AF443521

Studies in Systems, Decision and Control

Volume 482

Series Editor

Janusz Kacprzyk, Systems Research Institute, Polish Academy of Sciences, Warsaw, Poland

The series "Studies in Systems, Decision and Control" (SSDC) covers both new developments and advances, as well as the state of the art, in the various areas of broadly perceived systems, decision making and control–quickly, up to date and with a high quality. The intent is to cover the theory, applications, and perspectives on the state of the art and future developments relevant to systems, decision making, control, complex processes and related areas, as embedded in the fields of engineering, computer science, physics, economics, social and life sciences, as well as the paradigms and methodologies behind them. The series contains monographs, textbooks, lecture notes and edited volumes in systems, decision making and control spanning the areas of Cyber-Physical Systems, Autonomous Systems, Sensor Networks, Control Systems, Energy Systems, Automotive Systems, Biological Systems, Vehicular Networking and Connected Vehicles, Aerospace Systems, Automation, Manufacturing, Smart Grids, Nonlinear Systems, Power Systems, Robotics, Social Systems, Economic Systems and other. Of particular value to both the contributors and the readership are the short publication timeframe and the world-wide distribution and exposure which enable both a wide and rapid dissemination of research output.

Indexed by SCOPUS, DBLP, WTI Frankfurt eG, zbMATH, SCImago.

All books published in the series are submitted for consideration in Web of Science.

Andreas Varga

Solving Fault Diagnosis Problems

Linear Synthesis Techniques with Julia Code Examples

Second Edition

 Springer

Andreas Varga
Gilching, Germany

ISSN 2198-4182   ISSN 2198-4190 (electronic)
Studies in Systems, Decision and Control
ISBN 978-3-031-35766-4   ISBN 978-3-031-35767-1 (eBook)
https://doi.org/10.1007/978-3-031-35767-1

To my wife, Neli, with love

Preface to the Second Edition

Six years passed since the publication of the first edition, which served as the basis to implement a whole new collection of free software tools, **FDITools**, to solve the addressed synthesis problems of fault detection and model detection filters. **FDITools** is a MATLAB[1] toolbox, which is based on the **DSTools** collection to manipulate linear time-invariant systems in their most general form (also known as descriptor systems). The first edition of the book together with the associated software tools served as the basis of a series of graduate courses on "Model-Based Fault Diagnosis—A Linear Synthesis Framework Using MATLAB" organized in the framework of the International Graduate School on Control (IGSC) Program under the auspices of the European Embedded Control Institute (ESCI). Intensive 21 hours/week courses have been held jointly with Daniel Ossmann at Paris-Saclay University in 2018, University of Padova in 2019, University of Applied Sciences of Munich in 2021, as well as at the Institute of Flight System Dynamics within the Technical University of Munich in 2018. Although **FDITools** and **DSTools** have been developed as open software tools, which can be freely downloaded from **GitHub**, these tools must run within MATLAB, a widely popular but only commercially available software environment. This fact was one of the main motivations to develop alternative, entirely freely accessible software tools, which fully cover the functionality of **FDITools** and **DSTools**.

In the second edition, the changes in the theoretical part of fault and model detection are minor, the detailed presentation of algorithms has been kept unchanged to make the book as self-contained as possible, the bibliography and index have been updated and all known errors and typos have been corrected. A notational change worth mentioning is the new ordering used in the grouping of inputs in the system models, which now matches the employed ordering in the software implementations. The goal of this book is the same as for the first edition, "to address the fault detection and isolation topics from a computational perspective", by covering the same important aspects, namely (1) providing a completely general theoretical treatment of fault and model detection problems for linear time-invariant systems; (2) presenting the

---

[1] MATLAB is a registered trademark of The MathWorks Inc.

best suited numerical approaches to solve the specific computational problems; (3) providing supporting software to solve the analysis and filter synthesis problems. The major difference to the first edition is in the underlying computational support, which is now based on software developed in a relatively new language called Julia (originally released in 2012). The presentation of synthesis procedures and examples is similar to the first edition, but it is now interlaced with Julia codes, which can be used to reproduce all computational examples and figures presented in the book. An Appendix has been added to cover some basic issues related to using Julia and the new **FaultDetectionTools** and **DescriptorSystems** packages.

The question naturally arises: why Julia? First of all, Julia is a free and open-source environment, which solves the so-called "two language problem". This means that besides Julia's user-friendly operation capability, which is similar to that of MATLAB, Python or R, no other programming language is needed to write high-performance code, which is competitive with native C or Fortran-based programs. Some language features as multiple dispatch, dynamic and optional typing, ease significantly the interactive usage of Julia functions. The built-in easy interoperability with other languages such as Python, R, C, Fortran and even MATLAB allows an easy transfer or usage of codes and data from these languages. Not least, Julia enables fully reproducible computing results, which is an essential demand for the long-term usability of research results.

The development of Julia tools to be used as basic computational ingredients in implementing high-performance software to solve fault diagnosis problems started in 2019 with the implementation of the first package called **MatrixEquations**, to solve classes of Lyapunov, Sylvester and Riccati matrix equations. My main goal was to demonstrate that Julia programs can achieve high computational performance, which is comparable with the performance of efficient structure exploiting Fortran implementations, as those available in the Systems and Control Library SLICOT. This goal has been fully achieved for the Lyapunov and Sylvester equation solvers for which the codes for both real and complex data perform at practically the same performance level as similar functions available in the *Control System Toolbox* of MATLAB (which rely on SLICOT). The functionally complete and fully documented version v1.0 of the package has been released in November 2019.

The second major implementation effort has been directed to develop high-performance numerical software to compute a range of so-called Kronecker-like forms (KLFs) of linear matrix pencils, to reveal their Kronecker structure. The numerically reliable computation of various KLFs is a major ingredient to address a wide range of computational problems of linear time-invariant systems in their most general representations. Version v1.0 of the resulting package, called **MatrixPencils**, has been released in May 2020.

The transition from MATLAB to Julia of the **DSTools** and **FDITools** collections involved a substantial implementation effort, because all underlying codes, previously relying on SLICOT-based *mex*-interfaces and benefiting from the functions and system objects of the *Control System Toolbox*, had to be implemented anew. The implementations of versions v1.0 of **DescriptorSystems** and **FaultDetectionTools** have been finished in June 2021 and February 2022, respectively. The developments still continued during 2022 with some functional enhancements, being essentially completed just before receiving the proposal from Springer to prepare the second edition.

For the usage of the Julia supporting tools previous experience with Julia is certainly useful, but, in my opinion, not absolutely necessary. MATLAB users will find the transition to Julia relatively easy, and I consider myself a good example of this. The open-source software development model, typical for the Julia community, could be stimulating for students and young researchers to start implementing software in Julia, which extends or complements the existing tools.

The intended audience of this book is primarily formed of researchers and advanced graduate students in the areas of fault diagnosis and fault-tolerant control. Chapters 1–6 and 8 fit well into advanced fault diagnosis curricula relying on computer-aided design tools. Chapters 7 and 10 will appeal to mathematicians with interests in control-oriented numerics.

Acknowledgements. The ESCI graduate courses held jointly with Daniel Ossmann during 2018–2021 have been an excellent opportunity to validate the computational approaches presented in the first edition together with the accompanying MATLAB software implementations. Many thanks to Françoise Lamnabhi-Lagarrigue and colleagues from the ESCI for the invitation to hold these courses. I also thank Florian Holzapfel from the Institute of Flight System Dynamics of the Technische Universität München for the invitation to hold the same course for his doctoral students in 2018. Many thanks to Daniel Ossmann for the good collaboration in jointly holding all graduate courses and for his readiness to test the course examples, now implemented in Julia. My special thanks go to Cristina Stoica for her curiosity to explore this new language by performing the whole software installation process, testing all code examples in Appendix and for many valuable remarks on it.

Several people encouraged and supported me in my implementation endeavour during 2018–2022. I thank Martin Otter, a former colleague at the German Aerospace Center (DLR) and the main coauthor of the **Modia** package, for the first motivating discussions on Julia and for his kind support in guiding my first steps in the Julia community and the **GitHub** world. I am glad to have the opportunity to thank several new collaborators (I never met them personally), who invested their time and passion in providing support in different stages of my implementation efforts. John Verzani, the main author of the **Polynomials** package, was particularly responsive to my questions related to polynomial computations and receptive to my proposals to extend the **Polynomials** package with tools for handling rational functions (which later became the basis of implementing the rational transfer function object in the **DescriptorSystems**). Daniel Karrasch, the main author of **LinearMaps**, a Julia package for defining and working with linear maps, helped me a lot in the fine-tuning of sensitivity estimation functions of **MatrixEquations**, which are based on **LinearMaps**. Fredrik Bagge Carlson, the main author of the **ControlSystems** package, was one of the first users of **DescriptorSystems** and provided valuable suggestions and contributions to enhance the definition of the basic descriptor system object, which led to a substantial reduction of compilation times of many functions. I thank him also for reading the final version of the new Appendix. Finally, my thanks go to the members of the Julia community, who answered many of the issues, which I raised, and helped me to understand many subtleties of Julia.

I thank the staff of Springer Nature for making the production of this book possible and I am especially grateful for the assistance of Dr. Zainab Liaqat (Associate Editor) and Mrs. Shakila Sundarraman (Project Coordinator - Book Production). Special thanks go to Mr. Mano Priya Saravanan (Project Manager - Book Production) and his team for the high professionalism in the formatting and fine-tuning of the final manuscript.

This book edition has been prepared under Windows 11 and has been typeset in LaTeX 2ε using the **svmono** document style provided by Springer Verlag. For displaying and highlighting the Julia code, the **listings** package has been used, with a custom definition of basic Julia language commands and special Unicode characters. I used the free and open-source tools **TeXStudio**, as the LaTeX editor, and MiKTeX, as the underlying TEX distribution. To perform copy-paste operations directly from the Julia code listings provided in the book, the free PDF viewer **SumatraPDF** is a good choice under Windows, because of its support of Unicode characters (also used in Julia). The numerical examples have been computed using Julia 1.8.3 and the figures used in this book have been generated using the **CairoMakie** plotting package for Julia. For the automatic generation of the documentation of the implemented supporting packages, the **Documenter** package, integrated in Julia, has been employed.

Gilching, Germany   Andreas Varga
February 2023

Preface to the First Edition

The model based approach to fault detection and diagnosis has been the subject of continuing research for several decades. A number of monographs, several edited books, and many conference proceedings cover a multitude of aspects of two closely related important fields: fault diagnosis and fault tolerant control. This book addresses from a procedural point of view a basic aspect of these techniques: the synthesis of residual generators for fault detection and isolation. Although the need for specialized computational procedures has been earlier recognized by several authors, the computational aspects have been largely ignored in the existing literature or they are addressed only superficially without a clear understanding of the main numerical issues. Therefore, the aim of this book, to address the fault detection and isolation topics from a computational perspective, contrasts with most existing literature. This book is an attempt to close the gap between the existing well developed theoretical results and the realm of reliable computational synthesis procedures.

The book addresses several important aspects, which make it unique in the fault detection literature. A *first aspect* is the solution of standard synthesis problems in the most general setting. Consequently, the presented synthesis procedures can determine a solution to a specific problem whenever a solution exists, in accordance with the stated general existence conditions. Although this feature is a legitimate goal when developing computational approaches, the existing literature still abounds with technical assumptions, which, although they often facilitate establishing of particular theoretical results, are not necessary for the solution of the problems. A distinctive feature of the presented synthesis methods is their general applicability to both continuous- and discrete-time systems, regardless of whether the underlying system is proper or not.

The *second aspect* is the focus on the best suited numerical algorithms to solve the formulated filter synthesis problems. In contrast to the opinions of some authors that cultural aspects (e.g., familiarity with one approach or another) may influence the choice of appropriate algorithms, I firmly believe that only state-space description based algorithms are a viable choice when solving relatively high order problems. Therefore, I completely dismissed computational procedures based on polynomial or rational matrix manipulations, and exclusively rely on state-space representation based numerically reliable computational methods. An extra bonus is the availability of a huge arsenal of numerically reliable linear algebra software tools that facilitate the implementation of dedicated robust numerical software.

The *third aspect* emphasized is the development of so-called integrated computational procedures, where the resulting filters are determined by successive updating of partial synthesis results addressing specific requirements. Since each partial synthesis may represent a valid fault detection filter, this approach is highly flexible when using or combining different synthesis techniques. A common feature of all synthesis methods is the use of the nullspace method as the first synthesis step to reduce all synthesis problems to a simple standard form that allows easy checking of solvability conditions and addressing least order synthesis problems.

The *fourth aspect* is the provision of a comprehensive set of supporting software tools, which accompany this book. This software allows the easy implementation of all synthesis procedures presented in the book and facilitates performing rapid prototyping experiments in the computational environment MATLAB. The software tools rely on numerically reliable algorithms for solving computational problems for systems in a generalized state-space form, also known as descriptor systems. The provided collection of MATLAB functions, called DESCRIPTOR SYSTEM TOOLS, has been entirely implemented during the preparation of this book. There are also numerous MATLAB scripts, which allow the recalculation of all worked examples, and of several case studies. Since virtually all numerical results in the book can be reproduced by the readers using these scripts, this book is one of the first contributions to reproducible research in the field of fault detection.

Two alternative models are used in the book to describe systems with faults: an input-output description based on transfer function matrices, and state-space descriptions in standard or generalized (descriptor) forms. It is important to clarify from the beginning the roles of these two model types used in the book. The input output models underlie the theoretical developments in the book to solve various fault detection problems. Therefore, they serve to formulate the specific fault detection problems to be solved, to establish algebraic existence conditions, and even to describe high level conceptual solution procedures. In this way, it is possible to hide in a first reading most of the involved computational details by focusing mainly on conceptual aspects. However, when entering the realm of developing reliable numerical algorithms, there is an exclusive reliance on the equivalent state-space representations. There are several important reasons for this decision. Firstly, state space models are better suited to numerical computations than are the potentially highly sensitive polynomial based models. Secondly, state-space models allow the formulation of integrated algorithms, where successive steps are closely connected and structural features can be fully exploited. Finally, by developing explicit state space representation based updating formulas, the resulting algorithms lead to minimal order filters, by implicitly performing all hidden pole-zero cancelations. This significantly increases the reliability of computations and confers a sound numerical basis for all developed algorithms.

The treatment of fault detection problems is limited to linear time-invariant systems for which both the theory and computational tools are well developed.

Although the proposed linear synthesis methodologies offer satisfactory solutions for many practical applications, performance limitations (e.g., lack of sufficient robustness) may still occur when facing more complicated systems. Nevertheless, linear techniques are frequently used to address even problems where the underlying system models are nonlinear and parameter dependent. Besides providing guidance for problem solvability and performance limitations, linear synthesis approaches form the basis of complex gains-scheduling-based synthesis methodologies that are able to fully address robustness aspects. Also, using linear fault detection filters in conjunction with signal processing techniques for on-line identification of various types of faults often represents a viable approach to enhance the robustness of the fault detection and to provide useful information for control reconfiguration purposes. The considered case studies illustrate these aspects.

The book includes a substantial amount of background material on rational matrices, descriptor systems and computational algorithms. The presentation of theoretical backgrounds on rational matrices and descriptor systems exhibits a certain parallelism meant to ease introducing or recalling to readers the main theoretical concepts. The algorithmic details are presented only in the final chapter. This may help readers, especially those not familiar with or not interested in numerical aspects, to focus primarily on the main synthesis steps of the conceptual synthesis procedures, by blending out all non-essential technicalities. Nevertheless, the presentation of the underlying algorithms is a main part of this book, and the final chapter even includes several algorithmic improvements, which are presented for the first time here.

This book is primarily aimed at researchers and advanced graduate students in the areas of fault diagnosis and fault tolerant control. The Chaps. 1–6 and 8 fit well into an advanced fault diagnosis curricula relying on computer aided design tools. The Chaps. 7 and 10 will appeal to mathematicians with interests in control oriented numerics.

The present book is largely based on my own research, started in 2002, on developing reliable numerical methods for the synthesis of fault detection filters. Many colleagues working in the field of fault diagnosis recognized the need to develop efficient and reliable computational methods. This was a constant stimulus for me to understand all the subtleties of problem formulations, to discover the technical and numerical limitations of many of the existing computational approaches, and to pursue research to develop efficient and reliable algorithms, which are able to provide a solution whenever one exists. A particular impulse came from the organizers of the SAFEPROCESS'2012 conference in Mexico City, Professor Jan Lunze (Program Chair) and Professor Cristina Verde (General Chair), who invited me to hold a semi-plenary lecture at this conference. In my talk I presented a systematic account of linear time-invariant synthesis techniques of fault detection filters. The underlying plenary paper was later published, in an extended and revised form, in Annual Reviews in Control (ARC, 2013). This paper, which is simultaneously a survey of synthesis methods and a presentation of several new ideas and research results, laid the methodical foundation for this book. During its preparation, I realized the importance of the availability of numerical software suitable for the easy implementation of the presented synthesis methods. Therefore, implementing the free software tools that accompany this book was a natural extension of my original plans.

Acknowledgements. I am particularly grateful to Professor Janos Gertler for his encouragement during my first contacts with the fault detection field and for a continuous exchange of information during many years. I also highly benefitted from his constructive criticism and skilful moderation as Chief Editor of ARC during the publication process of the previously mentioned article. And finally, many thanks for plenty of detailed comments and useful suggestions for improvements of the Part I of this book. I would like to thank my employer of the last 23 years, the German Aerospace Center (DLR) in Oberpfaffenhofen, for the creative work environment and for the excellent support of my research career. In particular, I am grateful to Professor Georg Grübel, the former head of the Control Group in Oberpfaffenhofen, for his visionary interest in and stimulation of my efforts to develop algorithms and software for descriptor systems. Most of this software is nowadays freely available as part of the SLICOT software library. My thanks go also to Dr. Johann Bals, the Director of the Institute of System Dynamics and Control, for his constant support of my various research ambitions and professional involvements. I thank my former colleague and friend Dr. Vasile Sima for his time spent for reading of the entire material in a preliminary version of this book, for many detailed comments and suggestions for corrections. I also acknowledge his expert support in preparing the SLICOT-based interfaces, which underlie the accompanying software tools. I thank the staff of Springer Science for making the production of this book possible and I am especially grateful for the assistance of Dr. Thomas Ditzinger (Executive Editor) and Mr. Narayanasamy Prasanna Kumar (Project Coordinator). Special thanks go to Mr. Sukanya Servai (Production Editor) and his team for the high professionalism in the formatting and fine tuning of the final manuscript. Finally, I dedicate this book to my loving wife Cornelia Varga (Neli). Without her encouraging support over so many years, and more recently, without her motivating pressure on me "to find a meaningful preoccupation after my retirement", this book would not have been possible.

Gilching, Germany   Andreas Varga
November 2016

Contents

Acronyms

AFDP	Approximate fault detection problem
AFDIP	Approximate fault detection and isolation problem
AMDP	Approximate model detection problem
AMMP	Approximate model-matching problem
EFDP	Exact fault detection problem
EFEP	Exact fault estimation problem
EFDIP	Exact fault detection and isolation problem
EMDP	Exact model detection problem
EMMP	Exact model-matching problem
FDD	Fault detection and diagnosis
FDI	Fault detection and isolation
GCARE	Generalized continuous-time algebraic Riccati equation
GDARE	Generalized discrete-time algebraic Riccati equation
GRSD	Generalized real Schur decomposition
GRSF	Generalized real Schur form
LCF	Left coprime factorization
LDP	Least distance problem
LFT	Linear fractional transformation
LPV	Linear parameter-varying
LTI	Linear time invariant
MIMO	Multiple-input multiple-output
MMP	Model-matching problem
RCF	Right coprime factorization
RSD	Real Schur decomposition
RSF	Real Schur form
SISO	Single-input single-output
SVD	Singular value decomposition
TFM	Transfer function matrix

Synthesis Procedures

EFD   Exact synthesis of fault detection filters
AFD   Approximate synthesis of fault detection filters
EFDI   Exact synthesis of FDI filters
AFDI   Approximate synthesis of FDI filters
EMM   Exact model-matching synthesis of FDI filters
EMMS   Exact model-matching synthesis of strong FDI filters
AMMS   Approximate model-matching synthesis of FDI filters
EMD   Exact synthesis of model detection filters
AMD   Approximate synthesis of model detection filters

Notations and Symbols

General Notations

$\mathbb{C}$	Field of complex numbers
$\mathbb{R}$	Field of real numbers
$\mathbb{C}_s$	Stability domain (i.e., open left complex half-plane in continuous time or open unit disc centred in the origin in discrete time)
$\partial\mathbb{C}_s$	Boundary of stability domain (i.e., extended imaginary axis with infinity included in continuous time, or unit circle centred in the origin in discrete time)
$\overline{\mathbb{C}}_s$	Closure of $\mathbb{C}_s : \overline{\mathbb{C}}_s = \mathbb{C}_s \cup \partial\mathbb{C}_s$
$\mathbb{C}_u$	Open instability domain: $\mathbb{C}_u := \mathbb{C}\backslash\overline{\mathbb{C}}_s$
$\overline{\mathbb{C}}_u$	Closure of $\mathbb{C}_u : \overline{\mathbb{C}}_u := \mathbb{C}_u \cup \partial\mathbb{C}_s$
$\mathbb{C}_g$	"Good" domain of $\mathbb{C}$
$\mathbb{C}_b$	"Bad" domain of $\mathbb{C} : \mathbb{C}_b = \mathbb{C}\backslash\mathbb{C}_g$
s	Complex frequency variable in the Laplace transform: $s = \sigma + i\omega$
z	Complex frequency variable in the z-transform: $z = e^{sT}$, T-sampling time
λ	Complex frequency variable: $\lambda = s$ in continuous time or $\lambda = z$ in discrete time
λ_s	DC-gain frequency (i.e., $\lambda_s = 0$ in continuous time or $\lambda_s = 1$ in discrete time)
$\mathbb{R}(\lambda)$	Field of real rational functions in indeterminate λ
$\mathbb{R}(\lambda)$	Set of rational matrices in indeterminate λ with real coefficients and unspecified dimensions
$\mathbb{R}(\lambda)^{p\times m}$	Set of $p \times m$ rational matrices in indeterminate λ with real coefficients
$\mathbb{R}[\lambda]$	Ring of real rational polynomials in indeterminate λ
$\mathbb{R}[\lambda]$	Set of polynomial matrices in indeterminate λ with real coefficients and unspecified dimensions

$\mathbb{R}[\lambda]^{p\times m}$	Set of $p \times m$ polynomial matrices in indeterminate λ with real coefficients
$\mathbb{R}_s(\lambda)$	Set of stable and proper rational transfer functions with real coefficients
$\delta(G(\lambda))$	McMillan degree of the rational matrix $G(\lambda)$
$G^\sim(\lambda)$	Conjugate of $G(\lambda) \in \mathbb{R}(\lambda) : G^\sim(s) = G^T(-s)$ in continuous time and $G^\sim(z) = G^T(1/z)$ in discrete time
ℓ_2	Banach space of square-summable sequences
$\mathcal{L}_2$	Lebesgue space of square-integrable functions
$\mathcal{H}_2$	Hardy space of square-integrable complex-valued functions analytic in $\mathbb{C}_u$
$\mathcal{L}_\infty$	Space of complex-valued functions bounded and analytic in $\partial\mathbb{C}_s$
$\mathcal{H}_\infty$	Hardy space of complex-valued functions bounded and analytic in $\mathbb{C}_u$
$\|G\|_2$	$\mathcal{H}_2$- or $\mathcal{L}_2$-norm of the transfer function matrix $G(\lambda)$
$\|G\|_\infty$	$\mathcal{H}_\infty$- or $\mathcal{L}_\infty$-norm of the transfer function matrix $G(\lambda)$
$\|G\|_{\infty/2}$	either the $\mathcal{H}_\infty$- or $\mathcal{H}_2$-norm of the transfer function matrix $G(\lambda)$
$\|G\|_H$	Hankel norm of the transfer function matrix $G(\lambda)$
$\|G\|_-$	$\mathcal{H}_-$-index of the transfer function matrix $G(\lambda)$
$\|G\|_{\Omega-}$	Modified $\mathcal{H}_-$-index of the transfer function matrix $G(\lambda)$
$\|G\|_{2-}$	$\mathcal{H}_{2-}$-index of the transfer function matrix $G(\lambda)$
$\|G\|_{\infty-}$	$\mathcal{H}_{\infty-}$-index of the transfer function matrix $G(\lambda)$
$\|G\|_{2/\infty-}$	Either the $\mathcal{H}_{2-}$- or $\mathcal{H}_{\infty-}$-index of the transfer function matrix $G(\lambda)$
$\|u\|_2$	Norm of $u \in \mathcal{H}_2$, $u \in \mathcal{L}_2$, or $u \in \ell_2$
$\|u(t)\|_2$	Euclidean norm of the real vector $u(t)$
$M(i_1 : i_2, j_1 : j_2)$	Submatrix formed of rows $i_1, i_1 + 1, \cdots, i_2$, and columns $j_1, j_1 + 1, \cdots, j_2$ of the matrix M
$M(i_1 : i_2, :)$	Submatrix formed of rows $i_1, i_1 + 1, \cdots, i_2$ and all columns of matrix M
$M(:, j_1 : j_2)$	Submatrix formed of all rows and columns $j_1, j_1 + 1, \cdots, j_2$ of the matrix M
M^T	Transpose of the matrix T
M^P	Pertranspose of the matrix M
M^{-1}	Inverse of the matrix M
M^{-T}	Transpose of the inverse matrix M^{-1}
$M^\dagger$	Pseudo-inverse of the matrix M
$\overline{\sigma}(M)$	Largest singular value of the matrix M
$\underline{\sigma}(M)$	Least singular value of the matrix M
$\mathcal{N}(M)$	Kernel (or right nullspace) of the matrix M
$\mathcal{R}(M)$	Range (or image space) of the matrix M
$LFT_u(M, \Delta)$	Upper LFT defined by M and Δ
$\mathcal{N}_L(G(\lambda))$	Left kernel (or left nullspace) of $G(\lambda) \in \mathbb{R}(\lambda)$

$\mathcal{N}_R(G(\lambda))$	Right kernel (or right nullspace) $G(\lambda) \in \mathbb{R}(\lambda)$
$\Lambda(A)$	Set of eigenvalues of the matrix A
$\Lambda(A, E)$	Set of generalized eigenvalues of the pair (A, E)
$\Lambda(A - \lambda E)$	Set of eigenvalues of the pencil $A - \lambda E$
$\mathbf{u}$	Unit roundoff of the floating-point representation
$\mathcal{O}(\varepsilon)$	Quantity of order of ε
I_n or I	Identity matrix of order n or of an order resulting from context
e_i	The i-th column of the (known size) identity matrix
$0_{m \times n}$ or 0	Zero matrix of size $m \times n$ or of a size resulting from context
span M	Span (or linear hull) of the columns of the matrix M

Fault Diagnosis-Related Notations

$y(t)$	Measured output vector: $y(t) \in \mathbb{R}^p$
$\mathbf{y}(\lambda)$	Laplace- or $\mathcal{Z}$-transformed measured output vector
$u(t)$	Control input vector: $u(t) \in \mathbb{R}^{m_u}$
$\mathbf{u}\lambda)$	Laplace- or $\mathcal{Z}$-transformed control input vector
$d(t)$	Disturbance input vector: $d(t) \in \mathbb{R}^{m_d}$
$\mathbf{d}(\lambda)$	Laplace- or $\mathcal{Z}$-transformed disturbance input vector
$f(t)$	Fault input vector: $f(t) \in \mathbb{R}^{m_f}$
$\mathbf{f}(\lambda)$	Laplace- or $\mathcal{Z}$-transformed fault input vector
$w(t)$	Noise input vector: $w(t) \in \mathbb{R}^{m_w}$
$\mathbf{w}(\lambda)$	Laplace- or $\mathcal{Z}$-transformed noise input vector
$x(t)$	State vector: $x(t) \in \mathbb{R}^n$
$G_u(\lambda)$	Transfer function matrix from u to y
$G_d(\lambda)$	Transfer function matrix from d to y
$G_f(\lambda)$	Transfer function matrix from f to y
$G_{f_j}(\lambda)$	Transfer function matrix from the j-th fault input f_j to y
$G_w(\lambda)$	Transfer function matrix from w to y
A	System state matrix
E	System descriptor matrix
B_u, B_d, B_f, B_w	System input matrices from u, d, f, w
C	System output matrix
D_u, D_d, D_f, D_w	System feedthrough matrices from u, d, f, w
$r(t)$	Residual vector: $r(t) \in \mathbb{R}^q$
$\mathbf{r}(\lambda)$	Laplace- or $\mathcal{Z}$-transformed residual vector
n_b	Number of components of residual vector r
$r^{(i)}(t)$	i-th residual vector component: $r^{(i)}(t) \in \mathbb{R}^{q_i}$
$\mathbf{r}^{(i)}(\lambda)$	Laplace- or $\mathcal{Z}$-transformed i-th residual vector component
$Q(\lambda)$	Transfer function matrix of the implementation form of the residual generator from y and u to r
$Q_y(\lambda)$	Transfer function matrix of residual generator from y to r

$Q_u(\lambda)$	Transfer function matrix of residual generator from u to r
$Q^{(i)}(\lambda)$	Transfer function matrix of the implementation form of the i-th residual generator from y and u to $r^{(i)}$
$R(\lambda)$	Transfer function matrix of the internal form of the residual generator from u, d, w and f to r
$R_u(\lambda)$	Transfer function matrix from u to r
$R_d(\lambda)$	Transfer function matrix from d to r
$R_f(\lambda)$	Transfer function matrix from f to r
$R_{f_j}(\lambda)$	Transfer function matrix from the j-th fault input f_j to r
$R_{f_j}^{(i)}(\lambda)$	Transfer function matrix from the j-th fault input f_j to $r^{(i)}$
$R_w(\lambda)$	Transfer function matrix from w to r
S	Binary structure matrix
S_{R_f}	Binary structure matrix corresponding to $R_f(\lambda)$
$M_r(\lambda)$	Transfer function matrix of a reference model from f to r
$\theta(t)$	Residual evaluation vector
$\iota(t)$	Binary decision vector
τ, τ_i	Decision thresholds

Model Detection-Related Notations

N	Number of component models of the multiple model
$y(t)$	Measured output vector: $y(t) \in \mathbb{R}^p$
$\mathbf{y}(\lambda)$	Laplace- or $\mathcal{Z}$-transformed measured output vector
$u(t)$	Control input vector: $u(t) \in \mathbb{R}^{m_u}$
$\mathbf{u}(\lambda)$	Laplace- or $\mathcal{Z}$-transformed control input vector
$u^{(j)}(t)$	Control input vector of j-th model: $u^{(j)}(t) := u(t) \in \mathbb{R}^{m_u}$
$\mathbf{u}^{(j)}(\lambda)$	Laplace- or $\mathcal{Z}$-transformed control input vector of j-th model
$d^{(j)}(t)$	Disturbance input vector of j-th model: $d^{(j)}(t) \in \mathbb{R}^{m_d^{(j)}}$
$\mathbf{d}^{(j)}(\lambda)$	Laplace- or $\mathcal{Z}$-transformed disturbance input vector of j-th model
$w^{(j)}(t)$	Noise input vector of j-th model: $w^{(j)}(t) \in \mathbb{R}^{m_w^{(j)}}$
$\mathbf{w}^{(j)}(\lambda)$	Laplace- or $\mathcal{Z}$-transformed noise input vector of j-th model
$y^{(j)}(t)$	Output vector of j-th model: $y^{(j)}(t) \in \mathbb{R}^p$
$\mathbf{y}^{(j)}(\lambda)$	Laplace- or $\mathcal{Z}$-transformed output vector of j-th model
$x^{(j)}(t)$	State vector of j-th model: $x^{(j)}(t) \in \mathbb{R}^{n_j}$
$G_u^{(j)}(\lambda)$	Transfer function matrix of j-th model from $u^{(j)}$ to $y^{(j)}$
$G_d^{(j)}(\lambda)$	Transfer function matrix of j-th model from $d^{(j)}$ to $y^{(j)}$
$G_w^{(j)}(\lambda)$	Transfer function matrix of j-th model from $w^{(j)}$ to $y^{(j)}$
$A^{(j)}$	System state matrix of j-th model
$E^{(j)}$	System descriptor matrix of j-th model
$B_u^{(j)}, B_d^{(j)}, B_w^{(j)}$	System input matrices of j-th model from $u^{(j)}, d^{(j)}, w^{(j)}$
$C^{(j)}$	System output matrix of j-th model

$D_u^{(j)}, D_d^{(j)}, D_w^{(j)}$	System feedthrough matrices of j-th model from $u^{(j)}, d^{(j)}, w^{(j)}$
$r^{(i)}(t)$	i-th residual vector component: $r^{(i)}(t) \in \mathbb{R}^{q_i}$
$\mathbf{r}^{(i)}(\lambda)$	Laplace- or $\mathcal{Z}$-transformed i-th residual vector component
$r(t)$	Overall residual vector: $r(t) \in \mathbb{R}^q$, $q = \sum_{i=1}^{N} q_i$
$\mathbf{r}(\lambda)$	Laplace- or $\mathcal{Z}$-transformed overall residual vector
$Q^{(i)}(\lambda)$	Transfer function matrix of the implementation form of the i-th residual generator from y and u to $r^{(i)}$
$Q_y^{(i)}(\lambda)$	Transfer function matrix of residual generator from y to $r^{(i)}$
$Q_u^{(i)}(\lambda)$	Transfer function matrix of residual generator from u to $r^{(i)}$
$Q(\lambda)$	Transfer function matrix of the implementation form of the overall residual generator from y and u to r
$R^{(i,j)}(\lambda)$	Transfer function matrix of the internal form of the overall residual generator from $\left(u^{(j)}, d^{(j)}, w^{(j)}\right)$ to $r^{(i)}$
$R_u^{(i,j)}(\lambda)$	Transfer function matrix of the internal form of the overall residual generator from $u^{(j)}$ to $r^{(i)}$
$R_d^{(i,j)}(\lambda)$	Transfer function matrix of the internal form of the overall residual generator from $d^{(j)}$ to $r^{(i)}$
$R_w^{(i,j)}(\lambda)$	Transfer function matrix of the internal form of the overall residual generator from $w^{(j)}$ to $r^{(i)}$
$\theta(t)$	N-dimensional residual evaluation vector
$\iota(t)$	N-dimensional binary decision vector
τ_i	Decision threshold for i-th component of the residual vector

List of Figures

List of Tables

Listings

Part I
Basics of Fault Diagnosis

In this part, the main concepts related to model-based fault diagnosis are introduced, and several fault detection and isolation problems, as well as model detection problems, are formulated. After a short introductory chapter, the modelling aspects of linear systems with additive faults and physical faults are discussed. Two model types are used to describe systems with additive faults: an input–output description based on transfer function matrices, and state-space descriptions in standard or generalized (descriptor) forms. The input–output models underlie the main theoretical developments in the book, such as the formulation of synthesis problems and the development of suitable conceptual synthesis algorithms. The state-space model-based description is the basis of developing numerically reliable computational algorithms. To address robustness aspects when solving synthesis problems for systems described by linear parametric models or multiple linear models, it is necessary to convert such models to the standard synthesis form used throughout the book. Several fault detection and isolation as well as model detection problems are formulated in two separate chapters. The solvability of these problems is characterized in terms of appropriately defined concepts such as fault detectability, fault isolability and model detectability. An important aspect of the employed problem formulations is that they are independent of any particular solution method, which can be potentially used for their solution. This led to general solvability conditions, which form the basis of developing general synthesis procedures, whose primary aim is to produce a satisfactory solution whenever the existence conditions are fulfilled.

Chapter 1
Introduction

1.1 Linear Synthesis Techniques for Fault Diagnosis

Fault diagnosis is a widely used term across many application domains. In this book, we restrict the meaning of this term to designate the usage of specific techniques to discover anomalous behaviours occurring in physical plants (known as *fault detection*) and the more challenging aspect of locating a fault within an industrial equipment (known as *fault isolation*). The subsequent characterization of the type, size and nature of occurring faults (known as *fault identification*) is also often a part of fault diagnosis. Among many approaches for fault diagnosis, we focus on the model-based framework, where plant models are used to provide the required redundancy, also called *analytical redundancy*, to execute the fault detection and isolation tasks. A further restriction we purposely made is to restrict our focus to the class of *linear time-invariant* (LTI) plant models for which a reasonably complete theory for the synthesis of *fault detection and isolation* (FDI) filters exists.

The focus on linear system techniques may appear as a strong limitation, taking into account that most technical processes are nonlinear systems and their dynamical behaviour depends on parameters, which may vary during the plant operation or may have uncertain values. Besides that, unknown external signals acting as physical disturbances (e.g., external loads), as well as the ubiquitous presence of measurement noises, often increase the complexity of problem solving. To account (to some extent) for these inherent modelling deficiencies, linear models depending on parameters, the so-called *linear parameter-varying* (LPV) models, or collections of linearized models, the so-called LTI *multiple models*, can be used to serve as approximate plant models. These models can then be put in a standard form, which underlies all developments in this book.

The main emphasis of the book is on procedural aspects, by presenting general synthesis procedures of FDI filters to address six "canonical" problems, termed as fault detection, fault isolation and reference model matching problems, aiming for both exact and approximate solutions. The main goal of the book is to provide a comprehensive presentation of the synthesis algorithms for the formulated problems, both at a conceptual level (using frequency-domain concepts) as well as at a detailed implementable algorithm level (using state-space description-based computational methods). Although the discussion of procedural aspects is a recurring theme in several textbooks, in our opinion the first edition of this book described, for the first time, a complete collection of numerically viable methods, which can serve as basis for robust numerical software implementations. Most of the "computational" methods described in the fault diagnosis literature are not adequate for this purpose. The reasons are simple: the basic requirements for satisfactory numerical algorithms (for example, as those to solve linear algebra problems) are not fulfilled. Most of the time, the important aspect of numerical reliability is completely ignored, and therefore many computational algorithms are provably numerically unstable. Also, the use of ill-conditioned coordinate transformations (e.g., to compute certain canonical forms) may drastically worsen the conditioning of the problem, by increasing the sensitivity of the solution to variations in problem data. Thus, the effects of inherent roundoff errors are amplified and the accuracy of the solution is diminished.

We also present the somewhat new topic of the model detection, which consists of finding in a collection of available models, that one, which best matches the current plant behaviour. The solution to this problem is highly relevant to solving special classes of fault detection problems (e.g., as those due to extreme variations of plant parameters) or in the multiple-model-based adaptive control approaches.

The first edition of this book [161] used computational examples to be executed in the popular MATLAB environment. An important novelty of the second edition is that the same computational examples are now implemented in a relatively new programming language called Julia, which is particularly well suited for numerical computational science applications. A rich Julia software ecosystem includes several thousands of contributed free software packages among which are the computational tools developed by the author used in this edition of the book. As in the first edition, all synthesis procedures are accompanied by synthesis examples for which listings of Julia scripts (instead of MATLAB scripts) are included in the book. These scripts are simple examples intended to illustrate the applicability of the synthesis procedures and can serve as starting points for more advanced applications, for example, the case study examples presented in Chap. 8. All implemented software tools for both MATLAB and Julia, including the scripts associated with the worked out examples and case studies, are freely available on the collaborative version control platform **GitHub**[1] (see also the author's homepage dedicated to this book project[2]).

To a lesser extent, we addressed the decision-making aspects, which are however crucial for the use of fault diagnosis systems. The presented norm-based decision schemes can be easily replaced by statistical methods based on change detection techniques.

There are several issues which are not included in the presentation, for example, synthesis methods based on the parity-space approach, polynomial representation and

---

[1] https://github.com/andreasvarga.

[2] https://sites.google.com/view/andreasvarga/home/book2ed.

unknown-input observers. The reasons for this are either the lack of the generality of an approach (e.g., the observer-based method) or the intrinsic numerical instability of the associated computations (e.g., parity-space and polynomial approaches). Furthermore, we exclusively use a deterministic framework, thus leaving out a rich collection of statistical approaches. Note, however, that the synthesis approaches described in the book may prove useful also for a stochastic framework, since the main difference lies in the employed decision-making tools (i.e., statistical methods instead of norm-based approaches).

1.2 Outline of the Book

The book naturally falls into two parts, while a third part contains extensive background material. We briefly comment on the contents of these parts.

Part I—Basics of Fault Diagnosis. This part contains four chapters, which serve to introduce the fault diagnosis topic and to formulate the basic synthesis problems associated with it. This chapter is introductory. Chapter 2 introduces the standard forms of LTI models used throughout the book for the synthesis of fault detection filters. Both input–output and state-space representations are used, where the former mainly serves to simplify the problem formulations and describe conceptual synthesis procedures, while the latter serves for developing reliable and efficient computational algorithms. To mathematically describe systems with faults, both additive fault models as well as physical fault models can be used. To address the robustness aspects, LPV models (with explicit dependence of varying parameters) or multiple LTI models (with implicit dependence of parameters) can be recast such that (fictitious) noise inputs account for the effects of parameter variations. Chapter 3 discusses the main aspects related to the fault diagnosis topic, as residual generation and evaluation, definition and characterization of the basic fault detectability and isolability concepts, formulation of six "canonical" problems for exact or approximate synthesis of fault detection filters, and the selection of appropriate thresholds for decision-making. For all six formulated fault detection problems, conditions for the existence of a solution are given. In Chap. 4, the model detection topic is discussed, by covering the generation of a structured set of residuals, definition and characterization of the model detectability concept, formulation of exact and approximate model detection problems, and selection of thresholds for decision-making.

Part II—Synthesis of Residual Generators. The second part of the book is concerned with the synthesis procedures of residual generators. Chapter 5 is the central part of the book and presents conceptual synthesis procedures for the solution of the FDI problems formulated in Chap. 3. A common synthesis paradigm of all procedures is the use of the nullspace method as the first synthesis step, to reduce all synthesis problems to a simple standard form, which allows for easily checking the solvability conditions and to address least-order synthesis aspects. The selection of thresholds suitable for decision-making is discussed for each of the approximate synthesis methods. Chapter 6 presents the synthesis procedures for solving the model detection problems formulated in Chap. 4. Once again, the nullspace method is used at the first step of both the exact and approximate synthesis procedures to simplify the formulation of the model detection problems and allow checking the existence conditions of a solution. The main computational aspects of the presented synthesis procedures are discussed in Chap. 7, focusing more on an informative presentation rather than on algorithmic details. The discussion of numerical aspects of the synthesis algorithms, such as the numerical stability or reliability, algorithmic performance, choice of underlying synthesis models and model conditioning, is rare in the fault detection-related literature. A common basic procedural framework of all presented procedures is the use of updating techniques of the different representation forms of fault detection filters (i.e., implementation, internal). As a consequence, each synthesis procedure produces a factored representation of the fault detection filter, where partial syntheses achieved at intermediate synthesis steps may represent valid solutions satisfying partial synthesis goals. The main computational aspects, as the application of the nullspace method, least-order synthesis, coprime factorization techniques and the solution of exact or approximate model-matching problems, are presented from the fault detection perspective. Chapter 8 presents several case studies related to flight control applications, as the isolation and identification of flight actuator faults and reliable isolation of air data sensor failures.

Part III—Background Material. This part includes a substantial amount of background material on advanced system theoretical concepts and specialized computational algorithms. Chapter 9 deals with the presentation of basic concepts and results on rational and polynomial matrices, followed by the discussion, in parallel, of similar aspects in terms of equivalent descriptor system representations. Chapter 10 presents in detail the main algorithms for descriptor systems, which underlie the computational procedures of this book. For readers interested in the algorithmic details, this will allow the understanding in depth of algorithmic subtleties of the basic computations. Software implementations as Julia functions are available for all of the presented algorithms and are included in several packages accompanying this book. The presentation of the underlying algorithms is intentionally done only in the final chapter of the book, to relieve the casual readers of the need to understand highly sophisticated numerical algorithms.

Appendix. The novelty of Julia as a programming language and the large number of packages available in the Julia software ecosystem may be challenging for users not yet familiar with Julia. Therefore, to guide the interested readers to start using Julia and to select some of the relevant packages, we included a minimal amount of information on how to install and run Julia, how to install and run packages and how to execute the code examples presented in the book. Furthermore, we provide some basic information on the available functions in two of the underlying collections: **DescriptorSystems**, for the manipulation of descriptor systems, and **FaultDetectionTools**, for solving synthesis problems of fault detection and model detection filters. Simple examples are used to illustrate the main functions.

1.3   Notes and References

Several monographs and textbooks partly overlap with our book, especially in the formulation of the main synthesis problems [15, 23, 29, 51, 68]. Statistical approaches for decision-making are considered in [7, 15, 51]. Data-driven methods for fault diagnosis are presented in [30]. The use of sliding mode control techniques to address both fault diagnosis and fault-tolerant control problems is the subject of the monograph [1].

Chapter 2
Modelling Systems with Faults

This book is intended as a guide to synthesis methods of residual generators for various fault detection problems. The main nontrivial task to be fulfilled before employing any of these methods is the development of adequate models, which fit into the envisaged design methodologies. Our interest lies primarily in developing synthesis methods for the class of linear time-invariant (LTI) plant models for which the resulting residual generators are themselves LTI systems or filters. In spite of this apparent limitation, a wide class of fault detection problems can be addressed either by appropriate adjustments of the underlying models or by suitable reformulations of the design objectives.

In this chapter, we address the basic aspects of modelling systems with faults. Two basic approaches are described to model systems with faults. The first approach involves models with *additive faults*, where the faults are explicitly defined as fictive inputs, which act on the system similarly to the unknown external disturbance inputs. The main advantage of this modelling approach is that, by avoiding the explicit modelling of different fault modes, a single model can be used to account for many possible physical faults. For example, models with additive faults are widely used to describe systems with various types of actuator and sensor faults. Models with *multiplicative faults* often describe systems with *parametric faults* (i.e., abnormal variations of some model parameters).

The second approach is based on *physical fault models*, where to each fault mode corresponds a dedicated model, which is usually derived by adjusting appropriately the non-faulty system model. For example, parametric faults can alternatively be modelled using physical models by setting the model parameters to some abnormal values. In other cases, physical fault models can be derived by removing some of the system control inputs in the case of total loss of control or defining new disturbance inputs to account for specific fault effects.

The physical fault modelling approach typically involves *multiple models*, representing a collection of individual models where each model corresponds to a specific fault situation. The multiple-model-based approach for fault modelling is well suited for certain fault-tolerant control applications, where the detection of the "right" fault model automatically triggers the reconfiguration of the controller. Another important application field is the multiple-model adaptive control, where switching among several controllers is done after "recognizing" the best matching model at each time moment.

In this chapter, we first review the main types of faults and then introduce the input–output and state-space forms of LTI plant models with additive faults. These models represent the basis of all synthesis approaches presented in this book. The underlying fault-free models often include fictive noise inputs, which account for the effects of parametric uncertainties in the matrices of the state-space representations. We describe two approaches to arrive at such models starting from linear parameter-varying models and multiple LTI models. Finally, we present physical fault models described by a collection of LTI models, which form the basis of model detection-based approaches.

2.1 Types of Faults

A typical setup for the modelling of a system with faults is presented in Fig. 2.1.

The main system variables are the control inputs u, the unknown disturbance inputs d, the noise inputs w and the output measurements y. The control inputs u are assumed to be known (measurable) and, in general, can have arbitrary bounded variations. The output y and control input u are the only measurable signals, which therefore can be used for fault monitoring purposes. The disturbance inputs d and noise inputs w are non-measurable "unknown" input signals, which act adversely on the system performance. For example, the unknown disturbance inputs d may represent physical disturbance inputs, for example, wind turbulence acting on an aircraft or external loads acting on a plant. Typical noise inputs are sensor noise signals as well as process input noise. However, fictive noise inputs can also account for the cumulative effects of unmodelled system dynamics or for the effects of parametric uncertainties. In general, there is no clear-cut separation between disturbances and noise, and, therefore, the appropriate definition of the disturbance and noise inputs is a challenging aspect when modelling systems for solving fault detection problems.

We define a *fault* as any unexpected variation of some physical parameters or variables of a plant causing an unacceptable violation of certain specification limits for normal operation. For example, a change in the characteristics of an actuator

Fig. 2.1 Plant model with faults

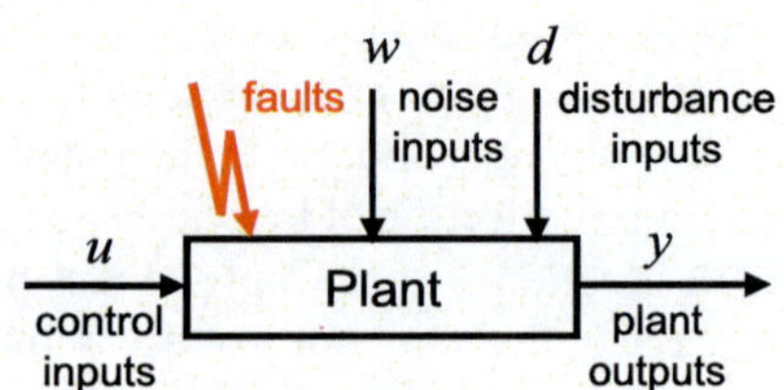

leading to a loss of efficiency or even to a complete breakdown is termed an *actuator fault*, while erroneous measurements (e.g., corrupted by bias or drift) obtained with a defective sensor are caused by a *sensor fault*. The malfunction of an internal component (e.g., leakage or shortcut) is often assimilated with a *parametric fault*. A *failure* designates a permanent interruption of the plant operation (e.g., complete breakdown) and may be caused by one or more faults.

Besides this physical classification, faults are often classified on the basis of their perceived effects. *Additive faults* are fictive inputs acting independently on the plant outputs. These inputs are zero in a fault-free situation and nonzero if a fault occurs. *Multiplicative faults* produce effects on the plant outputs, which depend on the magnitude of some internal signals or known inputs. Sensor faults and several types of actuator faults are usually considered as additive faults, while parametric faults are considered as multiplicative faults.

Faults can be also differentiated on the basis of their behaviour over time. *Intermittent faults* are short duration malfunctions, which can still induce long-lasting effects. For example, actuator saturations caused by excessive loads fall often in this category of faults. *Persistent faults* have a long-range time evolution, and often manifest as slowly evolving incipient faults or abrupt changes, with permanent character, of physical parameters or system structure.

2.2 Plant Models with Additive Faults

The synthesis methods presented in this book primarily deal with LTI systems described by input–output relations of the form

$$\mathbf{y}(\lambda) = G_u(\lambda)\mathbf{u}(\lambda) + G_d(\lambda)\mathbf{d}(\lambda) + G_f(\lambda)\mathbf{w}(\lambda) + G_w(\lambda)\mathbf{f}(\lambda), \qquad (2.1)$$

where $\mathbf{y}(\lambda)$, $\mathbf{u}(\lambda)$, $\mathbf{d}(\lambda)$, $\mathbf{f}(\lambda)$ and $\mathbf{w}(\lambda)$, with boldface notation, denote the Laplace-transformed (in the continuous-time case) or Z-transformed (in the discrete-time case) time-dependent vectors, namely the p-dimensional system output vector $y(t)$, m_u-dimensional control input vector $u(t)$, m_d-dimensional disturbance vector $d(t)$, m_f-dimensional fault vector $f(t)$ and m_w-dimensional noise vector $w(t)$, respectively. $G_u(\lambda)$, $G_d(\lambda)$, $G_f(\lambda)$ and $G_w(\lambda)$ are the transfer function matrices (TFMs) from the control inputs u, disturbance inputs d, fault inputs f and noise inputs w to the outputs y, respectively. According to the system type, $\lambda = s$, the complex variable in the Laplace transform in the case of a continuous-time system, or $\lambda = z$, the complex variable in the Z-transform in the case of a discrete-time system. For most of the practical applications, the TFMs $G_u(\lambda)$, $G_d(\lambda)$, $G_f(\lambda)$ and $G_w(\lambda)$ are proper rational matrices. However, for complete generality of our problem settings, we will allow that these TFMs are general improper rational matrices for which we will not *a priori* assume any further properties (e.g., stability or full rank).

Remark 2.1 Throughout this book, the main difference between the disturbance input $d(t)$ and noise input $w(t)$ arises from the formulation of the fault monitoring goals. In this respect, when synthesizing devices to serve for fault diagnosis purposes, we will generally target the *exact* decoupling of the effects of disturbance inputs. Since generally the exact decoupling of effects of noise inputs is not achievable, we will simultaneously try to attenuate their effects, to achieve an *approximate* decoupling. Consequently, we will try to solve synthesis problems exactly or approximately, in accordance with the absence or presence of noise inputs in the underlying plant model, respectively. $\square$

An equivalent *descriptor* state-space realization of the input–output model (2.1) has the form

$$
\begin{aligned}
E\lambda x(t) &= Ax(t) + B_u u(t) + B_d d(t) + B_f f(t) + B_w w(t)\,, \\
y(t) &= Cx(t) + D_u u(t) + D_d d(t) + D_f f(t) + D_w w(t)\,,
\end{aligned}
\tag{2.2}
$$

with the n-dimensional state vector $x(t)$, where $\lambda x(t) = \dot{x}(t)$ or $\lambda x(t) = x(t+1)$ depending on the type of the system, continuous- or discrete-time, respectively. In general, the square matrix E can be singular, but we will assume that the linear pencil $A - \lambda E$ is regular. For systems with proper TFMs in (2.1), we can always choose a *standard* state-space realization where $E = I$. In general, we can also assume that the representation (2.2) is minimal, that is, the pair $(A - \lambda E, C)$ is *observable* and the pair $(A - \lambda E, [\,B_u\ B_d\ B_f\ B_w\,])$ is *controllable*. The corresponding TFMs of the model in (2.1) are

$$
\begin{aligned}
G_u(\lambda) &= C(\lambda E - A)^{-1} B_u + D_u, \\
G_d(\lambda) &= C(\lambda E - A)^{-1} B_d + D_d, \\
G_f(\lambda) &= C(\lambda E - A)^{-1} B_f + D_f, \\
G_w(\lambda) &= C(\lambda E - A)^{-1} B_w + D_w
\end{aligned}
\tag{2.3}
$$

or in an equivalent notation

$$
\big[\,G_u(\lambda)\ G_d(\lambda)\ G_f(\lambda)\ G_w(\lambda)\,\big] := \left[\begin{array}{c|cccc} A - \lambda E & B_u & B_d & B_f & B_w \\ \hline C & D_u & D_d & D_f & D_w \end{array}\right].
$$

Remark 2.2 Although the overall model (2.2) can always be chosen minimal (i.e., controllable and observable), the state-space realizations of individual channels of the input–output model (2.1) may not be minimal. For example, the pair $(A - \lambda E, B_u)$ (which is part of the state-space realization of $G_u(\lambda)$) may be uncontrollable and even not stabilizable. In spite of this apparent deficiency, the chosen form (2.2) of the system model is instrumental for the development of all computational procedures presented in this book. $\square$

An important class of models with additive faults arises when defining the fault signals for two main categories of faults, namely actuator and sensor faults. Modelling actuator faults can be done by replacing $u(t)$ by a perturbed input $u(t) + S_a f_a(t)$, where $f_a(t)$ is the actuator fault signal and S_a is a fault distribution matrix. S_a is usually a full column rank matrix formed from distinct columns of an identity matrix of appropriate order. Thus, the corresponding fault-to-output TFM is defined as $G_f(\lambda) := G_u(\lambda)S_a$. Similarly, sensor faults can be modelled by replacing $y(t)$ by $y(t) + S_s f_s(t)$, where $f_s(t)$ is the sensor fault signal and S_s is an appropriate fault distribution matrix. The corresponding TFM is simply $G_f(\lambda) := S_s$. In the case when both actuator and sensor faults are present, then for the fault signal $f(t) := \begin{bmatrix} f_a(t) \\ f_s(t) \end{bmatrix}$, the corresponding fault-to-output TFM is

$$G_f(\lambda) := \begin{bmatrix} G_u(\lambda)S_a & S_s \end{bmatrix}. \tag{2.4}$$

In general, the matrices S_a and S_s can be chosen to also ensure a certain uniform range of magnitudes of the expected fault signals via appropriate scaling of fault inputs. The corresponding state-space realization (2.2) is obtained with $B_f := [\, B_u S_a \;\; 0 \,]$ and $D_f := [\, D_u S_a \;\; S_s \,]$. An important aspect of this approach is that the resulting models with additive faults can simultaneously cover several categories of actuator and sensor faults.

Example 2.1 Flight actuators with faults are often modelled as continuous-time LTI models, whose transfer function representation is

$$\mathbf{y}(s) = G_u(s)\mathbf{u}(s) + G_f(s)\mathbf{f}(s)\,,$$

where $u(t)$ and $y(t)$ are, respectively, the commanded and achieved surface positions and $f(t)$ is a fault signal. For an input (actuator) fault we can take $G_f(s) = G_u(s)$, while for an output (sensor) fault $G_f(s) = 1$. If both types of faults are present, then $f(t)$ is a two-dimensional vector and $G_f(s) = [\, G_u(s)\; 1\,]$. First- or second-order actuator models are frequently used for fault detection purposes, where the effects of the load (e.g., air resistance) are included in the actuator parameters. A first-order actuator model has the transfer function

$$G_u(s) = \frac{k}{s + k}\,,$$

where k is a constant gain. A second-order actuator model can have the transfer function

$$G_u(s) = \frac{\omega^2}{s^2 + 2\zeta\omega s + \omega^2}\,,$$

where ζ is the damping ratio and ω is the natural oscillation frequency. These simple faulty system models are suitable to serve for monitoring several categories of actuator faults, which can be considered as additive faults, such as jamming, runaway, oscillatory failure, and certain types of loss of efficiency. $\diamond$

The underlying plant models (i.e., without noise and fault inputs) often represent linearizations of nonlinear dynamic plant models in specific operation points and for fixed values of plant parameters. To cope with variabilities in operating conditions and plant parameters, alternative representations are often used, which cover a whole family of linearized models. To use such models for solving fault diagnosis problems by employing the linear system techniques described in this book, we have to convert them into LTI state-space representations with additional noise or fault inputs, where the noise inputs account for the effects of existing variabilities in operating points and parameters, or for extreme parameter variations due to parametric faults. In what follows, we show for two classes of plant models, namely state-space models with parameter-dependent matrices and families of linearized state-space models (i.e., multiple LTI models), how they can be recast as standard LTI models with additional noise or fault inputs.

2.2.1 *Models with Parametric Uncertainties*

The fault-free LTI model without noise inputs, which underlies the additive fault model of the form (2.1) or (2.2), often represents an approximation via linearization of a nonlinear system model in a certain nominal operating point for a particular combination of some model parameter values. Therefore, the validity of approximations by linearized models is often restricted to small variations around some nominal operating points and parameter values. To extend the range of validity of linear models, so-called *linear parameter-varying* (LPV) models have been introduced, where the dependence of (time-varying) operating conditions and parameters is explicitly reflected in the model (i.e., in the matrices of the state-space model). LPV models are therefore useful to represent nonlinear systems in terms of a family of linear models. The existing explicit parametric dependence can be exploited in various ways both in robust synthesis methods as well as in robustness analysis. In this section, we describe a useful technique to recast LPV models with parametric uncertainties into LTI models with fictitious noise inputs, which account for the effects of parametric variations. These models can thus serve to arrive at additive fault models of the general form (2.1) or (2.2).

There exist various techniques to determine LPV models. These techniques encompass (1) the symbolic manipulation of the nonlinear model equations leading to so-called quasi-LPV models, where the matrices of the state-space model depend on a parameter vector, whose components include not only plant parameters but also components of the state or output vectors of the nonlinear model; (2) direct parameter estimation using special global identification experiments; (3) interpolation of a set of local models (e.g., obtained via linearizations) using regression-based parameter fitting techniques. We will not further discuss various existing techniques, but note that this research field is still very active as documented by a rapidly increasing amount of the literature dedicated to this topic.

Let ρ be a time-varying parameter vector and consider a state-space realization of the fault-free system in the LPV form

$$E(\rho)\lambda x(t) = A(\rho)x(t) + B_u(\rho)u(t) + B_d(\rho)d(t)\,,$$
$$y(t) = C(\rho)x(t) + D_u(\rho)u(t) + D_d(\rho)d(t)\,. \tag{2.5}$$

Consider the parameter-dependent matrix

$$S(\rho) := \begin{bmatrix} E(\rho) & A(\rho) & B_u(\rho) & B_d(\rho) \\ 0 & C(\rho) & D_u(\rho) & D_d(\rho) \end{bmatrix} \tag{2.6}$$

and express $S(\rho)$ in the form

$$S(\rho) = S^{(0)} + \Delta_S \Gamma_S(\rho)\,, \tag{2.7}$$

where $S^{(0)}$ is the (nominal) value of $S(\lambda)$ defined for a constant value $\rho = \rho_0$ as

$$S^{(0)} := S(\rho_0) = \begin{bmatrix} E^{(0)} & A^{(0)} & B_u^{(0)} & B_d^{(0)} \\ 0 & C^{(0)} & D_u^{(0)} & D_d^{(0)} \end{bmatrix}, \tag{2.8}$$

$\Gamma_S(\rho)$ satisfies $\Gamma_S(\rho_0) = 0$ and Δ_S is a constant matrix. The system (2.5) can be alternatively expressed in the form

$$E^{(0)}\lambda x(t) = A^{(0)}x(t) + B_u^{(0)}u(t) + B_d^{(0)}d(t) + \Delta x(t,\rho),$$
$$y(t) = C^{(0)}x(t) + D_u^{(0)}u(t) + D_d^{(0)}d(t) + \Delta y(t,\rho), \tag{2.9}$$

where $\Delta x(t,\rho)$ and $\Delta y(t,\rho)$ can be interpreted as input and output noise terms and are given by

$$\begin{bmatrix} \Delta x(t,\rho) \\ \Delta y(t,\rho) \end{bmatrix} := \Delta_S \Gamma_S(\rho) \begin{bmatrix} -\lambda x(t) \\ x(t) \\ u(t) \\ d(t) \end{bmatrix}. \tag{2.10}$$

If we denote with $\mathcal{R}(\cdot)$ the range (or image) of a matrix, then we have for all values of ρ

$$\begin{bmatrix} \Delta x(t,\rho) \\ \Delta y(t,\rho) \end{bmatrix} \in \mathcal{R}(\Delta_S)\,.$$

We can define a LTI model of the form

$$E^{(0)}\lambda x(t) = A^{(0)}x(t) + B_u^{(0)}u(t) + B_d^{(0)}d(t) + B_w^{(0)}w(t),$$
$$y(t) = C^{(0)}x(t) + D_u^{(0)}u(t) + D_d^{(0)}d(t) + D_w^{(0)}w(t)\,,$$

to replace the LPV model (2.5), provided we can determine the two matrices $B_w^{(0)}$ and $D_w^{(0)}$ to satisfy the range condition

$$\mathcal{R}\left(\begin{bmatrix} B_w^{(0)} \\ D_w^{(0)} \end{bmatrix}\right) = \mathcal{R}(\Delta_S)\,,$$

where $w(t)$ is a fictitious "noise" signal, whose dimension m_w is equal to the column dimension of $B_w^{(0)}$ and $D_w^{(0)}$. Such a LTI model can be considered an "exact" (thus not a conservative) replacement of the LPV model (2.5).

The determination of $B_w^{(0)}$ and $D_w^{(0)}$ can be done from the following singular value decomposition (SVD):

$$\Delta_S = \begin{bmatrix} U_1 & U_2 \end{bmatrix} \begin{bmatrix} \Sigma & 0 \\ 0 & 0 \end{bmatrix} \begin{bmatrix} V_1 & V_2 \end{bmatrix}^T = U_1 \Sigma V_1^T,$$

where $\begin{bmatrix} U_1 & U_2 \end{bmatrix}$ and $\begin{bmatrix} V_1 & V_2 \end{bmatrix}$ are orthogonal matrices, and Σ is a nonsingular diagonal matrix with the decreasingly ordered nonzero singular values on its diagonal. From linear algebra, we know that the columns of U_1 form an orthogonal basis for the range of Δ_S. Therefore, we can choose $\begin{bmatrix} B_w^{(0)} \\ D_w^{(0)} \end{bmatrix} = U_1$ or $\begin{bmatrix} B_w^{(0)} \\ D_w^{(0)} \end{bmatrix} = U_1 \Sigma$. The latter choice includes different scalings of noise inputs.

The representation of $S(\rho)$ in the form (2.7) can be easily obtained for LPV models whose matrices depend rationally on the components of ρ. For such models, $S(\rho)$ can always be expressed using an upper *linear fractional transformation*[1] (LFT)-based equivalent representation

$$S(\rho) = LFT_u(M, \Delta), \tag{2.11}$$

where $M = \begin{bmatrix} M_{11} & M_{12} \\ M_{21} & M_{22} \end{bmatrix}$ is a certain constant matrix with M_{11} square and $\Delta = \Delta(\rho)$ is a diagonal matrix depending on the components of ρ such that $\Delta(\rho_0) = 0$. Straightforward algorithms are available to obtain the above representation. The above LFT-based representation of $S(\rho)$ allows to immediately obtain $S^{(0)} = M_{22}$, $\Delta_S = M_{21}$ and $\Gamma_S(\rho) = \Delta(I - \Delta M_{11})^{-1} M_{12}$.

Example 2.2 We consider an LPV model with a standard state-space realization (2.5) with $E = I$, $B_d = 0$, $D_d = 0$ and

$$A(\rho_1, \rho_2) = \begin{bmatrix} -0.8 & 0 & 0 \\ 0 & -0.5(1 + \rho_1) & 0.6(1 + \rho_2) \\ 0 & -0.6(1 + \rho_2) & -0.5(1 + \rho_1) \end{bmatrix}, \quad B_u = \begin{bmatrix} 1 & 1 \\ 1 & 0 \\ 0 & 1 \end{bmatrix},$$

$$C = \begin{bmatrix} 0 & 1 & 1 \\ 1 & 1 & 0 \end{bmatrix}, \quad D_u = \begin{bmatrix} 0 & 0 \\ 0 & 0 \end{bmatrix}.$$

In the expression of $A(\rho_1, \rho_2)$, ρ_1 and ρ_2 are uncertainties in the real and imaginary parts of the two complex conjugate eigenvalues $\lambda_{1,2} = -0.5 \pm j0.6$ of the nominal value $A^{(0)} = A(0, 0)$.

---

[1] An upper LFT for a partitioned matrix $M = \begin{bmatrix} M_{11} & M_{12} \\ M_{21} & M_{22} \end{bmatrix}$ and a compatible Δ is defined as $LFT_u(M, \Delta) := M_{22} + M_{21} \Delta (I - \Delta M_{11})^{-1} M_{12}$.

We can recast the effects of uncertain parameters ρ_1 and ρ_2 as fictitious noise inputs. Since all system matrices, excepting the state matrix $A(\rho_1, \rho_2)$, are constant, we only have to represent $A(\rho_1, \rho_2)$ as

$$A(\rho_1, \rho_2) = A^{(0)} + \Delta_A \Gamma_A(\rho),$$

with

$$A^{(0)} = \begin{bmatrix} -0.8 & 0 & 0 \\ 0 & -0.5 & 0.6 \\ 0 & -0.6 & -0.5 \end{bmatrix}$$

and Δ_A and $\Gamma_A(\rho)$ given by

$$\Delta_A = \begin{bmatrix} 0 & 0 \\ 0 & 1 \\ 1 & 0 \end{bmatrix}, \quad \Gamma_A(\rho) = \begin{bmatrix} 0 & -0.6\rho_2 & -0.5\rho_1 \\ 0 & -0.5\rho_1 & 0.6\rho_2 \end{bmatrix}. \tag{2.12}$$

Thus, the noise terms in (2.9) are $\Delta x(t, \rho) = \Delta_A \Gamma_A(\rho) x(t)$ and $\Delta y(t, \rho) = 0$ and therefore, Δ_S has the reduced form

$$\Delta_S = \begin{bmatrix} \Delta_A \\ 0 \end{bmatrix}, \tag{2.13}$$

which can be used for range computation. For this simple LPV model, we can define the noise vector as $w(t) := \Gamma_A(\rho) x(t)$ and the corresponding noise matrices result as

$$B_w^{(0)} := \Delta_A, \quad D_w^{(0)} = 0.$$

The resulting equivalent LTI model with noise inputs is

$$\begin{aligned} \dot{x}(t) &= A^{(0)} x(t) + B_u u(t) + B_w^{(0)} w(t), \\ y(t) &= Cx(t) + D_u u(t) + D_w^{(0)} w(t). \end{aligned} \tag{2.14}$$

Using the LFT-based representation $A(\rho) = LFT_u(M, \Delta)$ with $\Delta = \mathrm{diag}(\rho_1, \rho_1, \rho_2, \rho_2)$ and

$$M = \left[\begin{array}{c|c} M_{11} & M_{12} \\ \hline M_{21} & M_{22} \end{array} \right] = \left[\begin{array}{cccc|cccccccc} 0 & 0 & 0 & 0 & 0 & 0 & 0 & 0 & 1 & 0 & 0 & 0 \\ 0 & 0 & 0 & 0 & 0 & 0 & 0 & 0 & 0 & 1 & 0 & 0 \\ 0 & 0 & 0 & 0 & 0 & 0 & 0 & 0 & 0 & 1 & 0 & 0 \\ 0 & 0 & 0 & 0 & 0 & 0 & 0 & 0 & 1 & 0 & 0 & 0 \\ \hline 0 & 0 & 0 & 0 & 1 & 0 & 0 & -0.8 & 0 & 0 & 1 & 1 \\ -0.5 & 0 & 0.6 & 0 & 0 & 1 & 0 & 0 & -0.5 & 0.6 & 1 & 0 \\ 0 & -0.5 & 0 & -0.6 & 0 & 0 & 1 & 0 & -0.6 & -0.5 & 0 & 1 \\ 0 & 0 & 0 & 0 & 0 & 0 & 0 & 0 & 1 & 1 & 0 & 0 \\ 0 & 0 & 0 & 0 & 0 & 0 & 0 & 1 & 1 & 0 & 0 & 0 \end{array} \right],$$

we can check that $\mathcal{R}(M_{21}) = \mathcal{R}(\Delta_S)$, with Δ_S given in (2.13). Thus, we obtain the same $B_w^{(0)}$ (though with permuted columns) and $D_w^{(0)}$ as above.

The Julia script **Ex2_4.jl** in Listing 2.1 generates the matrices $B_w^{(0)}$ and $D_w^{(0)}$ of the LTI model (2.14) from the LPV model considered in this example. $\diamond$

Listing 2.1 Script **Ex2_4.jl** to convert the LPV model of Example 2.2 to LTI form

```julia
using FaultDetectionTools, DescriptorSystems, Measurements,
      LinearAlgebra, Test

# Example 2.4 - Generation of a synthesis model with noise inputs
# from a LPV uncertain model
println("Example 2.4")
# define ρ1 and ρ2 as uncertain parameters
ρ1 = measurement(0, rand()); ρ2 = measurement(0, rand());

# define A(ρ1,ρ2), Bu, C, Du
n = 3; mu = 2; p = 2;          # enter dimensions
A = [ -.8  0  0;
       0 -0.5*(1+ρ1)  0.6*(1+ρ2);
       0 -0.6*(1+ρ2)  -0.5*(1+ρ1) ];
Bu = [ 1 1; 1 0; 0 1]; C = [0 1 1;1 1 0]; Du = zeros(p,mu);

# build ΔA and ΔC
ΔA = getfield.(A,:err)
ΔC = zeros(p,n)

# determine the number of noise inputs
mw = rank(ΔA)

# compute orthogonal range of [ΔA;0]
U = svd([ΔA;ΔC]).U[:,1:mw]

# determine Bw and Dw
Bw = U[1:n,:]
Dw = U[n+1:end,:]

# setup of the synthesis model with additive actuator faults
# with Bf = Bu, Df = Du
sysf = fdimodset(dss(getfield.(A,:val),[Bu Bw],C,[Du Dw]),
        c = 1:mu, n = mu .+ (1:mw), f = 1:mu);
display(sysf)
```

2.2.2 Models with Parametric Faults

Multiplicative faults are frequently used as synonym for parametric faults. Let $\rho(t)$ be a time-varying parameter vector and consider a state-space realization of the system in the LPV form (2.5). The components of ρ consist of those system parameters, whose extreme variations represent the parametric faults for the given plant. In this section, we discuss the conversion of LPV models of the form (2.5) into equivalent LTI models with additive faults of the form (2.2). This conversion can be done using similar techniques as those described in Sect. 2.2.1.

The parameter-dependent matrix $S(\rho)$ in (2.6) can be expressed in the form $S(\rho) = S^{(0)} + \Delta_S \Gamma_S(\rho)$, where $S^{(0)}$ is the fault-free (nominal) value of $S(\lambda)$ defined for a (constant) normal value $\rho = \rho_0$ as in (2.8); $\Gamma_S(\rho)$ satisfies $\Gamma_S(\rho_0) = 0$ and therefore is zero in the fault-free case; and Δ_S is a constant matrix. The system (2.5) can be expressed in the alternative form (2.9), where $\Delta x(t, \rho)$ and $\Delta y(t, \rho)$ in (2.10) can be now interpreted as fault input terms. These terms defined in (2.10) depend on the magnitudes of state and input vectors, which justifies the characterization of parametric faults as multiplicative faults. The fault input terms are therefore zero if the system state, the state derivative and the system inputs are zero, and they are also zero if the plant operates in its normal condition corresponding to $\rho = \rho_0$.

Let $B_f^{(0)}$ and $D_f^{(0)}$ be two matrices, which satisfy the range condition

$$\mathcal{R}\left(\begin{bmatrix} B_f^{(0)} \\ D_f^{(0)} \end{bmatrix} \right) = \mathcal{R}(\Delta_S) \, .$$

Since for all values of ρ we have

$$\begin{bmatrix} \Delta x(t, \rho) \\ \Delta y(t, \rho) \end{bmatrix} \in \mathcal{R}(\Delta_S) \, ,$$

we can define a fictitious "fault" signal $f(t)$, whose dimension m_f is equal to the column dimension of $B_f^{(0)}$ and $D_f^{(0)}$, and build the equivalent LTI model of the form

$$\begin{aligned} E^{(0)} \lambda x(t) &= A^{(0)} x(t) + B_u^{(0)} u(t) + B_d^{(0)} d(t) + B_f^{(0)} f(t), \\ y(t) &= C^{(0)} x(t) + D_u^{(0)} u(t) + D_d^{(0)} d(t) + D_f^{(0)} f(t). \end{aligned} \tag{2.15}$$

Remark 2.3 In general, the parameter vector ρ can be split in two components, $\rho = [\, \rho_1^T \ \rho_2^T \,]^T$, where ρ_1 includes those model parameters, which are susceptible to parametric faults, while ρ_2 is the part of parameters, which have to be exclusively handled as uncertainties. This separation should be also reflected in the equivalent LTI model (2.15), where, besides the additive fault input terms $B_f^{(0)} f(t)$ and $D_f^{(0)} f(t)$, which correspond to the parametric faults in ρ_1, noise input terms $B_w^{(0)} w(t)$ and $D_w^{(0)} w(t)$ have to be added, respectively, which correspond to the parametric uncertainties in ρ_2. Such a model can be easily determined, if the parametric matrix $S(\rho)$ in (2.6) exhibits an additive separability property with respect to the two components of ρ of the form $S(\rho) = S_1(\rho_1) + S_2(\rho_2)$. In this case, the fault and noise input matrices can be generated separately for each term of $S(\rho)$, employing the approaches described in this and previous sections. An important class of parametric models for which this condition is fulfilled is formed by LPV models with affine dependence of system matrices of parameters. $\square$

Although the LTI model (2.15) can be considered an "exact" (thus not a conservative) replacement of the LPV model (2.5), still it hides the complex dependence of the fault input terms on the system parameters and variables exhibited in (2.10). For example, a direct correspondence between the components of ρ and the components of f is not explicitly provided, which could make the fault isolation task difficult. By taking into account the explicit dependence of the fault input terms of the current values of the state and input vectors, a more detailed (structured) representation of parametric faults is possible, which, however, leads to time-varying matrices in the fault input channel of the additive fault model (2.15) (i.e., using time-varying matrices $B_f(t)$ and $D_f(t)$ instead of the constant matrices $B_f^{(0)}$ and $B_f^{(0)}$, respectively).

For some particular LPV models, such as those with the system matrices having affine dependence on the components of ρ, it is possible to exploit this feature to obtain a direct correspondence between the components of the additive fault vector $f(t)$ and the components of the parameter vector ρ. To show this, let us assume that ρ has k components and the LPV system matrices have the following affine representations:

$$E(\rho) = E^{(0)} + \sum_{i=1}^{k} E^{(i)} \rho_i, \quad A(\rho) = A^{(0)} + \sum_{i=1}^{k} A^{(i)} \rho_i,$$

$$B_u(\rho) = B_u^{(0)} + \sum_{i=1}^{k} B_u^{(i)} \rho_i, \quad B_d(\rho) = B_d^{(0)} + \sum_{i=1}^{k} B_d^{(i)} \rho_i,$$

$$C(\rho) = C^{(0)} + \sum_{i=1}^{k} C^{(i)} \rho_i, \quad D_u(\rho) = D_u^{(0)} + \sum_{i=1}^{k} D_u^{(i)} \rho_i, \quad D_d(\rho) = D_d^{(0)} + \sum_{i=1}^{k} D_d^{(i)} \rho_i.$$

This allows to express $S(\rho)$ in (2.6) as $S(\rho) = S^{(0)} + \sum_{i=1}^{k} S^{(i)} \rho_i$, with

$$S^{(i)} := \left[\begin{array}{c|ccc} E^{(i)} & A^{(i)} & B_u^{(i)} & B_d^{(i)} \\ \hline 0 & C^{(i)} & D_u^{(i)} & D_d^{(i)} \end{array} \right] =: \left[\frac{S_x^{(i)}}{S_y^{(i)}} \right] \tag{2.16}$$

for $i = 0, 1, \ldots, k$. The fault input terms in (2.9) can be now expressed as

$$\left[\begin{array}{c} \Delta x(t, \rho) \\ \Delta y(t, \rho) \end{array} \right] := \left(\sum_{i=1}^{k} S^{(i)} \rho_i \right) z(t) = \left[\, S^{(1)} z(t) \ \cdots \ S^{(k)} z(t) \, \right] \rho,$$

where $z(t) := \left[-\lambda x^T(t) \ x^T(t) \ u^T(t) \ d^T(t) \right]^T$. Using the partitioning of $S^{(i)}$ in (2.16), we can express the fault input terms as

$$\Delta x(t, \rho) = B_f(t)\rho(t), \qquad \Delta y(t, \rho) = B_f(t)\rho(t),$$

where $B_f(t) = \left[\, S_x^{(1)} z(t) \ \cdots \ S_x^{(k)} z(t) \, \right]$ and $D_f(t) = \left[\, S_y^{(1)} z(t) \ \cdots \ S_y^{(k)} z(t) \, \right]$. By defining $f(t) := \rho(t)$, we obtain the equivalent additive fault model

$$E^{(0)}\lambda x(t) = A^{(0)}x(t) + B_u^{(0)}u(t) + B_d^{(0)}d(t) + B_f(t)f(t)\,,$$
$$y(t) = C^{(0)}x(t) + D_u^{(0)}u(t) + D_d^{(0)}d(t) + D_f(t)f(t)\,, \qquad (2.17)$$

with the fault input channel containing time-varying matrices with a special structure. Although the synthesis methods presented in this book are mainly intended for LTI models with additive faults, still some of these methods can be extended to handle models of the form (2.17) (see Remark 7.4).

The following example illustrates the model conversion techniques presented in this section on the basis of the model considered in Example 2.2.

Example 2.3 Consider the same model as that used in Example 2.2, where only the state matrix $A(\rho_1, \rho_2)$ depends on parameters. Large variations of these parameters are considered parametric faults. We can convert this LPV model to a LTI model with additive faults, using the calculations already done in Example 2.2. The resulting LTI model with additive fault inputs can be set up as

$$\dot{x}(t) = A^{(0)}x(t) + B_u u(t) + B_f^{(0)} f(t)\,,$$
$$y(t) = Cx(t) + D_u u(t)\,, \qquad (2.18)$$

with $B_f^{(0)} := \Delta_A$, where $A(\rho_1, \rho_2) = A^{(0)} + \Delta_A \Gamma_A(\rho)$, and Δ_A and $\Gamma_A(\rho)$ are given in (2.12). The equivalent fault input is defined as $f(t) := \Gamma_A(\rho)x(t)$.

For the fault vector $f(t)$, we can use a more structured representation using the alternative affine representation of $A(\rho)$ as $A(\rho) = A^{(0)} + A^{(1)}\rho_1 + A^{(2)}\rho_2$, which leads to

$$A(\rho)x(t) = A^{(0)}x(t) + \left[\, A^{(1)}x(t)\ A^{(2)}x(t) \,\right]\rho\,.$$

It follows that with $f(t) := \rho(t)$, we obtain a time-varying input matrix $B_f(t) = \left[\, A^{(1)}x(t)\ A^{(2)}x(t) \,\right]$ to replace $B_f^{(0)}$ in (2.18). $\diamond$

2.2.3 Multiple Linear Models

A frequent situation, which occurs in practical applications, is that we only have at our disposal N plant models (i.e., a multiple model) of the form

$$E^{(i)}\lambda x^{(i)}(t) = A^{(i)}x^{(i)}(t) + B_u^{(i)}u(t) + B_d^{(i)}d(t)\,,$$
$$y^{(i)}(t) = C^{(i)}x^{(i)}(t) + D_u^{(i)}u(t) + D_d^{(i)}d(t)\,, \qquad (2.19)$$

where, for $i = 1, \ldots, N$, $x^{(i)}(t) \in \mathbb{R}^n$ and $y^{(i)}(t) \in \mathbb{R}^p$ are the state vector and output vector of the i-th system, respectively. For simplicity, we assume that in all models, the dimensions of the state, output and input vectors are the same. Typically, (2.19) describes a family of linearized models for N relevant combinations of plant operating points and plant parameters. In what follows, we describe a simple method to recast such a multiple model into a unique LTI model with additional fictitious noise inputs, which account for the effects of variations in operating points and parameters. The resulting LTI model can then serve to build models with additive faults of the form (2.1) or (2.2).

The matrices of each component model can be expressed in the form

$$E^{(i)} = E^{(0)} + \Delta_E^{(i)}, \quad A^{(i)} = A^{(0)} + \Delta_A^{(i)}, \quad B_u^{(i)} = B_u^{(0)} + \Delta_{B_u}^{(i)}, \quad , \dots$$

where $E^{(0)}$, $A^{(0)}$, $B_u^{(0)}$, ... are some nominal values (or simply the mean values of the corresponding matrices), while $\Delta_E^{(i)}$, $\Delta_A^{(i)}$, $\Delta_{B_u}^{(i)}$, ... are the deviations from the nominal (or mean) values. If we denote

$$\Delta_S^{(i)} := \begin{bmatrix} -\Delta_E^{(i)} & \Delta_A^{(i)} & \Delta_{B_u}^{(i)} & \Delta_{B_d}^{(i)} \\ 0 & \Delta_C^{(i)} & \Delta_{D_u}^{(i)} & \Delta_{D_d}^{(i)} \end{bmatrix},$$

then each model of the form (2.19) can be equivalently represented in the form

$$\begin{aligned} E^{(0)}\lambda x^{(i)}(t) &= A^{(0)}x^{(i)}(t) + B_u^{(0)}u(t) + B_d^{(0)}d(t) + \Delta_x^{(i)}(t), \\ y^{(i)}(t) &= C^{(0)}x^{(i)}(t) + D_u^{(0)}u(t) + D_d^{(0)}d(t) + \Delta_y^{(i)}(t), \end{aligned} \qquad (2.20)$$

where $\Delta_x^{(i)}(t)$ and $\Delta_y^{(i)}(t)$ are the noise terms specific to the i-th model, given by

$$\begin{bmatrix} \Delta_x^{(i)}(t) \\ \Delta_y^{(i)}(t) \end{bmatrix} := \Delta_S^{(i)} \begin{bmatrix} \lambda x^{(i)}(t) \\ x^{(i)}(t) \\ u(t) \\ d(t) \end{bmatrix}.$$

Therefore, for each component model we have

$$\begin{bmatrix} \Delta_x^{(i)}(t) \\ \Delta_y^{(i)}(t) \end{bmatrix} \in \mathcal{R}(\Delta_S^{(i)}).$$

We can try to define a unique model of the form

$$\begin{aligned} E^{(0)}\lambda x(t) &= A^{(0)}x(t) + B_u^{(0)}u(t) + B_d^{(0)}d(t) + B_w^{(0)}w(t), \\ y(t) &= C^{(0)}x(t) + D_u^{(0)}u(t) + D_d^{(0)}d(t) + D_w^{(0)}w(t) \end{aligned}$$

to approximate the collection of N models in (2.20), provided the two matrices $B_w^{(0)}$ and $D_w^{(0)}$ satisfy the range condition

$$\mathcal{R}\left(\begin{bmatrix} B_w^{(0)} \\ D_w^{(0)} \end{bmatrix}\right) = \mathcal{R}(\Delta_S^{(1)}) \cup \cdots \cup \mathcal{R}(\Delta_S^{(N)}) = \mathcal{R}\left(\begin{bmatrix} \Delta_S^{(1)} & \cdots & \Delta_S^{(N)} \end{bmatrix}\right)$$

and $w(t)$ is a fictitious "noise" signal, which formally matches the column dimension of $B_w^{(0)}$ and $D_w^{(0)}$. Such an unique model is certainly conservative, because it includes in a single model the effects of *all* possible parametric variations. Nevertheless, the degree of conservatism may be acceptable in practice, because often the component models share common structural features, which are reflected in the matrices $\Delta_S^{(i)}$ (e.g., constant rank or fixed zero entries).

The determination of $B_w^{(0)}$ and $D_w^{(0)}$ can be done (as in the preceding section) from the SVD

$$\Delta_S := \begin{bmatrix} \Delta_S^{(1)} & \cdots & \Delta_S^{(N)} \end{bmatrix} = \begin{bmatrix} U_1 & U_2 \end{bmatrix} \begin{bmatrix} \Sigma & 0 \\ 0 & 0 \end{bmatrix} \begin{bmatrix} V_1 & V_2 \end{bmatrix}^T = U_1 \Sigma V_1^T, \qquad (2.21)$$

where $\begin{bmatrix} U_1 & U_2 \end{bmatrix}$ and $\begin{bmatrix} V_1 & V_2 \end{bmatrix}$ are orthogonal matrices, and Σ is a diagonal matrix with the decreasingly ordered nonzero singular values on its diagonal. Therefore, we can choose $\begin{bmatrix} B_w^{(0)} \\ D_w^{(0)} \end{bmatrix} = U_1$ or $\begin{bmatrix} B_w^{(0)} \\ D_w^{(0)} \end{bmatrix} = U_1 \Sigma$. The latter choice includes different scalings of noise inputs. Often, we can even use instead of U_1, only a few of its leading columns, which correspond to the most significant singular values.

Remark 2.4 The determination of U_1 in the SVD (2.21) involves the computation of the SVD of the potentially large matrix Δ_S with $n + p$ rows and $N(2n + m_u + m_d)$ columns. This computation may require a tremendous computational effort for large N or large n if Δ_S is explicitly formed. Fortunately, this can be avoided by a suitable preprocessing of Δ_S. The proposed computational approach below leads to significant saving in the computational effort if $N(2n + m_u + m_d) \gg n + p$ (which is usually the case). Let Q be an orthogonal matrix, which compresses the columns of Δ_S to a $(n + p) \times (n + p)$ matrix R_S (upper triangular) according to the following orthogonal RQ-decomposition of Δ_S as

$$\Delta_S = \begin{bmatrix} R_S & 0 \end{bmatrix} Q .$$

Then, the SVDs of the large matrix Δ_S and of the compressed matrix R_S provide the same U_1 matrix as basis for $\mathcal{R}(\Delta_S)$. Since the computation of the right transformation matrix $\begin{bmatrix} V_1 & V_2 \end{bmatrix}$ is not necessary, it is possible to determine R_S without determining explicitly Q. Moreover, we can compute R_S even without the need to form Δ_S explicitly, using the following recursion based on successive low-dimensional RQ-decompositions

$$\begin{bmatrix} R_i & 0 \end{bmatrix} Q_i = \begin{bmatrix} R_{i-1} & \Delta_S^{(i)} \end{bmatrix}, \quad i = 1, \ldots, N ,$$

where $R_0 = 0_{(n+p) \times (n+p)}$. Here, each R_i is an $(n + p) \times (n + p)$ (upper triangular) matrix and Q_i is an orthogonal matrix of order $3n + p + m_u + m_d$, which need not be computed. At the end, we set $R_S = R_N$ and the SVD of R_S provides the orthogonal basis matrix U_1 of $\mathcal{R}(S_\Delta)$ from the SVD (2.21).

Example 2.4 We consider once again the LPV system with a standard state-space realization (2.5) used in Example 2.2. Consider a set of parameter values $(\rho_1^{(i)}, \rho_2^{(i)})$, for $i = 1, \ldots, N$. For each value $(\rho_1^{(i)}, \rho_2^{(i)})$, we define

$$A^{(i)} := A(\rho_1^{(i)}, \rho_2^{(i)}) = A^{(0)} + \Delta_A^{(i)},$$

where $A^{(0)} = A(0, 0)$ is the nominal value of $A(\rho_1, \rho_2)$

$$A^{(0)} = \begin{bmatrix} -0.8 & 0 & 0 \\ 0 & -0.5 & 0.6 \\ 0 & -0.6 & -0.5 \end{bmatrix}$$

and $\Delta_A^{(i)}$ is given by

$$\Delta_A^{(i)} = \begin{bmatrix} 0 & 0 & 0 \\ 0 & -0.5\rho_1^{(i)} & 0.6\rho_2^{(i)} \\ 0 & -0.6\rho_2^{(i)} & 0.5\rho_1^{(i)} \end{bmatrix}.$$

With the reduced $\Delta_S^{(i)}$ defined as

$$\Delta_S^{(i)} := \begin{bmatrix} \Delta_A^{(i)} \\ 0 \end{bmatrix},$$

we have that $\mathcal{R}(\Delta_S) = \mathcal{R}(\Delta_S^{(i)})$ for all simultaneously nonzero $\rho_1^{(i)}$ and $\rho_2^{(i)}$. For convenience, we take $\rho_1^{(1)} = \sqrt{0.5}$ and $\rho_2^{(1)} = \sqrt{0.5}$, and we only compute the SVD of the nonzero part $\Delta_A^{(1)}$ as

$$\Delta_A^{(1)} = [U_1 | U_2] \begin{bmatrix} \Sigma & 0 \\ 0 & 0 \end{bmatrix} [V_1 | V_2]^T = \begin{bmatrix} 0 & 0 & 1 \\ -0.7071 & -0.7071 & 0 \\ -0.7071 & 0.7071 & 0 \end{bmatrix} \begin{bmatrix} I_2 & 0 \\ 0 & 0 \end{bmatrix} \begin{bmatrix} 0 & 0 & 1 \\ -1 & 0 & 0 \\ 0 & 1 & 0 \end{bmatrix}^T.$$

By defining the noise vector as $w(t) := \Sigma V_2^T x(t)$ and the corresponding matrices

$$B_w^{(0)} := \begin{bmatrix} 0 & 0 \\ -0.7071 & -0.7071 \\ -0.7071 & 0.7071 \end{bmatrix}, \quad D_w^{(0)} = \begin{bmatrix} 0 & 0 \\ 0 & 0 \end{bmatrix},$$

we arrive at a LTI model similar to (2.14). It is easy to see that although $B_w^{(0)}$ in Examples 2.2 and 2.4 are different, their ranges are the same.   ◇

2.3 Physical Fault Models

For physically modelled faults, each fault mode leads to a distinct model. Assume that we have N LTI models describing the fault-free and faulty systems, and for $i = 1, \ldots, N$ the i-th model is specified in the input–output form

$$\mathbf{y}^{(i)}(\lambda) = G_u^{(i)}(\lambda)\mathbf{u}^{(i)}(\lambda) + G_d^{(i)}(\lambda)\mathbf{d}^{(i)}(\lambda) + G_w^{(i)}(\lambda)\mathbf{w}^{(i)}(\lambda), \qquad (2.22)$$

where $y^{(i)}(t) \in \mathbb{R}^{p^{(i)}}$ is the output vector of the i-th system with control input $u^{(i)}(t) \in \mathbb{R}^{m_u^{(i)}}$, disturbance input $d^{(i)}(t) \in \mathbb{R}^{m_d^{(i)}}$ and noise input $w^{(i)}(t) \in \mathbb{R}^{m_w^{(i)}}$, respectively, and where $G_u^{(i)}(\lambda)$, $G_d^{(i)}(\lambda)$ and $G_w^{(i)}(\lambda)$ are the TFMs from the corresponding plant inputs to outputs. The significance of disturbance and noise inputs, and the basic difference between them, have already been discussed in Sect. 2.1. The state-space realizations corresponding to the multiple model (2.22) are for $i = 1, \ldots, N$ of the form

$$\begin{aligned}
E^{(i)}\lambda x^{(i)}(t) &= A^{(i)}x^{(i)}(t) + B_u^{(i)}u^{(i)}(t) + B_d^{(i)}d^{(i)}(t) + B_w^{(i)}w^{(i)}(t)\,, \\
y^{(i)}(t) &= C^{(i)}x^{(i)}(t) + D_u^{(i)}u^{(i)}(t) + D_d^{(i)}d^{(i)}(t) + D_w^{(i)}w^{(i)}(t)\,,
\end{aligned} \qquad (2.23)$$

where $x^{(i)}(t) \in \mathbb{R}^{n^{(i)}}$ is the state vector of the i-th system and, generally, can have different dimensions for different systems.

The multiple-model description represents a very general way to describe plant models with various faults. For example, extreme variations of parameters representing the so-called parametric faults can be easily described by multiple models. Let ρ be a parameter vector, which includes a set of model parameters whose extreme values characterize the different fault cases. We assume that the system model depending on ρ has the form

$$\mathbf{y}(\lambda) = G_u(\lambda, \rho)\mathbf{u}(\lambda) + G_d(\lambda, \rho)\mathbf{d}(\lambda) + G_w(\lambda, \rho)\mathbf{w}(\lambda)\,. \qquad (2.24)$$

Let $\rho^{(i)}, i = 1, \ldots, N$ be a set of values, which characterize both the normal operation as well as the fault cases. Then, the multiple model (2.22) for $i = 1, \ldots, N$ can be defined for $u^{(i)} = u$, $d^{(i)} = d$ and $w^{(i)} = w$ as

$$G_u^{(i)}(\lambda) := G_u(\lambda, \rho^{(i)}), \quad G_d^{(i)}(\lambda) := G_d(\lambda, \rho^{(i)}), \quad G_w^{(i)}(\lambda) := G_w(\lambda, \rho^{(i)})\,. \qquad (2.25)$$

Similarly, if the state-space realization of the system model has the LPV form

$$\begin{aligned}
E(\rho)\lambda x(t) &= A(\rho)x(t) + B_u(\rho)u(t) + B_d(\rho)d(t) + B_w(\rho)w(t)\,, \\
y(t) &= C(\rho)x(t) + D_u(\rho)u(t) + D_d(\rho)d(t) + D_w(\rho)w(t)\,,
\end{aligned} \qquad (2.26)$$

then a multiple model of the form (2.23) for $i = 1, \ldots, N$ can be defined with

$$E^{(i)} = E\big(\rho^{(i)}\big), \quad A^{(i)} = A\big(\rho^{(i)}\big), \quad \ldots \qquad (2.27)$$

and $x^{(i)} = x$, $u^{(i)} = u$, $d^{(i)} = d$ and $w^{(i)} = w$.

As an example, consider the modelling of a category of loss of efficiency actuator faults for a system of the form

$$\mathbf{y}(\lambda) = G_u(\lambda)\mathbf{u}(\lambda) + G_d(\lambda)\mathbf{d}(\lambda),$$

without noise input. The loss of efficiency of the i-th actuator can be modelled by defining

$$G_u^{(i)}(\lambda) := G_u(\lambda) F_a^{(i)}, \qquad G_d^{(i)}(\lambda) := G_d(\lambda), \tag{2.28}$$

where $F_a^{(i)}$ is a diagonal matrix with unit diagonal entries excepting the i-th diagonal entry, which is set to a non-negative subunitary value. Several values for $F_a^{(i)}$ can be employed to cope with different degrees of failures of a single actuator. A complete failure of the i-th actuator can be easily modelled by setting the i-th diagonal element to zero.

Different categories of sensor faults (e.g., bias, drift and frozen value) can be modelled by adding fictive disturbances, which act on the respective outputs. For example, for a fault in the i-th output sensor we can define the fault model with

$$G_u^{(i)}(\lambda) = F_s^{(i)} G_u(\lambda), \qquad G_d^{(i)}(\lambda) = [\, F_s^{(i)} G_d(\lambda), \ e_i \,],$$

where $F_s^{(i)}$ is chosen as a diagonal matrix (similar to $F_a^{(i)}$) to account for the i-th sensor fault and e_i is the i-th column of the identity matrix. Simultaneous sensor faults can be also easily modelled in this way.

The multiple-model approach to fault modelling offers a simple framework to model faults, by associating a distinct model to each fault or combination of several simultaneous faults. Occasionally, this involves defining additional disturbance inputs, which account for the effects of modelled faults. However, employing this approach for complex systems with many potential actuator and sensor faults can easily lead to a large number of models. The situation can be even worse if additionally parametric faults can occur. Since the number of required models increases exponentially with the number of faults, the applicability of this modelling framework is restricted to a system with a relatively small number of faults. Often, the single-fault-at-time assumption is used to limit the number of considered faults and thus of the associated models.

Example 2.5 Flight actuators can be often modelled as first-order parameter-dependent linear continuous-time models, whose transfer function representation is (assuming constant parameter)

$$\mathbf{y}(s) = G_u(s, k)\mathbf{u}(s),$$

with

$$G_u(s, k) = \frac{k}{s + k}.$$

The input $u(t)$ is usually the demanded position (e.g., angle) of the attached control surface and the output $y(t)$ is the actual surface position. Here, k is the effective actuator gain, which generally depends on flight parameters such as the current weight-balance, current flight conditions, and the current deflection of the attached control surface. High-accuracy models may also cover the dependence of the effects of air resistance on the associated control surface.

Several categories of actuator faults are best modelled as parametric faults and described by several special values of the gain parameter k. Assume that the normal operation of a flight actuator is well approximated by a LTI model with $G_u^{(1)}(s) := G_u(s, k_0)$, where k_0 is a mean value of k in normal operation (e.g., a typical value may be $k_0 = 14$). The actuator disconnection fault due to a broken rod between the actuator and corresponding control surface is considered a severe fault (although very improbable). Physically, this fault is equivalent to the lack of any air resistance and hinge moments, because the actuator rod can move practically without encountering any resistance from the control surface. Therefore, this fault can be modelled by a LTI model with $G_u^{(2)}(s) := G_u(s, k_{max})$, where k_{max} is the highest achievable gain (e.g., a typical value satisfies $k_{max} > 50$). On the opposite side, highly deflected control surfaces produce large air resistance and therefore, the actuators are susceptible to intermittent position saturations. This intermittent fault is called *stall load* and represents a sudden change of gain to its lowest value k_{min} (e.g., a typical value may be $k_{min} = 0.01$). Thus, this fault can be modelled by a LTI with $G_u^{(3)}(s) := G_u(s, k_{min})$. Finally, a sluggish behaviour of the actuator can be associated with a second type of loss-of-effectiveness fault and can be modelled as a LTI with $G_u^{(4)}(s) := G_u(s, \gamma k_0)$, where $0 < \gamma < 1$ is a parameter, which indicates the degradation of the actuator dynamics (e.g., $\gamma = 0.5$ for a 50% sluggishness). Several values of γ can be used to characterize different degradation levels. $\Diamond$

2.4 Notes and References

Background material on input–output representations via TFMs is given in Sect. 9.1. The structural properties of descriptor system representations are discussed, for example, in [26]. See also Sect. 9.2 for background material on descriptor systems.

The LTI model with additive faults, control, disturbance and noise inputs has been already used in the textbook of Gertler [51], while other authors such as Chen and Patton [23] and more recently Ding [29] employ LTI models with faults without making the difference between disturbance and noise inputs, when solving fault detection problems. However, the distinction between disturbance and noise inputs is the basis of the synthesis methods presented in this book and allows the exploitation of all existing structural features related to the unknown inputs acting on the system. The presence or absence of noise inputs in the underlying LTI synthesis models determines a direct correspondence with the synthesis methods labelled as "approximate" or "exact", respectively. This systematics has been introduced in a recent survey of synthesis methods [157].

Several methods to recast uncertain models into models with noise inputs are described by Chen and Patton [23]. For the derivation of LPV models, there are many approaches proposed in the literature (see, for example, the special issue of the IEEE Transactions on Control Systems Technology [83]). The determination of a high-fidelity first-order LPV flight actuator model approximation has been described in [158].

The use of multiple models is a standard way to address robust synthesis problems in the presence of parametric uncertainties. Several applications of multiple-model-based approaches are presented in [94]. The use of multiple models for fault detection and fault-tolerant control has been proposed by Maybeck [89] and Boškovic and Mehra [17].

Techniques for handling models with parametric (multiplicative) faults are described by Gertler in [51]. The special case of handling affine LPV models is considered by Ding in [29].

Chapter 3
Fault Diagnosis

In this chapter, we describe first the basic fault monitoring tasks, such as fault detection, fault isolation, fault estimation and fault identification. The concepts of fault detectability and fault isolability are then introduced and characterized. Six "canonical" fault detection problems are formulated for the class of LTI systems with additive faults. The formulations of the exact and approximate synthesis problems are independent of any possible solution method and allow the derivation of general solvability conditions in terms of ranks of certain transfer function matrices. An important aspect to emphasize is that the formulations of approximate problems include, as particular cases, the formulations of exact problems. This chapter is concluded with a discussion of performance requirements for fault diagnosis systems and the selection of thresholds to be used for decision-making.

3.1 Basic Fault Monitoring Tasks

A fault represents a deviation from the normal behaviour of a system due to an unexpected event (e.g., physical component failure or supply breakdown). In safety-critical technical systems, a fault must be detected as early as possible to prevent any serious consequence. For this purpose, fault diagnosis techniques are used to allow the detection of occurrence of faults (fault detection), the localization of detected faults (fault isolation), the reconstruction of the fault signal (fault estimation) and a precise classification of the detected faults and their characteristics (fault identification). In a specific practical application, the term *fault detection and diagnosis* (FDD) may include, besides fault detection, also further aspects such as fault isolation, fault estimation and fault identification. An FDD system is a device (usually based on a collection of real-time processing algorithms) suitably set up to fulfil the above tasks. The minimal functionality of any FDD system is illustrated in Fig. 3.1.

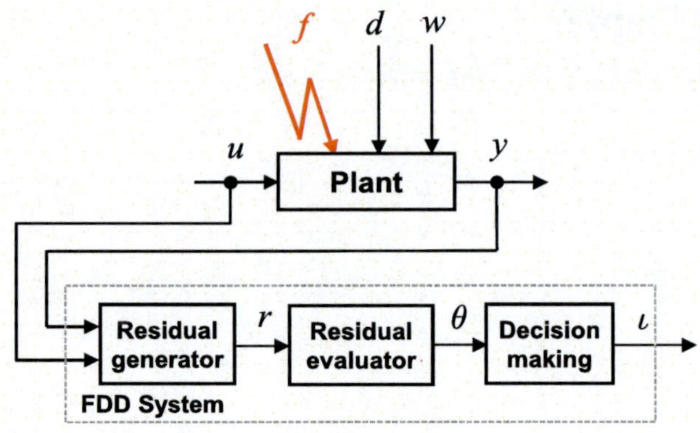

Fig. 3.1 Basic fault diagnosis setup

The main component of any FDD system (as that in Fig. 3.1) is the *residual generator* (or *fault detection filter*, or simply *fault detector*), which produces residual signals grouped in a q-dimensional vector r by processing the available measurements y and the known values of control inputs u. The role of the residual signals is to indicate the presence or absence of faults, and therefore the residual r must be equal (or close) to zero in the absence of faults and significantly different from zero after a fault occurs. For decision-making, suitable measures of the residual magnitudes (e.g., signal norms) are generated in a vector θ, which is then used to produce the corresponding decision vector ι. In what follows, several fault monitoring tasks are formulated and discussed. The discussion, which follows, also serves to fix the basic terminology used throughout this book.

Fault detection is simply a binary decision on the presence of any fault ($f \neq 0$) or the absence of all faults ($f = 0$). Typically, $\theta(t)$ is a scalar evaluation signal, which approximates $\|r\|_2$, the $\mathcal{L}_2$- or ℓ_2-norms of signal r, while $\iota(t)$ is a scalar decision-making signal defined as $\iota(t) = 1$ if $\theta(t) > \tau$ (fault occurrence) or $\iota(t) = 0$ if $\theta(t) \leq \tau$ (no fault), where τ is a suitable threshold quantifying the gap between the "small" and "large" magnitudes of the residual. The decision on the occurrence or absence of faults must be done in the presence of arbitrary control inputs u, disturbance inputs d and noise inputs w acting simultaneously on the system. The effects of the control inputs on the residual can be always decoupled by a suitable choice of the residual generation filter. In the ideal case, when no noise inputs are present ($w \equiv 0$), the residual generation filter must additionally be able to *exactly* decouple the effects of the disturbance inputs in the residual and ensure, simultaneously, the sensitivity of the residual to all faults (i.e., *complete fault detectability*; see Sect. 3.3). In this case, $\tau = 0$ can be (ideally) used. However, in the general case when $w \not\equiv 0$, only an *approximate* decoupling of w can be achieved (at best) and a sufficient gap must exist between the magnitudes of residuals in fault-free and faulty situations. Therefore, an appropriate choice of $\tau > 0$ must avoid false alarms and missed detections.

Fault isolation concerns with the exact localization of occurred faults and involves for each component f_j of the fault vector f the decision on the presence of j-th fault ($f_j \neq 0$) or its absence ($f_j = 0$). Ideally, this must be achieved regardless of the faults that occur one at a time or several faults occurring simultaneously. Therefore, the fault isolation task is significantly more difficult than the simpler fault detection. For fault isolation purposes, we will assume a partitioning of the q-dimensional residual vector r in n_b stacked q_i-dimensional subvectors $r^{(i)}$, $i = 1, \ldots, n_b$, in the form

$$r = \begin{bmatrix} r^{(1)} \\ \vdots \\ r^{(n_b)} \end{bmatrix}, \tag{3.1}$$

where $q = \sum_{i=1}^{n_b} q_i$. A typical fault evaluation setup used for fault isolation is to define $\theta_i(t)$, the i-th component of $\theta(t)$, as a real-time computable approximation of $\|r^{(i)}\|_2$. The i-th component of $\iota(t)$ is set to $\iota_i(t) = 1$ if $\theta_i(t) > \tau_i$ (i-th residual fired) or $\iota_i(t) = 0$ if $\theta_i(t) \leq \tau_i$ (i-th residual not fired), where τ_i is a suitable threshold for the i-th subvector $r^{(i)}(t)$. If a sufficiently large number of measurements are available, then it can be aimed that $r^{(i)}$ is influenced only by the i-th fault signal f_i. This setting, with $n_b = m_f$, allows *strong fault isolation*, where an arbitrary number of simultaneous faults can be isolated. The isolation of the i-th fault is achieved if $\iota_i(t) = 1$, while for $\iota_i(t) = 0$ the i-th fault is not present. In many practical applications, the lack of a sufficiently large number of measurements impedes strong isolation of simultaneous faults. Therefore, often only *weak fault isolation* can be performed under simplifying assumptions, for example, that the faults occur one at a time or no more than two faults may occur simultaneously. The fault isolation schemes providing weak fault isolation compare the resulting n_b-dimensional binary decision vector $\iota(t)$ with a predefined set of binary fault signatures. If each individual fault f_j has associated a distinct signature s_j, the j-th fault can be isolated by simply checking that $\iota(t)$ matches the associated signature s_j. Similar to fault detection, besides the decoupling of the control inputs u from the residual r (always possible), the exact decoupling of the disturbance inputs d from r can be strived in the case when $w \equiv 0$. However, in the general case when $w \not\equiv 0$, only approximate decoupling of w can be achieved (at best) and a careful selection of tolerances τ_i is necessary to perform fault isolation without false alarms and missed detections.

The combined *fault detection and isolation* (FDI) is the basis of the fault diagnosis, and it is nowadays fully accepted that FDI is a must for almost all practical applications. This justifies partly the main emphasis of this book on residual generation for this class of problems.

Fault estimation represents the next level of complexity in fault diagnosis and addresses the reconstruction of fault signals from the available measurements. The fault estimation can be interpreted as a more challenging strong fault isolation, where additionally, the fault signal must be reconstructed. Fault estimation can serve, for example, as a basis for control law reconfiguration for specific classes of faults or for virtual sensor development. Exact fault estimation can be seldom achieved (even if sufficient number of measurements are available), and often the only meaningful requirement is the approximate reconstruction of the fault signal (e.g., via asymptotic estimation).

Fault identification is intended to fully characterize the type, size and nature of the faults, and therefore is often a part of fault diagnosis, usually subsequent to fault isolation. The fault identification relies on a range of techniques, which include parameter estimation, active fault detection, signal analysis and classification, pattern recognition, etc. These techniques will not be addressed in this book, but some aspects related to characterizing fault modes (e.g., detection of classes of persistent faults) can be explicitly considered in formulating concrete design requirements for residual generators.

3.2 Residual Generation

Recall the additive fault model (2.1) introduced in Chap. 2, described by input–output representations of the form

$$\mathbf{y}(\lambda) = G_u(\lambda)\mathbf{u}(\lambda) + G_d(\lambda)\mathbf{d}(\lambda) + G_f(\lambda)\mathbf{f}(\lambda) + G_w(\lambda)\mathbf{w}(\lambda)\,, \qquad (3.2)$$

where $\mathbf{y}(\lambda)$, $\mathbf{u}(\lambda)$, $\mathbf{d}(\lambda)$, $\mathbf{f}(\lambda)$ and $\mathbf{w}(\lambda)$ are Laplace-transformed (in the continuous-time case) or Z-transformed (in the discrete-time case) vectors of the p-dimensional system output vector $y(t)$, m_u-dimensional control input vector $u(t)$, m_d-dimensional disturbance vector $d(t)$, m_f-dimensional fault vector $f(t)$ and m_w-dimensional noise vector $w(t)$, respectively, and where $G_u(\lambda)$, $G_d(\lambda)$, $G_f(\lambda)$ and $G_w(\lambda)$ are the *transfer function matrices* (TFMs) from the control inputs to outputs, disturbance inputs to outputs, fault inputs to outputs and noise inputs to outputs, respectively. For complete generality of our problem formulations, we will allow that these TFMs are general rational matrices (proper or improper) for which we will not a priori assume any further properties.

A linear residual generator (or fault detection filter) processes the measurable system outputs $y(t)$ and known control inputs $u(t)$ and generates the residual signals $r(t)$, which serve for decision-making on the presence or absence of faults. The input–output form of this filter is

$$\mathbf{r}(\lambda) = Q(\lambda)\begin{bmatrix} \mathbf{y}(\lambda) \\ \mathbf{u}(\lambda) \end{bmatrix} = Q_y(\lambda)\mathbf{y}(\lambda) + Q_u(\lambda)\mathbf{u}(\lambda)\,, \qquad (3.3)$$

with $Q(\lambda) = [\, Q_y(\lambda)\ Q_u(\lambda) \,]$, and is called the *implementation form*. The TFM $Q(\lambda)$ for a physically realizable filter must be *proper* (i.e., only with finite poles) and *stable* (i.e., only with poles having negative real parts for a continuous-time system or magnitudes less than one for a discrete-time system). The dimension q of the residual vector $r(t)$ depends on the fault detection problem to be addressed.

The residual signal $r(t)$ in (3.3) generally depends on all system inputs $u(t)$, $d(t)$, $f(t)$ and $w(t)$ via the system output $y(t)$. The *internal form* of the filter is obtained by replacing in (3.3) $\mathbf{y}(\lambda)$ by its expression in (3.2), and is given by

$$\mathbf{r}(\lambda) = R(\lambda) \begin{bmatrix} \mathbf{u}(\lambda) \\ \mathbf{d}(\lambda) \\ \mathbf{f}(\lambda) \\ \mathbf{w}(\lambda) \end{bmatrix} = R_u(\lambda)\mathbf{u}(\lambda) + R_d(\lambda)\mathbf{d}(\lambda) + R_f(\lambda)\mathbf{f}(\lambda) + R_w(\lambda)\mathbf{w}(\lambda) , \quad (3.4)$$

with $R(\lambda) = [\, R_u(\lambda) \; R_d(\lambda) \; R_f(\lambda) \; R_w(\lambda) \,]$ defined as

$$\left[\, R_u(\lambda) \,\middle|\, R_d(\lambda) \,\middle|\, R_f(\lambda) \,\middle|\, R_w(\lambda) \,\right] := Q(\lambda) \left[\begin{array}{c|c|c|c} G_u(\lambda) & G_d(\lambda) & G_f(\lambda) & G_w(\lambda) \\ I_{m_u} & 0 & 0 & 0 \end{array}\right].$$
$$(3.5)$$

For a properly designed filter $Q(\lambda)$, the corresponding internal representation $R(\lambda)$ is also a proper and stable system, and additionally fulfils specific fault detection and isolation requirements.

3.3 Fault Detectability

The concepts of fault detectability and complete fault detectability deal with the sensitivity of the residual to an individual fault and to all faults, respectively. For the discussion of these concepts, we will assume that no noise input is present in the system model (3.2) ($w \equiv 0$).

Definition 3.1 For the system (3.2), the j-th fault f_j is *detectable* if there exists a fault detection filter $Q(\lambda)$ such that for all control inputs u and all disturbance inputs d, the residual $r \neq 0$ if $f_j \neq 0$ and $f_k = 0$ for all $k \neq j$.

Definition 3.2 The system (3.2) is *completely fault detectable* if there exists a fault detection filter $Q(\lambda)$ such that for each j, $j = 1, \ldots, m_f$, all control inputs u and all disturbance inputs d, the residual $r \neq 0$ if $f_j \neq 0$ and $f_k = 0$ for all $k \neq j$.

We have the following result to characterize fault detectability.

Theorem 3.1 *For the system (3.2), the j-th fault is detectable if and only if*

$$\operatorname{rank} \left[\, G_d(\lambda) \; G_{f_j}(\lambda) \,\right] > \operatorname{rank} G_d(\lambda), \qquad (3.6)$$

where $G_{f_j}(\lambda)$ is the j-th column of $G_f(\lambda)$ and rank $(\cdot)$ *is the normal rank (i.e., over rational functions) of a rational matrix.*

Proof To prove necessity, we show that the assumption of the detectability of f_j implies condition (3.6). If f_j is detectable, then there exists a filter $Q(\lambda)$ such that if $\mathbf{f}_j(\lambda) \neq 0$ and $\mathbf{f}_k(\lambda) = 0$ for $k \neq j$, the corresponding residual in (3.4)

$$\mathbf{r}(\lambda) = R_{f_j}(\lambda)\mathbf{f}_j(\lambda) + R_u(\lambda)\mathbf{u}(\lambda) + R_d(\lambda)\mathbf{d}(\lambda) \qquad (3.7)$$

is nonzero for all $\mathbf{u}(\lambda)$ and $\mathbf{d}(\lambda)$, where $R_{f_j}(\lambda)$ is the j-th column of $R_f(\lambda)$. In particular, there always exists a scalar output filter $Q(\lambda)$, which fulfils the above conditions. From (3.7), it follows that it is necessary that the corresponding $R_u(\lambda)$ and $R_d(\lambda)$ in (3.4) satisfy $R_u(\lambda) = 0$ and $R_d(\lambda) = 0$, otherwise there always exist control inputs and disturbances, which make $\mathbf{r}(\lambda) = 0$ for any $\mathbf{f}_j(\lambda)$. It follows that

$$\mathbf{r}(\lambda) = R_{f_j}(\lambda)\mathbf{f}_j(\lambda). \tag{3.8}$$

To guarantee the sensitivity of r to f_j, $R_{f_j}(\lambda)$ must be nonzero.

Condition (3.8) together with $R_u(\lambda) = 0$ and $R_d(\lambda) = 0$ can be transcribed into a linear rational matrix equation fulfilled by $Q(\lambda)$

$$Q(\lambda) \begin{bmatrix} G_{f_j}(\lambda) & G_u(\lambda) & G_d(\lambda) \\ 0 & I_{m_u} & 0 \end{bmatrix} = \begin{bmatrix} R_{f_j}(\lambda) & 0 & 0 \end{bmatrix}. \tag{3.9}$$

This equation has a solution $Q(\lambda)$ if and only if the following rank condition for the compatibility of the linear system (3.9) is fulfilled

$$\mathrm{rank} \begin{bmatrix} G_{f_j}(\lambda) & G_u(\lambda) & G_d(\lambda) \\ 0 & I_{m_u} & 0 \end{bmatrix} = \mathrm{rank} \begin{bmatrix} G_{f_j}(\lambda) & G_u(\lambda) & G_d(\lambda) \\ 0 & I_{m_u} & 0 \\ R_{f_j}(\lambda) & 0 & 0 \end{bmatrix}.$$

Since rank $R_{f_j}(\lambda) = 1$, by equating the ranks of the left- and right-hand side matrices above, we obtain

$$m_u + \mathrm{rank} \begin{bmatrix} G_{f_j}(\lambda) & G_d(\lambda) \end{bmatrix} = m_u + 1 + \mathrm{rank}\, G_d(\lambda),$$

which is equivalent to (3.6).

To prove the sufficiency of (3.6), we determine a filter $Q(\lambda)$ such that if $\mathbf{f}_j(\lambda) \neq 0$ and $\mathbf{f}_k(\lambda) = 0$ for $k \neq j$, the residual $\mathbf{r}(\lambda)$ in (3.7) is nonzero for all $\mathbf{u}(\lambda)$ and $\mathbf{d}(\lambda)$. For this, we can choose an arbitrary $R_{f_j}(\lambda) \neq 0$ and determine $Q(\lambda)$ by solving the equation (3.9), whose solvability is guaranteed by condition (3.6). ∎

The following extension of Theorem 3.1 characterizes the complete fault detectability.

Theorem 3.2 *The system (3.2) is completely fault detectable if and only if*

$$\mathrm{rank} \begin{bmatrix} G_d(\lambda) & G_{f_j}(\lambda) \end{bmatrix} > \mathrm{rank}\, G_d(\lambda), \quad j = 1, \ldots, m_f. \tag{3.10}$$

Proof The necessity follows simply by applying Theorem 3.1 for each individual fault f_j. To prove sufficiency, we need to show that there exists a single filter $Q(\lambda)$ such that the residual r is sensitive to all faults. This is equivalent to asking that all columns of the corresponding $R_f(\lambda)$ in (3.5) are nonzero. The conditions (3.10) guarantee that each fault is detectable, and, therefore, we can build for each fault f_j

a filter $Q^{(j)}(\lambda)$ such that the corresponding $R_{f_j}(\lambda)$ is nonzero and $Q^{(j)}(\lambda)$ satisfies the linear rational equation

$$Q^{(j)}(\lambda) \begin{bmatrix} G_{f_j}(\lambda) & G_u(\lambda) & G_d(\lambda) \\ 0 & I_{m_u} & 0 \end{bmatrix} = \begin{bmatrix} R_{f_j}^{(j)}(\lambda) & 0 & 0 \end{bmatrix}.$$

We define

$$Q(\lambda) := \begin{bmatrix} Q^{(1)}(\lambda) \\ \vdots \\ Q^{(m_f)}(\lambda) \end{bmatrix},$$

for which the corresponding $R_f(\lambda)$ has the following partitioned form:

$$R_f(\lambda) = \begin{bmatrix} R_{f_1}^{(1)}(\lambda) & \cdots & * \\ \vdots & \ddots & \vdots \\ * & \cdots & R_{f_{m_f}}^{(m_f)}(\lambda) \end{bmatrix}.$$

Each of the m_f columns of $R_f(\lambda)$ is nonzero because of the presence of the nonzero diagonal blocks $R_{f_j}^{(j)}(\lambda)$ for $j = 1, \ldots, m_f$. ∎

Remark 3.1 The complete fault detectability condition (3.10) is generically fulfilled if $p > m_d$, which, for a full column rank $G_d(\lambda)$, is also a necessary condition for complete fault detectability. This imposes that there must be more output measurements than disturbance inputs. □

Remark 3.2 For the case $m_d = 0$, condition (3.10) reduces to the *input observability* conditions of all faults

$$G_{f_j}(\lambda) \neq 0, \quad j = 1, \ldots m_f. \tag{3.11}$$

Note that in several works, this simpler condition is used for the definition of complete fault detectability. While these conditions are necessary for complete fault detectability, they are not sufficient for the existence of a filter guaranteeing that all columns of the corresponding $R_f(\lambda)$ are nonzero. □

Remark 3.3 For the characterization of complete fault detectability, the stability and properness of the filter $Q(\lambda)$ and of the corresponding $R_f(\lambda)$ play no role. However, it is easy to see that we can always impose that both $Q(\lambda)$ and $R_f(\lambda)$ are proper and stable, because any filter $Q(\lambda)$ can be replaced by an updated filter $M(\lambda)Q(\lambda)$, where $M(\lambda)$ is a stable, invertible TFM, which can be always chosen such that $M(\lambda)[\, Q(\lambda)\ R_f(\lambda)\,]$ is stable and proper (see Sect. 9.1.6). □

Strong fault detectability is a concept related to the reliability and easiness of performing fault detection. The main idea behind this concept is the ability of the residual generators to produce persistent residual signals in the case of persistent fault excitation. For example, for reliable fault detection it is advantageous to have an asymptotically non-vanishing residual signal in the case of persistent faults as step or sinusoidal signals. On the contrary, the lack of strong fault detectability may make the detection of these types of faults more difficult, because their effects manifest in the residual only during possibly short transients, thus the effect disappears in the residual after enough long time although the fault itself still persists.

The definitions of strong fault detectability and complete strong fault detectability cover several classes of persistent fault signals. Let $\Omega \subset \partial \mathbb{C}_s$ be a set of complex frequencies, which characterize the classes of persistent fault signals in question. Common choices in a continuous-time setting are $\Omega = \{0\}$ for a step signal or $\Omega = \{i\omega\}$ for a sinusoidal signal of frequency ω. However, Ω may contain several such frequency values or even a whole interval of frequency values, such as $\Omega = \{i\omega \mid \omega \in [\omega_1, \omega_2]\}$. We denote by $\mathcal{F}_\Omega$ the class of persistent fault signals characterized by Ω.

Definition 3.3 For the system (3.2) and a given set of frequencies $\Omega \subset \partial \mathbb{C}_s$, the j-th fault f_j is *strong fault detectable* with respect to Ω if there exists a stable fault detection filter $Q(\lambda)$ such that for all control inputs u and all disturbance inputs d, the residual $r(t) \neq 0$ for $t \to \infty$ if $f_j \in \mathcal{F}_\Omega$ and $f_k = 0$ for all $k \neq j$.

Definition 3.4 The system (3.2) is *completely strong fault detectable* with respect to a given set of frequencies $\Omega \subset \partial \mathbb{C}_s$, if there exists a stable fault detection filter $Q(\lambda)$ such that for each $j = 1, \ldots, m_f$, all control inputs u and all disturbance inputs d, the residual $r(t) \neq 0$ for $t \to \infty$ if $f_j \in \mathcal{F}_\Omega$ and $f_k = 0$ for all $k \neq j$.

For a given stable filter $Q(\lambda)$, checking the strong detection property of the filter for the j-th fault f_j involves to check that $R_{f_j}(\lambda)$ has no zeros in Ω. A characterization of strong detectability as a system property is given in what follows.

Theorem 3.3 *Let $\Omega \subset \partial \mathbb{C}_s$ be a given set of frequencies. For the system (3.2), f_j is strong fault detectable with respect to Ω if and only if f_j is fault detectable and the rational matrices $G_{e,j}(\lambda)$ and $\begin{bmatrix} G_{e,j}(\lambda) \\ F_e(\lambda) \end{bmatrix}$ have the same zero structure for each $\lambda_z \in \Omega$, where*

$$G_{e,j}(\lambda) := \begin{bmatrix} G_{f_j}(\lambda) & G_u(\lambda) & G_d(\lambda) \\ 0 & I_{m_u} & 0 \end{bmatrix}, \quad F_e(\lambda) := [\, 1 \;\; 0_{1 \times m_u} \;\; 0_{1 \times m_d} \,]. \tag{3.12}$$

Proof First we prove the necessity. Strong fault detectability trivially implies fault detectability. Along the lines of reasoning in the proof of necessity of Theorem 3.1, the condition (3.8) together with $R_u(\lambda) = 0$ and $R_d(\lambda) = 0$ can be transcribed into the matrix equation (3.9) to be fulfilled by any scalar output filter $Q(\lambda)$

$$Q(\lambda)G_{e,j}(\lambda) = F_{e,j}(\lambda), \tag{3.13}$$

where $F_{e,j}(\lambda) := [\, R_{f_j}(\lambda) \;\; 0 \;\; 0 \,]$ is a stable TFM. According to Lemma 9.5, this equation has a stable solution $Q(\lambda)$ if and only if rank $\begin{bmatrix} G_d(\lambda) & G_{f_j}(\lambda) \end{bmatrix} >$ rank $G_d(\lambda)$

(fault detectability of f_j), and $G_{e,j}(\lambda)$ and $\begin{bmatrix} G_{e,j}(\lambda) \\ F_{e,j}(\lambda) \end{bmatrix}$ have the same pole–zero structure for all $\lambda_z \in \overline{\mathbb{C}}_u$. Since $F_{e,j}(\lambda)$ is assumed stable, the condition on poles is trivially fulfilled, while the condition on zeros includes the requirement for having the same zero structure for all $\lambda_z \in \Omega$ (because $\Omega \subset \overline{\mathbb{C}}_u$). Since $R_{f_j}(\lambda)$ must not have zeros in any $\lambda_z \in \Omega$, the condition on the lack of zeros in Ω must be fulfilled, in particular, for $R_{f_j}(\lambda) = 1$.

To prove sufficiency, we determine a stable scalar output filter $Q(\lambda)$ such that if $\mathbf{f}_j(\lambda) \neq 0$ and $\mathbf{f}_k(\lambda) = 0$ for $k \neq j$, the residual $\mathbf{r}(\lambda)$ in (3.7) is nonzero for all $\mathbf{u}(\lambda)$ and $\mathbf{d}(\lambda)$, and the corresponding $R_{f_j}(\lambda)$ has no zeros in Ω. For this, we solve for $\widetilde{Q}(\lambda)$ the linear rational matrix equation $\widetilde{Q}(\lambda)G_{e,j}(\lambda) = F_e(\lambda)$, whose solvability is guaranteed by the fault detectability condition of f_j. Furthermore, the condition on the zero structures in $\lambda_z \in \Omega$ (see Lemma 9.5) ensures that the resulting solution $\widetilde{Q}(\lambda)$ has no poles in Ω. However, $\widetilde{Q}(\lambda)$ may still have poles in $\mathbb{C}_u$ or at infinity. In this case, we can choose a stable proper transfer function $M(\lambda)$, without zeros in Ω, such that $Q(\lambda) := M(\lambda)\widetilde{Q}(\lambda)$ is stable and proper (see Sect. 9.1.6), and the corresponding $R_{f_j}(\lambda) = M(\lambda)$ is stable and has no zeros in Ω. $\blacksquare$

Remark 3.4 A necessary condition for strong fault detectability of f_j is that $G_{f_j}(\lambda)$ has no zeros in Ω. This follows from the condition on zeros in Theorem 3.3, which guarantees the existence of a solution $Q(\lambda)$, without poles in Ω, of the equation $Q(\lambda)G_{e,j}(\lambda) = F_e(\lambda)$. This includes the solvability of the equation

$$Q(\lambda)\begin{bmatrix} G_{f_j}(\lambda) \\ 0 \end{bmatrix} = 1,$$

with $Q(\lambda)$ having no poles in Ω, which requires that $G_{f_j}(\lambda)$ has no zeros in Ω. $\square$

Remark 3.5 Strong fault detectability implies fault detectability, which can be thus assimilated with a kind of *weak* fault detectability property. For the characterization of the strong fault detectability, we can impose a weaker condition, involving only the existence of a filter $Q(\lambda)$ without poles in Ω (instead imposing stability). For such a filter $Q(\lambda)$, the stability can always be achieved by replacing $Q(\lambda)$ by $M(\lambda)Q(\lambda)$, where $M(\lambda)$ is a stable and invertible TFM without zeros in Ω. Such an $M(\lambda)$ can be determined from a left coprime factorization with least-order denominator of $[\, Q(\lambda)\ R_f(\lambda)\,]$ (see Sect. 9.1.6). $\square$

Example 3.1 Consider the continuous-time system of the form (3.2) with

$$G_u(s) = \begin{bmatrix} \frac{1}{s} \\ \frac{1}{s} \end{bmatrix}, \quad G_d(s) = \begin{bmatrix} 0 \\ \frac{s}{s+3} \end{bmatrix}, \quad G_f(s) = \begin{bmatrix} \frac{s+1}{s+2} \\ \frac{1}{s+2} \end{bmatrix}$$

and $\Omega = \{0\}$. Since $G_{e,1}(s)$ has a double zero in 0, while $\begin{bmatrix} G_{e,1}(\lambda) \\ F_e(\lambda) \end{bmatrix}$ has only a single zero in 0, it follows that this system is not strongly fault detectable for constant faults. $\diamondsuit$

Example 3.2 Consider a slightly different continuous-time system of the form (3.2) with

$$
G_u(s) = \begin{bmatrix} \frac{1}{s+1} \\ \frac{1}{s+1} \end{bmatrix}, \quad G_d(s) = \begin{bmatrix} 0 \\ \frac{s}{s+3} \end{bmatrix}, \quad G_f(s) = \begin{bmatrix} \frac{s+1}{s+2} \\ \frac{1}{s+2} \end{bmatrix}
$$

and $\Omega = \{0\}$. Since both $G_{e,1}(s)$ and $\begin{bmatrix} G_{e,1}(\lambda) \\ F_e(\lambda) \end{bmatrix}$ have a single zero in 0, it follows that the system is strongly fault detectable for constant fault inputs. For example, the fault detection filter

$$
Q(s) = \begin{bmatrix} 1 & 0 & -\frac{1}{s+1} \end{bmatrix}
$$

achieves

$$
R_u(s) = 0, \quad R_d(s) = 0, \quad R_f(s) = \frac{s+1}{s+2} \, .
$$

$\Diamond$

For complete strong fault detectability, the strong fault detectability of each individual fault is necessary, however, it is not a sufficient condition. This fact is illustrated in the following example.

Example 3.3 Consider the system with

$$
G_u(s) = 0, \quad G_d(s) = 0, \quad G_f(s) = \begin{bmatrix} 1 & \frac{1}{s} \end{bmatrix}, \quad G_w(s) = 0 \, .
$$

The checks of Theorem 3.3 indicate strong detectability of both fault inputs f_1 and f_2 for constant fault inputs. However, it is easy to see that there is no stable filter $Q(s)$ such that the corresponding $R_f(s) = Q(s)G_f(s)$ is stable and both $R_{f_1}(0)$ and $R_{f_2}(0)$ are nonzero. For example, with $Q(s) = s/(s+1)$ the corresponding $R_f(s)$ is

$$
R_f(s) = \begin{bmatrix} \frac{s}{s+1} & \frac{1}{s+1} \end{bmatrix} \, .
$$

Since $R_{f_1}(0) = 0$, strong fault detectability is not achieved. This fact has a simple explanation. Any filter, which makes $R_f(s)$ stable, must have a zero at $s = 0$, which therefore becomes a zero of $R_{f_1}(s)$ too. $\Diamond$

The following theorem gives a general characterization of the complete strong fault detectability as a system property.

Theorem 3.4 *Let Ω be the set of frequencies, which characterize the persistent fault signals. The system (3.2) with $w \equiv 0$ is completely strong fault detectable with respect to Ω if and only if each fault f_j, for $j = 1, \ldots, m_f$, is strong fault detectable with respect to Ω and all $G_{f_j}(\lambda)$, for $j = 1, \ldots, m_f$, have the same pole structure in λ_p for all $\lambda_p \in \Omega$.*

Proof To prove necessity, we observe first that the complete strong fault detectability with respect to Ω trivially implies the strong fault detectability with respect to Ω of all fault inputs f_j, for $j = 1, \ldots, m_f$. According to Remark 3.4, this also implies that $G_{f_j}(\lambda)$ has no zeros in Ω. Let $Q(\lambda)$ be any scalar output fault detection filter such that $Q(\lambda)$ and the corresponding $R_f(\lambda)$ are stable, and, for all $\lambda_p \in \Omega$, $R_{f_j}(\lambda_p) \neq 0$, $j = 1, \ldots, m_f$. Each $G_{f_j}(\lambda)$, for $j = 1, \ldots, m_f$, can be expressed in the form

$$G_{f_j}(\lambda) = \frac{1}{\beta^{(j)}(\lambda)} \widetilde{G}_{f_j}(\lambda) \,,$$

where $\beta^{(j)}(\lambda)$ is a *monic polynomial*[1] whose roots are the poles of $G_{f_j}(\lambda)$ in Ω, and $\widetilde{G}_{f_j}(\lambda)$ has no poles (and also no zeros) in Ω. Since $Q(\lambda)$ is a stable filter, which satisfies the linear equation

$$Q(\lambda) \begin{bmatrix} G_{f_j}(\lambda) \\ 0 \end{bmatrix} = R_{f_j}(\lambda) \,,$$

it follows that $Q(\lambda)$ must have the form $Q(\lambda) = \beta^{(j)}(\lambda) \widetilde{Q}^{(j)}(\lambda)$, where $\widetilde{Q}^{(j)}(\lambda)$ is a stable TFM, which satisfies

$$\widetilde{Q}^{(j)}(\lambda) \begin{bmatrix} \widetilde{G}_{f_j}(\lambda) \\ 0 \end{bmatrix} = R_{f_j}(\lambda) \,.$$

Moreover, since $\widetilde{G}_{f_j}(\lambda_z) \neq 0$ for all $\lambda_z \in \Omega$, it follows that $\widetilde{Q}^{(j)}(\lambda_z) \neq 0$ as well; thus $\widetilde{Q}^{(j)}(\lambda)$ has no zeros in Ω. For any two distinct values i and j, we have $\beta^{(i)}(\lambda) \widetilde{Q}^{(i)}(\lambda) = \beta^{(j)}(\lambda) \widetilde{Q}^{(j)}(\lambda)$, and, therefore, we can express, for example, $\widetilde{Q}^{(i)}(\lambda)$ as

$$\widetilde{Q}^{(i)}(\lambda) = \frac{\beta^{(j)}(\lambda)}{\beta^{(i)}(\lambda)} \widetilde{Q}^{(j)}(\lambda) \,.$$

Since both $\widetilde{Q}^{(i)}(\lambda)$ and $\widetilde{Q}^{(j)}(\lambda)$ are stable and have no zeros in Ω, the above relation can be fulfilled only if $\beta^{(i)}(\lambda) = \beta^{(j)}(\lambda)$. It follows that all $G_{f_j}(\lambda)$ must have the same set of poles in Ω.

To prove sufficiency, we show that we can construct a stable filter $Q(\lambda)$ such that the corresponding $R_f(\lambda)$ is stable and none of its columns $R_{f_j}(\lambda)$, for $j = 1, \ldots, m_f$, has zeros in Ω. Since all $G_{f_j}(\lambda)$, for $j = 1, \ldots, m_f$, have the same pole structure for all $\lambda_p \in \Omega$, we can express each $G_{f_j}(\lambda)$ in the form $G_{f_j}(\lambda) = \widetilde{G}_{f_j}(\lambda)/\beta(\lambda)$, where $\beta(\lambda)$ is the monic polynomial whose roots are the common poles in Ω of $G_{f_j}(\lambda)$, for $j = 1, \ldots, m_f$, and, consequently, $\widetilde{G}_{f_j}(\lambda)$, for $j = 1, \ldots, m_f$, have no poles in Ω. The strong fault detectability of each fault component f_j allows to determine a stable scalar output filter $Q^{(j)}(\lambda)$ such that

$$Q^{(j)}(\lambda) \begin{bmatrix} G_{f_j}(\lambda) & G_u(\lambda) & G_d(\lambda) \\ 0 & I_{m_u} & 0 \end{bmatrix} = \begin{bmatrix} R_{f_j}^{(j)}(\lambda) & 0 & 0 \end{bmatrix} \,,$$

---

[1] A monic polynomial has its leading coefficient equal to 1.

with $R_{f_j}^{(j)}(\lambda)$ having no poles in Ω and $R_{f_j}^{(j)}(\lambda_z) \neq 0$ for all $\lambda_z \in \Omega$. Since $Q^{(j)}(\lambda)$ also satisfies

$$Q^{(j)}(\lambda) \begin{bmatrix} \widetilde{G}_{f_j}(\lambda) \\ 0 \end{bmatrix} /\beta(\lambda) = R_{f_j}^{(j)}(\lambda),$$

it must have the form $Q^{(j)}(\lambda) = \beta(\lambda)\widetilde{Q}^{(j)}(\lambda)$, with $\widetilde{Q}^{(j)}(\lambda)$ stable and without zeros in Ω. For all $i = 1, \ldots, m_f$, we have

$$R_{f_i}^{(j)}(\lambda) := \widetilde{Q}^{(j)}(\lambda) \begin{bmatrix} \widetilde{G}_{f_i}(\lambda) \\ 0 \end{bmatrix}$$

and, therefore, all $R_{f_i}^{(j)}(\lambda)$, for $i = 1, \ldots, m_f$, have no poles in Ω. To enforce the stability of all $R_{f_i}^{(j)}(\lambda)$, $i = 1, \ldots, m_f$, we can replace $\widetilde{Q}^{(j)}(\lambda)$ by $\chi^{(j)}(\lambda)\widetilde{Q}^{(j)}(\lambda)$, where $\chi^{(j)}(\lambda)$ is a suitably chosen proper stable rational function without zeros in Ω (see Sect. 9.1.6). Therefore, the resulting updated $R_{f_j}^{(j)}(\lambda) \leftarrow \chi^{(j)}(\lambda)R_{f_j}^{(j)}(\lambda)$ is guaranteed to have no zeros in Ω.

We obtain the overall stable filter $Q(\lambda)$ and the corresponding stable $R_f(\lambda)$ in the block-structured forms

$$Q(\lambda) := \begin{bmatrix} Q^{(1)}(\lambda) \\ \vdots \\ Q^{(m_f)}(\lambda) \end{bmatrix}, \qquad R_f(\lambda) = \begin{bmatrix} R_{f_1}^{(1)}(\lambda) & \cdots & * \\ \vdots & \ddots & \vdots \\ * & \cdots & R_{f_{m_f}}^{(m_f)}(\lambda) \end{bmatrix}.$$

We see that for all $\lambda_z \in \Omega$, each of the m_f columns of $R_f(\lambda_z)$ is nonzero because of the presence of the nonzero diagonal blocks $R_{f_j}^{(j)}(\lambda_z)$ for $j = 1, \ldots, m_f$. $\blacksquare$

Example 3.4 Consider the system with

$$G_u(s) = 0, \quad G_d(s) = 0, \quad G_f(s) = \begin{bmatrix} \frac{1}{s} & \frac{1}{s(s+2)} \end{bmatrix}, \quad G_w(s) = 0,$$

where both $G_{f_1}(s)$ and $G_{f_2}(s)$ have a pole in the origin. For $Q(s) = s/(s+1)$, the corresponding $R_f(s)$ is

$$R_f(s) = Q(s)G_f(s) = \begin{bmatrix} \frac{1}{s+1} & \frac{1}{(s+1)(s+2)} \end{bmatrix}$$

and therefore the system is completely strong fault detectable for constant faults. $\Diamond$

The following straightforward consequence of Theorem 3.4 states that, when $G_f(\lambda)$ has no poles in Ω, the strong fault detectability of all individual faults is a necessary and sufficient condition for complete strong fault detectability.

Corollary 3.1 *Let $\Omega \subset \partial\mathbb{C}_s$ be a given set of frequencies. The system (3.2) with $G_f(\lambda)$ having no poles in Ω is completely strong fault detectable with respect to Ω if and only if each fault f_j, for $j = 1, \ldots, m_f$, is strong fault detectable with respect to Ω.*

3.4 Fault Isolability

While the detectability of a fault can be individually defined and checked, for the definition of fault isolability, we need to deal with the interactions among all fault inputs. Therefore for fault isolation, we assume a structuring of the residual vector r into n_b subvectors as in (3.1), where each individual q_i-dimensional subvector $r^{(i)}$ is differently sensitive to faults. We assume that each fault f_j is characterized by a distinct pattern of zeros and ones in a n_b-dimensional vector s_j called the *signature* of the j-th fault. Then, fault isolation consists of recognizing which signature matches the resulting decision vector ι generated by the FDD system in Fig. 3.1 according to the partitioning of r in (3.1).

For the discussion of fault isolability, we will assume that no noise input is present in the model (3.2) ($w \equiv 0$). The structure of the residual vector in (3.1) corresponds to a $q \times m_f$ TFM $Q(\lambda)$ ($q = \sum_{i=1}^{n_b} q_i$) of the residual generation filter, built by stacking a bank of n_b filters $Q^{(1)}(\lambda), \ldots, Q^{(n_b)}(\lambda)$ as

$$Q(\lambda) = \begin{bmatrix} Q^{(1)}(\lambda) \\ \vdots \\ Q^{(n_b)}(\lambda) \end{bmatrix}. \tag{3.14}$$

Thus, the i-th subvector $r^{(i)}$ is the output of the i-th filter with the $q_i \times m_f$ TFM $Q^{(i)}(\lambda)$

$$\mathbf{r}^{(i)}(\lambda) = Q^{(i)}(\lambda) \begin{bmatrix} \mathbf{y}(\lambda) \\ \mathbf{u}(\lambda) \end{bmatrix}. \tag{3.15}$$

Let $R_f(\lambda)$ be the corresponding $q \times m_f$ fault-to-residual TFM in (3.4) and we denote $R_{f_j}^{(i)}(\lambda) := Q^{(i)}(\lambda) \begin{bmatrix} G_{f_j}(\lambda) \\ 0 \end{bmatrix}$, the $q_i \times 1$ (i, j)-th block of $R_f(\lambda)$, which describes how the j-th fault f_j influences the i-th residual subvector $r^{(i)}$. Thus, $R_f(\lambda)$ is an $n_b \times m_f$ block-structured TFM of the form

$$R_f(\lambda) = \begin{bmatrix} R_{f_1}^{(1)}(\lambda) & \cdots & R_{f_{m_f}}^{(1)}(\lambda) \\ \vdots & \ddots & \vdots \\ R_{f_1}^{(n_b)}(\lambda) & \cdots & R_{f_{m_f}}^{(n_b)}(\lambda) \end{bmatrix}. \tag{3.16}$$

We associate with such a structured $R_f(\lambda)$ the $n_b \times m_f$ *structure matrix* S_{R_f} whose (i, j)-th element is defined as

$$
\begin{aligned}
S_{R_f}(i, j) &= 1 \quad \text{if} \quad R_{f_j}^{(i)}(\lambda) \neq 0, \\
S_{R_f}(i, j) &= 0 \quad \text{if} \quad R_{f_j}^{(i)}(\lambda) = 0.
\end{aligned}
\tag{3.17}
$$

If $S_{R_f}(i, j) = 1$, then we say that the residual component $r^{(i)}$ is sensitive to the j-th fault f_j, while if $S_{R_f}(i, j) = 0$ then the j-th fault f_j is decoupled from $r^{(i)}$.

Fault isolability is a property which involves all faults. This is reflected in the following definition, which relates the fault isolability property to a certain structure matrix S. For a given structure matrix S, we refer to the i-th row of S as the *specification* associated with the i-th residual component $r^{(i)}$, while the j-th column of S is called the *signature* (or *code*) associated with the j-th fault f_j.

Definition 3.5 For a given $n_b \times m_f$ structure matrix S, the model (3.2) is *S-fault isolable* if for all control inputs u and all disturbance inputs d there exists a fault detection filter $Q(\lambda)$ such that $S_{R_f} = S$.

When solving fault isolation problems, the choice of a suitable structure matrix S is an important aspect. This choice is, in general, not unique and several choices may lead to satisfactory synthesis results. We discuss shortly only some basic aspects regarding the choice of S and for a pertinent discussion, the literature mentioned at the end of this chapter can be consulted. In this context, the availability of the maximally achievable structure matrix is of paramount importance, because it allows to construct any S by simply selecting a (minimal) number of achievable specifications (i.e., rows of this matrix). The **Procedure GENSPEC**, presented in Sect. 5.4, allows to compute the maximally achievable structure matrix for a given system.

The choice of S should usually reflect the fact that complete fault detectability must be a necessary condition for the S-fault isolability. This requirement is fulfilled if S is chosen without zero columns. Also, for the unequivocal isolation of the j-th fault, the corresponding j-th column of S must be different from all other columns. Structure matrices having all columns pairwise distinct are called *weakly isolating*. Fault signatures, which result as (logical OR) combinations of two or more columns of the structure matrix, can be occasionally employed to isolate simultaneous faults, provided they are distinct from all columns of S. In this sense, a structure matrix S, which allows the isolation of an arbitrary number of simultaneously occurring faults, is called *strongly isolating*. It is important to mention in this context that a system, which is not fault isolable for a given S, may still be fault isolable for another choice of the structure matrix.

Before we establish formal conditions for fault isolability, we consider several examples of structure matrices for a system with $m_f = 4$ faults to illustrate several important types of fault isolability.

1. The structure matrix

$$S = \begin{bmatrix} 1 & 1 & 1 & 1 \end{bmatrix}$$

characterizes the *complete fault detectability* property and allows the detection (but no isolation) of any fault or an arbitrary number of simultaneous faults.

2. The structure matrix

$$S = \begin{bmatrix} 0 & 1 & 1 & 1 \\ 1 & 0 & 1 & 1 \\ 1 & 1 & 0 & 1 \\ 1 & 1 & 1 & 0 \end{bmatrix}$$

characterizes a kind of *weak fault isolability*, which allows to isolate any individual fault occurring one at a time. This structure has the remarkable property that it can cope with partial firings of the residual components, without producing false alarms.[2]

3. The structure matrix

$$S = \begin{bmatrix} 1 & 0 & 0 & 0 \\ 0 & 1 & 0 & 0 \\ 0 & 0 & 1 & 0 \\ 0 & 0 & 0 & 1 \end{bmatrix}$$

characterizes the *strong fault isolability*. This structure matrix allows to isolate an arbitrary number of simultaneous faults, because any (logical OR) combination of two or more columns of S leads to a fault signature, which can be unequivocally associated with a certain occurrence of simultaneous faults.

4. The block-diagonal structure matrix

$$S = \left[\begin{array}{c|cc|c} 1 & 0 & 0 & 0 \\ \hline 0 & 1 & 1 & 0 \\ \hline 0 & 0 & 0 & 1 \end{array} \right]$$

characterizes a weaker form of strong isolability called *strong block fault isolability*. It allows to isolate simultaneous faults belonging to three different groups of faults.

5. The structure matrix

$$S = \left[\begin{array}{c|cc|c} 0 & 1 & 1 & 1 \\ 1 & 0 & 0 & 1 \\ 1 & 1 & 1 & 0 \end{array} \right]$$

characterizes a kind of *weak block fault isolability*, which allows to isolate three groups of faults, with faults in different groups occurring one at a time.

To characterize the fault isolability property, we observe that each block row $Q^{(i)}(\lambda)$ of the TFM $Q(\lambda)$ is itself a fault detection filter, which must achieve the

---

[2] This structure matrix is called in [51] *strongly isolating*.

specification contained in the i-th row of S. Thus, the isolability conditions will consist of a set of n_b independent conditions, each of them characterizing the complete detectability of particular subsets of faults. We have the following straightforward characterization of fault isolability.

Theorem 3.5 *For a given $n_b \times m_f$ structure matrix S, the model (3.2) is S-fault isolable if and only if for $i = 1, \ldots, n_b$*

$$\text{rank}\,[\,G_d(\lambda)\ \widehat{G}_d^{(i)}(\lambda)\ G_{f_j}(\lambda)\,] > \text{rank}[\,G_d(\lambda)\ \widehat{G}_d^{(i)}(\lambda)\,],\quad \forall j,\ \ S_{ij} \neq 0, \quad (3.18)$$

where $\widehat{G}_d^{(i)}(\lambda)$ is formed from the columns $G_{f_j}(\lambda)$ of $G_f(\lambda)$ for which $S_{ij} = 0$.

Proof According to the specification encoded in the i-th row of S, the corresponding residual $r^{(i)}$ must be decoupled from all faults for which $S_{ij} = 0$. We simply redefine these faults as fictive disturbances and build the corresponding TFM $\widehat{G}_d^{(i)}(\lambda)$. Now, we apply the results of Theorem 3.1 for a system with an extended disturbance set characterized by the TFM $[\,G_d(\lambda)\ \widehat{G}_d^{(i)}(\lambda)\,]$ to obtain condition (3.18) for the complete detectability of all faults f_j for which $S_{ij} \neq 0$. ∎

The conditions (3.18) of Theorem 3.5 give a very general characterization of isolability of faults. An important particular case is *strong fault isolability* for which $S = I_{m_f}$, and thus diagonal. The following result characterizes the strong isolability.

Theorem 3.6 *The model (3.2) is strongly fault isolable if and only if*

$$\text{rank}\,[\,G_d(\lambda)\ G_f(\lambda)\,] = \text{rank}\,G_d(\lambda) + m_f\,. \tag{3.19}$$

Proof Since $S = I_{m_f}$, $R_f(\lambda)$ must be a full column rank block-diagonal TFM and

$$\mathbf{r}(\lambda) = R_f(\lambda)\mathbf{f}(\lambda)\,. \tag{3.20}$$

This condition together with $R_u(\lambda) = 0$ and $R_d(\lambda) = 0$ can be expressed as the matrix equation to be satisfied by $Q(\lambda)$

$$Q(\lambda)\begin{bmatrix} G_f(\lambda) & G_u(\lambda) & G_d(\lambda) \\ 0 & I_{m_u} & 0 \end{bmatrix} = [\,R_f(\lambda)\ 0\ 0\,]\,.$$

This equation has a solution $Q(\lambda)$ if and only if the following rank condition for compatibility is fulfilled:

$$\text{rank}\begin{bmatrix} G_f(\lambda) & G_u(\lambda) & G_d(\lambda) \\ 0 & I_{m_u} & 0 \end{bmatrix} = \text{rank}\begin{bmatrix} G_f(\lambda) & G_u(\lambda) & G_d(\lambda) \\ 0 & I_{m_u} & 0 \\ R_f(\lambda) & 0 & 0 \end{bmatrix}\,.$$

Since rank $R_f(\lambda) = m_f$, by equating the ranks of the left- and right-hand side matrices above, we obtain

$$m_u + \text{rank}\left[\, G_f(\lambda)\ G_d(\lambda)\,\right] = m_u + m_f + \text{rank}\, G_d(\lambda)\,,$$

which is equivalent to (3.19). $\blacksquare$

Note that this result can be also derived as a corollary of Theorem 3.5 for the case $S = I_{m_f}$.

Remark 3.6 Generically, the strong fault isolability condition (3.19) is fulfilled if $p \geq m_f + m_d$, which imposes that the system must have at least as many measurable outputs as fault and disturbance inputs counted together. $\square$

Remark 3.7 In the case $m_d = 0$, the strong fault isolability condition reduces to the left invertibility condition

$$\text{rank}\, G_f(\lambda) = m_f\,. \tag{3.21}$$

This condition is a necessary condition even in the case $m_d \neq 0$ (otherwise $R_f(\lambda)$ would not have full column rank). $\square$

Remark 3.8 The definition of the structure matrix S_{R_f} associated with a given TFM $R_f(\lambda)$ can be extended to cover the strong fault detectability requirement defined by $\Omega \subset \partial \mathbb{C}_s$, where Ω is the set of relevant frequencies. This comes down to a straightforward modification of the definition (3.17) of the structure matrix

$$\begin{aligned}
S_{R_f}(i,j) &= 1 \quad \text{if} \quad R_{f_j}^{(i)}(\lambda_z) \neq 0 \text{ for all } \lambda_z \in \Omega\,,\\
S_{R_f}(i,j) &= 0 \quad \text{if} \quad R_{f_j}^{(i)}(\lambda_z) = 0 \text{ for some } \lambda_z \in \Omega\,.
\end{aligned} \tag{3.22}$$

$\square$

3.5 Fault Detection and Isolation Problems

In this section, we formulate several synthesis problems of fault detection and isolation filters for LTI systems. These problems can be considered as a minimal (canonical) set to cover the needs of most practical applications. For the solution of these problems, we seek linear residual generators (or fault detection filters) of the form (3.3), which process the measurable system outputs $y(t)$ and known control inputs $u(t)$ and generate the residual signals $r(t)$, which serve for decision-making on the presence or absence of faults. The standard requirements on all TFMs appearing in the implementation form (3.3) and internal form (3.4) of the fault detection filter are *properness* and *stability*, to ensure physical realizability of the filter $Q(\lambda)$ and to guarantee a stable behaviour of the FDD system. The *order* of the filter $Q(\lambda)$ is its *McMillan degree*, that is, the dimension of the state vector of a minimal state-space realization of $Q(\lambda)$. For practical purposes, lower order filters are preferable to larger order ones, and, therefore, determining *least-order residual generators* is also a desirable synthesis goal. Finally, while the dimension q of the residual vector $r(t)$ depends on the fault detection problem to be solved, filters with the *least number of outputs* are always of interest for practical usage.

For the solution of fault detection and isolation problems, it is always possible to completely decouple the control input $u(t)$ from the residual $r(t)$ by requiring $R_u(\lambda) = 0$. Regarding the disturbance input $d(t)$ and noise input $w(t)$, we aim to impose a similar condition on the disturbance input $d(t)$ by requiring $R_d(\lambda) = 0$, while minimizing simultaneously the effect of noise input $w(t)$ on the residual (e.g., by minimizing the norm of $R_w(\lambda)$). Thus, from a practical synthesis point of view, the distinction between $d(t)$ and $w(t)$ lies solely in the way these signals are treated when solving the residual generator synthesis problem.

In all fault detection problems formulated in what follows, we require that by a suitable choice of a stable fault detection filter $Q(\lambda)$, we achieve that the residual signal $r(t)$ is fully decoupled from the control input $u(t)$ and disturbance input $d(t)$. Thus, the following *decoupling conditions* must be fulfilled for the filter synthesis

$$
\begin{aligned}
&(i)\ \ \ R_u(\lambda) = 0\,,\\
&(ii)\ R_d(\lambda) = 0\,.
\end{aligned}
\tag{3.23}
$$

In the case when condition (ii) cannot be fulfilled (e.g., due to a lack of sufficient number of measurements), we can redefine some (or even all) components of $d(t)$ as noise inputs and include them in $w(t)$.

For each fault detection problem formulated in what follows, specific requirements have to be fulfilled, which are formulated as additional synthesis conditions. For all formulated problems, we also give the existence conditions of the solutions of these problems.

3.5.1 *Exact Fault Detection Problem*

For the *exact fault detection problem* (EFDP), the basic additional requirement is simply to achieve by a suitable choice of a stable and proper fault detection filter $Q(\lambda)$ that, in the absence of noise input (i.e., $w \equiv 0$), the residual $r(t)$ is sensitive to all fault components $f_j(t)$, $j = 1, \ldots, m_f$. Thus, the following *detection condition* has to be fulfilled:

$$
(iii)\ \ R_{f_j}(\lambda) \neq 0,\ \ j = 1, \ldots, m_f \ \text{ with } \ R_f(\lambda) \text{ stable.}
\tag{3.24}
$$

This is precisely the complete fault detectability requirement (see also Remark 3.3) and leads to the following solvability condition:

Theorem 3.7 *For the system (3.2) with $w \equiv 0$, the EFDP is solvable if and only if the system (3.2) is completely fault detectable.*

Proof On the basis of Theorem 3.2, the synthesis conditions (3.23) and (3.24), without enforcing the stability of $Q(\lambda)$ and $R_f(\lambda)$, can be fulfilled provided the system (3.2) is completely fault detectable. To ensure the stability of both $Q(\lambda)$ and $R_f(\lambda)$, the filter $Q(\lambda)$ and its internal form $R_f(\lambda)$ can be replaced by $M(\lambda)Q(\lambda)$ and $M(\lambda)R_f(\lambda)$, respectively, where

$$M^{-1}(\lambda)N(\lambda) = [\, Q(\lambda) \ \ R_f(\lambda) \,]$$

is a stable left coprime factorization.   ∎

Let $\Omega \subset \partial \mathbb{C}_s$ be a given set of frequencies, which characterize the relevant persistent faults. We can give a similar result in the case when the EFDP is solved with the following *strong detection condition* to be fulfilled:

$$(iii)' \ \ R_{f_j}(\lambda_z) \neq 0, \ \ \forall \lambda_z \in \Omega, \ j = 1, \ldots, m_f \ \text{ with } \ R_f(\lambda) \text{ stable.} \qquad (3.25)$$

The solvability condition of the EFDP with the strong detection condition above is precisely the complete strong fault detectability requirement as stated by the following theorem.

Theorem 3.8 *Let Ω be the set of frequencies, which characterize the persistent fault signals. For the system (3.2) with $w \equiv 0$, the EFDP with the strong detection condition (3.25) is solvable if and only if the system (3.2) is completely strong fault detectable with respect to Ω.*

Proof On the basis of Theorem 3.4, the synthesis conditions (3.23), (3.24) and (3.25) can be fulfilled provided the system (3.2) with $w \equiv 0$ is complete strong fault detectable with respect to Ω.   ∎

3.5.2 Approximate Fault Detection Problem

The effects of the noise input $w(t)$ can usually not be fully decoupled from the residual $r(t)$. In this case, the basic requirements for the choice of $Q(\lambda)$ can be expressed to achieve that the residual $r(t)$ is influenced by all fault components $f_j(t)$ and the influence of the noise signal $w(t)$ is negligible. For the *approximate fault detection problem* (AFDP), the following two additional conditions have to be fulfilled:

$$\begin{aligned}
&(iii) \ \ R_{f_j}(\lambda) \neq 0, \ \ j = 1, \ldots, m_f \ \text{ with } \ R_f(\lambda) \text{ stable;} \\
&(iv) \ \ R_w(\lambda) \approx 0, \ \ \text{ with } \ R_w(\lambda) \text{ stable.}
\end{aligned} \qquad (3.26)$$

Here, (iii) is the *detection condition* of all faults employed also in the EFDP, while (iv) is the *attenuation condition* for the noise input. The condition $R_w(\lambda) \approx 0$

expresses the requirement that the transfer gain $\|R_w(\lambda)\|$ (measured by any suitable norm) can be made arbitrarily small.

The solvability conditions of the formulated AFDP can be easily established.

Theorem 3.9 *For the system (3.2), the AFDP is solvable if and only if the EFDP is solvable.*

Proof We can always determine a solution $Q(\lambda)$ of the EFDP such that additionally the resulting $R_w(\lambda)$ is stable. Moreover, by rescaling $Q(\lambda)$ with a constant factor γ, the norm of $R_w(\lambda)/\gamma$ can be made arbitrarily small. The necessity is trivial, because any solution of the AFDP is also a solution of the EFDP.   ∎

Remark 3.9 The proof of the theorem relies on the fact that any solution $Q(\lambda)$ of the EFDP can be also used as a solution of the AFDP. While mathematically this is true, still the employed scaled filter $Q(\lambda)/\gamma$, which makes $\|R_w(\lambda)/\gamma\|$ "small", reduces simultaneously $\|R_f(\lambda)/\gamma\|$, while preserving the fault detectability property. In practical applications, the usefulness of a solution $Q(\lambda)$ of the AFDP must be judged by taking into account the maximum size w_{max} of the noise signal and the desired minimum detectable sizes of faults. Since $Q(\lambda)$ automatically ensures that $R_u(\lambda) = 0$ and $R_d(\lambda) = 0$, the corresponding residual is given by

$$\mathbf{r}(\lambda) = R_f(\lambda)\mathbf{f}(\lambda) + R_w(\lambda)\mathbf{w}(\lambda)\,.$$

If $f_{j,min}$ is the desired minimum detectable size of the j-th fault f_j, then a filter $Q(\lambda)$ is satisfactory provided the least size faults can be detected in the presence of the worst-case (i.e., maximum size) noise inputs. This is achieved if

$$\min_j(\|R_{f_j}(\lambda)\|f_{j,min}) > \|R_w(\lambda)\|w_{max}\,.$$

It follows that for assessing the usefulness of a design, not the size of $\|R_w(\lambda)\|$ is the relevant measure, but the ratio

$$\frac{\min_j(\|R_{f_j}(\lambda)\|f_{j,min})}{\|R_w(\lambda)\|w_{max}}\,.$$

Therefore, a meaningful goal for the synthesis methods, which solve the AFDP, is to determine a filter $Q(\lambda)$, which maximizes this ratio.   □

3.5.3 *Exact Fault Detection and Isolation Problem*

For a row-wise block-structured fault detection filter $Q(\lambda)$ as in (3.14), let $R_f(\lambda)$ be the corresponding block-structured fault-to-residual TFM as defined in (3.16) with $n_b \times m_f$ blocks, and let S_{R_f} be the corresponding $n_b \times m_f$ structure matrix defined in (3.17) (see Sect. 3.4). Let s_j, $j = 1, \ldots, m_f$ be a set of n_b-dimensional binary signature vectors associated with the faults f_j, $j = 1, \ldots, m_f$, which form the desired structure matrix $S := [\, s_1 \ \ldots \ s_{m_f} \,]$. The *exact fault detection and isolation problem* (EFDIP) requires to determine for a given $n_b \times m_f$ structure matrix S, a stable and proper filter $Q(\lambda)$ of the form (3.14) such that the following condition is additionally fulfilled:

$$(iii) \quad S_{R_f} = S, \quad \text{with } R_f(\lambda) \text{ stable.} \tag{3.27}$$

We have the following straightforward solvability condition:

Theorem 3.10 *For the system (3.2) with $w \equiv 0$ and a given structure matrix S, the EFDIP is solvable if and only if the system (3.2) is S-fault isolable.*

Proof The synthesis conditions (3.23) and (3.27), without enforcing the stability of $Q(\lambda)$ and $R_f(\lambda)$, can be fulfilled provided the system (3.2) is S-fault isolable. To ensure the stability of both $Q(\lambda)$ and $R_f(\lambda)$, the filter $Q(\lambda)$ and its internal form $R_f(\lambda)$ can be replaced by $M(\lambda)Q(\lambda)$ and $M(\lambda)R_f(\lambda)$, respectively, where

$$M^{-1}(\lambda)N(\lambda) = [\, Q(\lambda) \ \ R_f(\lambda) \,]$$

is a stable left fractional factorization with $M(\lambda)$ diagonal (see Sect. 9.1.6). $\blacksquare$

A similar result can be established for the case when S is the m_f-th order identity matrix $S = I_{m_f}$. We call the associated synthesis problem the *strong* EFDIP. The proof is similar to that of Theorem 3.10.

Theorem 3.11 *For the system (3.2) with $w \equiv 0$ and $S = I_{m_f}$, the EFDIP is solvable if and only if the system (3.2) is strongly fault isolable.*

3.5.4   Approximate Fault Detection and Isolation Problem

Let S be the desired $n_b \times m_f$ structure matrix targeted to be achieved by using a structured fault detection filter $Q(\lambda)$ with n_b row blocks as in (3.14). The $n_b \times m_f$ block-structured fault-to-residual TFM $R_f(\lambda)$, corresponding to $Q(\lambda)$ that is defined in (3.16), can be additively decomposed as $R_f(\lambda) = \widetilde{R}_f(\lambda) + \overline{R}_f(\lambda)$, where $\widetilde{R}_f(\lambda)$ and $\overline{R}_f(\lambda)$ have the same block structure as $R_f(\lambda)$ and have their (i, j)-th blocks defined as

$$\widetilde{R}_{f_j}^{(i)}(\lambda) = S_{ij} R_{f_j}^{(i)}(\lambda), \quad \overline{R}_{f_j}^{(i)}(\lambda) = (1 - S_{ij}) R_{f_j}^{(i)}(\lambda) \,.$$

To address the approximate fault detection and isolation problem, we will target to enforce for the part $\widetilde{R}_f(\lambda)$ of $R_f(\lambda)$ the desired structure matrix S, while the part $\overline{R}_f(\lambda)$ must be (ideally) negligible. The *approximate fault detection and isolation problem* (AFDIP) can be formulated as follows. For a given $n_b \times m_f$ structure matrix

S, determine a stable and proper filter $Q(\lambda)$ in the form (3.14) such that the following conditions are additionally fulfilled:

$$
\begin{aligned}
(iii) \quad & S_{\widetilde{R}_f} = S, \ \overline{R}_f(\lambda) \approx 0, \ \text{with } R_f(\lambda) \text{ stable,} \\
(iv) \quad & R_w(\lambda) \approx 0, \ \text{with } R_w(\lambda) \text{ stable.}
\end{aligned}
\tag{3.28}
$$

It is straightforward to show that a necessary and sufficient condition for the solvability of the AFDIP is the solvability of the EFDP.

Theorem 3.12 *For the system (3.2) and a given structure matrix S without zero columns, the AFDIP is solvable if and only if the EFDP is solvable.*

Proof The necessity trivially follows for any structure matrix S without zero columns, since the solvability of the EFDP in the case $w \equiv 0$ corresponds to a $1 \times m_f$ binary structure matrix with all elements equal to one. Such a structure matrix can be built as the (logical) sum of all rows of the given S (e.g., using the elementary binary operations $1 \oplus 0 = 1$, $1 \oplus 1 = 1$ and $0 \oplus 0 = 0$).

To prove the sufficiency, we show that if the EFDP for $w \equiv 0$ is solvable, then we can determine a filter $Q(\lambda)$ in the form (3.14) such that conditions (3.28) are fulfilled. Since the EFDP is solvable, we can determine for each row i of S a filter $Q^{(i)}(\lambda)$ such that the blocks $R_{f_j}^{(i)}(\lambda)$ of the i-th block row of the corresponding $R_f(\lambda)$ in (3.16) are nonzero for all j for which $S_{ij} \neq 0$. Additionally, we can choose $Q^{(i)}(\lambda)$ to also enforce the stability of the block rows

$$
R_f^{(i)}(\lambda) := Q^{(i)}(\lambda) \begin{bmatrix} G_f(\lambda) \\ 0 \end{bmatrix}, \quad R_w^{(i)}(\lambda) := Q^{(i)}(\lambda) \begin{bmatrix} G_w(\lambda) \\ 0 \end{bmatrix}.
$$

We collect all blocks $R_{f_j}^{(i)}(\lambda)$ for which $S_{ij} = 0$ into a matrix $\overline{R}_f^{(i)}$. By rescaling $Q^{(i)}(\lambda)$ with a constant factor γ_i, arbitrary small norms of $R_w^{(i)}(\lambda)/\gamma_i$ and $\overline{R}_f^{(i)}(\lambda)/\gamma_i$ can be achieved. The overall $Q(\lambda)$, $R_f(\lambda)$ and $R_w(\lambda)$ result as

$$
Q(\lambda) = \begin{bmatrix} Q^{(1)}(\lambda)/\gamma_1 \\ \vdots \\ Q^{(n_b)}(\lambda)/\gamma_{n_b} \end{bmatrix}, \quad
R_f(\lambda) = \begin{bmatrix} R_f^{(1)}(\lambda)/\gamma_1 \\ \vdots \\ R_f^{(n_b)}(\lambda)/\gamma_{n_b} \end{bmatrix}, \quad
R_w(\lambda) = \begin{bmatrix} R_w^{(1)}(\lambda)/\gamma_1 \\ \vdots \\ R_w^{(n_b)}(\lambda)/\gamma_{n_b} \end{bmatrix}
$$

and $R_f(\lambda)$ and $R_w(\lambda)$ fulfil the conditions (3.28). $\blacksquare$

Remark 3.10 If the given structure matrix S has zero columns, then all faults corresponding to the zero columns of S can be redefined as additional noise inputs. In this case, Theorem 3.12 can be applied to a modified system with a reduced set of faults and increased set of noise inputs. $\square$

Example 3.5 The somehow surprising result of Theorem 3.12 allows to address the solution of approximate fault detection and isolation problems in the cases when no sufficient measurements are available to solve the EFDIP. Consider the static model with two additive faults

$$y(t) = f_1(t) + f_2(t) \, ,$$

for which the EFDP is solvable, but the EFDIP with $S = I_2$ is not solvable. However, the AFDIP for this system with $S = I_2$ is solvable according to Theorem 3.12. Still, it is clear that the isolation of either of faults is generally a futile task, unless additional information on the nature of faults is available. Let us assume, for example, that $f_1(t)$ and $f_2(t)$ represent two narrowband signals, with well-separated centre frequencies ω_1 and ω_2. Then we can build a bank of two filters

$$\begin{bmatrix} \mathbf{r}^{(1)}(\lambda) \\ \mathbf{r}^{(2)}(\lambda) \end{bmatrix} = \begin{bmatrix} Q^{(1)}(\lambda) \\ Q^{(2)}(\lambda) \end{bmatrix} \mathbf{y}(\lambda),$$

where $Q^{(1)}(\lambda)$ is a bandpass filter for the frequency band around ω_1, and $Q^{(2)}(\lambda)$ is a bandpass filter for the frequency band around ω_2. It follows that $r^{(1)}(t) \approx f_1(t)$ and $r^{(2)}(t) \approx f_2(t)$, and therefore, the isolation of these faults is easily achievable. $\diamond$

The solvability of the EFDIP is clearly a sufficient condition for the solvability of the AFDIP, but is not, in general, also a necessary condition, unless we impose in the formulation of the AFDIP the stronger condition $\overline{R}_f(\lambda) = 0$ (instead $\overline{R}_f(\lambda) \approx 0$). This is equivalent to require $S_{R_f} = S$. Therefore, we can alternatively formulate the AFDIP to fulfil the conditions:

$$\begin{aligned} (iii)' \quad & S_{R_f} = S, \ \text{with } R_f(\lambda) \text{ stable,} \\ (iv)' \quad & R_w(\lambda) \approx 0, \ \text{with } R_w(\lambda) \text{ stable.} \end{aligned} \tag{3.29}$$

In this case, we have the straightforward result:

Theorem 3.13 *For the system (3.2) and a given structure matrix S, the AFDIP is solvable with $S_{R_f} = S$ if and only if the EFDIP is solvable.*

Proof We can always determine a solution $Q(\lambda)$ of the EFDIP, which also ensures that $R_w(\lambda)$ is stable. By rescaling $Q(\lambda)$ with a constant factor γ, the norm of $R_w(\lambda)/\gamma$ can be made arbitrarily small. The necessity is trivial, because any solution of the AFDIP with $S_{R_f} = S$ is also a solution of the EFDIP.   ∎

3.5.5 Exact Model-Matching Problem

Let $M_r(\lambda)$ be a given $q \times m_f$ TFM of a stable and proper reference model specifying the desired input–output behaviour from the faults to residuals as $\mathbf{r}(\lambda) = M_r(\lambda)\mathbf{f}(\lambda)$. Thus, we want to achieve by a suitable choice of a stable and proper $Q(\lambda)$ satisfying (i) and (ii) in (3.23) that we have additionally $R_f(\lambda) = M_r(\lambda)$. For example, a typical choice for $M_r(\lambda)$ is an $m_f \times m_f$ diagonal and invertible TFM, which ensures that each residual $r_i(t)$ is influenced only by the fault $f_i(t)$. The choice $M_r(\lambda) = I_{m_f}$ targets the solution of an *exact fault estimation problem* (EFEP).

To determine $Q(\lambda)$, we have to solve the linear rational equation (3.5), with the settings $R_u(\lambda) = 0$, $R_d(\lambda) = 0$ and $R_f(\lambda) = M_r(\lambda)$ ($R_w(\lambda)$ and $G_w(\lambda)$ are assumed empty matrices). The choice of $M_r(\lambda)$ may lead to a solution $Q(\lambda)$, which is not proper or is unstable or has both these undesirable properties. Therefore, besides determining $Q(\lambda)$, we also consider the determination of a suitable updating factor $M(\lambda)$ of $M_r(\lambda)$ to ensure the stability and properness of the solution $Q(\lambda)$ for $R_f(\lambda) = M(\lambda)M_r(\lambda)$. Obviously, $M(\lambda)$ must be chosen a proper, stable and invertible TFM. Additionally, by choosing $M(\lambda)$ diagonal, the zero and nonzero entries of $M_r(\lambda)$ can be also preserved in $R_f(\lambda)$ (see also Sect. 3.5.3).

The *exact model-matching problem* (EMMP) can be formulated as follows: given a stable and proper $M_r(\lambda)$, it is required to determine a stable and proper filter $Q(\lambda)$ and a diagonal, proper, stable and invertible TFM $M(\lambda)$ such that the following condition is additionally fulfilled:

$$(iii) \quad R_f(\lambda) = M(\lambda)M_r(\lambda). \tag{3.30}$$

The solvability condition of the EMMP is the standard solvability condition of systems of linear equations:

Theorem 3.14 *For the system* (3.2) *with* $w \equiv 0$ *and a given* $M_r(\lambda)$, *the EMMP is solvable if and only if the following condition is fulfilled*

$$\text{rank}\,[\,G_f(\lambda)\ G_d(\lambda)\,] = \text{rank}\begin{bmatrix} G_f(\lambda) & G_d(\lambda) \\ M_r(\lambda) & 0 \end{bmatrix}. \tag{3.31}$$

Proof The necessity is immediate by writing down the condition for the existence of a stable solution $Q(\lambda)$ for a suitable invertible $M(\lambda)$. To prove the sufficiency, we determine first $\widetilde{Q}(\lambda)$, a solution of the linear equations $R_u(\lambda) = 0$, $R_d(\lambda) = 0$ and $R_f(\lambda) = M_r(\lambda)$. The existence of such a solution is guaranteed by the solvability condition (3.31) for this linear system. A stable solution $Q(\lambda)$ satisfying condition (3.30) is obtained by expressing $\widetilde{Q}(\lambda)$ in a left factorized form $\widetilde{Q}(\lambda) = M^{-1}(\lambda)Q(\lambda)$, with $M(\lambda)$ and $Q(\lambda)$ proper and stable TFMs, and $M(\lambda)$ diagonal (see Sect. 9.1.6). ∎

Remark 3.11 When $M_r(\lambda)$ has full column rank m_f, the solvability condition (3.31) of the EMMP reduces to the strong isolability condition (3.19) (see also Theorem 3.11). □

The solvability conditions become more involved if we strive for a stable proper solution $Q(\lambda)$ for a given reference model $M_r(\lambda)$ without allowing its updating. For example, this is the case when solving the EFEP for $M_r(\lambda) = I_{m_f}$. For a slightly more general case, we have the following result.

Theorem 3.15 *For the system* (3.2) *with* $w \equiv 0$ *and a given stable and minimum-phase* $M_r(\lambda)$ *of full column rank, the EMMP is solvable with* $M(\lambda) = I$ *if and only if the system is strongly fault isolable and* $G_f(\lambda)$ *is minimum phase.*

Proof We prove first the necessity. Since $M_r(\lambda)$ has full column rank m_f, the solvability condition for $Q(\lambda)$ satisfying $R_u(\lambda) = 0$, $R_d(\lambda) = 0$ and $R_f(\lambda) = M_r(\lambda)$ is precisely the strong fault isolability of the system (3.2) (see condition (3.19) in Theorem 3.6). The underlying linear system satisfied by $Q(\lambda)$ is $Q(\lambda)G(\lambda) = F(\lambda)$, where

$$G(\lambda) = \begin{bmatrix} G_f(\lambda) & G_u(\lambda) & G_d(\lambda) \\ 0 & I_{m_u} & 0 \end{bmatrix}, \quad F(\lambda) = \begin{bmatrix} M_r(\lambda) & 0 & 0 \end{bmatrix}. \tag{3.32}$$

According to Lemma 9.5 and taking into account that $F(\lambda)$ is stable, a stable and proper filter $Q(\lambda)$ satisfying $Q(\lambda)G(\lambda) = F(\lambda)$ exists if additionally the TFMs $G(\lambda)$ and $\begin{bmatrix} G(\lambda) \\ F(\lambda) \end{bmatrix}$ have the same unstable zero structure. Equivalently,

$$\begin{bmatrix} G_f(\lambda) & G_d(\lambda) \end{bmatrix} \text{ and } \begin{bmatrix} G_f(\lambda) & G_d(\lambda) \\ M_r(\lambda) & 0 \end{bmatrix} \tag{3.33}$$

must have the same unstable zero structure. Since rank $G_f(\lambda) = m_f$, any zero of $G_f(\lambda)$ is also a zero of $[\, G_f(\lambda) \; G_d(\lambda) \,]$. Therefore, $G_f(\lambda)$ must be minimum-phase, because otherwise any unstable zero of $G_f(\lambda)$ would violate the existence conditions of Lemma 9.5 of a stable and proper solution (i.e., the same zero structure of the TFMs in (3.33) for all unstable zeros). This proves the necessity.

   To prove the sufficiency, observe that the strong fault isolability condition (3.19) and the minimum-phase property of $G_f(\lambda)$ and $M_r(\lambda)$ ensure that the TFMs in (3.33) have the same zero structure for all unstable zeros (any such unstable zero must be a zero of $G_d(\lambda)$). According to Lemma 9.5, this also guarantees the existence of a stable solution $Q(\lambda)$ of the system $Q(\lambda)G(\lambda) = F(\lambda)$, with $G(\lambda)$ and $F(\lambda)$ defined in (3.32). ∎

Remark 3.12 If $G_f(\lambda)$ has unstable or infinite zeros, the solvability of the EMMP with $M(\lambda) = I$ is possible provided $M_r(\lambda)$ is chosen such that the condition (3.33) is fulfilled. For this it is necessary that $M_r(\lambda)$ has the same unstable and infinity zeros structure as $G_f(\lambda)$. □

3.5.6   Approximate Model-Matching Problem

Similar to the formulation of the EMMP, we include the determination of an updating factor of the reference model in the formulation of the *approximate model-matching problem* (AMMP). Specifically, for a given stable and proper TFM $M_r(\lambda)$, it is required to determine a stable and proper filter $Q(\lambda)$ and a diagonal, proper, stable and invertible TFM $M(\lambda)$ such that the following conditions are additionally fulfilled:

$$
\begin{aligned}
&(iii) \ \ R_f(\lambda) \approx M(\lambda)M_r(\lambda), \ \ \text{with} \ \ R_f(\lambda) \ \text{stable};\\
&(iv) \ \ R_w(\lambda) \approx 0, \ \ \text{with} \ \ R_w(\lambda) \ \text{stable}.
\end{aligned}
\tag{3.34}
$$

The condition (iii) means that we strive to achieve that $\| R_f(\lambda) - M(\lambda)M_r(\lambda)\| \approx 0$.

A sufficient condition for the solvability of AMMP is simply the solvability of the EMMP:

Proposition 3.1 *For the system (3.2) and a given $M_r(\lambda)$, the AMMP is solvable if the EMMP is solvable.*

Proof Let $Q(\lambda)$ and $M(\lambda)$ be a solution of the EMMP, which also ensures that $R_w(\lambda)$ is stable. By simultaneously rescaling both $Q(\lambda)$ and $M(\lambda)$ with the same constant factor γ, arbitrary small norms of $(R_f(\lambda) - M(\lambda)M_r(\lambda))/\gamma$ and $R_w(\lambda)/\gamma$ can be achieved. ∎

Alternative sufficient conditions for the solvability of the AMMP exist for appropriately formulated norm minimization-based solution approaches (see Sect. 5.7).

3.6 Threshold Selection

A well-designed FDD system as that in Fig. 3.1 must fulfil standard performance requirements such as operation without producing false alarms, operation without missing the detection of relevant faults and generation of timely decisions on the presence of faults. A false alarm is when a fault is declared although no fault exists, whereas a missed detection is when a fault occurs but is not detected. The origins of false alarms and missed detections lie in the presence of uncertainties. In the linear setting of this book, these uncertainties are exclusively due to the presence of noise inputs (which also may account for the presence of model uncertainties; see Sects. 2.2.1 and 2.2.3). Critical parameters to choose to avoid false alarms and missed detections are the decision threshold τ used for fault detection or the set of thresholds τ_i, $i = 1, \ldots, n_b$ used in FDI schemes. For example, consider the case of a fault detection setup of the FDD system in Fig. 3.1, where $\theta(t)$ is the corresponding residual evaluation signal (usually an approximation of $\|r\|_2$, the $\mathcal{L}_2$- or ℓ_2-norm of the time signal $r(t)$). Then, false alarms may result at time $t_{\bar{a}}$ if for a given decision threshold τ, the condition for the occurrence of a fault $\theta(t_{\bar{a}}) > \tau$ is fulfilled, in the absence of any fault. This may occur in the case when the unknown noise inputs w have too large magnitudes. Likewise, missed detections of small amplitude faults can occur, if the detection condition $\theta(t) > \tau$ is not fulfilled for any t in the maximum allowed detection interval $[t_f, t_f + \Delta t_d]$, where t_f is the fault occurrence time and Δt_d is the allowed maximal duration of fault detection. Missed detections may occur, for example, if the value of τ is set too high (e.g., to avoid false alarms due to large noise input signals).

A related performance aspect of FDD systems is the timely decision on the occurrence of faults. For a fault occurring at time t_f, let t_d be the *fault detection time* defined as the least value of time $t \geq t_f$ for which $\theta(t) \geq \tau$. For timely decisions, the difference $t_d - t_f$ must be less than a maximal allowed duration Δt_d, thus the following constraint on the detection time must be fulfilled

$$t_d - t_f \leq \Delta t_d . \tag{3.35}$$

The detection time can be reduced, for example, by imposing faster dynamics for the residual generator. Also, a careful selection of the decision threshold τ is instrumental in achieving satisfactory detection times.

In what follows, we present a general approach for choosing suitable thresholds to avoid both false alarms and missed detections. Let $\mathcal{W}$ and $\mathcal{F}$ be the classes of noise inputs w and fault inputs f, respectively, which are relevant for a fault monitoring application. For example, $\mathcal{W}$ may be the class of white noise signals of given maximal amplitude and covariance, or may be the class of disturbance signals with bounded variations, while $\mathcal{F}$ usually includes several categories of fault signals of given minimal amplitudes. For our discussion on the selection of thresholds, we only consider the selection of the threshold τ for the simpler case of fault detection. However, the same approach can be employed to select the individual thresholds τ_i for the components of the residual vector used for an FDI setup.

Let τ be the threshold used for decision-making. The requirement for no false alarms in the absence of faults leads to a lower bound for τ, representing the *false alarm bound*

$$\tau_f := \sup_{\substack{t \in [0, t_m] \\ f = 0 \\ w \in \mathcal{W}}} \theta(t), \tag{3.36}$$

where t_m is the maximum signal monitoring time. The requirement for no missed detections leads to an upper bound for τ, representing the *detection bound*

$$\tau_d := \inf_{\substack{t \in [t_f, t_f + \Delta t_d] \\ f \in \mathcal{F}, w \in \mathcal{W}}} \theta(t) \geq \tau . \tag{3.37}$$

To ensure simultaneously the lack of false alarms and missed detections, the condition $\tau_f \leq \tau_d$ must be fulfilled, which ensures that the threshold τ can be chosen such that

$$\tau_f \leq \tau \leq \tau_d .$$

If however $\tau_f > \tau_d$, then either the requirement for the lack of false alarms or the requirement for the lack of missed detections cannot be fulfilled. To deal with such cases in practical applications, often the detection bound has to be increased by suitably adjusting the requirements on the least size of the detectable faults.

The lack of false alarms, of missed detection, and the fulfilment of the constraint (3.35) on the detection time are standard requirements when setting up any FDD system. The appropriate selection of the detection threshold τ is of paramount importance for the fulfilment of the above aims. In practice, the value of τ is determined from the requirement of the lack of false alarms. This value is then used to determine the minimum amplitude detectable faults. For this purpose, the false alarm bound τ_f can be used to determine the least size of detectable faults δ_f, which satisfy

$$\tau_f = \inf_{\substack{t \in [t_f, t_f + \Delta t_d] \\ w \in \mathcal{W} \\ f \in \mathcal{F} \\ \|f\| = \delta_f}} \theta(t). \tag{3.38}$$

Alternatively, a conservative estimate of δ_f can be determined from

$$2\tau_f = \inf_{\substack{t \in [t_f, t_f + \Delta t_d] \\ w = 0, f \in \mathcal{F} \\ \|f\| = \delta_f}} \theta(t). \tag{3.39}$$

In what follows, we illustrate the selection of the decision threshold τ for the AFDP formulated previously. For this, we will make some assumptions. First, we assume that the fault detection filter has the form (3.3) and the corresponding internal form of the residual generator is

$$\mathbf{r}(\lambda) = R_w(\lambda)\mathbf{w}(\lambda) + R_f(\lambda)\mathbf{f}(\lambda).$$

This form always results when solving the AFDP, provided the decoupling conditions $R_u(\lambda) = 0$ and $R_d(\lambda) = 0$ are satisfied by the resulting filter. Second, we assume that we use to compute the residual evaluation signal $\theta(t)$, finite time approximations of the usual signal norms as $\|r\|_2$, where we assume $r(t) = 0$ for $t < 0$. For example, in a continuous-time setting, we can use

$$\theta(t) = \|r\|_2^{[0,\,t]} := \left(\int_0^t r^T(\tilde{t}) r(\tilde{t}) \mathrm{d}\tilde{t} \right)^{1/2},$$

while in a discrete-time setting we can use

$$\theta(t) = \|r\|_2^{[0,\,t]} := \left(\sum_{\tilde{t}=0}^t r^T(\tilde{t}) r(\tilde{t}) \right)^{1/2},$$

where discrete values $\tilde{t}, \tilde{t}+1, \ldots, t$ are assumed for the time $\tilde{t}$. Note that $\|r\|_2 = \lim_{t\to\infty} \|r\|_2^{[0,\,t]}$ and in the frequency domain, we have similar norms $\|\mathbf{r}\|_2$ of the Fourier-transformed continuous- or discrete-time signals (see Sect. 9.1.7). According to Plancherel's theorem, we have

$$\|r\|_2 = \|\mathbf{r}\|_2 ,$$

which allows to use time-domain and frequency-domain signal norms interchangeably. Finally, we assume that $w(t)$ has bounded energy and $\|w\|_2 \leq \delta_w$ holds.

For the AFDP, we can determine the false alarm bound as

$$\tau_f = \sup_{\|w\|_2 \leq \delta_w} \|R_w(\lambda)\mathbf{w}(\lambda)\|_2 = \|R_w(\lambda)\|_\infty \delta_w$$

and set the threshold to $\tau := \tau_f$. With these values, we can estimate the minimum size detectable fault using, for example, (3.39) (see Remark 5.9 in Sect. 5.3). The determination of the thresholds for the AFDIP and AMMP are discussed in Sect. 5.5 (see Remark 5.11) and Sect. 5.7 (see Remark 5.15), respectively.

Remark 3.13 The computation of the residual evaluation signal $\theta(t)$ must be performed in real time. For fast detection of faults, often only the instantaneous value of the residual can ensure timely detections. In this case, we can simply use the Euclidean-norm of the vector $r(t)$ as evaluation signal

$$\theta(t) = \|r(t)\|_2 .$$

It is also possible to use a finite-interval approximation of the norm over a sliding time window $[t - T,\ t]$ as

$$\theta(t) = \left(\int_{t-T}^{t} r^T(\tilde{t})r(\tilde{t})\mathrm{d}\tilde{t} \right)^{1/2} .$$

In several applications, a weighted combination of instantaneous and low pass filtered values represents the best compromise, as provided by a Narendra-type evaluation scheme

$$\theta(t) = \alpha\|r(t)\|_2 + \beta \left(\int_{0}^{t} e^{-\gamma(t-\tilde{t})}r^T(\tilde{t})r(\tilde{t})\mathrm{d}\tilde{t} \right)^{1/2} , \tag{3.40}$$

where $\alpha \geq 0$ and $\beta > 0$ are suitable weighting factors, while $e^{-\gamma(t-\tilde{t})}$ with $\gamma > 0$ is an exponential forgetting factor used to reduce the influence of old data. For example, the last approximation can be simply implemented as a first-order filter

$$\dot{\xi}(t) = -\gamma\xi(t) + r^T(t)r(t) ,$$
$$\theta(t) = \alpha\|r(t)\|_2 + \beta\sqrt{\xi(t)} .$$

Similar evaluation schemes can be used in the discrete-time setting. $\qquad\qquad\square$

3.7 Notes and References

The material in this chapter is partially covered in textbooks [15, 23, 29, 51, 68]. The used terminology largely agrees on terms as *fault detection*, *fault isolation* and *fault identification* with the published literature, while there are slight variations in the coverage of the term *fault diagnosis*.

The notions of *implementation form* and *internal forms* of a fault detection filter have been introduced by Gertler [51]. The definition of *structure matrix* and the related nomenclature used in Sect. 3.4 stem partly also from [51].

The definition of the *fault detectability* concept as a property of the underlying system has been introduced and characterized by Nyberg in [98], where conditions for *strong fault detectability* are also provided. The characterizations of strong fault detectability and complete strong fault detectability in terms of the original system matrices, as given in Theorems 3.3 and 3.4, respectively, appear to be new results. The definition of the *fault isolability* concept used in this chapter differs from those introduced in the literature, although the ideas behind this concept are similar to the discussions in several textbooks [15, 24, 51]. The two main novelties introduced are (1) the characterization of fault isolability in the presence of unknown disturbance inputs (intended to be decoupled by the fault detector) and (2) the characterization with respect to an arbitrary given structure matrix. The main appeal of employed definitions is that the conditions for fault detectability and fault isolability coincide with the conditions for the existence of the solution of the related fault detection and isolation problems.

The formulation of several basic synthesis problems is done in terms of the TFMs of the input–output description of the internal form of the residual generator filter. These formulations lead in a straightforward manner to intuitive equivalent algebraic synthesis conditions. An important aspect is that our formulations are independent of any concrete solution approach and the solvability conditions are derived in the most general setting. Similar formulations (of fourteen) fault detection and isolation problems, and derivations of solvability conditions have been done by Saberi et al. [113].

The solvability conditions of the EFDP have been established for proper systems, see, for example [35, 98], and form the basis for establishing the solvability conditions for the EFDIP. The solvability condition for the strong EFDIP in Theorem 3.11 has been established in [47]. For a comprehensive discussion of choosing suitable fault signatures, see [51].

The model-matching approach is a widely employed formulation to solve approximate FDI filter synthesis problems (see, for example, [15, 23, 51] and the references cited therein). The earliest use of the model-matching approach for the synthesis of FDI filters is apparently due to Viswanadham and Minto [170]. The formulation of the AMMP used in this book extends the standard model-matching approach, by also including the determination of an updating factor of the reference model.

The general solvability condition of the EMMP has been discussed in [141]. When $M_r(\lambda)$ has full column rank m_f, the solvability condition of the EMMP coincides with those derived in [47] for diagonal $M_r(\lambda)$. The solvability conditions of the EMMP in the case when the reference model $M_r(\lambda)$ is fixed and no updating is allowed can be derived from the general conditions for solving linear rational equations with stability constraints. See [72] for a complete treatment of this case. The choice of a diagonal reference model in the EMMP corresponds to the so-called *directional residuals*-based FDI approach used by Gertler [51]. Here, also the use of a diagonal $M(\lambda)$ for updating purposes has been proposed (see also earlier papers cited in [51]).

The concepts of *false alarm bound*, *detection bound* and *least size detectable fault* have been introduced in [42]. The conservative estimation of the least size of detectable faults based on the equation (3.39) has been also suggested in [42], and is the basis for several attempts to determine explicit analytic expressions for the least norm of detectable faults; see, for example [32, 112]. The Narendra-type residual evaluation scheme has been proposed in [95] and successfully used in recent applications (see, for example [159]).

Chapter 4
Model Detection

In this chapter, we first formulate the basic model detection task to discover among N given LTI models, that particular model, which best matches the current plant behaviour. Then, the concept of model detectability is introduced and characterized. The exact and approximate model detection problems are formulated. These problems target the synthesis of a bank of N LTI model detection filters, which generate a structured set of residuals allowing the discrimination of models in the case of absence or presence of noise inputs, respectively. The discussion of specific performance requirements for model detection and the selection of thresholds to be used for decision-making conclude the chapter.

4.1 Basic Model Detection Task

Multiple models, which describe various fault situations, have been frequently used for fault detection purposes. In such applications, the detection of the occurrence of a fault comes down to identifying, using the available measurements from the measurable outputs and control inputs, that model (from a collection of models), which best matches the dynamical behaviour of the faulty plant. Another typical application is the multiple-model-based adaptive control, where the adaptation of the control law (for example, by switching from one controller to another) is based on the recognition of that model, which best approximates the current dynamical behaviour of the plant. In this book, we will use the (not yet standard) term *model detection* to describe the model identification task consisting of the selection of a model from a collection of N models, which best matches the current dynamical behaviour of a plant.

A related term used in the literature is *model validation*, which covers an arsenal of statistical methods to assess the adequacy of a model to a set of measurements. Often model validation also includes the identification of suitable uncertainty bounds, which account for the unmodelled dynamics, initial condition uncertainty

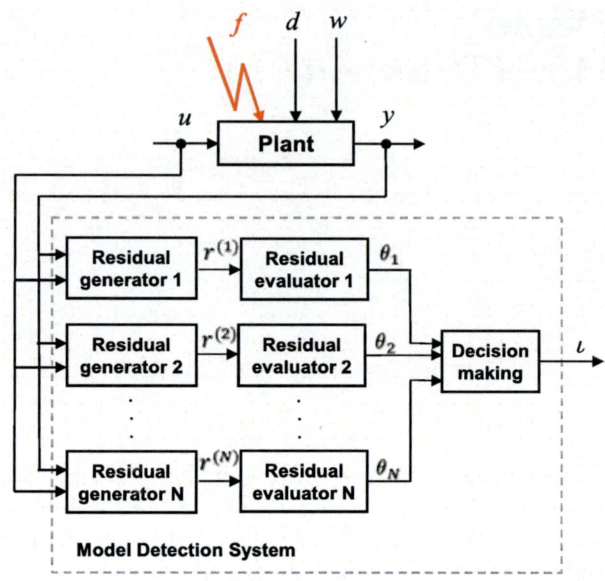

Fig. 4.1 Basic model detection setup

and measurement noise. Strictly speaking, model validation is generally impossible (or at least very challenging), because it would involve to checking that the model can describe *any* input–output behaviour of the physical plant. Therefore, a closer related term to model detection is *model invalidation*, which relies on the trivial fact that a model can be invalidated (i.e., it does not fit with the input and output data) on the basis of a *single* input–output data set. In this sense, the invalidation of $N - 1$ models can be seen as part of the model detection task.

A typical model detection setting is shown in Fig. 4.1. A bank of N residual generation filters (or residual generators) is used, with $r^{(i)}(t)$ being the output of the i-th residual generator. The i-th component θ_i of the N-dimensional evaluation vector θ usually represents an approximation of $\|r^{(i)}\|_2$, the $\mathcal{L}_2$- or ℓ_2-norm of $r^{(i)}$. The i-th component of the N-dimensional decision vector ι is set to 0 if $\theta_i \leq \tau_i$ and 1 otherwise, where τ_i is a suitable threshold. The j-th model is "detected" if $\iota_j = 0$ and $\iota_i = 1$ for all $i \neq j$. It follows that model detection can be interpreted as a particular type of week fault isolation with N signature vectors, where the N-dimensional j-th signature vector has all elements set to one, excepting the j-th entry, which is set to zero. An alternative decision scheme can also be devised if θ_i can be associated with a distance function from the current model to the i-th model. In this case, ι is a scalar, set to $\iota = j$, where j is the index for which $\theta_j = \min_{i=1:N} \theta_i$. Thus, the decision scheme selects that model j, which best fits with the current model characterized by the measured input and output data.

The underlying synthesis techniques of model detection systems rely on multiple-model descriptions of physical fault cases of the form (2.22). Since different degrees of performance degradations can be easily described via multiple models, model detection techniques have potentially the capability to address certain fault identification aspects too.

4.2 Residual Generation

Assume we have N LTI models of the form (2.22), where for $j = 1, ..., N$, the j-th model is specified in the input–output form

$$\mathbf{y}^{(j)}(\lambda) = G_u^{(j)}(\lambda)\mathbf{u}^{(j)}(\lambda) + G_d^{(j)}(\lambda)\mathbf{d}^{(j)}(\lambda) + G_w^{(j)}(\lambda)\mathbf{w}^{(j)}(\lambda). \tag{4.1}$$

We further assume that the N models originate from a common underlying system with $y(t) \in \mathbb{R}^p$, the measurable output vector, and $u(t) \in \mathbb{R}^{m_u}$, the known control input. Therefore, $y^{(j)}(t) \in \mathbb{R}^p$ is the output vector of the j-th system with the control input $u^{(j)}(t) \in \mathbb{R}^{m_u}$, disturbance input $d^{(j)}(t) \in \mathbb{R}^{m_d^{(j)}}$ and noise input $w^{(j)}(t) \in \mathbb{R}^{m_w^{(j)}}$, respectively, and $G_u^{(j)}(\lambda)$, $G_d^{(j)}(\lambda)$ and $G_w^{(j)}(\lambda)$ are the TFMs from the corresponding plant inputs to outputs. We assume that all models are controlled with the same control inputs $u^{(j)}(t) := u(t)$, but the disturbance and noise inputs $d^{(j)}(t)$ and $w^{(j)}(t)$, respectively, may differ for each component model. For complete generality of our problem formulations, we will allow that these TFMs are general rational matrices (proper or improper) for which we will not *a priori* assume any further properties.

Residual generation for model detection is performed using N linear residual generators, which process the measurable system outputs $y(t)$ and known control inputs $u(t)$ and generate N residual signals $r^{(i)}(t)$, $i = 1, \ldots, N$, which serve for decision-making on which model best matches the current input–output measurement data. As already mentioned, model detection can be interpreted as a week fault isolation problem with an $N \times N$ structure matrix S having all its elements equal to one, excepting those on its diagonal, which are zero. The task of model detection is thus to find out the model, which best matches the measurements of outputs and inputs, by comparing the resulting decision vector ι with the set of signatures associated with each model and coded in the columns of S. The residual generation filters in their implementation form are described by the input–output relations

$$\mathbf{r}^{(i)}(\lambda) = Q^{(i)}(\lambda) \begin{bmatrix} \mathbf{y}(\lambda) \\ \mathbf{u}(\lambda) \end{bmatrix}, \ i = 1, \ldots, N, \tag{4.2}$$

where y is the *actual* measured system output, being one of the system outputs generated by the multiple model (4.1). The TFMs $Q^{(i)}(\lambda)$, for $i = 1, \ldots, N$, must be proper and stable. The overall model detection filter has the form

$$Q(\lambda) = \begin{bmatrix} Q^{(1)}(\lambda) \\ \vdots \\ Q^{(N)}(\lambda) \end{bmatrix}. \tag{4.3}$$

The dimension q_i of the residual vector component $r^{(i)}(t)$ can be chosen always one, but occasionally values $q_i > 1$ may provide better sensitivity to model mismatches.

Assuming $y(t) = y^{(j)}(t)$, the residual signal component $r^{(i)}(t)$ in (4.2) generally depends on all system inputs $u^{(j)}(t)$, $d^{(j)}(t)$ and $w^{(j)}(t)$ via the system output $y^{(j)}(t)$. The *internal form* of the i-th filter driven by the j-th model is obtained by replacing in (4.2) $\mathbf{y}(\lambda)$ with $\mathbf{y}^{(j)}(\lambda)$ from (4.1) and $\mathbf{u}(\lambda)$ with $\mathbf{u}^{(j)}(\lambda)$. To make explicit the dependence of $r^{(i)}$ on the j-th model, we will use $\widetilde{r}^{(i,j)}$, to denote the i-th residual output for the j-th model. After replacing in (4.2), $\mathbf{y}(\lambda)$ with $\mathbf{y}^{(j)}(\lambda)$ from (4.1), and $\mathbf{u}(\lambda)$ with $\mathbf{u}^{(j)}(\lambda)$, we obtain

$$\begin{aligned} \widetilde{\mathbf{r}}^{(i,j)}(\lambda) &:= R^{(i,j)}(\lambda) \begin{bmatrix} \mathbf{u}^{(j)}(\lambda) \\ \mathbf{d}^{(j)}(\lambda) \\ \mathbf{w}^{(j)}(\lambda) \end{bmatrix} \\ &= R_u^{(i,j)}(\lambda)\mathbf{u}^{(j)}(\lambda) + R_d^{(i,j)}(\lambda)\mathbf{d}^{(j)}(\lambda) + R_w^{(i,j)}(\lambda)\mathbf{w}^{(j)}(\lambda), \end{aligned} \tag{4.4}$$

with $R^{(i,j)}(\lambda) := \left[\, R_u^{(i,j)}(\lambda)\,\big|\,R_d^{(i,j)}(\lambda)\,\big|\,R_w^{(i,j)}(\lambda)\,\right]$ defined as

$$\left[\, R_u^{(i,j)}(\lambda)\,\big|\,R_d^{(i,j)}(\lambda)\,\big|\,R_w^{(i,j)}(\lambda)\,\right] := Q^{(i)}(\lambda)\begin{bmatrix} G_u^{(j)}(\lambda) & \big| & G_d^{(j)}(\lambda) & \big| & G_w^{(j)}(\lambda) \\ I_{m_u} & \big| & 0 & \big| & 0 \end{bmatrix}. \tag{4.5}$$

For a successfully designed set of filters $Q^{(i)}(\lambda)$, $i = 1, \ldots, N$, the corresponding internal representations $R^{(i,j)}(\lambda)$ in (4.4) are also a proper and stable.

4.3 Model Detectability

The concept of model detectability concerns with the sensitivity of the components of the residual vector to individual models from a given collection of models. Assume that we have N models, with the j-th model specified in the input–output form (4.1). For the discussion of the model detectability concept, we will assume that no noise inputs are present in the models (4.1) (i.e., $w^{(j)} \equiv 0$ for $j = 1, \ldots, N$). For model detection purposes, N filters of the form (4.2) are employed. It follows from (4.4) that the i-th component $r^{(i)}$ of the residual r is sensitive to the j-th model provided

$$R^{(i,j)}(\lambda) := \left[\, R_u^{(i,j)}(\lambda)\ \ R_d^{(i,j)}(\lambda)\,\right] \neq 0. \tag{4.6}$$

We can associate to the $N \times N$ blocks $R^{(i,j)}(\lambda)$ defined in (4.6), the $N \times N$ structure matrix S_R with the (i, j)-th element set to 1 if $R^{(i,j)}(\lambda) \neq 0$ and set to 0 if $R^{(i,j)}(\lambda) = 0$. As already mentioned, model detection can be interpreted as a week fault isolation problem with an $N \times N$ structure matrix S having all its elements equal to one, excepting those on its diagonal, which are zero. Having this analogy in mind, we introduce the following concept of model detectability.

Definition 4.1 The multiple model defined by the N component systems (4.1) with $w^{(j)} \equiv 0$ for $j = 1, \ldots, N$, is *model detectable* if there exist N filters of the form (4.2), such that $R^{(i,j)}(\lambda)$ defined in (4.6) fulfils $R^{(i,i)}(\lambda) = 0$ for $i = 1, \ldots, N$ and $R^{(i,j)}(\lambda) \neq 0$ for all $i, j = 1, \ldots, N$ such that $i \neq j$.

The following result characterizes the model detectability property.

Theorem 4.1 *The multiple model defined by the N component systems (4.1) with $w^{(j)} \equiv 0$ for $j = 1, \ldots, N$, is model detectable if and only if for $i = 1, \ldots, N$*

$$\text{rank} \, [\, G_d^{(i)}(\lambda) \ \ G_d^{(j)}(\lambda) \ \ G_u^{(i)}(\lambda) - G_u^{(j)}(\lambda) \,] > \text{rank} \, G_d^{(i)}(\lambda) \ \ \forall j \neq i \, . \qquad (4.7)$$

Proof For the proof of necessity, assume the model detectability of the multiple model (4.1) and, for $i = 1, \ldots, N$, let $Q^{(i)}(\lambda)$ be a corresponding set of filters for model detection. Let us partition the columns of each $Q^{(i)}(\lambda)$ as

$$Q^{(i)}(\lambda) = [\, Q_y^{(i)}(\lambda) \ \ Q_u^{(i)}(\lambda) \,] \, ,$$

to correspond to the two filter inputs $y(t)$ and $u(t)$ in (4.2). The conditions to achieve the i-th specification are $R^{(i,i)}(\lambda) = 0$ and $R^{(i,j)}(\lambda) \neq 0$ for all $j \neq i$. With the above partitioning of $Q^{(i)}(\lambda)$, this comes down to

$$Q_y^{(i)}(\lambda)G_u^{(i)}(\lambda) + Q_u^{(i)}(\lambda) = 0 \, ,$$
$$Q_y^{(i)}(\lambda)G_d^{(i)}(\lambda) = 0$$

and

$$\left[\, Q_y^{(i)}(\lambda)G_u^{(j)}(\lambda) + Q_u^{(i)}(\lambda) \quad Q_y^{(i)}(\lambda)G_d^{(j)}(\lambda) \,\right] \neq 0, \ \ \forall j \neq i \, .$$

Since $Q_u^{(i)}(\lambda) = -Q_y^{(i)}(\lambda)G_u^{(i)}(\lambda)$, after some manipulations, we obtain the conditions to be satisfied by $Q_y^{(i)}(\lambda)$

$$Q_y^{(i)}(\lambda)G_d^{(i)}(\lambda) = 0 \, ,$$
$$Q_y^{(i)}(\lambda) \left[\, G_u^{(j)}(\lambda) - G_u^{(i)}(\lambda) \quad G_d^{(j)}(\lambda) \,\right] \neq 0, \ \ \forall j \neq i \, .$$

For each $j \neq i$, the second condition requires that there exists at least one column in $\left[\, G_u^{(j)}(\lambda) - G_u^{(i)}(\lambda) \ \ G_d^{(j)}(\lambda) \,\right]$, say $g(\lambda)$, for which $Q_y^{(i)}(\lambda)g(\lambda) \neq 0$. This condition together with $Q_y^{(i)}(\lambda)G_d^{(i)}(\lambda) = 0$ is equivalent with the fault detectability condition (see Theorem 3.1)

$$\text{rank}\,[\,G_d^{(i)}(\lambda)\ \ g(\lambda)\,] > \text{rank}\ G_d^{(i)}(\lambda)\,.$$

It is easy to observe that this condition implies (4.7).

To prove the sufficiency of (4.7), we determine a bank of N filters $Q^{(i)}(\lambda)$, $i = 1, \ldots, N$ to solve the model detection problem. For this, we construct the i-th filter $Q^{(i)}(\lambda)$ such that the corresponding

$$R^{(i,j)}(\lambda) := Q^{(i)}(\lambda) \begin{bmatrix} G_u^{(j)}(\lambda)\ G_d^{(j)}(\lambda) \\ I_{m_u} \qquad 0 \end{bmatrix}$$

satisfies $R^{(i,i)}(\lambda) = 0$ and $R^{(i,j)}(\lambda) \neq 0 \ \forall j \neq i$. We show that we can determine $Q^{(i)}(\lambda)$ in the stacked form

$$Q^{(i)}(\lambda) = \begin{bmatrix} Q_1^{(i)}(\lambda) \\ \vdots \\ Q_N^{(i)}(\lambda) \end{bmatrix}, \tag{4.8}$$

where each row $Q_j^{(i)}(\lambda)$ is a stable scalar output filter, which satisfies

$$Q_j^{(i)}(\lambda) \begin{bmatrix} G_u^{(i)}(\lambda)\ G_d^{(i)}(\lambda) \\ I_{m_u} \qquad 0 \end{bmatrix} = 0 \tag{4.9}$$

and additionally for $j \neq i$

$$Q_j^{(i)}(\lambda) \begin{bmatrix} G_u^{(j)}(\lambda)\ G_d^{(j)}(\lambda) \\ I_{m_u} \qquad 0 \end{bmatrix} \neq 0\,. \tag{4.10}$$

For convenience, we set $Q_i^{(i)}(\lambda) = 0$ (a null row vector). This construction of $Q^{(i)}(\lambda)$ in (4.8) ensures with the help of the condition (4.10) that the corresponding $R^{(i,j)}(\lambda) \neq 0 \ \forall j \neq i$.

To determine $Q_j^{(i)}(\lambda)$ for $j \neq i$, we observe that the condition (4.7) can be interpreted as an extended fault detectability condition for (fictive) fault inputs corresponding to an input–output faulty system defined by the triple of TFMs

$$\{G_u^{(i)}(\lambda), G_d^{(i)}(\lambda), [\,G_d^{(j)}(\lambda)\ \ G_u^{(i)}(\lambda) - G_u^{(j)}(\lambda)\,]\}$$

from suitably defined control, disturbance and fault inputs, respectively. It follows that there exists $Q_j^{(i)}(\lambda)$ such that (4.9) is fulfilled and

$$Q_j^{(i)}(\lambda) \begin{bmatrix} G_d^{(j)}(\lambda)\ G_u^{(i)}(\lambda) - G_u^{(j)}(\lambda) \\ 0 \qquad\qquad 0 \end{bmatrix} \neq 0\,.$$

Taking into account (4.9), this condition can be rewritten in the equivalent form (4.10), which in turn implies that $R^{(i,j)}(\lambda) \neq 0$ for $j \neq i$. ∎

4.4 Model Detection Problems

In this section, we formulate the exact and approximate synthesis problems of model detection filters for the collection of N LTI systems (4.1). As in the case of the EFDIP or AFDIP, we seek N linear residual generators (or model detection filters) of the form (4.2), which process the measurable system outputs $y(t)$ and known control inputs $u(t)$ and generate the N residual signals $r^{(i)}(t)$ for $i = 1, \ldots, N$. These signals serve for decision-making by comparing the pattern of fired and not fired residuals with the signatures coded in the columns of the associated standard $N \times N$ structure matrix S with zeros on the diagonal and ones elsewhere. The standard requirements for the TFM of the overall filter $Q(\lambda)$ in (4.3) are *properness* and *stability*. For practical purposes, the order of the overall filter $Q(\lambda)$ must be as small as possible. A least order $Q(\lambda)$ can be usually achieved by employing N scalar output least order filters (see Sect. 6.2).

In analogy to the formulations of the EFDIP and AFDIP, we use the internal form of the i-th residual generator (4.4) to formulate the basic model detection requirements. Independently of the presence of the noise inputs $w^{(j)}$, we will target that the i-th residual is exactly decoupled from the i-th model if $w^{(i)} \equiv 0$ and sensitive to the j-th model, for all $j \neq i$. These requirements can be easily translated into algebraic conditions using the internal form (4.4) of the i-th residual generator:

$$
\begin{aligned}
&(i)\ \ [R_u^{(i,i)}(\lambda)\ \ R_d^{(i,i)}(\lambda)] = 0, \ \ i = 1, \ldots, N, \\
&(ii)\ [R_u^{(i,j)}(\lambda)\ \ R_d^{(i,j)}(\lambda)] \neq 0, \ \ \forall j \neq i, \text{ with } [R_u^{(i,j)}(\lambda)\ \ R_d^{(i,j)}(\lambda)] \text{ stable.}
\end{aligned}
\tag{4.11}
$$

Here, (i) is the *model decoupling condition* for the i-th model in the i-th residual component, while (ii) is the *model sensitivity condition* of the i-th residual component to all models, excepting the i-th model. In the case when condition (i) cannot be fulfilled (e.g., due to lack of sufficient measurements), some (or even all) components of $d^{(i)}(t)$ can be redefined as noise inputs and included in $w^{(i)}(t)$.

In what follows, we formulate two model detection problems, which are addressed in this book.

4.4.1 Exact Model Detection Problem

The standard requirement for solving the *exact model detection problem* (EMDP) is to determine for the multiple model (4.1), in the absence of noise input (i.e., $w^{(j)} \equiv 0$ for $j = 1, \ldots, N$), a set of N proper and stable filters $Q^{(i)}(\lambda)$ such that, for

$i = 1, \ldots, N$, the conditions (4.11) are fulfilled. These conditions are similar to the model detectability requirement and lead to the following solvability condition:

Theorem 4.2 *For the multiple model (4.1) with $w^{(j)} \equiv 0$ for $j = 1, \ldots, N$, the EMDP is solvable if and only if the multiple model (4.1) is model detectable.*

Proof For each i, the conditions (4.11) can be fulfilled provided the multiple model (4.1) is model detectable. To ensure the stability of $Q^{(i)}(\lambda)$, $R_u^{(i,j)}(\lambda)$ and $R_d^{(i,j)}(\lambda)$, the filter with TFM $Q^{(i)}(\lambda)$ can be replaced by $M^{(i)}(\lambda)Q^{(i)}(\lambda)$, where

$$(M^{(i)}(\lambda))^{-1} N^{(i)}(\lambda) = [\, Q^{(i)}(\lambda) \;\; R_u^{(i,1)}(\lambda) \;\; R_d^{(i,1)}(\lambda) \;\; \cdots \;\; R_u^{(i,N)}(\lambda) \;\; R_d^{(i,N)}(\lambda) \,]$$

is a stable left coprime factorization.   ∎

4.4.2   Approximate Model Detection Problem

The effects of the noise input $w^{(i)}(t)$ can usually not be fully decoupled from the residual $r^{(i)}(t)$. In this case, the basic requirements for the choice of $Q^{(i)}(\lambda)$ can be expressed as achieving that the residual $r^{(i)}(t)$ is influenced by all models in the multiple model (4.1), while the influence of the i-th model is only due to the noise signal $w^{(i)}(t)$ and is negligible. For the *approximate model detection problem* (AMDP) the following additional conditions to (4.11) have to be fulfilled:

$$
\begin{aligned}
&(iii) \;\; R_w^{(i,i)}(\lambda) \approx 0, \;\; \text{with} \;\; R_w^{(i,i)}(\lambda) \;\; \text{stable};\\
&(iv) \;\; R_w^{(i,j)}(\lambda) \;\; \text{stable} \;\; \forall j \neq i.
\end{aligned}
\tag{4.12}
$$

Here, (iii) is the *attenuation condition* of the noise input.

The solvability conditions of the formulated AMDP can be easily established:

Theorem 4.3 *For the multiple model (4.1) the AMDP is solvable if and only the EMDP is solvable.*

Proof We can always determine a solution of the EMDP with $Q(\lambda)$ in the form (4.3), such that additionally the resulting $R_w^{(i,j)}(\lambda)$ are stable for $i, j = 1, \ldots, N$. Moreover, by rescaling $Q^{(i)}(\lambda)$ with a constant factor γ_i, the norm of $R_w^{(i,i)}(\lambda)/\gamma_i$ can be made arbitrarily small. The necessity is trivial, because any solution of the AMDP is also a solution of the EMDP.   ∎

4.5 Threshold Selection

Similar to the performance requirements for FDD systems, a well-designed model detection system as that in Fig. 4.1, must fulfil standard performance requirements as timely and unequivocal identification of a *single* model out of N candidate models, which best fits with the input–output measurements. Assume that we use N residual evaluation signals $\theta_i(t), i = 1, \ldots, N$, where $\theta_i(t)$ is an approximation of $\|r^{(i)}\|_2$ (see Sect. 3.6), and for each i let τ_i be the corresponding threshold. For the unequivocal identification of the i-th model, we must have $\theta_i(t) \leq \tau_i$ and $\theta_j(t) > \tau_j$ for all $j \neq i$, which corresponds to a binary signature with $N - 1$ ones and single zero in the i-th element. A false alarm occurs when, due to the effects of noise inputs, the j-th model (a "false" one) is identified as the best matching one instead of the i-th model (the "true" one). A missed detection occurs, for example, when $\theta_i(t) > \tau_i$ for all $i = 1, \ldots, N$, or when the resulting binary signature contains several zero entries. In both of these cases, no unequivocal model identification can take place.

In what follows, we discuss the choice of the decision thresholds $\tau_i, i = 1, \ldots, N$ to be used in the model detection schemes, such that false alarms and missed detections can be avoided. For $j = 1, \ldots, N$, let $\mathcal{U}^{(j)}$, $\mathcal{D}^{(j)}$ and $\mathcal{W}^{(j)}$ be the classes of control inputs $u^{(j)}$, disturbance inputs $d^{(j)}$ and noise inputs $w^{(j)}$, respectively, which are relevant for a model detection application. For example, $\mathcal{U}^{(j)}$ is the class of nonzero control inputs with bounded variations, $\mathcal{D}^{(j)}$ may be the class of disturbance signals with bounded variations for the j-th model, while $\mathcal{W}^{(j)}$ may be the class of white noise signals of given maximal amplitude and covariance for the j-th model. We consider the selection of the threshold τ_i, which is instrumental for the discrimination of the i-th model from the rest of models.

To account for the dependence of the evaluation signal $\theta(t)$ of the input variables $u^{(j)} \in \mathcal{U}^{(j)}, d^{(j)} \in \mathcal{D}^{(j)}$, and $w^{(j)} \in \mathcal{W}^{(j)}$ and of the corresponding time response of the output signal $y^{(j)}$ of the j-th model up to the time moment t, we will indicate this dependence explicitly as $\theta(t, u^{(j)}, d^{(j)}, w^{(j)}, y^{(j)})$. Assume that the i-th model is the current model (to be detected) and $y^{(i)}$ is the corresponding time response of the i-th model output. The requirement for no false alarms in recognizing the i-th model leads to a lower bound for τ_i, representing the i-th *false alarm bound*

$$
\tau_f^{(i)} := \sup_{\substack{t \in [0, t_m] \\ u^{(i)} \in \mathcal{U}^{(i)} \\ d^{(i)} \in \mathcal{D}^{(i)} \\ w^{(i)} \in \mathcal{W}^{(i)}}} \theta_i(t, u^{(i)}, d^{(i)}, w^{(i)}, y^{(i)}) , \tag{4.13}
$$

where t_m is the maximum signal monitoring time. We can define the i-th *detection bound* as the least of the $N - 1$ lower bounds of the evaluation signal for any other current model different from the i-th model:

$$\tau_d^{(i)} := \min_{j \neq i} \quad \inf_{\substack{t \in [0, t_m] \\ u^{(j)} \in \mathcal{U}^{(j)} \\ d^{(j)} \in \mathcal{D}^{(j)} \\ w^{(j)} \in \mathcal{W}^{(j)}}} \theta_i(t, u^{(j)}, d^{(j)}, w^{(j)}, y^{(j)}). \tag{4.14}$$

It is usually assumed that the choice of the i-th filter $Q^{(i)}(\lambda)$ can be done such that $\tau_f^{(i)} < \tau_d^{(i)}$, which ensures that the threshold τ_i can be chosen such that

$$\tau_f^{(i)} < \tau_i \leq \tau_d^{(i)}.$$

With such a choice for all N threshold values τ_i, $i = 1, \ldots, N$, it is possible to guarantee the lack of false alarms and missed detections, and thus ensure the unequivocal identification of any of the N models. Note that, with a suitable rescaling of the N component filters $Q^{(i)}(\lambda)$, $i = 1, \ldots, N$, it is possible to arrange that all thresholds can be taken equal to a common value $\tau_i = \tau$, for $i = 1, \ldots, N$. If the condition $\tau_f^{(i)} < \tau_d^{(i)}$ cannot be enforced, then no unequivocal identification of the i-th model is possible. A possible remedy in such cases is to redefine the set of models, by including only models, which are sufficiently "far" from each other.

Remark 4.1 In practical applications the chosen N models usually form a representative set of models, but frequently do not cover the entire set of possible models, which can even be infinite due to continuous ranges of variation of fault parameters (e.g., loss of efficiency degree). Thus, a typical operation mode for any model detection setup is with the current model lying "in between" two candidate models. To handle this situation and to avoid false alarms and missed detections, an alternative decision scheme can be employed, where the i-th model is selected, provided the corresponding evaluation signal $\theta_i < \theta_j$ for all $j \neq i$. Although this decision scheme "always" works, still wrong identifications may result, because of the difficulty to correctly map (via a set of N filters $Q^{(i)}(\lambda)$, $i = 1, \ldots, N$), the "nearness" of two models, as for example, the i-th and j-th models, into the "nearness" of the corresponding evaluations θ_i and θ_j. $\square$

4.6 Notes and References

The term *model detection* has been apparently used for the first time in [148]. Model validation and also model invalidation have been discussed in [110, 117] in the context of model identification for robust control. The definition of *model detectability* appears to be new.

Two *model selection* problems have been formulated, in a stochastic setting, by Baram in [4], which are very similar to the model detection problem considered in this book. These problems consist of the selection of a model out of N given models, which is the closest to or exactly matches the "true" model. Stochastic measures of closeness are used to discriminate between two models. The use of Kalman filters to perform model selection has been discussed by Willsky [174]. The selection of adequate models for the purpose of *multiple-model adaptive control* (MMAC) is discussed in [44].

The use of multiple model techniques for fault detection and isolation has been considered in several publications, see—for example, [17, 171]. The exact model detection problem has been formulated and solved in [148]. The formulation of the approximate model detection problem is similar to several formulations based on the use of Kalman filters as model detection filters, where all unknown inputs (noise and disturbances) are assumed to be white noise signals [89, 95, 174]. The model detection approach discussed in this book is a viable alternative to Kalman filter-based approaches used for switching or interpolating among different controllers for MMAC (see—for example, [2]) or in *interacting multiple model* (IMM) Kalman filters-based reconfiguration schemes [73]. The main advantages of using model detection filters over various Kalman filter-based techniques are (1) the ability of model detection filters to exactly decouple the influence of nonstochastic disturbances from the residual signals, and (2) the significantly lower dynamical orders of model detection filters compared to Kalman filters. The first of these advantages has been also noted in a related approach based on unknown-input observers proposed in [171].

The decision scheme based on the choice of that model for which the corresponding evaluation signal has the least value among all evaluation signals has been advocated in [95], where the Narendra-type residual evaluation filter has also been introduced.

Part II
Synthesis of Residual Generators

Part II is dedicated to the development of synthesis procedures of fault detection and model detection filters, which solve the fault and model detection problems formulated in Part I of this book. The proposed synthesis procedures are completely general and deliver solutions whenever the general existence conditions (also established in Part I) are fulfilled. Although this goal for generality appears as perfectly legitimate, it rules out several "consecrated" synthesis methods (e.g., unknown-input observer-based synthesis), whose applicability requires the fulfilment of not necessary (technical) assumptions. A distinctive feature of the presented synthesis methods is their general applicability to both continuous- and discrete-time systems, regardless of whether the underlying system is proper or not.

The developed synthesis procedures of fault detection and isolation filters and of model detection filters are presented in two separate chapters. For each synthesis procedure, we present a general derivation of the underlying synthesis method and give the pseudocode of a conceptual synthesis procedure which makes precise the involved computational steps. A detailed synthesis example, accompanied by the listing of a Julia script, is also given. This allows to reproduce virtually all results of the synthesis examples presented in the book. The numerical aspects of the developed procedures are discussed in Chap. 7, where state-space representation-based formulas are derived for each main computational step of the developed synthesis procedures. The selection of computational procedures on the basis of widely accepted criteria for satisfactory computational algorithms such as generality, numerical reliability and computational efficiency largely contributed to dismissal of the so-called parity-space methods, which, although general, cannot be considered reliable computational methods due to their intrinsic numerical instability.

Several common computational paradigms emerge when contemplating the presented synthesis procedures. A first paradigm is the product form representation of all synthesized filters, in both implementation and internal forms. Each factor of the resulting filter is derived by enforcing new features for an already available partial synthesis. However, the explicit forming of the product is completely avoided by employing suitable updating formulas. Therefore, the derivation of the explicit updating formulas is an important part of the presented algorithmic developments.

Besides ensuring implicit pole–zero cancellations, the use of updating techniques leads to highly integrated steps, where the structural features achieved at the termination of a computational step are fully exploited by the computations in the subsequent step. Other computational paradigms worth to be mentioned are the use of the nullspace method as the first synthesis step to reduce all synthesis problems to a simple standard form which allows to easily check solvability conditions; the use of minimum dynamic cover algorithms to address least-order synthesis problems; the use of coprime factorization techniques to enforce the desired filter dynamics.

Two case studies illustrate the use of linear synthesis techniques to address the robustness aspects in the synthesis of fault detection and isolation filters for two fault monitoring applications in aeronautics. The first case study considers the robust monitoring of flight actuator faults using global and local monitoring approaches. The second case study considers the robust monitoring of air data sensor faults.

Chapter 5
Synthesis of Fault Detection and Isolation Filters

This chapter presents general synthesis procedures of fault detection filters, which solve the fault detection problems formulated in Chap. 3. The synthesis procedures are described in terms of input–output models, which allow simpler conceptual presentations. Numerically reliable state-space representation-based synthesis algorithms, well suited for robust software implementations, are described in Chap. 7.

In the recently developed computational procedures for the synthesis of fault detection filters, two important computational paradigms emerged, which are instrumental in developing generally applicable, numerically reliable and computationally efficient synthesis methods. The first paradigm is the use of factorization-based synthesis methods. Accordingly, for all presented synthesis procedures, it is possible to express the TFM of the final filter $Q(\lambda)$ in a factored form as

$$Q(\lambda) = Q_K(\lambda) \cdots Q_2(\lambda) Q_1(\lambda), \tag{5.1}$$

where $Q_1(\lambda)$, $Q_2(\lambda) Q_1(\lambda)$, ..., can be interpreted as partial syntheses addressing specific requirements. Since each partial synthesis may represent a valid fault detection filter, this approach has a high flexibility in using or combining different synthesis techniques. The factorization-based synthesis approach naturally leads to the so-called integrated computational algorithms, with strongly coupled successive computational steps. For a K-step synthesis procedure to determine $Q(\lambda)$ in the factored form (5.1), K updating operations of the form $Q(\lambda) \leftarrow Q_i(\lambda) Q(\lambda)$ are performed for $i = 1, \ldots, K$, where $Q_i(\lambda)$ is the factor computed at the i-th synthesis step. The state-space description-based filter updating formulas are described in Chap. 7 for specific synthesis steps.

The second paradigm is the use of the nullspace method as a first synthesis step to reduce all synthesis problems to simpler problems, which easily allows to check the solvability conditions and address least order synthesis problems. The nullspace-based synthesis approach is described in Sect. 5.1. In Sects. 5.2–5.7, specific synthesis procedures, relying on the nullspace method, are presented for each of the fault detection and isolation problems formulated in Chap. 3.

5.1 Nullspace-Based Synthesis

A useful parametrization of all fault detection filters can be obtained on the basis of conditions $R_u(\lambda) = 0$ and $R_d(\lambda) = 0$ in (3.23). For any fault detection filter $Q(\lambda)$, the condition $[\, R_u(\lambda)\ R_d(\lambda)\,] = 0$ is equivalent to

$$Q(\lambda) \begin{bmatrix} G_u(\lambda)\ G_d(\lambda) \\ I_{m_u} \qquad 0 \end{bmatrix} = 0 \,.$$

Thus, any fault detection filter $Q(\lambda)$ must be a left annihilator of the TFM

$$G(\lambda) := \begin{bmatrix} G_u(\lambda)\ G_d(\lambda) \\ I_{m_u} \qquad 0 \end{bmatrix} . \tag{5.2}$$

Assume $r_d < p$ is the normal rank of $G_d(\lambda)$ (i.e., the maximal rank over all λ). Using standard linear algebra results (see Sect. 9.1.3), there exists a maximal full row rank left annihilator $N_l(\lambda)$ of size $(p - r_d) \times (p + m_u)$ such that $N_l(\lambda)G(\lambda) = 0$. Any such $N_l(\lambda)$ represents a *basis* of $\mathcal{N}_L(G(\lambda))$, the left nullspace of $G(\lambda)$. Using this fact, we have the following straightforward parametrization of all fault detection filters:

Theorem 5.1 *Let $N_l(\lambda)$ be a basis of $\mathcal{N}_L(G(\lambda))$, with $G(\lambda)$ defined in (5.2). Then, any fault detection filter $Q(\lambda)$ satisfying (3.23) can be expressed in the form*

$$Q(\lambda) = V(\lambda)N_l(\lambda), \tag{5.3}$$

where $V(\lambda)$ is a suitable TFM.

Proof Let $q^{(i)}(\lambda)$ be the i-th row of $Q(\lambda)$. Since $q^{(i)}(\lambda)G(\lambda) = 0$, it follows that $q^{(i)}(\lambda) \in \mathcal{N}_L(G(\lambda))$, and therefore, there exists a vector $v^{(i)}(\lambda)$ such that $q^{(i)}(\lambda) = v^{(i)}(\lambda)N_l(\lambda)$, representing a linear combination of the nullspace basis vectors. Thus, we build $V(\lambda)$ in (5.3) as a TFM whose i-th row is $v^{(i)}(\lambda)$. ∎

Remark 5.1 For any non-singular polynomial or rational matrix $M(\lambda)$ of appropriate dimension, $\widetilde{N}_l(\lambda) := M(\lambda)N_l(\lambda)$ is also a nullspace basis. Frequently, $M(\lambda)$ is the denominator matrix of a *left coprime factorization* (LCF) of an original basis $N_l(\lambda)$ in the form

$$N_l(\lambda) = M(\lambda)^{-1}\widetilde{N}_l(\lambda), \tag{5.4}$$

where the factors $M(\lambda)$ and $\widetilde{N}_l(\lambda)$ are determined to satisfy special requirements such as properness or to have only poles in a certain "good" region of the complex plane (e.g., in the stability region), or both. In this case, if $N_l(\lambda)$ is a basis, then $\widetilde{N}_l(\lambda) = M(\lambda)N_l(\lambda)$ is a basis as well. Moreover, the zeros of $M(\lambda)$ are the poles of $N_l(\lambda)$ lying outside of the "good" region. For more details on coprime factorizations, see Sect. 9.1.6. □

An interesting property of nullspace bases is the following elementary fact. Consider a column partitioning of $G(\lambda)$ as $G(\lambda) = \begin{bmatrix} G_1(\lambda) & G_2(\lambda) \end{bmatrix}$ and let $N_{l,1}(\lambda)$ be a basis of $\mathcal{N}_L(G_1(\lambda))$ and $N_{l,2}(\lambda)$ be a basis of $\mathcal{N}_L(N_{l,1}(\lambda)G_2(\lambda))$. Then, $N_{l,2}(\lambda)N_{l,1}(\lambda)$ is a basis of $\mathcal{N}_L(G(\lambda))$. Using this fact with the following partitioning

$$G(\lambda) = \begin{bmatrix} G_1(\lambda) \big| G_2(\lambda) \end{bmatrix} := \left[\begin{array}{c|c} G_u(\lambda) & G_d(\lambda) \\ I_{m_u} & 0 \end{array} \right],$$

we immediately obtain the left nullspace basis $N_l(\lambda)$ in the factorized form

$$N_l(\lambda) = N_{l,d}(\lambda)\begin{bmatrix} I_p & -G_u(\lambda) \end{bmatrix}, \tag{5.5}$$

where $N_{l,d}(\lambda)$ is a $(p - r_d) \times p$ TFM representing a basis of $\mathcal{N}_L(G_d(\lambda))$. This form leads to simple expressions of $N_l(\lambda)$ for particular cases as $N_l(\lambda) = N_{l,d}(\lambda)$ if $m_u = 0$, or $N_l(\lambda) = \begin{bmatrix} I_p & -G_u(\lambda) \end{bmatrix}$ if $m_d = 0$, or $N_l(\lambda) = I_p$ if $m_u + m_d = 0$.

A proper and stable representation of $N_l(\lambda)$ for arbitrary rational or polynomial matrices $G_u(\lambda)$, $G_d(\lambda)$, $G_f(\lambda)$ and $G_w(\lambda)$ can be obtained from the LCF

$$\begin{bmatrix} G_u(\lambda) & G_d(\lambda) & G_f(\lambda) & G_w(\lambda) \end{bmatrix} = \widehat{M}^{-1}(\lambda)\begin{bmatrix} \widehat{G}_u(\lambda) & \widehat{G}_d(\lambda) & \widehat{G}_f(\lambda) & \widehat{G}_w(\lambda) \end{bmatrix}, \tag{5.6}$$

where $\widehat{M}(\lambda)$ and $\begin{bmatrix} \widehat{G}_u(\lambda) & \widehat{G}_d(\lambda) & \widehat{G}_f(\lambda) & \widehat{G}_w(\lambda) \end{bmatrix}$ are proper and stable factors. With obvious replacements, the left nullspace basis $N_l(\lambda)$ can be chosen as

$$N_l(\lambda) = \widehat{N}_{l,d}(\lambda)\begin{bmatrix} \widehat{M}(\lambda) & -\widehat{G}_u(\lambda) \end{bmatrix}, \tag{5.7}$$

where $\widehat{N}_{l,d}(\lambda)$ is a $(p - r_d) \times p$ proper and stable TFM representing a basis of $\mathcal{N}_L(\widehat{G}_d(\lambda))$. If $m_u = m_d = 0$, then we can formally set $N_l(\lambda) := \widehat{M}(\lambda)$.

For the particular form of the nullspace basis in (5.7), we have the following straightforward corollary of Theorem 5.1:

Corollary 5.1 *Let $\widehat{G}_d(\lambda)$ and $\widehat{G}_u(\lambda)$ be the TFMs defined in (5.6) and let $\widehat{N}_{l,d}(\lambda)$ be a basis of $\mathcal{N}_L(\widehat{G}_d(\lambda))$. Then, any fault detection filter $Q(\lambda)$ satisfying (3.23) can be expressed in the form*

$$Q(\lambda) = W(\lambda)\widehat{N}_{l,d}(\lambda)\begin{bmatrix} \widehat{M}(\lambda) & -\widehat{G}_u(\lambda) \end{bmatrix}, \tag{5.8}$$

where $W(\lambda)$ is a suitable TFM.

The parametrization result of Theorem 5.1 underlies the nullspace method based synthesis procedures of fault detection filters, which form the main focus of this book. All synthesis procedures of the fault detection filters rely on the initial factored form

$$Q(\lambda) = \overline{Q}_1(\lambda)Q_1(\lambda), \tag{5.9}$$

where $Q_1(\lambda) = N_l(\lambda)$ is a basis of $\mathcal{N}_L(G(\lambda))$, while $\overline{Q}_1(\lambda)$ is a factor to be subsequently determined. The nullspace-based first step allows to reduce all synthesis

problems formulated for the system (3.2) to simpler problems, which make straightforward to check the solvability conditions.

Using the factored form (5.9), the fault detection filter in (3.3) can be rewritten in
the alternative form

$$\mathbf{r}(\lambda) = \overline{Q}_1(\lambda) Q_1(\lambda) \begin{bmatrix} \mathbf{y}(\lambda) \\ \mathbf{u}(\lambda) \end{bmatrix} = \overline{Q}_1(\lambda) \overline{\mathbf{y}}(\lambda) \,, \tag{5.10}$$

where

$$\overline{\mathbf{y}}(\lambda) := Q_1(\lambda) \begin{bmatrix} \mathbf{y}(\lambda) \\ \mathbf{u}(\lambda) \end{bmatrix} = \overline{G}_f(\lambda)\mathbf{f}(\lambda) + \overline{G}_w(\lambda)\mathbf{w}(\lambda) \,, \tag{5.11}$$

with

$$[\,\overline{G}_f(\lambda)\ \overline{G}_w(\lambda)\,] := Q_1(\lambda) \begin{bmatrix} G_f(\lambda)\ G_w(\lambda) \\ 0 \qquad 0 \end{bmatrix} . \tag{5.12}$$

With this first preprocessing step, we reduced the original problems, formulated
for the system (3.2), to simpler ones, which can be formulated for the reduced system
(5.11) (without control and disturbance inputs) for which we have to determine the
TFM $\overline{Q}_1(\lambda)$ of the simpler fault detection filter (5.10).

Remark 5.2 At this stage, we can assume that both $Q_1(\lambda)$ and the TFMs of the
reduced system (5.11) are proper and even stable. This can always be achieved
by replacing any basis $N_l(\lambda)$, with a stable basis $Q_1(\lambda) = M(\lambda)N_l(\lambda)$, where
$M(\lambda)$ is an invertible, stable and proper TFM, of least McMillan degree, such that
$M(\lambda)[\,N_l(\lambda)\ \overline{G}_f(\lambda)\ \overline{G}_w(\lambda)\,]$ is stable and proper. Such an $M(\lambda)$ can be determined
as the minimum-degree denominator of a stable and proper LCF of $[\,N_l(\lambda)\ \overline{G}_f(\lambda)$
$\overline{G}_w(\lambda)\,]$ (see Sect. 9.1.6). Even if $N_l(\lambda)$ is a minimal basis, the resulting stable basis
$Q_1(\lambda)$ is, in general, not a minimal basis.   □

We conclude this section with the derivation of simpler conditions for checking
the fault detectability conditions studied in the Sect. 3.3. The following result characterizes the complete fault detectability of the system (3.2) as the complete fault
input observability property of the reduced system (5.11).

Proposition 5.1 *For the system (3.2) with $w \equiv 0$, let $Q_1(\lambda) = N_l(\lambda)$ be a rational
basis of $\mathcal{N}_L(G(\lambda))$, where $G(\lambda)$ is defined in (5.2), and let (5.11) be the corresponding
reduced system with $w \equiv 0$. Then, the system (3.2) is completely fault detectable if
and only if*

$$\overline{G}_{f_j}(\lambda) \neq 0, \quad j = 1, \ldots m_f \,. \tag{5.13}$$

Proof To prove the necessity, we show that if the original system is completely
fault detectable, then the reduced system (5.11) is also completely fault detectable
(i.e., conditions (5.13) are fulfilled). For the completely fault detectable system

(3.2), let $Q(\lambda)$ be a filter such that $R_{f_j}(\lambda) \neq 0$ for $j = 1, \ldots, m_f$. According to Theorem 5.1, for a given nullspace basis $N_l(\lambda)$, any filter $Q(\lambda)$ can be expressed in the form $Q(\lambda) = W(\lambda)N_l(\lambda)$, where $W(\lambda)$ is a suitable rational matrix. It follows that $R_{f_j}(\lambda) = W(\lambda)\overline{G}_{f_j}(\lambda)$, and therefore, $R_{f_j}(\lambda) \neq 0$ only if $\overline{G}_{f_j}(\lambda) \neq 0$.

The proof of sufficiency is trivial, since with $Q(\lambda) := N_l(\lambda)$, the corresponding $R_f(\lambda) = \overline{G}_f(\lambda)$, and thus satisfies $R_{f_j}(\lambda) \neq 0$ for $j = 1, \ldots, m_f$. ∎

The following result is a general characterization of the complete strong fault detectability of the system (3.2) in terms of a particular reduced system (5.11) and can serve as an easy check of this property.

Proposition 5.2 *Let Ω be the set of frequencies, which characterize the persistent fault signals. For the system (3.2) with $w \equiv 0$ and for $G(\lambda)$ defined in (5.2), let $Q_1(\lambda)$ be a least order rational basis of $\mathcal{N}_L(G(\lambda))$, such that $Q_1(\lambda)$ and $\overline{G}_f(\lambda)$ in (5.12) have no poles in Ω. Then, the system (3.2) is completely strong fault detectable with respect to Ω if and only if*

$$\overline{G}_{f_j}(\lambda_z) \neq 0, \quad j = 1, \ldots m_f, \quad \forall \lambda_z \in \Omega . \tag{5.14}$$

Proof To prove necessity, we note that complete strong fault detectability implies that there exists a stable filter $Q(\lambda)$ such that the corresponding $R_f(\lambda)$ is stable, and $R_{f_j}(\lambda)$, the j-th column of $R_f(\lambda)$, has no zeros in Ω. According to Theorem 5.1, any filter $Q(\lambda)$ satisfying $Q(\lambda)G(\lambda) = 0$ can be expressed in the form $Q(\lambda) = W(\lambda)Q_1(\lambda)$, where $W(\lambda)$ is a suitable rational matrix. It follows that $R_{f_j}(\lambda) = W(\lambda)\overline{G}_{f_j}(\lambda)$. Assume $\lambda_z \in \Omega$ is a zero of $\overline{G}_{f_j}(\lambda)$, such that $\overline{G}_{f_j}(\lambda_z) = 0$. However, this implies that $R_{f_j}(\lambda_z) = 0$, which contradicts the assumption of complete strong detectability. Therefore, $\overline{G}_{f_j}(\lambda)$ cannot have zeros in Ω. This requirement is expressed, for $j = 1, \ldots, m_f$, by the conditions (5.14).

To prove sufficiency, we show that for any given basis $Q_1(\lambda)$ without poles in Ω and for $\overline{G}_{f_j}(\lambda)$ without poles and zeros in Ω, we can build a stable filter $Q(\lambda)$ such that $R_{f_j}(\lambda)$ has no zeros in Ω as well. For this, we take $Q(\lambda) = M(\lambda)Q_1(\lambda)$, where $[\, Q_1(\lambda)\ \overline{G}_f(\lambda)\,] = M^{-1}(\lambda)[\, Q(\lambda)\ R_f(\lambda)\,]$ is a stable left coprime factorization. The zeros of $M(\lambda)$ are the unstable poles of $[\, Q_1(\lambda)\ \overline{G}_f(\lambda)\,]$. Since by assumption, this TFM has no poles in Ω, it follows that $M(\lambda)$ has no zeros in Ω. Therefore, for any $\lambda_z \in \Omega$, $\det M(\lambda_z) \neq 0$. For each f_j follows that if $\overline{G}_{f_j}(\lambda_z) \neq 0$, then $R_{f_j}(\lambda_z) = M(\lambda_z)\overline{G}_{f_j}(\lambda_z) \neq 0$. This proves the complete strong fault detectability with respect to Ω. ∎

Remark 5.3 The conditions on the poles of $Q_1(\lambda)$ and $\overline{G}_f(\lambda)$ imposed in Proposition 5.2 are essential to check the complete strong fault detectability. In Example 3.3 with $G_u(s) = 0$, $G_d(s) = 0$ and $G_f(s) = [\,1\ \ 1/s\,]$, we can choose $Q_1(s) = s/(s+1)$ to obtain $\overline{G}_f(s) = Q_1(s)G_f(s) = [\,s/(s+1)\ \ 1/(s+1)\,]$. This system is not completely strong fault detectable with respect to constant faults because $\overline{G}_{f_1}(0) = 0$. The following example shows that the check of strong fault detectability may lead to an erroneous result if the condition on the poles of $Q_1(\lambda)$ is not fulfilled. $\qquad\qquad\square$

Example 5.1 Consider the continuous-time system (3.2) from Example 3.1 with

$$
G_u(s) = \begin{bmatrix} \frac{1}{s} \\ \frac{1}{s} \end{bmatrix}, \quad
G_d(s) = \begin{bmatrix} 0 \\ \frac{s}{s+3} \end{bmatrix}, \quad
G_f(s) = \begin{bmatrix} \frac{s+1}{s+2} \\ \frac{1}{s+2} \end{bmatrix}
$$

and $\Omega = \{0\}$. This system is not strongly fault detectable. To see this, we employ the check based on Proposition 5.2.

A stable rational (minimal) basis is

$$
Q_1(s) = \begin{bmatrix} \dfrac{s}{s+1} & 0 & -\dfrac{1}{s+1} \end{bmatrix},
$$

which leads to

$$
\overline{G}_f(s) = \frac{s}{s+2}.
$$

Since $\overline{G}_f(s)$ has a zero in 0, the system is not strongly fault detectable for constant faults.

However, if we use, instead, the rational (minimal) basis with a pole in the origin

$$
Q_1(s) = \begin{bmatrix} 1 & 0 & -\dfrac{1}{s} \end{bmatrix},
$$

we obtain

$$
\overline{G}_f(s) = \frac{s+1}{s+2},
$$

for which the zeros-based check indicates, erroneously, strong fault detectability. $\diamond$

5.2 Solving the Exact Fault Detection Problem

Using Proposition 5.1, the solvability conditions of the *exact fault detection problem* (EFDP) formulated in Sect. 3.5.1 for the system (3.2) with $w \equiv 0$ can be expressed as fault input observability conditions for the reduced system (5.11) with $w \equiv 0$ according to the following corollary to Theorem 3.7:

Corollary 5.2 *For the system (3.2) with $w \equiv 0$ the EFDP is solvable if and only if the reduced system (5.11) with $w \equiv 0$ is completely fault detectable, or equivalently, the following input observability conditions hold*

$$\overline{G}_{f_j}(\lambda) \neq 0, \quad j = 1, \ldots m_f. \tag{5.15}$$

Using Proposition 5.2, the solvability conditions of the EFDP with the strong detection condition (3.25) can be equivalently expressed as conditions on the lack of zeros in Ω for all columns of the TFM $\overline{G}_f(\lambda)$ of reduced system (5.11) according to the following corollary to Theorem 3.8:

Corollary 5.3 *Let $\Omega \subset \partial \mathbb{C}_s$ be a given set of frequencies, and assume that the reduced system (5.11) has been obtained by choosing $Q_1(\lambda)$ without poles in Ω and such that also $\overline{G}_f(\lambda)$ in (5.12) has no poles in Ω. Then, for $w \equiv 0$, the EFDP with the strong detection condition (3.25) is solvable if and only if the reduced system (5.11) with $w \equiv 0$ is completely strong fault detectable with respect to Ω, or equivalently, the following conditions hold:*

$$\overline{G}_{f_j}(\lambda_z) \neq 0, \quad j = 1, \ldots m_f, \quad \forall \lambda_z \in \Omega. \tag{5.16}$$

When solving the EFDP, it is obvious that any stable and proper rational nullspace basis $Q_1(\lambda)$ already represents a solution, provided the complete fault detectability conditions (5.15) or the complete strong fault detectability conditions (5.16) are fulfilled and $\overline{G}_f(\lambda)$ is stable. According to Remark 5.2, the dynamics of both $Q_1(\lambda)$ and $\overline{G}_f(\lambda)$ (i.e., their poles) can be arbitrarily assigned. Moreover, fault detection filters with an arbitrary number of outputs $q \leq p - r_d$ can be easily obtained, by building linear combinations of the rows of $Q_1(\lambda)$.

Example 5.2 Consider a continuous-time system with the transfer function matrices

$$G_u(s) = \begin{bmatrix} \dfrac{s+1}{s+2} \\ \dfrac{s+2}{s+3} \end{bmatrix}, \quad G_d(s) = \begin{bmatrix} \dfrac{1}{s+2} \\ 0 \end{bmatrix}, \quad G_f(s) = \begin{bmatrix} \dfrac{s+1}{s+2} & 0 \\ 0 & 1 \end{bmatrix}, \quad G_w(s) = 0.$$

A minimal left nullspace basis of $G(\lambda)$ defined in (5.2) for $\lambda = s$ can be obtained in the form (5.5) as $N_l(s) = N_{l,d}(s)\begin{bmatrix} I_2 & -G_u(s) \end{bmatrix}$, with $N_{l,d}(s) = \begin{bmatrix} 0 & 1 \end{bmatrix}$. We obtain $Q_1(s) = N_l(s)$ as

$$Q_1(s) = \begin{bmatrix} 0 & 1 & -\dfrac{s+2}{s+3} \end{bmatrix}$$

and the TFMs of the reduced system (5.11) are

$$\overline{G}_f(s) = [\, 0 \ \ 1 \,], \qquad \overline{G}_w(s) = 0 \,.$$

The presence of a zero column in $\overline{G}_f(s)$ indicates that the EFDP has no solution, because the fault f_1 and the disturbance d share the same signal space. By appropriately redefining d and w, we will address this problem in Example 5.5 and show that an approximate solution of this problem is still possible. Note that the filter with $Q(\lambda) = N_l(s)$ can be still used for the detection of f_2. $\qquad\qquad\qquad\Diamond$

We can exploit in various ways the existing freedom in determining fault detection filters, which solve the EFDP. For practical use, it is sometimes advantageous to impose for the number of residual signals q a certain low value, as for example, $q = 1$, which leads to *scalar output* fault detection filters. Of both theoretical and practical interest are fault detection filters, which have the least possible order (i.e., least McMillan degree). For example, least order scalar output fault detection filters can be employed to build banks of scalar output filters with global least orders to solve the more involved FDIPs.

For the computation of a least order solution, we can choose the factor $\overline{Q}_1(\lambda)$ in (5.9) in the factored form

$$\overline{Q}_1(\lambda) = Q_3(\lambda)Q_2(\lambda),$$

where $Q_2(\lambda)$ is a $q \times (p - r_d)$ proper TFM determined such that $Q_2(\lambda)Q_1(\lambda)$ has least order, while $Q_3(\lambda)$ is a $q \times q$ proper, stable and invertible TFM determined such that both the overall filter $Q(\lambda) = Q_3(\lambda)Q_2(\lambda)Q_1(\lambda)$ and $R_f(\lambda) = Q_3(\lambda)Q_2(\lambda)\overline{G}_f(\lambda)$ are stable. The least possible order of the fault detection filter $Q(\lambda)$ is uniquely determined by the fulfilment of a certain admissibility condition. When solving the EFDP, we say that the filter $Q(\lambda)$ is *admissible*, if the fault detection conditions (3.24) are fulfilled by the corresponding $R_f(\lambda)$. Thus, an admissible choice of $Q_2(\lambda)$ must guarantee the admissibility of $Q(\lambda)$. Since $Q_3(\lambda)$ is invertible, its choice plays no role in ensuring admissibility. Interestingly, a least order filter synthesis can be always achieved by a scalar output fault detection filter.

The **Procedure EFD** given in what follows summarizes the main computational steps of the synthesis of least order fault detection filters. In view of potential applications of **Procedure EFD**, we devised this procedure to be applicable to the complete faulty system (2.1), including also the noise inputs.

Procedure EFD: Exact synthesis of fault detection filters

Inputs : $\{G_u(\lambda), G_d(\lambda), G_f(\lambda), G_w(\lambda)\}, q$
Outputs: $Q(\lambda), R_f(\lambda), R_w(\lambda)$

(1) Compute a $(p - r_d) \times (p + m_u)$ minimal basis matrix $Q_1(\lambda)$ for the left nullspace of $G(\lambda)$ defined in (5.2), where $r_d := \operatorname{rank} G_d(\lambda)$;
set $Q(\lambda) = Q_1(\lambda)$ and compute $\left[\, R_f(\lambda)\,\middle|\,R_w(\lambda)\,\right] = Q_1(\lambda) \begin{bmatrix} G_f(\lambda) & G_w(\lambda) \\ 0 & 0 \end{bmatrix}$.
Exit if exists $j \in \{1, \ldots, m_f\}$ such that $R_{f_j}(\lambda) = 0$ (no solution exists).

(2) Choose a $\min(q, p - r_d) \times (p - r_d)$ rational matrix $Q_2(\lambda)$ such that $Q_2(\lambda)Q(\lambda)$ has least McMillan degree and $Q_2(\lambda)R_{f_j}(\lambda) \neq 0$, $j = 1, \ldots, m_f$; compute $Q(\lambda) \leftarrow Q_2(\lambda)Q(\lambda)$, $R_f(\lambda) \leftarrow Q_2(\lambda)R_f(\lambda)$ and $R_w(\lambda) \leftarrow Q_2(\lambda)R_w(\lambda)$.

(3) Choose a proper and stable invertible rational matrix $Q_3(\lambda)$ such that $Q_3(\lambda)Q(\lambda)$, $Q_3(\lambda)R_f(\lambda)$ and $Q_3(\lambda)R_w(\lambda)$ have desired stable dynamics; compute $Q(\lambda) \leftarrow Q_3(\lambda)Q(\lambda)$, $R_f(\lambda) \leftarrow Q_3(\lambda)R_f(\lambda)$, $R_w(\lambda) \leftarrow Q_3(\lambda)R_w(\lambda)$.

This procedure illustrates several computational paradigms common to all synthesis algorithms presented in this book such as the use of product form representations of the filter and the use of the associated filter updating techniques, the use of nullspace method as the first computational step, the determination of least order of the resulting filter on the basis of suitable admissibility conditions, or the arbitrary assignment of filter dynamics by using coprime factorization techniques.

The computational details of the above procedure differ according to the type of the employed nullspace basis at Step (1). We consider first the case when at Step (1) of **Procedure EFD**, $Q(\lambda) = Q_1(\lambda)$ is a minimal polynomial basis and the corresponding $R_f(\lambda)$ satisfies $R_{f_j}(\lambda) \neq 0$ for $j = 1, \ldots, m_f$. For simplicity, we determine a least order fault detection filter with scalar output (i.e., for $q = 1$). At Step (2), we have to determine $Q_2(\lambda) = \phi(\lambda)$, where $\phi(\lambda)$ is a polynomial vector, such that $\phi(\lambda)Q_1(\lambda)$ has least degree and $\phi(\lambda)R_{f_j}(\lambda) \neq 0$ for $j = 1, \ldots, m_f$. Assume $Q_1(\lambda)$ is formed of $p - r_d$ row vectors $v_i(\lambda)$, where $v_i(\lambda)$ is a polynomial basis vector of degree n_i. We assume that the basis vectors $v_i(\lambda)$ are ordered such that $n_1 \leq n_2 \leq \ldots \leq n_{p-r_d}$. We can easily construct linear combinations of basis vectors of final degree n_i, for $i = 1, \ldots, p - r_d$, by choosing $\phi(\lambda) = \phi^{(i)}(\lambda)$, with

$$\phi^{(i)}(\lambda) = [\, \phi_1^{(i)}(\lambda) \; \ldots \; \phi_i^{(i)}(\lambda) \; 0 \; \ldots \; 0\,], \tag{5.17}$$

where $\phi_j^{(i)}(\lambda)$ is a polynomial of maximum degree $n_i - n_j$ and $\phi_i^{(i)}(\lambda)$ is a nonzero constant value. The achievable least order can be determined by successively constructing linear combinations of polynomials with increasing degrees $n_1, n_2, \ldots,$ n_{p-r_d} (e.g., with randomly generated coefficients). For each trial degree n_i, the condition $\phi^{(i)}(\lambda)R_{f_j}(\lambda) \neq 0$ for $j = 1, \ldots, m_f$ is checked. The search stops for the first value of i for which this condition is fulfilled. At Step (3), we can often choose $Q_3(\lambda) = 1/d(\lambda)$, with $d(\lambda)$ a polynomial of degree n_i with only stable roots. However, if the resulting $Q_3(\lambda)R_f(\lambda)$ is not stable or not proper, then $Q_3(\lambda)$ must be computed to also enforce the stability of $Q_3(\lambda)R_f(\lambda)$ as well as of $Q_3(\lambda)R_w(\lambda)$. This can be achieved by replacing $Q(\lambda)$, $R_f(\lambda)$ and $R_w(\lambda)$ resulting at Step (2) with the proper and stable factors $\widetilde{Q}(\lambda)$, $\widetilde{R}_f(\lambda)$ and $\widetilde{R}_w(\lambda)$, respectively, resulting from a LCF with proper and stable factors

$$[\, Q(\lambda)\ R_f(\lambda)\ R_w(\lambda)\,] = Q_3^{-1}(\lambda)[\, \widetilde{Q}(\lambda)\ \widetilde{R}_f(\lambda)\ \widetilde{R}_w(\lambda)\,]\,, \qquad (5.18)$$

where the poles of the scalar transfer function $Q_3(\lambda)$ can be freely assigned.

The polynomial nullspace approach allows to easily solve the least order synthesis problem of fault detection filters with scalar outputs. The least order is bounded below by n_i, the degree of the i-th basis vector, where i is the first index for which there exists a $\phi^{(i)}(\lambda)$ of the form (5.17) such that $\phi^{(i)}(\lambda)R_{f_j}(\lambda) \neq 0$ for $j = 1, \ldots, m_f$ (with $R_f(\lambda)$ computed at Step (1) of **Procedure EFD**). The value n_i for the McMillan degree of the final filter $Q(\lambda)$ can be often achieved, as for example, when $G_f(\lambda)$ and $G_w(\lambda)$ are already stable and proper.

Remark 5.4 Step (2) of this synthesis procedure can be significantly simplified by determining directly the least degree of candidate polynomial vectors suited to solve the EFDP, instead of iterating with candidate vectors of increasing orders. For this purpose, we can use the $(p - r_d) \times m_f$ structure matrix S_{R_f} associated to the resulting $R_f(\lambda)$ at Step (1) of **Procedure EFD**. Let S_{R_f} be the binary matrix (see Sect. 3.4), whose the (i, j)-th element is set to 1 if the (i, j)-th element of $R_f(\lambda)$ is nonzero, and otherwise is set to 0. Let i be the least row index such that the leading i rows of S_{R_f} contain at least one nonzero element in all columns. It follows that we can build, using a polynomial vector $\phi^{(i)}(\lambda)$ of the form (5.17), a linear combination of the first i basis vectors of least degree n_i, such that all faults can be detected. A straightforward simplification is to use, instead of the polynomial vector $\phi^{(i)}(\lambda)$ in (5.17), a constant vector (with the same structure)

$$h^{(i)} = [\, h_1, \ldots, h_i, 0, \ldots, 0\,]\,, \qquad (5.19)$$

with $h_j \neq 0$, $j = 1, \ldots, i$, to build a linear combination of basis vectors up to degree n_i (e.g., using randomly generated values). The nonzero components of $h^{(i)}$ can be interpreted as weighting factors of the individual basis vectors. Therefore, an optimal choice of these weights can maximize the overall sensitivity of residual to faults. Suitable fault sensitivity measures for this purpose are discussed in Remark 5.6. $\square$

Remark 5.5 Although the EFDP can always be solved using a scalar output fault detection filter of least dynamical order, there may exist advantages when using filters with more than one output. A first advantage is the possibility to enforce with a residual vector with several components a more uniform sensitivity of the residual vector to individual fault components. This aspect is related to an increased number of free parameters, which can thus be optimally chosen (see Remark 5.6). A second potential advantage is that, with several residual outputs, it may be possible to also achieve a certain block isolation of a group of faults. For example, suitable combinations of individual basis vectors in $Q_1(\lambda)$ can be easily constructed using the binary information coded in the structure matrix S_{R_f} associated with the resulting $R_f(\lambda)$ at Step (1) of the **Procedure EFD**. This can be advantageous, especially in the case when the expected magnitudes of the residual signals may significantly vary for different groups of faults. A more involved synthesis procedure to achieve block isolation can be performed using several scalar output filters, where each filter is designed to be sensitive to a group of faults and insensitive to the rest of faults (see **Procedure EFDI** in Sect. 5.4). $\square$

When using a proper rational basis instead a polynomial one at Step (1) of the **Procedure EFD**, a synthesis approach leading directly to a proper filter can be devised.   Assume $Q_1(\lambda)$ is a simple minimal proper rational basis (see Sect. 9.1.3 for the definition of simple bases) formed of $p - r_d$ rational row vectors $v_i(\lambda)/d_i(\lambda)$, where $v_i(\lambda)$ is a polynomial vector of degree n_i and $d_i(\lambda)$ is a stable polynomial of degree n_i. We assume that the vectors $v_i(\lambda)$ are the basis vectors of a minimal polynomial basis, ordered such that $n_1 \leq n_2 \leq \ldots \leq n_{p-r_d}$, and each denominator $d_i(\lambda)$ divides $d_j(\lambda)$ for $i < j$. It follows immediately, that a linear combination $h^{(i)} Q_1(\lambda)$ of the first i rows with $h^{(i)}$ of the form (5.19) has a McMillan degree n_i. At Step (2), choosing the least index i such that $h^{(i)} R_{f_j}(\lambda) \neq 0$ for $j = 1, \ldots, m_f$, allows to take $Q_2(\lambda) := h^{(i)}$. Often the choice $Q_3(\lambda) = 1$ at Step (3) solves the synthesis problem. However, if $R_f(\lambda)$ is unstable or not proper, then the same computational approach, based on the LCF in (5.18), can be used as in the case of a polynomial basis.

The **Procedure EFD** employing polynomial or simple proper nullspace bases involves polynomial manipulations and therefore is not a reliable computational approach for large order systems due to the intrinsic high sensitivity of polynomial-based representations. A numerically reliable alternative algorithm employs minimal (non-simple) proper bases and is based on state-space computations described in detail in Sect. 7.4 (see also Sect. 10.3.2). The importance of **Procedure EFD**, and especially of the synthesis with least order scalar fault detection filters, lies in being the basic computational procedure, which allows to solve the more involved fault detection and isolation problem formulated in Sect. 3.5.3.

Remark 5.6 Steps (2) and (3) of **Procedure EFD** can be easily embedded into an optimization-based tuning procedure to determine an optimal $Q_2(\lambda)$, which ensures a more uniform sensitivity of the detector to individual faults. The free parameters to be tuned are the polynomial coefficients of $\phi^{(i)}(\lambda)$ in (5.17) or the nonzero components of the real vector $h^{(i)}$ in (5.19). It is assumed that for given values of these parameters at Step (2), the computations at Step (3) follow automatically to produce a stable candidate solution $Q(\lambda)$. For optimal tuning of parameters, the *fault sensitivity condition* can be used as a criterion to be maximized. For a given $R_f(\lambda)$, this criterion is defined as

$$\xi := \min_j \| R_{f_j}(\lambda) \|_\infty \, / \, \max_j \| R_{f_j}(\lambda) \|_\infty \tag{5.20}$$

and can be interpreted as a scaling-independent measure of the *complete fault detectability*.[1] For strong fault detectability-based tuning, a similar *fault sensitivity condition* can be defined in terms of the gains at a selected frequency λ_s as

$$\xi^s := \min_j \| R_{f_j}(\lambda_s) \|_2 / \max_j \| R_{f_j}(\lambda_s) \|_2 . \tag{5.21}$$

A small value of the fault sensitivity condition ξ (or ξ^s) indicates potential difficulties in detecting faults due to a substantial gap between the maximum and minimum gains. In such cases, employing fault detection filters with several outputs ($q > 1$) could be advantageous. $\square$

Example 5.3 Consider a continuous-time system with the TFMs

$$G_u(s) = \begin{bmatrix} \dfrac{s+1}{s+2} \\ \dfrac{s+2}{s+3} \end{bmatrix}, \quad G_d(s) = \begin{bmatrix} \dfrac{s-1}{s+2} \\ 0 \end{bmatrix}, \quad G_f(s) = \begin{bmatrix} \dfrac{s+1}{s+2} & 0 \\ \dfrac{s+2}{s+3} & 1 \end{bmatrix}, \quad G_w(s) = 0 .$$

The fault f_1 corresponds to an additive actuator fault, while f_2 describes an additive sensor fault in the second output y_2. The TFM $G_d(s)$ is non-minimum phase, having an unstable zero at 1.

At Step (1) of the **Procedure EFD**, a proper minimal left nullspace basis can be determined, consisting of a single row vector, which we can choose, for example,

$$Q_1(s) = \begin{bmatrix} 0 & 1 & -\dfrac{s+2}{s+3} \end{bmatrix} .$$

For the reduced system (5.11) computed at Step (1), we obtain

$$R_f(s) = \overline{G}_f(s) = \begin{bmatrix} \dfrac{s+2}{s+3} & 1 \end{bmatrix} ,$$

which shows that according to Corollary 5.2, the EFDP has a solution. Since this basis is already stable, $Q(s) = Q_1(s)$ is a least order solution of the EFDP. $\Diamond$

Example 5.4 Consider an unstable continuous-time system with the TFMs

$$G_u(s) = \begin{bmatrix} \dfrac{s+1}{s-2} \\ \dfrac{s+2}{s-3} \end{bmatrix}, \quad G_d(s) = \begin{bmatrix} \dfrac{s-1}{s+2} \\ 0 \end{bmatrix}, \quad G_f(s) = \begin{bmatrix} \dfrac{s+1}{s-2} & 0 \\ \dfrac{s+2}{s-3} & 1 \end{bmatrix}, \quad G_w(s) = 0 ,$$

where as before, the fault f_1 corresponds to an additive actuator fault, while f_2 describes an additive sensor fault in the second output y_2, with the difference that

---

[1] In [161], the reciprocal of fault sensitivity condition has been defined instead.

the underlying system is unstable. The TFM $G_d(s)$ is non-minimum phase, having an unstable zero at 1.

At Step (1) of the **Procedure EFD**, a proper minimal left nullspace basis can be determined, consisting of a single row vector, which we can choose, for example,

$$Q_1(s) = \left[\, 0 \; 1 \; -\frac{s+2}{s-3} \,\right].$$

For the reduced system (5.11) computed at Step (1), we obtain

$$R_f(s) = \overline{G}_f(s) = \left[\, \frac{s+2}{s-3} \; 1 \,\right],$$

which shows that according to Corollary 5.2 the EFDP has a solution. Since $Q(s) = Q_1(s)$ is unstable, it must be suitably updated. With $Q_2(s) = 1$ at Step (2) and $Q_3(s) = \frac{s-3}{s+3}$ at Step (3), we finally obtain

$$Q(s) = \left[\, 0 \; \frac{s-3}{s+3} \; -\frac{s+2}{s+3} \,\right], \quad R_f(s) = \left[\, \frac{s+2}{s+3} \; \frac{s-3}{s+3} \,\right].$$

The script **Ex5_4c.jl** in Listing 5.1 solves the EFDP considered in this example and calls the synthesis function **efdsyn**, which implements the **Procedure EFD**. The script also includes the verification of results by checking the synthesis conditions. A more detailed version of this script is **Ex5_4.jl** (not listed), which closely follows the computational steps of the **Procedure EFD**. Some of computed intermediary results differ from those of this example. $\Diamond$

Listing 5.1 Script **Ex5_4c.jl** to solve the EFDP of Example 5.4 using **Procedure EFD**

```
using FaultDetectionTools, DescriptorSystems, Test

# Example 5.4 - Solution of an EFDP
println("Example 5.4")

# define s as an improper transfer function
s = rtf('s');
# define Gu(s) and Gd(s)
Gu = [(s+1)/(s-2); (s+2)/(s-3)];          # enter Gu(s)
Gd = [(s-1)/(s+2); 0];                     # enter Gd(s)
Gf = [(s+1)/(s-2) 0; (s+2)/(s-3) 1];      # enter Gf(s)
p = 2; mu = 1; md = 1; mf = 2;            # set dimensions
atol = 1.e-7;                              # tolerance for rank tests

# setup the synthesis model with faults with Gf(s) = [ Gu(s) [0;1]]
sysf = fdimodset(dss([Gu Gd]),c =1,d = 2,f = 1,fs = 2);

# call efdsyn with the options for stability degree -3 and
# the synthesis of a scalar output filter
Qt, Rft, info = efdsyn(sysf, sdeg = -3, rdim = 1);

# normalize Q and Rf to match example
```

```
scale = evalfr(Rft.sys[1,1],Inf)[1,1];
Q = Qt/scale;
Rf = Rft/scale;
println("Q = $(dss2rm(Q.sys; atol))")
println("Rf = $(dss2rm(Rf.sys; atol))")


# check synthesis results
R = fdIFeval(Q, sysf; minimal = true, atol)
@test iszero(R.sys[:,[R.controls; R.disturbances]]; atol) &&
      iszero(R.sys[:,R.faults]-Rf.sys) &&
      fdisspec(Rf) == Bool[1 1] &&
      gpole(Q.sys) ≈ [-3] && gpole(Rf.sys) ≈  [-3]
```

5.3   Solving the Approximate Fault Detection Problem

Using the factorized representation $Q(\lambda) = \overline{Q}_1(\lambda)Q_1(\lambda)$ in (5.9) with $Q_1(\lambda)$ chosen proper and stable, it follows that $Q(\lambda)$ solves the *approximate fault detection problem* (AFDP) formulated in Sect. 3.5.2 for the system (3.2) if and only if $\overline{Q}_1(\lambda)$ solves the AFDP for the reduced system (5.11). By a suitable choice of $Q_1(\lambda)$, we can always additionally enforce that both $\overline{G}_w(\lambda)$ and $\overline{G}_f(\lambda)$ in (5.11) are proper, which will be assumed throughout this section. The solvability conditions of the AFDP for the system (3.2) can be replaced by similar conditions for the reduced system (5.11) according to the following corollary to Theorem 3.9.

Corollary 5.4 *For the system (3.2), the AFDP is solvable if and only if the system (5.11) is completely fault detectable, or equivalently, the following input observability conditions hold*

$$\overline{G}_{f_j}(\lambda) \neq 0, \ j = 1, \ldots, m_f.$$

We have seen in the proof of Theorem 3.9 that a solution of the AFDP can be determined by solving the related EFDP with $w \equiv 0$, using, for example, **Procedure EFD**. The usefulness of such a solution can be assessed in terms of the magnitudes of the minimum size detectable fault inputs in the presence of noise inputs. While for small noise levels, such a solution may often be satisfactory, for large noise levels, a purposely designed fault detection filter, which maximizes the magnitudes of the minimum size detectable fault inputs for the given class of noise inputs, usually represents a better solution. Such a solution, which aims to maximize the sensitivity of residual to faults and, simultaneously, to minimize the effects of noise on the residual, can be targeted by solving a suitably formulated optimization problem.

Consider a fault detection filter $Q(\lambda)$, in the general parameterized form (5.9), which has the internal form

$$\mathbf{r}(\lambda) := R_f(\lambda)\mathbf{f}(\lambda) + R_w(\lambda)\mathbf{w}(\lambda).$$

Let $\gamma > 0$ be an admissible level for the effect of the noise signal $w(t)$ on the residual $r(t)$, which can be imposed, for example, as

$$\| R_w(\lambda) \|_{2/\infty} \leq \gamma \, , \tag{5.22}$$

where $\| \cdot \|_{2/\infty}$ denotes either the $\mathcal{H}_2$- or $\mathcal{H}_\infty$-norm. The $\mathcal{H}_2$-norm corresponds to the case when $w(t)$ is a white noise signal, while the $\mathcal{H}_\infty$-norm is better suited when $w(t)$ is an unknown signal with bounded energy (or power). The choice of γ usually reflects the desired robustness of the fault detection filter to reject the noise. The value $\gamma = 0$ can be used to formulate the EFDP as a particular AFDP. For $\gamma > 0$, it is always possible, via a suitable scaling of the filter, to use the normalized value $\gamma = 1$.

As measures of the sensitivity of residuals to faults, several "indices" have been proposed in the literature to characterize the least sensitivity in terms of $R_f(\lambda)$. Such an index, commonly denoted by $\| R_f(\lambda) \|_-$, has been defined in terms of the least singular value (denoted by $\underline{\sigma}(\cdot)$) of the frequency response of $R_f(\lambda)$ as

$$\| R_f(\lambda) \|_- := \inf_{\omega \in \Omega} \underline{\sigma} \left(R_f(\omega) \right) \, , \tag{5.23}$$

where $\Omega \subset \partial \mathbb{C}_s$ is a finite or infinite set of frequency values on the boundary of the appropriate stability domain. In view of Definition 3.4 (see Sect. 3.3), the requirement $\| R_f(\lambda) \|_- > 0$ can be interpreted as a complete strong fault detectability condition. In some works, the formulation of the AFDP involves the determination of a filter $Q(\lambda)$, which maximizes the index (5.23) such that the noise attenuation constraint (5.22) is simultaneously fulfilled. This binding of the formulation of the AFDP to a particular optimization-based solution method is generally not desirable, since it imposes additional constraints, usually of purely technical character, on the solvability of the AFDP. While the satisfaction of such constraints guarantees the solvability of the underlying mathematical optimization problem, these conditions are usually not necessary for the solvability of the AFDP (according to the formulation in Sect. 3.5.2). Two inherent weaknesses in the definition of the index $\| R_f(\lambda) \|_-$ worsen additionally the solvability of the optimization-based formulation of the AFDP.

A first issue is that the index (5.23) is meaningful only when $m_f \leq p$, because if $m_f > p$, only the detectability of p out of m_f faults can be assessed by this index, $m_f - p$ singular values being null. It was argued that the case $m_f > p$ can be addressed using a bank of filters, where each filter must be sensitive only to a subset of maximal p faults. However, this leads to an unnecessary increase of the global order of the resulting fault detection filter, and therefore, represents a strong technical limitation for practical use. The second issue is rather of conceptual nature. The definition (5.23) targets primarily the complete strong fault detectability aspect (see Definition 3.4), and therefore, appears to be less adequate to characterize the weaker property of complete fault detectability (see Definition 3.2), which merely requires that each column of $R_f(\lambda)$ must be nonzero. While this property can be still indirectly targeted, for example, by a suitable choice of Ω (e.g., $\Omega = \{\lambda_0\}$ with λ_0 a representative frequency value at which $R_{f_j}(\lambda_0)$ must be nonzero for $j = 1, \ldots, m_f$), an alternative index, discussed in what follows, is better suited to address directly the complete fault detectability aspect.

To overcome both these deficiencies, an alternative index will be used to characterize fault sensitivity. This index is defined as

$$\|R_f(\lambda)\|_{2/\infty-} := \min_{1 \le j \le m_f} \|R_{f_j}(\lambda)\|_{2/\infty}, \tag{5.24}$$

where $\| \cdot \|_{2/\infty}$ stands for either $\| \cdot \|_2$ or $\| \cdot \|_\infty$, while $\| \cdot \|_{2/\infty-}$ stands for either $\| \cdot \|_{2-}$ or $\| \cdot \|_{\infty-}$ indices defined in terms of $\mathcal{H}_2$ or $\mathcal{H}_\infty$ norms in (5.24), respectively. The requirement $\|R_f(\lambda)\|_{2/\infty-} > 0$ merely asks that all columns of $R_{f_j}(\lambda)$ are nonzero, and therefore, the index $\|R_f(\lambda)\|_{2/\infty-}$ characterizes the *complete fault detectability* (of an arbitrary number of faults) as defined in Definition 3.2. To characterize the *complete strong fault detectability with respect to* Ω, the modified index $\| \cdot \|_{\Omega-}$ can be used, which is defined as

$$\|R_f(\lambda)\|_{\Omega-} := \min_{1 \le j \le m_f} \left\{ \inf_{\omega \in \Omega} \|R_{f_j}(\omega)\|_2 \right\}. \tag{5.25}$$

For a particular problem, a combination of the two indices (5.24) and (5.25) can also be meaningful, by selecting for the j-th column of $R_f(\lambda)$ either $\|R_{f_j}(\lambda)\|_{2/\infty-}$ or $\|R_{f_j}(\lambda)\|_{\Omega-}$ as a problem specific fault sensitivity measure.

Using the above definitions of the $\| \cdot \|_{2/\infty-}$ and $\| \cdot \|_{\Omega-}$ indices, several optimization problems can be formulated to address the computation of a satisfactory solution of the AFDP for the reduced system (5.11) with $\overline{G}_w(\lambda)$ and $\overline{G}_f(\lambda)$ proper, using the parametrization (5.9) of the fault detection filter with stable $Q_1(\lambda)$. In what follows, we only discuss one of the most popular formulations, the $\mathcal{H}_{\infty-}/\mathcal{H}_\infty$ synthesis, for which we give a detailed computational procedure. The synthesis goal is to determine $\overline{Q}_1(\lambda)$, which maximizes the fault sensitivity for a given level of noise: Given $\gamma \ge 0$, determine the stable and proper optimal fault detection filter $\overline{Q}_1(\lambda)$ and the corresponding optimal fault sensitivity level $\beta > 0$ such that

$$\beta = \max_{\overline{Q}_1(\lambda)} \left\{ \left\|\overline{Q}_1(\lambda)\overline{G}_f(\lambda)\right\|_{\infty-} \,\middle|\, \left\|\overline{Q}_1(\lambda)\overline{G}_w(\lambda)\right\|_\infty \le \gamma \right\}. \tag{5.26}$$

An alternative formulation of an optimization-based solution, called the $\mathcal{H}_\infty/\mathcal{H}_{\infty-}$ synthesis, minimizes the effects of noise by imposing a certain fault sensitivity level: Given $\beta > 0$, determine $\gamma \ge 0$ and a stable and proper fault detection filter $\overline{Q}_1(\lambda)$ such that

$$\gamma = \min_{\overline{Q}_1(\lambda)} \left\{ \left\|\overline{Q}_1(\lambda)\overline{G}_w(\lambda)\right\|_\infty \,\middle|\, \left\|\overline{Q}_1(\lambda)\overline{G}_f(\lambda)\right\|_{\infty-} \ge \beta \right\}. \tag{5.27}$$

The two approaches may lead to different solutions, depending on the properties of the underlying transfer function matrices and problem dimensions. For both cases, the gap β/γ can be interpreted as a measure of the quality of fault detection. For

$\gamma = 0$, both formulations include the exact solution (i.e., of the EFDP for $w \equiv 0$) and the corresponding gap is infinite.

Before we discuss the computational issues, we consider a simple example, which highlights the roles of fault and noise input signals when solving an AFDP.

Example 5.5 This is the same as Example 5.2, where we redefined the noise input w as d and thus we have

$$
G_u(s) = \begin{bmatrix} \dfrac{s+1}{s+2} \\ \dfrac{s+2}{s+3} \end{bmatrix}, \quad G_d(s) = 0, \quad G_f(s) = \begin{bmatrix} \dfrac{s+1}{s+2} & 0 \\ 0 & 1 \end{bmatrix}, \quad G_w(s) = \begin{bmatrix} \dfrac{1}{s+2} \\ 0 \end{bmatrix}.
$$

A minimal basis is simply $N_l(s) = [\, I_2 \;\; -G_u(s) \,]$, which leads to $\overline{G}_f(s) = G_f(s)$ and $\overline{G}_w(s) = G_w(s)$. This basis is in fact a solution of an EFDP in the case $w \equiv 0$. Thus, this solution can be also employed to solve the AFDP, as pointed out in the proof of Theorem 3.9. To be useful for practical purposes, a fault detection filter must provide reliable detection of all faults in the presence of noise. This condition is evidently fulfilled by the fault input f_2, since with $Q(s) = N_l(s)$, the second component of the residual r_2 is simply $r_2 = f_2$, because there is no interaction between the noise input w and fault input f_2. However, because f_1 and w share the same input space, the minimal detectable size of f_1 will depend on the possible maximum size of noise input w. Assume $\|w\|_2 \le \delta_w$, thus for all w we have $\|G_w(s)\mathbf{w}(s)\|_2 \le \|\overline{G}_w(s)\|_\infty \|\mathbf{w}(s)\|_2 \le \delta_w/2$. Thus, the minimum size of detectable faults $f_{1,min}$ satisfies $\|G_{f_1}(s)\mathbf{f}_{1,min}(s)\|_2 > \delta_w/2$. The solution to this problem depends on the classes of faults considered. Assuming $\mathbf{f}_{1,min}(s) = \eta/s$ (thus a step input fault of amplitude η), the resulting asymptotic value of $G_{f_1}(s)\mathbf{f}_{1,min}(s)$ is $G_{f_1}(0)\eta = \eta/2$. It follows that we can reliably detect constant faults, provided their amplitude satisfies $\eta > \delta_w$. More generally, for step inputs in f_1, the condition $\eta > \|G_w(s)\|_\infty \delta_w / G_{f_1}(0)$ most be fulfilled for reliable detection. Similar conditions can be established in the case of sinusoidal fault inputs. $\Diamond$

To solve the $\mathcal{H}_{\infty-}/\mathcal{H}_\infty$ optimization problem (5.26), we devise a synthesis procedure based on successive simplifications of the original problem by reducing it to simpler problems with the help of a factorized representation of the fault detection filter. We start with the factorized representation (5.9) of the fault detection filter $Q(\lambda)$, where $Q_1(\lambda)$ is a left nullspace basis of $G(\lambda)$ in (5.2) and $\overline{Q}_1(\lambda)$ has to be determined. Let $\overline{G}_f(\lambda)$ and $\overline{G}_w(\lambda)$ be the TFMs of the reduced system (5.11) determined according to (5.12). We can immediately check the solvability conditions of the AFDP of Corollary 5.2 as $\|\overline{G}_f(\lambda)\|_{\infty-} > 0$. Assume that this test indicates the solvability of the AFDP. In this context, we introduce a useful concept to simplify the presentation. A fault detection filter $Q(\lambda)$ is called *admissible* if the corresponding $R_f(\lambda)$ satisfies $\|R_f(\lambda)\|_{\infty-} > 0$ (i.e., it has all its columns nonzero).

Let q be the desired number of residual components. As in the case of an EFDP, if a solution of the AFDP exists, then generally we can always use a scalar output fault detection filter (thus choose $q = 1$). However, larger values of q can be advantageous because they generally involve more free parameters, which can be appropriately tuned. In the proposed synthesis procedure (see **Procedure AFD**), the choice of q is restricted to $q \leq r_w \leq p - r_d$, where $r_w := \operatorname{rank} \overline{G}_w(\lambda)$ and $r_d := \operatorname{rank} G_d(\lambda)$. This choice is, however, only for convenience, because it leads to a simpler synthesis procedure. As shown in Remark 5.10, in practical applications, q must only satisfy $q \leq p - r_d$, which limits q to the maximum number of left nullspace basis vectors of $G(\lambda)$ in (5.2) (i.e., the number of rows of $Q_1(\lambda)$). This bound on q is the same as in the case of solving the EFDP.

At the next step, we use a factorized representation of $\overline{Q}_1(\lambda)$ in the form $\overline{Q}_1(\lambda) = \overline{Q}_2(\lambda) Q_2(\lambda)$, where the $r_w \times (p - r_d)$ factor $Q_2(\lambda)$ is determined such that $Q_2(\lambda)\overline{G}_w(\lambda)$ has full row rank r_w, and the product $Q_2(\lambda)Q_1(\lambda)$ is admissible and has least McMillan degree. If this latter requirement is not imposed, then a simple choice is $Q_2(\lambda) = H$, where H is a $r_w \times (p - r_d)$ full row rank constant matrix, which ensures admissibility (e.g., chosen as a randomly generated matrix with orthonormal columns). This choice corresponds to building $Q_2(\lambda)Q_1(\lambda)$ as r_w linear combinations of the left nullspace basis vectors contained in the rows of $Q_1(\lambda)$.

At this stage, the optimization problem to be solved falls in one of two categories. The *standard case* is when $Q_2(\lambda)\overline{G}_w(\lambda)$ has no zeros on the boundary of the stability domain $\partial \mathbb{C}_s$ (i.e., on the extended imaginary axis in the continuous-time case, or on the unit circle centred in the origin in the discrete-time case). The *nonstandard case* corresponds to the presence of such zeros. This categorization can be easily revealed at the next step, which also involves the computation of the respective zeros. For the full row rank TFM $Q_2(\lambda)\overline{G}_w(\lambda)$, we compute the quasi-co-outer–co-inner factorization

$$Q_2(\lambda)\overline{G}_w(\lambda) = G_{wo}(\lambda) G_{wi}(\lambda), \tag{5.28}$$

where the quasi-co-outer factor $G_{wo}(\lambda)$ is invertible, having only stable zeros excepting possible zeros on the boundary of the stability domain, and $G_{wi}(\lambda)$ is co-inner (i.e., $G_{wi}(\lambda)G_{wi}^{\sim}(\lambda) = I$ with $G_{wi}^{\sim}(s) = G_{wi}^T(-s)$ in the continuous-time case, and $G_{wi}^{\sim}(z) = G_{wi}^T(1/z)$ in the discrete-time case).

We choose $\overline{Q}_2(\lambda) = \overline{Q}_3(\lambda) Q_3(\lambda)$, with $Q_3(\lambda) = G_{wo}^{-1}(\lambda)$ and $\overline{Q}_3(\lambda)$ to be determined. Using (5.10)–(5.12), the fault detection filter in (3.3) can be rewritten as

$$\mathbf{r}(\lambda) = \overline{Q}_3(\lambda) Q_3(\lambda) Q_2(\lambda) \overline{\mathbf{y}}(\lambda) = \overline{Q}_3(\lambda) \widetilde{\mathbf{y}}(\lambda), \tag{5.29}$$

where

$$\widetilde{\mathbf{y}}(\lambda) := Q_3(\lambda) Q_2(\lambda) \overline{\mathbf{y}}(\lambda) = \widetilde{G}_f(\lambda)\mathbf{f}(\lambda) + G_{wi}(\lambda)\mathbf{w}(\lambda), \tag{5.30}$$

with

$$\widetilde{G}_f(\lambda) := Q_3(\lambda) Q_2(\lambda) \overline{G}_f(\lambda). \tag{5.31}$$

It follows that $\overline{Q}_3(\lambda)$ can be determined as the solution of

$$\beta = \max_{\overline{Q}_3(\lambda)} \left\{ \left\| \overline{Q}_3(\lambda)\widetilde{G}_f(\lambda) \right\|_{\infty-} \ \Big| \ \left\| \overline{Q}_3(\lambda) \right\|_{\infty} \le \gamma \right\}, \tag{5.32}$$

where we used that $\left\| \overline{Q}_3(\lambda)G_{wi}(\lambda) \right\|_{\infty} = \left\| \overline{Q}_3(\lambda) \right\|_{\infty}$.

In the standard case, we can always ensure that both the partial filter defined by the product of stable factors $Q_3(\lambda)Q_2(\lambda)Q_1(\lambda)$ and $\widetilde{G}_f(\lambda)$ are stable. Thus, $\overline{Q}_3(\lambda)$ is determined as $\overline{Q}_3(\lambda) = Q_4$, where Q_4 is a constant matrix representing the optimal solution of the reduced problem

$$\beta = \max_{Q_4} \left\{ \left\| Q_4\widetilde{G}_f(\lambda) \right\|_{\infty-} \ \Big| \ \left\| Q_4 \right\|_{\infty} \le \gamma \right\},$$

such that the resulting detector $Q(\lambda) = Q_4 Q_3(\lambda) Q_2(\lambda) Q_1(\lambda)$ is admissible. For square $Q_4(\lambda)$, $Q_4 = \gamma I$ is the simplest $\mathcal{H}_{\infty-}/\mathcal{H}_{\infty}$ optimal solution.

We give the following result without proof. For proofs in continuous- and discrete-time, see [81] and [82], respectively.

Theorem 5.2 *For the reduced system (5.11) and with a suitable choice of $Q_2(\lambda)$, assume that we have $\|Q_2(\lambda)\overline{G}_f(\lambda)\|_{\infty-} > 0$, $Q_2(\lambda)\overline{G}_w(\lambda)$ has full row rank and $Q_2(\lambda)\overline{G}_w(\lambda)$ has no zeros on the boundary of the stability domain. Then, for $\gamma > 0$, the $\mathcal{H}_{\infty-}/\mathcal{H}_{\infty}$ optimal solution of the optimization problem (5.26) is*

$$\overline{Q}_{1,opt}(\lambda) := \gamma G_{wo}^{-1}(\lambda)Q_2(\lambda) ,$$

where $G_{wo}(\lambda)$ is the co-outer factor of the co-outer–co-inner factorization (5.28).

In the nonstandard case, both the partial detector $\widetilde{Q}(\lambda) := Q_3(\lambda)Q_2(\lambda)Q_1(\lambda)$ and $\widetilde{G}_f(\lambda)$ can result unstable or improper due to the presence of poles on the boundary of the stability domain in the factor $Q_3(\lambda) = G_{wo}^{-1}(\lambda)$. In this case, we choose $\overline{Q}_3(\lambda) = Q_5 Q_4(\lambda)$, where $Q_4(\lambda)$ results from a LCF with stable and proper factors

$$[\, \widetilde{Q}(\lambda)\ \widetilde{G}_f(\lambda)\,] = Q_4^{-1}(\lambda)[\, \widehat{Q}(\lambda)\ \widehat{G}_f(\lambda)\,],$$

while Q_5 is a constant matrix, which solves

$$\beta = \max_{Q_5} \left\{ \left\| Q_5\widehat{G}_f(\lambda) \right\|_{\infty-} \ \Big| \ \|Q_5 Q_4(\lambda)\|_{\infty} \le \gamma \right\}.$$

Since $Q_4(\lambda)$ can be always chosen diagonal and such that its diagonal elements have $\mathcal{H}_{\infty}$-norms equal to 1, this choice will significantly simplify the solution of the above problem. For example, the choice $Q_5 = \gamma I$ is always a possibility to obtain a fault detection filter.

Remark 5.7 The presence of unstable zeros of $G_{wo}(\lambda)$ on the boundary of the stability domain prevents the computation of an "optimal" solution of the $\mathcal{H}_{\infty-}/\mathcal{H}_{\infty}$-optimization problem. When solving practical applications, this apparent limitation is superfluous because the presence of these zeros represents in fact an advantage rather than a disadvantage. For example, in the case of a continuous-time system, a zero at infinity (e.g., in the case when the original $G_w(s)$ is strictly proper) confers to $G_{wo}(s)$ a low-pass character as well, such that high-frequency noise will be attenuated in the noise input channel. Similarly, a zero in the origin will cancel all constant variations in the noise, thus will also attenuate slowly varying noise inputs. Finally, a pair of conjugated zeros on the imaginary axis will attenuate all sinusoidal noise signals of nearby frequencies. This behaviour is thus very similar to that of notch filters, which are purposely included in the feedback loops to address disturbance attenuation or rejection problems in control systems design. The above approach for the nonstandard case simply copes with the presence of zeros on the boundary of the stability domain. $\qquad\square$

Remark 5.8 In the nonstandard case, we can alternatively regularize the problem by replacing $G_{wo}(\lambda)$ in (5.28) by $G_{wo,\varepsilon}(\lambda)$, which, for $\varepsilon > 0$, is a minimum-phase spectral factor satisfying

$$G_{wo,\varepsilon}(\lambda)G_{wo,\varepsilon}^{\sim}(\lambda) = \varepsilon^2 I + G_{wo}(\lambda)G_{wo}^{\sim}(\lambda)\,.$$

By choosing $\overline{Q}_2(\lambda) = \overline{Q}_3(\lambda)Q_3(\lambda)$ with $Q_3(\lambda) = G_{wo,\varepsilon}^{-1}(\lambda)$, we arrive to the same optimization problem (5.32) for $\overline{Q}_3(\lambda)$ as for the standard case. The solution of the AFDP along this line has been discussed in [55]. $\qquad\square$

In the standard case, the dynamical order of the resulting residual generator is of the order of $Q_3(\lambda)$ if we choose $Q_4(\lambda)$ a constant matrix. This order results from the conditions that $Q_2(\lambda)\overline{G}_w(\lambda)$ has full row rank and $Q_2(\lambda)Q_1(\lambda)$ has least order and is admissible (i.e., $\|Q_2(\lambda)\overline{G}_f(\lambda)\|_{\infty-} > 0$). For each candidate $Q_2(\lambda)$, the corresponding optimal $Q_3(\lambda)$ results automatically, but the different "optimal" detectors for the same level γ of noise attenuation performance can have significantly differing fault detection performance levels (measured via the optimal cost β). Finding the best compromise between achieved order and the achieved performance (measured via the gap β/γ) should take into account that larger orders and larger number of detector outputs q potentially lead to better performance.

The **Procedure AFD**, given in what follows, allows the synthesis of least order fault detection filters to solve the AFDP employing an $\mathcal{H}_\infty-/\mathcal{H}_\infty$ optimization-based approach. This procedure also includes the **Procedure EFD** in the case when an exact solution exists. Similar synthesis procedures relying on alternative optimization-based formulations (e.g., $\mathcal{H}_\infty-/\mathcal{H}_2$, $\mathcal{H}_2-/\mathcal{H}_\infty$, $\mathcal{H}_2-/\mathcal{H}_2$, $\mathcal{H}_\Omega-/\mathcal{H}_\infty$, $\mathcal{H}_\Omega-/\mathcal{H}_2$ as well as their finite frequency range counterparts) can be devised by only adapting appropriately the last computational step of the **Procedure AFD**.

Procedure AFD: Approximate synthesis of fault detection filters

Inputs : $\{G_u(\lambda), G_d(\lambda), G_f(\lambda), G_w(\lambda)\}, q, \gamma$
Outputs: $Q(\lambda), R_f(\lambda), R_w(\lambda), \beta$

(1) Compute a $(p - r_d) \times (p + m_u)$ minimal proper stable basis $Q_1(\lambda)$ for the left nullspace of $G(\lambda)$ defined in (5.2), where $r_d = \text{rank}\, G_d(\lambda)$; set $Q(\lambda) = Q_1(\lambda)$ and compute

$$[\,R_f(\lambda)\ R_w(\lambda)\,] = Q_1(\lambda) \begin{bmatrix} G_f(\lambda) & G_w(\lambda) \\ 0 & 0 \end{bmatrix}.$$

Exit if exists $j \in \{1, \ldots, m_f\}$ such that $R_{f_j}(\lambda) = 0$ (no solution).

(2) Compute $r_w = \text{rank}\, R_w(\lambda)$; if $r_w = 0$, set $q_1 = \min(p - r_d, q)$; else, set $q_1 = r_w$; choose a $q_1 \times (p - r_d)$ rational matrix $Q_2(\lambda)$ such that $Q_2(\lambda)Q(\lambda)$ is admissible, has least McMillan degree and, if $r_w > 0$ then $Q_2(\lambda)R_w(\lambda)$ has full row rank r_w; compute $Q(\lambda) \leftarrow Q_2(\lambda)Q(\lambda)$, $R_f(\lambda) \leftarrow Q_2(\lambda)R_f(\lambda)$ and $R_w(\lambda) \leftarrow Q_2(\lambda)R_w(\lambda)$.

(3) If $r_w > 0$, compute the quasi-co-outer–co-inner factorization $R_w(\lambda) = R_{wo}(\lambda) R_{wi}(\lambda)$, where the quasi-co-outer factor $R_{wo}(\lambda)$ is invertible and has only stable zeros, excepting possible zeros on the boundary of the tability domain, and $R_{wi}(\lambda)$ is co-inner; with $Q_3(\lambda) = R_{wo}^{-1}(\lambda)$ compute $Q(\lambda) \leftarrow Q_3(\lambda)Q(\lambda)$, $R_f(\lambda) \leftarrow Q_3(\lambda)R_f(\lambda)$ and $R_w(\lambda) \leftarrow R_{wi}(\lambda)$.

(4) Compute $Q_4(\lambda)$ such that $Q_4(\lambda)Q(\lambda)$ and $Q_4(\lambda)R_f(\lambda)$ are proper and stable; compute $Q(\lambda) \leftarrow Q_4(\lambda)Q(\lambda)$, $R_f(\lambda) \leftarrow Q_4(\lambda)R_f(\lambda)$, and $R_w(\lambda) \leftarrow Q_4(\lambda) R_w(\lambda)$.

(5) If $r_w > 0$, choose $Q_5 \in \mathbb{R}^{\min(q, r_w) \times q_1}$ such that $\|Q_5 Q_4(\lambda)\|_\infty = \gamma$ and $\beta = \|Q_5 R_f(\lambda)\|_{\infty-} > 0$; update $Q(\lambda) \leftarrow Q_5 Q(\lambda)$, $R_f(\lambda) \leftarrow Q_5 R_f(\lambda)$, and $R_w(\lambda) \leftarrow Q_5 R_w(\lambda)$; else, set $\beta = \infty$.

Remark 5.9 The threshold selection approach of Sect. 3.6 can be applied to determine a threshold value τ, which guarantees the lack of false alarms. For any selected value of the threshold τ, we can estimate for $j = 1, \ldots, m_f$ the magnitude $\underline{\delta}_{f_j}$, of the minimum size detectable fault $f_j \neq 0$, provided $f_k = 0\ \forall k \neq j$. Consider the internal representation of the resulting fault detection filter in the form

$$\mathbf{r}(\lambda) = \sum_{j=1}^{m_f} R_{f_j}(\lambda)\mathbf{f}_j(\lambda) + R_w(\lambda)\mathbf{w}(\lambda).$$

By using (3.39) in the frequency domain (via Plancherel's theorem), $\underline{\delta}_{f_j}$ can be computed from

$$2\tau = \inf_{\|f_j\| = \underline{\delta}_{f_j}} \|R_{f_j}(\lambda)\mathbf{f}_j(\lambda)\|_2 = \underline{\delta}_{f_j} \|R_{f_j}(\lambda)\|_{\Omega-},$$

where we used the properties of the index defined in (5.25). For $w(t)$ having bounded energy and satisfying $\|w\|_2 \leq \delta_w$, we obtain

$$\underline{\delta}_{f_j} = \frac{2\|R_w(\lambda)\|_\infty \delta_w}{\|R_{f_j}(\lambda)\|_{\Omega-}}. \tag{5.33}$$

The resulting value of δ_{f_j} can be used to assess the "practical usefulness" of any solution. A small value of $\|R_{f_j}(\lambda)\|_{\Omega-}$ may indicate a large size of the minimal detectable faults for a particular choice of Ω. Therefore, various alternative choices of Ω may be used to arrive to more realistic estimates. For example, Ω can be defined as a relevant interval of frequency values, or only a finite set of relevant frequencies (e.g., the DC-gain frequency λ_s). $\qquad\square$

Example 5.6 If we apply **Procedure AFD** to solve the $\mathcal{H}_{\infty-}/\mathcal{H}_\infty$ synthesis problem for the system in Example 5.5, the resulting optimization problem is nonstandard, because $G_w(s)$ has a zero at infinity. Let us choose $\gamma = 1$. At Step (1), we set $Q_1(s) = N_l(s)$, with $N_l(s)$ determined in Example 5.5. We have that $R_w(s) = G_w(s)$ and $R_f(s) = G_f(s)$. Since each column of $R_f(s)$ is nonzero, the AFDP is solvable. Since $r_w = 1$, at Step (2), we can employ a constant vector $Q_2(\lambda) = [\,1\ 1\,]$ to obtain the updated quantities

$$Q(s) = \left[\,1\ 1\ -\frac{2s^2 + 8s + 7}{(s+2)(s+3)}\,\right], \quad R_f(s) = \left[\,\frac{s+1}{s+2}\ 1\,\right], \quad R_w(s) = \frac{1}{s+2}. \tag{5.34}$$

At Step (3), the quasi-outer factor $G_{wo}(s)$ is simply $R_w(s)$ and, being strictly proper, has thus a zero at infinity. With $Q_3(s) = R_w^{-1}(s)$, the resulting $Q(s)$ and $R_f(s)$ are, therefore, improper. At Step (4), we choose $Q_4(s)$ of unity $\mathcal{H}_\infty$-norm of the form $Q_4(s) = a/(s+a)$ with $a \geq 2$. For $\gamma = 1$, we obtain at Step (5) with $Q_5 = 1$ the final $Q(s)$, $R_f(s)$ and $R_w(s)$

$$Q(s) = \left[\,a\frac{s+2}{s+a}\ a\frac{s+2}{s+a}\ -a\frac{2s^2+8s+7}{(s+a)(s+3)}\,\right], \quad R_f(s) = \left[\,a\frac{s+1}{s+a}\ a\frac{s+2}{s+a}\,\right], \quad R_w(s) = \frac{a}{s+a}.$$

Since $\beta = \|R_f(s)\|_{\infty-} = a$, it follows that β can be arbitrarily large, and thus the $\mathcal{H}_{\infty-}/\mathcal{H}_\infty$ problem (5.26) has no optimal solution. Although not optimal, the resulting fault detection filter can be reliably employed for detecting faults, whose minimum amplitude is above a certain threshold. The value of this threshold can be easily determined using information on the size and waveform of the noise input.

The script **Ex5_6c.jl** in Listing 5.2 solves the AFDP considered in this example and calls the synthesis function **afdsyn**, which implements the **Procedure AFD**. The script also includes the verification of results by checking the synthesis conditions. A more detailed version of this script is **Ex5_6.jl** (not listed), which closely follows the computational steps of the **Procedure AFD**. $\qquad\diamond$

Listing 5.2 Script `Ex5_6c.jl` to solve the AFDP of Example 5.6 using **Procedure AFD**

```julia
using FaultDetectionTools, DescriptorSystems, Test

# Example 5.6 - Solution of an AFDP
println("Example 5.6")

# define s as an improper transfer function
s = rtf('s');
# define Gu(s), Gf(s) and Gw(s)
Gu = [(s+1)/(s+2); (s+2)/(s+3)];        # enter Gu(s)
Gf = [(s+1)/(s+2) 0; 0 1];              # enter Gf(s)
Gw = [1/(s+2); 0];                      # enter Gw(s)
p = 2; mu = 1; mw = 1; mf = 2;          # enter dimensions
atol = 1.e-7;                           # tolerance for rank tests

# setup the synthesis model
sysf = fdimodset(dss([Gu Gf Gw],minimal=true),c = 1:mu,
               f = mu .+ (1:mf), n = (mu+mf) .+ (1:mw));

# call afdsyn with the options for stability degree -3 and the synthesis
# of a scalar output filter
Q, Rfw, info = afdsyn(sysf; atol, minimal = false,
                   nullspace = false, smarg = -2, sdeg = -3,
                   rdim = 1, HDesign = [1 1]);

println("Q = $(dss2rm(Q.sys; atol))")
println("Rf = $(dss2rm(Rfw.sys[:,Rfw.faults]; atol))")
println("Rw = $(dss2rm(Rfw.sys[:,Rfw.noise]; atol))")
gap = fdif2ngap(Rfw)[1]; println("gap = $gap")
β = fdiscond(Rfw)[1]; println("scond = $β")
mmperf = fdimmperf(Rfw); println("mmperf = $mmperf")

# check synthesis results
R = fdIFeval(Q, sysf; minimal = true, atol)
@test iszero(R.sys[:,[R.controls; R.disturbances]]) &&
    iszero(R.sys[:,[R.faults;R.noise]]-Rfw.sys[:,[Rfw.faults;Rfw.noise]]) &&
    fdisspec(Rfw, block = true) == Bool[1 1] &&
    maximum(real(gpole(Q))) >= -3 && minimum(real(gpole(Q))) <= -3 &&
    gap ≈ info.gap
```

Example 5.7 We solve the problem in Example 5.6 using the alternative approach suggested in Remark 5.8. At Steps (1) and (2), we determine the same $Q(s)$, $R_w(s)$ and $R_f(s)$ as in (5.34). The quasi-outer factor is as before $G_{wo}(s) = R_w(s)$ and is strictly proper, having thus a zero at infinity. For $\varepsilon > 0$, we determine $G_{wo,\varepsilon}(s)$ such that $G_{wo,\varepsilon}(s)G_{wo,\varepsilon}^{\sim}(s) = \varepsilon^2 + G_{wo}(s)G_{wo}^{\sim}(s)$ and we obtain

$$G_{wo,\varepsilon}(s) = \frac{\varepsilon s + \sqrt{1 + 4\varepsilon^2}}{s + 2}.$$

With $Q_3(s) = G_{wo,\varepsilon}^{-1}(s)$, the optimal solution of the problem (5.32) is $\overline{Q}_3(s) = 1$ for which the final $Q(s)$, $R_f(s)$ and $R_w(s)$ are

$$Q(s) = \left[\frac{s+2}{\varepsilon s + \sqrt{1+4\varepsilon^2}} \quad \frac{s+2}{\varepsilon s + \sqrt{1+4\varepsilon^2}} \quad -\frac{2s^2 + 8s + 7}{(\varepsilon s + \sqrt{1+4\varepsilon^2})(s+3)} \right],$$

$$R_f(s) = \left[\frac{s+1}{\varepsilon s + \sqrt{1+4\varepsilon^2}} \quad \frac{s+2}{\varepsilon s + \sqrt{1+4\varepsilon^2}} \right], \quad R_w(s) = \frac{1}{\varepsilon s + \sqrt{1+4\varepsilon^2}}.$$

Since $\beta = \|R_f(s)\|_{\infty-} = 1/\varepsilon$, it follows that β becomes arbitrarily large as $\varepsilon \to 0$. Although the $\mathcal{H}_{\infty-}/\mathcal{H}_{\infty}$ problem (5.26) has no optimal solution, the resulting filter $Q(s)$ can be acceptable for a large range of values of ε. $\Diamond$

Example 5.8 If we solve the $\mathcal{H}_{\infty}/\mathcal{H}_{\infty-}$ synthesis problem for Example 5.5, the optimal solution $Q(s)$ and the corresponding $R_f(s)$ are simply

$$Q(s) = G_f^{-1}(s)N_l(s) = \begin{bmatrix} \frac{s+2}{s+1} & 0 & -1 \\ 0 & 1 & -\frac{s+2}{s+3} \end{bmatrix}, \qquad R_f(s) = \begin{bmatrix} 1 & 0 \\ 0 & 1 \end{bmatrix},$$

which lead to the optimal values $\beta = 1$ and $\gamma = 1$. In contrast to the filter in Example 5.7, this filter is optimal (in a certain sense) and able to perform fault isolation as well, by exactly reconstructing fault f_2 and approximately fault f_1. $\Diamond$

Remark 5.10 The solution of the AFDP can be refined in the case when $r_w < p - r_d$. In this case, it follows that there exists a left nullspace basis $\overline{N}_{l,w}(\lambda)$ such that $\overline{N}_{l,w}(\lambda)\overline{G}_w(\lambda) = 0$, thus the noise input can be exactly decoupled. Also, there exists a maximal subvector $f^{(1)}$ of fault inputs, which are completely fault detectable (i.e., the columns of the corresponding $\overline{N}_{l,w}(\lambda)\overline{G}_{f^{(1)}}(\lambda)$ are nonzero), while none of the components of its complementary part $f^{(2)}$ of f is fault detectable (i.e., all columns of the corresponding $\overline{N}_{l,w}(\lambda)\overline{G}_{f^{(2)}}(\lambda)$ are zero), and thus are completely decoupled. Here, we denoted with $\overline{G}_{f^{(1)}}(\lambda)$ and $\overline{G}_{f^{(2)}}(\lambda)$ the columns of $\overline{G}_f(\lambda)$ corresponding to $f^{(1)}$ and $f^{(2)}$, respectively. This allows the partitioning of the reduced system (5.11) as

$$\overline{\mathbf{y}}(\lambda) := \overline{G}_{f^{(1)}}(\lambda)\mathbf{f}^{(1)}(\lambda) + \overline{G}_{f^{(2)}}(\lambda)\mathbf{f}^{(2)}(\lambda) + \overline{G}_w(\lambda)\mathbf{w}(\lambda). \tag{5.35}$$

In general, we can construct $\overline{Q}_1(\lambda)$ and $Q(\lambda)$ in the forms

$$\overline{Q}_1(\lambda) = \begin{bmatrix} \overline{Q}_1^{(1)}(\lambda) \\ \overline{Q}_1^{(2)}(\lambda) \end{bmatrix}, \quad Q(\lambda) = \begin{bmatrix} Q^{(1)}(\lambda) \\ Q^{(2)}(\lambda) \end{bmatrix} := \begin{bmatrix} \overline{Q}_1^{(1)}(\lambda) \\ \overline{Q}_1^{(2)}(\lambda) \end{bmatrix} Q_1(\lambda), \tag{5.36}$$

where $\overline{Q}_1^{(1)}(\lambda)$ solves the EFDP for the reduced system (5.35) with respect to fault components $f^{(1)}$ and decouples $f^{(2)}$ and w in the leading components $r^{(1)}$ of the residual r, while $\overline{Q}_1^{(2)}(\lambda)$ solves the AFDP for the reduced system (5.35) for the fault components $f^{(2)}$ and generates the trailing components $r^{(2)}$ of the residual r. The maximum number of components of $r^{(1)}$ is $p - r_d - r_w$, while $r^{(2)}$ will have maximum r_w components. Thus, the number of components of r is limited to $p - r_d$. The case $f = f^{(1)}$ corresponds to the solution of an EFDP for which **Procedure EFD** can be used, while the case $f = f^{(2)}$ corresponds to the solution of an AFDP for which **Procedure AFD** can be used. $\square$

Example 5.9 Consider once again the solution of the $\mathcal{H}_{\infty-}/\mathcal{H}_\infty$ synthesis problem for Example 5.5. With $N_l(s)$ chosen as in Example 5.5, we have that the rank of $\overline{G}_w(s)$ (or $R_w(s)$ at Step (1) of **Procedure AFD**) is $r_w = 1$. With $\overline{N}_{l,w}(s) = \begin{bmatrix} 0 & 1 \end{bmatrix}$, we obtain $\overline{N}_{l,w}(s)\overline{G}_w(s) = 0$ and $\overline{N}_{l,w}(s)\overline{G}_f(s) = \begin{bmatrix} 0 & 1 \end{bmatrix}$. Thus, with $f^{(1)} = f_2$, $f^{(2)} = f_1$ and

$$\overline{G}_{f^{(1)}}(s) := \begin{bmatrix} 0 \\ 1 \end{bmatrix}, \qquad \overline{G}_{f^{(2)}}(s) := \begin{bmatrix} \dfrac{s+1}{s+2} \\ 0 \end{bmatrix},$$

we arrive to the partitioned subsystem (5.35). We determine $Q(s)$ in the partitioned form (5.36), where the solution of the EFDP with the above $\overline{G}_{f^{(2)}}(s)$ is simply

$$Q^{(1)}(s) = \overline{N}_{l,w}(s)N_l(s) = \begin{bmatrix} 0 & 1 & -\dfrac{s+2}{s+3} \end{bmatrix}.$$

We determine $Q^{(2)}(s)$ by solving the AFDP formulated with $\overline{G}_{f^{(1)}}(s)$ and $\overline{G}_w(s)$ using **Procedure AFD**. With $Q_1(s) = N_l(s)$ chosen in Example 5.5 and $Q_2(s) = \begin{bmatrix} 1 & 0 \end{bmatrix}$ we obtain at Step (2)

$$Q^{(2)}(s) = \begin{bmatrix} 1 & 0 & -\dfrac{s+1}{s+2} \end{bmatrix}, \quad R_{f^{(2)}}(s) = \dfrac{s+1}{s+2}, \quad R_w(s) = \dfrac{1}{s+2}.$$

With $Q_3(s) = R_w^{-1}(s)$ at Step (3), $Q_4(s) = Q_3^{-1}(s) = R_w(s)$ at Step (4), and $Q_5 = 2$ at Step (5), we obtain for $\gamma = 1$ the final $Q^{(2)}(s)$ and the corresponding $R_{f^{(2)}}(s)$

$$Q^{(2)}(s) = \begin{bmatrix} 2 & 0 & -2\dfrac{s+1}{s+2} \end{bmatrix}, \quad R_{f^{(2)}}(s) = 2\dfrac{s+1}{s+2},$$

for which $\beta = \|R_{f^{(2)}}(s)\|_{\infty-} = 2$. The combined solutions according to (5.36) give

$$Q(s) = \begin{bmatrix} Q^{(1)}(s) \\ Q^{(2)}(s) \end{bmatrix} = \begin{bmatrix} 0 & 1 & -\dfrac{s+2}{s+3} \\ 2 & 0 & -2\dfrac{s+1}{s+2} \end{bmatrix}, \quad R_f(s) = \begin{bmatrix} 0 & 1 \\ 2\dfrac{s+1}{s+2} & 0 \end{bmatrix}, \quad R_w(s) = \begin{bmatrix} 0 \\ \dfrac{2}{s+2} \end{bmatrix}.$$

The resulting filter is able to perform fault isolation as well, and even the exact reconstruction of the fault f_2. The optimal value $\beta = 1$ for $\gamma = 1$ is the same as for the "optimal" solution of Example 5.7. However, since the exact solution $Q^{(1)}(s)$ can be arbitrarily scaled, the effective value of β is 2, which is larger than for the "optimal" solution of Example 5.8.   ◊

5.4   Solving the Exact Fault Detection and Isolation Problem

Let S be a given $n_b \times m_f$ structure matrix to be achieved by the fault detection filter $Q(\lambda)$. Using the factorized representation $Q(\lambda) = \overline{Q}_1(\lambda)Q_1(\lambda)$ in (5.9), it follows that, to solve the *exact fault detection and isolation problem* (EFDIP) formulated in Sect. 3.5.3 for the system (3.2) with $w \equiv 0$, the same S must be achieved by $\overline{Q}_1(\lambda)$ for the reduced system (5.11) for $w \equiv 0$. For this, we consider $\overline{Q}_1(\lambda)$ partitioned with n_b block rows, in the form

$$\overline{Q}_1(\lambda) = \begin{bmatrix} \overline{Q}_1^{(1)}(\lambda) \\ \overline{Q}_1^{(2)}(\lambda) \\ \vdots \\ \overline{Q}_1^{(n_b)}(\lambda) \end{bmatrix}, \tag{5.37}$$

where the i-th block row $\overline{Q}_1^{(i)}(\lambda)$ generates the i-th component of the residual vector

$$\mathbf{r}^{(i)}(\lambda) := \overline{Q}_1^{(i)}(\lambda)\overline{\mathbf{y}}(\lambda) \tag{5.38}$$

and achieves the i-th specification contained in the i-th row of S.

The solvability conditions of the EFDIP given in Theorem 3.10 (also explicitly given in Theorem 3.5) can be replaced by simpler conditions for the reduced system (5.11). This comes down to checking for $i = 1, \ldots, n_b$, the solvability conditions for the i-th specification contained in the i-th row of S. For this purpose, we rewrite for each i, $i = 1, \ldots, n_b$, the reduced system (5.11) for $w \equiv 0$ as

$$\overline{\mathbf{y}}(\lambda) = \overline{G}_d^{(i)}(\lambda)\mathbf{d}^{(i)}(\lambda) + \overline{G}_f^{(i)}(\lambda)\mathbf{f}^{(i)}(\lambda), \tag{5.39}$$

where $d^{(i)}$ contains those components f_j of f for which $S_{ij} = 0$, $f^{(i)}$ contains those components f_j of f for which $S_{ij} \neq 0$, while $\overline{G}_d^{(i)}(\lambda)$ and $\overline{G}_f^{(i)}(\lambda)$ are formed from the corresponding sets of columns of $\overline{G}_f(\lambda)$, respectively. Thus, $d^{(i)}$ contains all fault components to be decoupled in the i-th component $r^{(i)}$ of the residual by the i-th filter $\overline{Q}_1^{(i)}(\lambda)$, while $f^{(i)}$ contains those faults, which need to be detected in the i-th component $r^{(i)}$ of the residual.

The following corollary to Theorem 3.10 provides the solvability conditions of the EFDIP in terms of the n_b reduced systems formed in (5.39).

Corollary 5.5 *For the system (3.2) with $w \equiv 0$ and a given structure matrix S, the EFDIP is solvable if and only if the system (5.11) with $w \equiv 0$ is S-fault isolable, or equivalently, for $i = 1, \ldots, n_b$*

$$\operatorname{rank} [\, \overline{G}_d^{(i)}(\lambda) \; \overline{G}_{f_j}(\lambda) \,] > \operatorname{rank} \overline{G}_d^{(i)}(\lambda), \quad \forall j, \; S_{ij} \neq 0 ,$$

where $\overline{G}_d^{(i)}(\lambda)$ is formed from the columns $\overline{G}_{f_j}(\lambda)$ of $\overline{G}_f(\lambda)$ for which $S_{ij} = 0$.

In other words, to check the fault isolability for the i-th specification, we have to simply check the complete fault detectability of the corresponding reduced system (5.39) with permuted inputs.

A similar corollary to Theorem 3.11 provides the solvability condition for the solution of the EFDIP with strong isolability.

Corollary 5.6 *For the system (3.2) with $w \equiv 0$ and $S = I_{m_f}$, the EFDIP is solvable if and only if the system (5.11) with $w \equiv 0$ is strongly fault isolable, or equivalently*

$$\operatorname{rank} \overline{G}_f(\lambda) = m_f .$$

To determine the i-th block row $\overline{Q}_1^{(i)}(\lambda)$ of $\overline{Q}_1(\lambda)$ in (5.37), we have to solve an EFDP for the corresponding reduced system (5.39). For this purpose, the **Procedure EFD** can be applied, which also checks the solvability conditions for the corresponding specification. The resulting overall detector $Q(\lambda)$ and the corresponding $R_f(\lambda)$ are

$$Q(\lambda) = \begin{bmatrix} Q^{(1)}(\lambda) \\ Q^{(2)}(\lambda) \\ \vdots \\ Q^{(n_b)}(\lambda) \end{bmatrix} = \begin{bmatrix} \overline{Q}_1^{(1)}(\lambda) \\ \overline{Q}_1^{(2)}(\lambda) \\ \vdots \\ \overline{Q}_1^{(n_b)}(\lambda) \end{bmatrix} Q_1(\lambda), \qquad R_f(\lambda) = \begin{bmatrix} R_f^{(1)}(\lambda) \\ R_f^{(2)}(\lambda) \\ \vdots \\ R_f^{(n_b)}(\lambda) \end{bmatrix}, \quad (5.40)$$

where the i-th block row $R_f^{(i)}(\lambda)$ achieves the i-th specification contained in the i-th row of S.

The **Procedure EFDI**, given in what follows, determines the n_b row blocks $Q^{(i)}(\lambda)$ and $R_f^{(i)}(\lambda)$, $i = 1, \ldots, n_b$, of $Q(\lambda)$ and $R_f(\lambda)$, respectively, with the i-th blocks having the desired row dimension q_i.

Procedure EFDI: Exact synthesis of fault detection and isolation filters

Inputs : $\{G_u(\lambda), G_d(\lambda), G_f(\lambda)\}$, $S \in \mathbb{R}^{n_b \times m_f}$, $\{q_1, \ldots, q_{n_b}\}$
Outputs: $Q^{(i)}(\lambda)$, $R_f^{(i)}(\lambda)$, $i = 1, \ldots, n_b$

(1) Compute a $(p - r_d) \times (p + m_u)$ minimal basis matrix $Q_1(\lambda)$ for the left
    nullspace of $G(\lambda)$ defined in (5.2), where $r_d := \operatorname{rank} G_d(\lambda)$;
    set $Q(\lambda) = Q_1(\lambda)$ and compute $R_f(\lambda) = Q_1(\lambda) \begin{bmatrix} G_f(\lambda) \\ 0 \end{bmatrix}$.

(2) For $i = 1, \ldots, n_b$

    (2.1) Define $\overline{G}_d^{(i)}(\lambda)$ as those columns $R_{f_j}(\lambda)$ of $R_f(\lambda)$ for which $S_{ij} = 0$
    and $\overline{G}_f^{(i)}(\lambda)$ as those columns $R_{f_j}(\lambda)$ for which $S_{ij} \neq 0$.

    (2.2) Apply the **Procedure EFD** to the system described by the quadru-
    ple $\{0, \overline{G}_d^{(i)}(\lambda), \overline{G}_f^{(i)}(\lambda), [\, Q(\lambda) \; R_f(\lambda) \,]\}$ to obtain the $q_i \times (p - r_d)$
    least order filter $\overline{Q}_1^{(i)}(\lambda)$. **Exit** if no solution exists.

    (2.3) Compute $Q^{(i)}(\lambda) = \overline{Q}_1^{(i)}(\lambda) Q(\lambda)$ and $R_f^{(i)}(\lambda) = \overline{Q}_1^{(i)}(\lambda) R_f(\lambda)$.

This synthesis procedure ensures that each block $\overline{Q}_1^{(i)}(\lambda)$ and the corresponding $R_f^{(i)}(\lambda)$ are stable. Thus, the overall $R_f(\lambda)$ in (5.40) is also stable. The stability of overall $Q(\lambda)$ in (5.40) can always be ensured, by choosing a stable left nullspace basis $Q_1(\lambda)$ at Step (1). As it will be shown in Sect. 7.4, this is not necessary because the computation of both $Q^{(i)}(\lambda) = \overline{Q}_1^{(i)}(\lambda) Q(\lambda)$ and $R_f^{(i)}(\lambda) = \overline{Q}_1^{(i)}(\lambda) R_f(\lambda)$ at Step (2.3) can be done by using state-space representation-based updating techniques, which always guarantee that $Q^{(i)}(\lambda)$ and $R_f^{(i)}(\lambda)$ have the state-space representations with the same state and descriptor matrices, and result simultaneously stable.

The applicability of **Procedure EFDI** for a given system relies on the assumption that the structure matrix S is achievable. Therefore, to select a minimal set of specifications, which cover all expected fault combinations, it is important to know all achievable specifications for a given system. For a system with m_f faults, the complete set of possible distinct specifications contains $2^{m_f} - 1$ elements. Thus, a brute force approach is based on an exhaustive search by trying to solve the EFDIP for each of these specifications to find out which specifications are feasible (i.e., the corresponding design was successful). The main problem with this approach is its lack of efficiency, as explained in what follows.

Each synthesis problem of a fault detection filter for a given specification can be reformulated as a standard EFDP, where all faults with zero signatures in the checked specification are redefined as disturbances. With this reformulation, the main computation is the determination of the nullspace basis of a TFM with $p + m_u$ rows and $m_u + m_d + k$ columns, where k denotes the number of null elements in the tested specification (i.e., $0 \leq k < m_f$) and represents the number of additional disturbance inputs, which results by recasting the fault inputs to be decoupled as disturbances. The nullspace computation must be performed for *all* $2^{m_f} - 1$ possible specifications, although this may not be necessary if $m_f > p - r_d$, where we recall that r_d is the rank of $G_d(\lambda)$. In what follows, we describe a more efficient approach, where the product representation of nullspace, mentioned in Sect. 5.1, is systematically exploited. The expected efficiency gain arises by replacing the above nullspace computations on matrices with $p + m_u$ rows and at least $m_u + m_d$ columns, with a succession of nullspace determinations on single column matrices with decreasing number of rows. This leads to a significant reduction of the total computational burden.

We now describe a recursive procedure to generate in a systematic and computationally efficient way suitable nullspace bases to serve for the determination of all achievable specifications. We illustrate the core computation with two generic $p_e \times m$ and $p_e \times m_f$ TFMs $G(\lambda)$ and $F(\lambda)$, respectively. The basic computational step consists of successively determining left nullspace bases $N_l(\lambda)$ of $G(\lambda)$ (i.e., $N_l(\lambda)G(\lambda) = 0$) such that the structure matrix of $N_l(\lambda)F(\lambda)$ has up to $\min(m_f, p_e - r) - 1$ zero columns, where $r = \operatorname{rank} G(\lambda)$. To initialize the procedure for the system (2.1), we initialize these TFMs as

$$G(\lambda) = \begin{bmatrix} G_u(\lambda) & G_d(\lambda) \\ I_{m_u} & 0 \end{bmatrix}, \quad F(\lambda) = \begin{bmatrix} G_f(\lambda) \\ 0 \end{bmatrix}, \tag{5.41}$$

with $p_e = p + m_u$ and $m = m_u + m_d$.

To describe the nullspace generation process in more detail, let $N_l^0(\lambda)$ be the $(p_e - r) \times p_e$ proper minimal left nullspace basis of $G(\lambda)$ and let S_{F^0} be the structure matrix of $F^0(\lambda) := N_l^0(\lambda)F(\lambda)$. This structure matrix is a $1 \times m_f$ row vector corresponding to $F^0(\lambda)$ seen as an $1 \times m_f$ block row (see the definition of structure matrix in (3.17) based on (3.16)), with the $(1, j)$-th block row element formed of the j-th column of $F^0(\lambda)$. If $\min(m_f, p_e - r) > 1$, then for each $i = 1, \ldots, m_f$, determine the left nullspace basis $N_l^i(\lambda)$ of the i-th column of $F^0(\lambda)$ and let S_{F^i} be the structure matrix corresponding to $F^i(\lambda) := N_l^i(\lambda)F^0(\lambda)$. Each S_{F^i} is a $1 \times m_f$ row vector with the i-th column element equal to zero. If the i-th column is zeroed with $N_l^i(\lambda)$, then $N_l^i(\lambda)$ is a $(p_e - r - 1) \times (p_e - r)$ TFM. If now $p_e - r - 1 > 1$, we continue by computing for each j-th column of $F^i(\lambda)$, $j > i$, the corresponding left nullspace $N_l^{j,i}(\lambda)$ and the corresponding structure matrix $S_{F^{j,i}}$ of $F^{j,i}(\lambda) := N_l^{j,i}(\lambda)F^i(\lambda)$. Each $S_{F^{j,i}}$ will have zeros in its i-th and j-th columns. This process continues in a similar way until all nonzero $S_{F^{k,\ldots,j,i}}$ have been generated. The resulting S is formed by concatenating row-wise the determined $S_{F^0}, S_{F^1}, \ldots, S_{F^{m_f}}, S_{F^{2,1}}, \ldots, S_{F^{m_f,1}}, \ldots,$ $S_{F^{m_f,m_f-1}}, \ldots$. The tree in Fig. 5.1 illustrates the performed computations for a system with $m_f = 3$ and $p_e - r = 3$.

If we denote with S the matrix formed of all achievable specifications, then, for the considered example, we have $S = [\, S_{F^0}^T \;\; S_{F^1}^T \;\; S_{F^{2,1}}^T \;\; S_{F^{3,1}}^T \;\; S_{F^2}^T \;\; S_{F^{3,2}}^T \;\; S_{F^3}^T \,]^T$, where each S_{F^i} has the i-th column zero, while each $S_{F^{j,i}}$ has the i-th and j-th columns zero. Note that in nongeneric cases, other elements may also be zero. It can be observed that the computation of $F^{1,2}(\lambda)$ is not necessary because the same information is provided by $F^{2,1}(\lambda)$. Similarly, the computation of both $F^{1,3}(\lambda)$ and $F^{2,3}(\lambda)$ is not necessary, because the corresponding information is provided by $F^{3,1}(\lambda)$ and $F^{3,2}(\lambda)$, respectively.

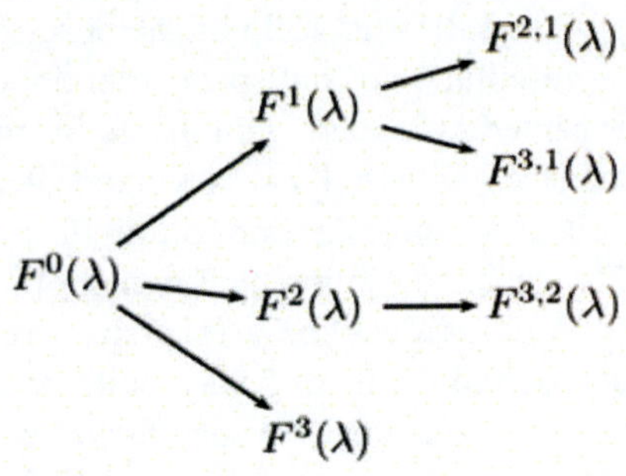

Fig. 5.1 Tree of performed computations of fault specifications

The computational process can be easily formulated as a recursive procedure, which, for the given matrices $G(\lambda)$ and $F(\lambda)$, computes the maximally achievable structure matrix S. This procedure can be formally called as $S = \text{GENSPEC}(G, F)$. For example, the maximally achievable structure matrix for the system (2.1) can be computed with $G(\lambda)$ and $F(\lambda)$ defined in (5.41).

Procedure GENSPEC: Generation of achievable fault specifications

Inputs : $G(\lambda) \in \mathbb{R}(\lambda)^{p_e \times m}$, $F(\lambda) \in \mathbb{R}(\lambda)^{p_e \times m_f}$
Output: $S \in \mathbb{R}^{q \times m_f}$

Function $S = \text{GENSPEC}(G, F)$

(1) Compute a left nullspace basis $N_l(\lambda)$ of $G(\lambda)$.

    Exit with $S = \emptyset$ if $N_l(\lambda)$ is empty.

(2) Compute $N_f(\lambda) = N_l(\lambda) F(\lambda)$.
(3) Compute the structure matrix S of $N_f(\lambda)$. **Exit** if $N_f(\lambda)$ is a row vector.
(4) **For** $i = 1, \ldots, m_f$

    (4.1) Form $\widetilde{G}_i(\lambda)$ as column i of $N_f(\lambda)$.
    (4.2) Form $\widetilde{F}_i(\lambda)$ from columns $1, \ldots, i-1, i+1, \ldots, m_f$ of $N_f(\lambda)$.
    (4.3) **Call** $\widetilde{S} = \text{GENSPEC}(\widetilde{G}_i, \widetilde{F}_i)$.
    (4.4) Partition $\widetilde{S} = [\,\widetilde{S}_1 \ \widetilde{S}_2\,]$ such that $\widetilde{S}_1$ has $i - 1$ columns.
    (4.5) Define $\widehat{S} = [\,\widetilde{S}_1 \ 0 \ \widetilde{S}_2\,]$ and update $S \leftarrow \begin{bmatrix} S \\ \widehat{S} \end{bmatrix}$.

The **Procedure GENSPEC** performs the minimum number of nullspace computations and updating. This number is given by $k_S = \sum_{i=0}^{i_{max}} \binom{m_f}{i}$, where $i_{max} = \min(m_f, p_e - r) - 1$ and r is the rank of the initial $G(\lambda)$. As it can be observed, k_S depends on the number of initial basis vectors $p_e - r$ and the number of faults m_f, and although the number of distinct specifications can be relatively low, still k_S can be a large number. For the example considered above, $m_f = 3$ and $p_e - r = 3$, thus the maximum number $k_S = 7(= 2^{m_f} - 1)$ nullspace computations are necessary. However, in contrast to the brute force approach, all but one of nullspace computations are performed for rational matrices with a single column (and varying number of rows), and therefore, a substantial saving in the computation effort can be expected.

Example 5.10 Consider a continuous-time system with triplex sensor redundancy on one of its measured output components, which we denote by y_1, y_2 and y_3. Each output is related to the control and disturbance inputs by the input–output relation

$$y_i(s) = G_u(s)\mathbf{u}(s) + G_d(s)\mathbf{d}(s), \quad i = 1, 2, 3,$$

where $G_u(s)$ and $G_d(s)$ are $1 \times m_u$ and $1 \times m_d$ TFMs, respectively. We assume all three outputs are susceptible to additive sensor faults. Thus, the input–output model of the system has the form

$$\mathbf{y}(s) := \begin{bmatrix} \mathbf{y}_1(s) \\ \mathbf{y}_2(s) \\ \mathbf{y}_3(s) \end{bmatrix} = \begin{bmatrix} G_u(s) \\ G_u(s) \\ G_u(s) \end{bmatrix} \mathbf{u}(s) + \begin{bmatrix} G_d(s) \\ G_d(s) \\ G_d(s) \end{bmatrix} \mathbf{d}(s) + \begin{bmatrix} \mathbf{f}_1(s) \\ \mathbf{f}_2(s) \\ \mathbf{f}_3(s) \end{bmatrix}.$$

The maximal achievable structure matrix obtained by applying the **Procedure GENSPEC** is

$$S_{max} = \begin{bmatrix} 1 & 1 & 1 \\ 0 & 1 & 1 \\ 1 & 0 & 1 \\ 1 & 1 & 0 \end{bmatrix}.$$

If we can assume that no simultaneous sensor failures occur, then we can target to solve a EFDIP for the structure matrix

$$S = \begin{bmatrix} 0 & 1 & 1 \\ 1 & 0 & 1 \\ 1 & 1 & 0 \end{bmatrix},$$

where the columns of S codify the desired fault signatures.

By using the **Procedure EFDI**, we compute first a left nullspace basis $N_l(s)$ of

$$G(s) = \begin{bmatrix} G_u(s) & G_d(s) \\ G_u(s) & G_d(s) \\ G_u(s) & G_d(s) \\ 1 & 0 \end{bmatrix},$$

in a product form similar to (5.5). We obtain

$$N_l(s) = \begin{bmatrix} 1 & -1 & 0 \\ 0 & 1 & -1 \end{bmatrix} \begin{bmatrix} 1 & 0 & 0 & -G_u(s) \\ 0 & 1 & 0 & -G_u(s) \\ 0 & 0 & 1 & -G_u(s) \end{bmatrix} = \begin{bmatrix} 1 & -1 & 0 & 0 \cdots 0 \\ 0 & 1 & -1 & 0 \cdots 0 \end{bmatrix}. \tag{5.42}$$

We set $Q_1(s) = N_l(s)$ and

$$R_f(s) = \begin{bmatrix} 1 & -1 & 0 \\ 0 & 1 & -1 \end{bmatrix}.$$ (5.43)

For example, to achieve the first specification $\begin{bmatrix} 0 & 1 & 1 \end{bmatrix}$, we redefine f_1 as a disturbance $d^{(1)} := f_1$ to be decoupled, $f^{(1)} := [\, f_2 \ f_3\,]^T$, $\overline{G}_d^{(1)}(s)$ as the first column of $R_f(s)$ and $\overline{G}_f^{(1)}(s)$ as the last two columns of $R_f(s)$. With **Procedure EFD**, we obtain $\overline{Q}_1^{(1)}(s) = [\, 0 \ 1\,]$ (as a constant basis of the left nullspace of $\overline{G}_d^{(1)}(s)$). Thus, the first row of the overall filter $Q(s)$ is given by

$$Q^{(1)}(s) = \overline{Q}_1^{(1)}(s)Q_1(s) = \begin{bmatrix} 0 & 1 & -1 & 0 & \cdots & 0 \end{bmatrix}.$$

The corresponding residual component is simply

$$r_1 = y_2 - y_3 = f_2 - f_3\,,$$

which is fully decoupled from f_1. Similarly, with $\overline{Q}_1^{(2)}(s) = [\, -1 \ -1\,]$ and $\overline{Q}_1^{(3)}(s) = [\, 1 \ 0\,]$, we obtain

$$Q^{(2)}(s) = \overline{Q}_1^{(2)}(s)Q_1(s) = \begin{bmatrix} -1 & 0 & 1 & 0 & \cdots & 0 \end{bmatrix}$$

and

$$Q^{(3)}(s) = \overline{Q}_1^{(3)}(s)Q_1(s) = \begin{bmatrix} 1 & -1 & 0 & 0 & \cdots & 0 \end{bmatrix}.$$

The TFM of the overall FDI filter is

$$Q(s) = \begin{bmatrix} Q^{(1)}(s) \\ Q^{(2)}(s) \\ Q^{(3)}(s) \end{bmatrix} = \begin{bmatrix} 0 & 1 & -1 & 0 & \cdots & 0 \\ -1 & 0 & 1 & 0 & \cdots & 0 \\ 1 & -1 & 0 & 0 & \cdots & 0 \end{bmatrix}$$ (5.44)

and the overall residual vector is

$$r = \begin{bmatrix} r_1 \\ r_2 \\ r_3 \end{bmatrix} := \begin{bmatrix} y_2 - y_3 \\ y_3 - y_1 \\ y_1 - y_2 \end{bmatrix} = \begin{bmatrix} 0 & 1 & -1 \\ -1 & 0 & 1 \\ 1 & -1 & 0 \end{bmatrix} \begin{bmatrix} f_1 \\ f_2 \\ f_3 \end{bmatrix}.$$

This fault detection filter implements the widely employed voting-based fault isolation scheme for the case when the assumption of a single sensor fault at a time is fulfilled. Its main appeal is its independence of the system dynamics. Thus, the constant filter (5.44) can be applied even in the case of a system with nonlinear dynamics. Since parametric variations have no effects on the residuals, a perfect robustness of this scheme is guaranteed. However, for applications to safety critical systems, the voting scheme is potentially unreliable in the (improbable) case of two simultaneous failures with a common value of faults (e.g., $f_2 = f_3 \neq 0$). In such a case, the faults remain undetected, and often the common fault value is wrongly used as the "valid" measurement.

The script **Ex5_10c.jl** in Listing 5.3 solves the EFDIP considered in this example and calls the synthesis function **efdisyn**, which implements the **Procedure EFDI**. The script also includes the verification of results by checking the synthesis conditions. A more detailed version of this script is **Ex5_10.jl** (not listed), which closely follows the computational steps of the **Procedure EFDI**.   ◊

Listing 5.3 Script **Ex5_10c.jl** to solve the EFDIP of Example 5.10 using **Procedure EFDI**

```julia
using FaultDetectionTools, DescriptorSystems, Test

# Example 5.10 - Solution of an EFDIP
println("Example 5.10")

# enter output and fault vector dimensions
p = 3; mf = 3;
# generate random dimensions for system order and input vectors
nu = Int(floor(1+4*rand())); mu = Int(floor(1+4*rand()));
nd = Int(floor(1+4*rand())); md = Int(floor(1+4*rand()));

# define random Gu(s) and Gd(s) with triplex sensor redundancy
# and Gf(s) for triplex sensor faults
Gu = ones(3,1)*rss(nu,1,mu);   # enter Gu(s) in state-space form
Gd = ones(3,1)*rss(nd,1,md);   # enter Gd(s) in state-space form
Gf = eye(3);                   # enter Gf(s) for sensor faults
atol = 1.e-7;                  # tolerance for rank tests

# build model with faults
sysf = fdimodset([Gu Gd Gf], c = 1:mu, d = mu.+(1:md),
                 f = (mu+md).+(1:mf))

SFDI = [ 0 1 1; 1 0 1; 1 1 0] .> 0 # set structure matrix

# call efdisyn with option for scalar output filters
Qt, Rft = efdisyn(sysf, SFDI; atol, rdim = 1)

# normalize Q and Rf to match example
scale = sign.([ Rft.sys[1].D[1,2], Rft.sys[2].D[1,3],
               Rft.sys[3].D[1,1]])
Q = FDIFilter(scale .* Qt.sys, p, mu)
R = FDIFilterIF(scale .* Rft.sys; mf)
println("Q = "); display(Q)
println("R = "); display(R)

# check synthesis conditions:
# Qt*[Gu Gd;I 0] = 0 and Qt*[Gf; 0] = Rf
Rt = fdIFeval(Qt,sysf) # form Qt*[Gu Gd Gf;I 0 0];
indcd = [Rt.controls;Rt.disturbances]
@test iszero(vcat(Rt.sys...)[:,indcd]; atol) &&
      iszero(vcat(Rt.sys...)[:,Rt.faults]-vcat(Rft.sys...); atol)

# check weak and strong fault detectability
@test fditspec(Rft) == fdisspec(Rft) == SFDI
```

5.5 Solving the Approximate Fault Detection and Isolation Problem

Let S be a given $n_b \times m_f$ structure matrix targeted to be achieved by the fault detection filter $Q(\lambda)$. Using the factorized representation $Q(\lambda) = \overline{Q}_1(\lambda) Q_1(\lambda)$ in (5.9), it follows that, to solve the *approximate fault detection and isolation problem* (AFDIP) formulated in Sect. 3.5.4, the same S has to be targeted by any $\overline{Q}_1(\lambda)$, which solves the AFDIP for the reduced system (5.11). For this, we consider $\overline{Q}_1(\lambda)$ partitioned with n_b block rows, in the form (5.37), where the i-th block row $\overline{Q}_1^{(i)}(\lambda)$ generates the i-th component $r^{(i)}$ of the residual vector r according to (5.38) and targets to achieve the i-th specification contained in the i-th row of S.

The solvability conditions of the AFDIP given in Theorems 3.12 and 3.13 can be replaced by simpler conditions for the reduced system (5.11). This comes down to checking for $i = 1, \ldots, n_b$, the solvability conditions for the i-th specification contained in the i-th row of S. To determine the filter $\overline{Q}_1^{(i)}(\lambda)$, an AFDP can be formulated for each i by suitably redefining the disturbance, fault and noise inputs of the reduced system (5.11).

The reduced system (5.11) can be rewritten for each $i = 1, \ldots, n_b$, in the form

$$\overline{\mathbf{y}}(\lambda) = \overline{G}_d^{(i)}(\lambda)\mathbf{d}^{(i)}(\lambda) + \overline{G}_f^{(i)}(\lambda)\mathbf{f}^{(i)}(\lambda) + \overline{G}_w(\lambda)\mathbf{w}(\lambda) , \qquad (5.45)$$

where $d^{(i)}$ contains those components f_j of f for which $S_{ij} = 0$, $f^{(i)}$ contains those components f_j of f for which $S_{ij} \neq 0$, while $\overline{G}_d^{(i)}(\lambda)$ and $\overline{G}_f^{(i)}(\lambda)$ are formed from the corresponding sets of columns of $\overline{G}_f(\lambda)$, respectively. The vector $f^{(i)}$ contains all faults, which need to be detected in the i-th component $r^{(i)}$ of the residual.

In the case when the AFDIP is formulated to fulfil the weaker conditions (3.28), $d^{(i)}$ contains all fault components, which have to be *approximately* decoupled in the i-th component $r^{(i)}$ of the residual by the i-th filter $\overline{Q}_1^{(i)}(\lambda)$, and therefore, $d^{(i)}$ have to be treated as additional noise inputs. The following corollary to Theorem 3.12 provides the solvability conditions of the AFDIP in terms of the reduced system (5.11) for an arbitrary structure matrix S (see also Remark 3.10).

Corollary 5.7 *For the system (3.2) and a given $n_b \times m_f$ structure matrix S with columns S_j, $j = 1, \ldots, m_f$, the AFDIP is solvable with conditions (3.28) if and only if the reduced system (5.11) is fault input observable for all faults f_j corresponding to nonzero columns of S, or equivalently,*

$$\overline{G}_{f_j}(\lambda) \neq 0 \ \ \forall j, \ \ S_j \neq 0 .$$

In the case when the AFDIP is formulated to fulfil the stronger conditions (3.29), $d^{(i)}$ contains all fault components to be *exactly* decoupled in the i-th component $r^{(i)}$ of the residual by the i-th filter $\overline{Q}_1^{(i)}(\lambda)$. The following corollary to Theorem 3.13 provides the solvability conditions of the AFDIP in terms of the reduced system (5.11):

Corollary 5.8 *For the system (3.2) and a given structure matrix S, the AFDIP is solvable with conditions (3.29) if and only if the reduced system (5.11) is S-fault isolable, or equivalently, for $i = 1, \ldots, n_b$*

$$\text{rank}\,[\,\overline{G}_d^{(i)}(\lambda)\ \overline{G}_{f_j}(\lambda)\,] > \text{rank}\,\overline{G}_d^{(i)}(\lambda), \quad \forall j, \ S_{ij} \neq 0\,,$$

where $\overline{G}_d^{(i)}(\lambda)$ is formed from the columns $\overline{G}_{f_j}(\lambda)$ of $\overline{G}_f(\lambda)$ for which $S_{ij} = 0$.

To determine $\overline{Q}^{(i)}(\lambda)$, we can always try first to achieve the i-th specification *exactly* by applying the **Procedure AFD** (see Sect. 5.3) to solve the AFDP for the reduced system (5.45), and determine a least order fault detection filter $\overline{Q}^{(i)}(\lambda)$ in (5.38), which fully decouples $d^{(i)}(t)$. If the AFDP for the reduced system (5.45) is not solvable, then the **Procedure AFD** can be applied to solve the AFDP for the same reduced system (5.45), but with the disturbance inputs $d^{(i)}(t)$ redefined as additional noise inputs.

The **Procedure AFDI**, given in what follows, determines for a given $n_b \times m_f$ structure matrix S, a bank of n_b least order fault detection filters $Q^{(i)}(\lambda)$, $i = 1, \ldots, n_b$, which solve the AFDIP. Additionally, the block rows of $R_f(\lambda)$ and $R_w(\lambda)$ corresponding to $Q^{(i)}(\lambda)$ are determined as

$$R_f^{(i)}(\lambda) := Q^{(i)}(\lambda)\begin{bmatrix} G_f(\lambda) \\ 0 \end{bmatrix}, \quad R_w^{(i)}(\lambda) := Q^{(i)}(\lambda)\begin{bmatrix} G_w(\lambda) \\ 0 \end{bmatrix}.$$

The existence conditions for the solvability of the AFDIP are implicitly tested when applying the **Procedure AFD** to solve the appropriate AFDP for the system (5.45), with specified number of components q_i of $r^{(i)}$ and noise signal gain level γ. For each filter $Q^{(i)}(\lambda)$, the achieved fault sensitivity level β_i is also computed by the **Procedure AFD**.

Procedure AFDI: Approximate synthesis of FDI filters

Inputs : $\{G_u(\lambda), G_d(\lambda), G_f(\lambda), G_w(\lambda)\}$, $S \in \mathbb{R}^{n_b \times m_f}$, $\{q_1, \ldots, q_{n_b}\}$, γ

Outputs: $Q^{(i)}(\lambda)$, $R_f^{(i)}(\lambda)$, $R_w^{(i)}(\lambda)$, β_i, $i = 1, \ldots, n_b$

(1) Compute a $(p - r_d) \times (p + m_u)$ minimal basis matrix $Q_1(\lambda)$ for the left nullspace of $G(\lambda)$ defined in (5.2), where $r_d := \operatorname{rank} G_d(\lambda)$;

     set $Q(\lambda) = Q_1(\lambda)$ and compute $[\, R_f(\lambda) \; R_w(\lambda) \,] = Q_1(\lambda) \begin{bmatrix} G_f(\lambda) & G_w(\lambda) \\ 0 & 0 \end{bmatrix}$.

(2) For $i = 1, \ldots, n_b$

    (2.1) Form $\overline{G}_d^{(i)}(\lambda)$ from the columns $R_{f_j}(\lambda)$ for which $S_{ij} = 0$ and $\overline{G}_f^{(i)}(\lambda)$ from the columns $R_{f_j}(\lambda)$ for which $S_{ij} \neq 0$.

    (2.2) Apply **Procedure AFD** to the system described by the quadruple $\{0, \overline{G}_d^{(i)}(\lambda), \overline{G}_f^{(i)}(\lambda), R_w(\lambda)\}$ to obtain the $q_i \times (p - r_d)$ least order filter $\overline{Q}_1^{(i)}(\lambda)$ and β_i. **Go to** Step (2.4) if successful.

    (2.3) Apply **Procedure AFD** to the system described by the quadruple $\{0, 0, \overline{G}_f^{(i)}(\lambda), [\, R_w(\lambda) \; \overline{G}_d^{(i)}(\lambda) \,]\}$ to obtain the $q_i \times (p - r_d)$ least order filter $\overline{Q}_1^{(i)}(\lambda)$ and β_i. **Exit** if no solution exists.

    (2.4) Compute $\quad Q^{(i)}(\lambda) = \overline{Q}_1^{(i)}(\lambda) Q(\lambda), \quad R_f^{(i)}(\lambda) = \overline{Q}_1^{(i)}(\lambda) R_f(\lambda) \quad$ and $R_w^{(i)}(\lambda) = \overline{Q}_1^{(i)}(\lambda) R_w(\lambda)$.

Remark 5.11 For the selection of the threshold τ_i for the component $r^{(i)}(t)$ of the residual vector, we can use a similar approach to that described in Remark 5.9. To determine the false alarm bound, we can use the corresponding internal representation of the resulting i-th fault detection filter in the form

$$\mathbf{r}^{(i)}(\lambda) = R_f^{(i)}(\lambda)\mathbf{f}(\lambda) + R_w^{(i)}(\lambda)\mathbf{w}(\lambda) . \tag{5.46}$$

If we assume, for example, a bounded energy noise input $w(t)$ such that $\|w\|_2 \leq \delta_w$, then the false alarm bound $\tau_f^{(i)}$ for the i-th residual vector component $r^{(i)}(t)$ can be computed as

$$\tau_f^{(i)} = \sup_{\|w\|_2 \leq \delta_w} \|R_w^{(i)}(\lambda)\mathbf{w}(\lambda)\|_2 = \|R_w^{(i)}(\lambda)\|_\infty \delta_w . \tag{5.47}$$

However, by simply setting $\tau_i = \tau_f^{(i)}$, we can only detect the presence of a fault in any of the components of f, but we ignore the additional structural information needed for fault isolation. Therefore, we need to take into account the partition of the components of f into two distinct vectors, namely $f^{(i)}$, which contains those components f_j of f for which $S_{ij} = 1$ (i.e., the faults to be detected in $r^{(i)}$) and $\bar{f}^{(i)}$, which contains those components f_j of f for which $S_{ij} = 0$ (i.e., the faults to be decoupled from $r^{(i)}$). By denoting $R_{f^{(i)}}^{(i)}(\lambda)$ and $R_{\bar{f}^{(i)}}^{(i)}(\lambda)$, the columns of $R_f^{(i)}(\lambda)$ corresponding to $f^{(i)}$ and $\bar{f}^{(i)}$, respectively, we can rewrite (5.46) in the form

$$\mathbf{r}^{(i)}(\lambda) = R^{(i)}_{f^{(i)}}(\lambda)\mathbf{f}^{(i)}(\lambda) + R^{(i)}_{\bar{f}^{(i)}}(\lambda)\bar{\mathbf{f}}^{(i)}(\lambda) + R^{(i)}_w(\lambda)\mathbf{w}(\lambda). \tag{5.48}$$

If we assume, for example, a bounded energy noise input $w(t)$ such that $\|w\|_2 \le \delta_w$ and, similarly, a bounded energy fault input $\bar{f}^{(i)}(t)$ such that $\|\bar{f}^{(i)}\|_2 \le \delta_{\bar{f}^{(i)}}$, then the false alarm bound for isolation $\tau^{(i)}_{fi}$ for the i-th residual vector component $r^{(i)}(t)$ can be bounded as follows:

$$\tau^{(i)}_{fi} = \sup_{\substack{\|w\|_2 \le \delta_w \\ \|\bar{f}^{(i)}\|_2 \le \delta_{\bar{f}^{(i)}}}} \|R^{(i)}_{\bar{f}^{(i)}}(\lambda)\bar{\mathbf{f}}^{(i)}(\lambda) + R^{(i)}_w(\lambda)\mathbf{w}(\lambda)\|_2$$

$$\le \|R^{(i)}_{\bar{f}^{(i)}}(\lambda)\|_\infty \delta_{\bar{f}^{(i)}} + \|R^{(i)}_w(\lambda)\|_\infty \delta_w := \tilde{\tau}^{(i)}_{fi}. \tag{5.49}$$

The setting of the threshold $\tau_i = \tilde{\tau}^{(i)}_{fi}$ ensures no false isolation alarms due to faults occurring in $\bar{f}^{(i)}$. A somewhat smaller (i.e., less conservative) threshold can be used if additionally the information on the maximum number of faults, which simultaneously may occur is included in bounding $\|R^{(i)}_{\bar{f}^{(i)}}(\lambda)\bar{\mathbf{f}}^{(i)}(\lambda)\|_2$. Note that if the i-th specification (coded in the i-th row of the structure matrix S) has been exactly achieved at Step (2.2) of the **Procedure AFDI**, then $R^{(i)}_{\bar{f}^{(i)}}(\lambda) = 0$, and therefore, $\tau^{(i)}_f = \tau^{(i)}_{fi} = \tilde{\tau}^{(i)}_{fi}$. In this case, we can set the threshold to the lowest value $\tau_i = \tau^{(i)}_f$ (i.e., the false alarm bound).

The least size $\underline{\delta}^{(i)}_{f_j}$ of the fault f_j, which can be detected in $r^{(i)}$ for $S_{ij} = 1$, can be estimated similarly as done in Remark 5.9 (see (5.33))

$$\underline{\delta}^{(i)}_{f_j} = \frac{2\|R^{(i)}_w(\lambda)\|_\infty \delta_w}{\|R^{(i)}_{f_j}(\lambda)\|_{\Omega-}}, \tag{5.50}$$

where Ω is a given set of relevant frequency values. Overall, $\underline{\delta}_{f_j}$, the least size of the isolable fault f_j, can be defined as

$$\underline{\delta}_{f_j} := \min_{i \in \mathcal{I}_j} \underline{\delta}^{(i)}_{f_j},$$

where $\mathcal{I}_j := \{i : i \in \{1, \ldots, n_b\} \wedge S_{ij} = 1\}$. $\qquad\qquad\square$

Example 5.11 Consider the solution of the AFDIP for the system

$$G_u(s) = \begin{bmatrix} \dfrac{s+1}{s+2} \\ \dfrac{s+2}{s+3} \end{bmatrix}, \quad G_d(s) = 0, \quad G_f(s) = \begin{bmatrix} \dfrac{s+1}{s+2} & 0 \\ 0 & 1 \end{bmatrix}, \quad G_w(s) = \begin{bmatrix} \dfrac{1}{s+2} \\ 0 \end{bmatrix}$$

used also in Examples 5.5 and 5.8. At Step (1) of the **Procedure AFDI**, we compute a minimal left nullspace basis of $G(\lambda)$ defined in (5.2) as $Q_1(s) = [\, I_2 \;\; -G_u(s) \,]$, which leads to $R_f(s) = G_f(s)$ and $R_w(s) = G_w(s)$. By inspecting $R_f(s)$, it follows that the strong isolability condition is fulfilled (i.e., rank $R_f(s) = 2$), thus we can target to solve an AFDIP with $S = I_2$.

To achieve the specification in the first row of S, we define the reduced model (5.45) with $\overline{G}_d^{(1)}(s) = R_{f_2}(s)$ and $\overline{G}_f^{(1)}(s) = R_{f_1}(s)$. We can apply the **Procedure AFD** to solve the AFDP for the quadruple $\{0, \overline{G}_d^{(1)}(s), \overline{G}_f^{(1)}(s), R_w(s)\}$. At Step (1), we compute a left nullspace basis of $\overline{G}_d^{(1)}(s)$ as $Q_1^{(1)}(s) = \begin{bmatrix} 1 & 0 \end{bmatrix}$ and determine

$$R_f^{(1)}(s) := Q_1^{(1)}(s)R_{f_1}(s) = \frac{s+1}{s+2}, \qquad R_w^{(1)}(s) := Q_1^{(1)}(s)R_w(s) = \frac{1}{s+2}.$$

Since $R_f^{(1)}(s) \neq 0$, it follows that the EFDP, and therefore, also the AFDP has a solution according to Theorem 3.9. At Step (2), we take $Q_2^{(1)}(s) = 1$, and at Step (3), the quasi-co-outer factor $G_{wo}(s)$ is simply $R_w^{(1)}(s)$, which is strictly proper and, thus, has a zero at infinity. With $Q_3^{(1)}(s) = \left(R_w^{(1)}(s)\right)^{-1}$, the resulting $\widetilde{Q}^{(1)}(s) := Q_3^{(1)}(s)Q_2^{(1)}(s)Q_1^{(1)}(s)$ and $\widetilde{R}_f^{(1)}(s) := Q_3^{(1)}(s)Q_2^{(1)}(s)Q_1^{(1)}(s)R_{f_1}(s)$ are, therefore, improper. At Step (4), we choose $Q_4^{(1)}(s)$ of unity $\mathcal{H}_\infty$-norm of the form $Q_4^{(1)}(s) = a/(s+a)$ with $a \geq 1$, such that $Q_4^{(1)}(s)[\ \widetilde{Q}^{(1)}(s)\ \ \widetilde{R}_f^{(1)}(s)\ ]$ is stable and proper. For $\gamma = 1$, we obtain at Step (5) with $Q_5^{(1)} = 1$ the final $\overline{Q}^{(1)}(s)$ as

$$\overline{Q}^{(1)}(s) = Q_5^{(1)}(s)Q_4^{(1)}(s)Q_3^{(1)}(s)Q_2^{(1)}(s)Q_1^{(1)}(s) = \begin{bmatrix} \dfrac{a(s+2)}{s+a} & 0 \end{bmatrix}.$$

At Step (2.4) of **Procedure AFDI**, we obtain

$$Q^{(1)}(s) = \overline{Q}^{(1)}(s)Q_1(s) = \begin{bmatrix} \dfrac{a(s+2)}{s+a} & 0 & -\dfrac{a(s+1)}{s+a} \end{bmatrix},$$

$$R_f^{(1)}(s) = \overline{Q}_1^{(1)}(s)R_f(s) = \begin{bmatrix} \dfrac{a(s+1)}{s+a} & 0 \end{bmatrix}, \qquad R_w^{(1)}(s) = \overline{Q}_1^{(1)}(s)R_w(s) = \frac{a}{s+a}.$$

Since $\beta_1 = \|R_{f_1}^{(1)}(s)\|_\infty = a$ can be arbitrarily large, the underlying $\mathcal{H}_{\infty-}/\mathcal{H}_\infty$ problem has no optimal solution. Still, the resulting $Q^{(1)}(s)$ is completely satisfactory by providing an arbitrary large gap $\beta_1/\gamma = a$.

To achieve the specification in the second row of S, we define $\overline{G}_d^{(2)}(s) = R_{f_1}(s)$ (the first column of $R_f(s)$) and $\overline{G}_f^{(2)}(s) = R_{f_2}(s)$. Again, we apply **Procedure AFD** to solve the AFDP for the quadruple $\{0, \overline{G}_d^{(2)}(s), \overline{G}_f^{(2)}(s), R_w(s)\}$. At Step (1), we compute a left nullspace basis of $\overline{G}_d^{(2)}(s)$ as $Q_1^{(2)}(s) = \begin{bmatrix} 0 & -1 \end{bmatrix}$ and determine

$$R_f^{(2)}(s) := Q_1^{(2)}(s)R_{f_2}(s) = -1, \qquad R_w^{(2)}(s) := Q_1^{(2)}(s)R_w(s) = 0.$$

Observe that we actually solved the AFDP as an EFDP by obtaining $\overline{Q}_1^{(2)}(s) = Q_1^{(2)}(s)$. At Step (2.4) of **Procedure AFDI**, we obtain

$$Q^{(2)}(s) = \overline{Q}^{(2)}(s)Q_1(s) = \left[\, 0 \;\; -1 \;\; \frac{s+2}{s+3} \,\right],$$

$$R_f^{(2)}(s) = \overline{Q}_1^{(2)}(s)R_f(s) = \left[\, 0 \;\; -1 \,\right], \quad R_w^{(2)}(s) = \overline{Q}_1^{(2)}(s)R_w(s) = 0\,.$$

Although $\beta_2 = \| R_{f_2}^{(2)}(s) \|_\infty = 1$, but β_2 can be arbitrary large by suitably rescaling $Q^{(2)}(s)$.

The script **Ex5_11c.jl** in Listing 5.4 solves the AFDIP considered in this example and calls the synthesis function **afdisyn**, which implements the **Procedure AFDI**. The script also includes the verification of results by checking the synthesis conditions. A more detailed version of this script is **Ex5_11.jl** (not listed), which closely follows the computational steps of the **Procedure AFDI**.

$\Diamond$

Listing 5.4 Script **Ex5_11c.jl** to solve the AFDIP of Example 5.11 using **Procedure AFDI**

```julia
using FaultDetectionTools, DescriptorSystems, Test

# Example 5.11 - Solution of an AFDIP
println("Example 5.11")

# define s as an improper transfer function
s = rtf('s');
# define Gu(s), Gf(s) and Gw(s)
Gu = [(s+1)/(s+2); (s+2)/(s+3)];        # enter Gu(s)
Gw = [1/(s+2); 0];                      # enter Gw(s)
Gf = [(s+1)/(s+2) 0; 0 1];              # enter Gf(s)
p = 2; mu = 1; mw = 1; mf = 2;          # enter dimensions
atol = 1.e-7;                           # tolerance for rank tests

SFDI = eye(mf) .> 0                     # set structure matrix to I

# build the synthesis model with additive faults
sysf = fdimodset(dss([Gu Gf Gw]), c = 1:mu,
                 f = mu .+ (1:mf), n = (mu+mf) .+ (1:mw));

# call afdisyn with the option for stability degree -3
Q, R, info = afdisyn(sysf, SFDI; smarg =-3, sdeg = -3)

# scale with -1 to match example
Q.sys[1] = -Q.sys[1]; R.sys[1] = -R.sys[1];

for i = 1:size(SFDI,1)
    println("Q[$i] = $(dss2rm(Q.sys[i]; atol))")
    println("Rf[$i] = $(dss2rm(R.sys[i][:,R.faults]; atol))")
    println("Rw[$i] = $(dss2rm(R.sys[i][:,R.noise]; atol))")
end

println("Q = "); display(Q)
```

```
println("R = "); display(R)
println("gap = $(fdif2ngap(R,SFDI)[1])")

# check synthesis conditions: Q*[Gu Gd;I 0] = 0 and Q*[Gf; 0] = Rf
Rt = fdIFeval(Q,sysf) # form Q*[Gu Gd Gf;I 0 0];
indcd = [Rt.controls;Rt.disturbances]
indfn = [Rt.faults;Rt.noise]
@test iszero(vcat(Rt.sys...)[:,indcd]; atol) &&
      iszero(vcat(Rt.sys...)[:,indfn]-vcat(R.sys...); atol) &&
      fdif2ngap(R,SFDI)[1] ≈ info.gap ≈ [3,Inf]
```

5.6   Solving the Exact Model-Matching Problem

Let $M_r(\lambda)$ be a given $q \times m_f$ TFM of a stable and proper reference model specifying the desired input–output behaviour from the faults to residuals as $\mathbf{r}(\lambda) = M_r(\lambda)\mathbf{f}(\lambda)$. Using the factorized representation $Q(\lambda) = \overline{Q}_1(\lambda)Q_1(\lambda)$ in (5.9), it follows that the *exact model-matching problem* (EMMP) formulated in Sect. 3.5.5 is solvable for the system (3.2) with $w \equiv 0$ if it is solvable for the reduced system (5.11) with $w \equiv 0$. The following corollary to Theorem 3.14 provides the solvability conditions of the EMMP in terms of the reduced system (5.11).

Corollary 5.9 *For the system (3.2) with $w \equiv 0$ and a given reference model $M_r(\lambda)$, the EMMP is solvable if and only if the EMMP is solvable for the reduced system (5.11) with $w \equiv 0$, or equivalently, the following condition is fulfilled:*

$$\text{rank } \overline{G}_f(\lambda) = \text{rank } \begin{bmatrix} \overline{G}_f(\lambda) \\ M_r(\lambda) \end{bmatrix}. \tag{5.51}$$

The case when $M_r(\lambda)$ is diagonal and invertible corresponds to a strong FDI requirement. The solvability condition for this case is the same as the solvability condition resulting from (5.51) for the case when $M_r(\lambda)$ has full column rank m_f.

Corollary 5.10 *For the system (3.2) with $w \equiv 0$ and a given reference model $M_r(\lambda)$ with rank $M_r(\lambda) = m_f$, the EMMP is solvable if and only if the reduced system (5.11) with $w \equiv 0$ is strongly isolable, or equivalently, the following condition is fulfilled:*

$$\text{rank } \overline{G}_f(\lambda) = m_f. \tag{5.52}$$

Remark 5.12 For a strongly isolable system (3.2) with $w \equiv 0$, the left invertibility condition (5.52) is a necessary and sufficient condition for the solvability of the EMMP for an arbitrary $M_r(\lambda)$. $\square$

For the solution of the EMMP, we present a synthesis procedure, which employs the factorized representation $Q(\lambda) = \overline{Q}_1(\lambda)Q_1(\lambda)$ in (5.9), where $Q_1(\lambda)$ is a minimal proper left nullspace basis of $G(\lambda)$ defined in (5.2). The factor $\overline{Q}_1(\lambda)$ can be determined in the product form

$$\overline{Q}_1(\lambda) = Q_3(\lambda)Q_2(\lambda),$$

where $Q_2(\lambda)$ is a solution, possibly of least McMillan degree, of the linear rational matrix equation

$$Q_2(\lambda)\overline{G}_f(\lambda) = M_r(\lambda), \tag{5.53}$$

while the diagonal updating factor $Q_3(\lambda) := M(\lambda)$ is determined such that

$$Q(\lambda) = Q_3(\lambda)Q_2(\lambda)Q_1(\lambda)$$

is stable and proper. The computation of $Q_3(\lambda)$ is necessary only if $Q_2(\lambda)Q_1(\lambda)$ is not proper or is unstable. The **Procedure EMM**, given below, summarizes the main computational steps for solving the EMMP.

Procedure EMM: Exact model-matching synthesis of FDI filters

Inputs : $\{G_u(\lambda), G_d(\lambda), G_f(\lambda)\}$, $M_r(\lambda)$
Outputs: $Q(\lambda)$, $M(\lambda)$

(1) Compute a minimal proper basis $Q_1(\lambda)$ for the left nullspace of $G(\lambda)$ defined in (5.2); set $Q(\lambda) = Q_1(\lambda)$ and compute $R_f(\lambda) = Q_1(\lambda)\begin{bmatrix} G_f(\lambda) \\ 0 \end{bmatrix}$.

(2) Solve, for the least McMillan degree solution $Q_2(\lambda)$, the linear rational matrix equation $Q_2(\lambda)R_f(\lambda) = M_r(\lambda)$. **Exit** if no solution exists. Otherwise update $Q(\lambda) \leftarrow Q_2(\lambda)Q(\lambda)$.

(3) Determine a diagonal, stable, proper and invertible $Q_3(\lambda) := M(\lambda)$ such that $M(\lambda)Q(\lambda)$ is stable and proper; update $Q(\lambda) \leftarrow Q_3(\lambda)Q(\lambda)$.

To perform the computation at Step (2), a state-space realization-based algorithm to compute least McMillan degree solutions of linear rational matrix equations is described in Sect. 10.3.7. For the determination of the diagonal updating factor $M(\lambda)$ at Step (3), coprime factorization techniques can be used, as described in Sect. 9.1.6. The underlying state-space realization based algorithms are presented in Sect. 10.3.5.

Remark 5.13 The solution of the EMMP can be alternatively performed by determining $Q(\lambda)$ as $Q(\lambda) = Q_2(\lambda)Q_1(\lambda)$, where $Q_1(\lambda)$ is a least McMillan degree solution of the linear rational matrix equation

$$Q_1(\lambda)\begin{bmatrix} G_u(\lambda) & G_d(\lambda) & G_f(\lambda) \\ I_{m_u} & 0 & 0 \end{bmatrix} = \begin{bmatrix} 0 & 0 & M_r(\lambda) \end{bmatrix}. \tag{5.54}$$

The diagonal updating factor $Q_2(\lambda) := M(\lambda)$ is determined to ensure that $Q(\lambda)$ is proper and stable. $\square$

Example 5.12 In Example 5.10, we solved an EFDIP for a system with triplex sensor redundancy. To solve an EMMP for the same system, we use the resulting $R_f(s)$ to define the reference model

$$M_r(s) := R_f(s) = \begin{bmatrix} 0 & 1 & -1 \\ -1 & 0 & 1 \\ 1 & -1 & 0 \end{bmatrix}.$$

Using **Procedure EMM**, we determine first a left nullspace basis $Q_1(s) = N_l(s)$, with $N_l(s)$ given in (5.42). The corresponding $R_f(s)$ (given in (5.43)) is

$$R_f(s) = \begin{bmatrix} 1 & -1 & 0 \\ 0 & 1 & -1 \end{bmatrix}.$$

The solvability condition can be easily checked

$$\operatorname{rank} R_f(s) = \operatorname{rank} \begin{bmatrix} R_f(s) \\ M_r(s) \end{bmatrix} = 2.$$

We solve for $Q_2(s)$

$$Q_2(s) R_f(s) = M_r(s)$$

and obtain

$$Q_2(s) = \begin{bmatrix} 0 & 1 \\ -1 & -1 \\ 1 & 0 \end{bmatrix}.$$

Finally, we have

$$Q(s) = Q_2(s) Q_1(s) = \begin{bmatrix} 0 & 1 & -1 & 0 & \cdots & 0 \\ -1 & 0 & 1 & 0 & \cdots & 0 \\ 1 & -1 & 0 & 0 & \cdots & 0 \end{bmatrix}.$$

We obtain the same result by solving directly (5.54) for $Q(s) = Q_1(s)$.

The script **Ex5_12c.jl** in Listing 5.5 solves the EMMP considered in this example and calls the synthesis function **emmsyn**, which implements the **Procedure EMM**. The script also includes the verification of results by checking the synthesis conditions. A more detailed version of this script is **Ex5_12.jl** (not listed), which closely follows the computational steps of the **Procedure EMM**.   ◊

Listing 5.5 Script **Ex5_12c.jl** to solve the EMMP of Example 5.12 using **Procedure EMM**

```julia
using FaultDetectionTools, DescriptorSystems, Test

# Example 5.12 - Solution of an EMMP
println("Example 5.12")

# enter output and fault vector dimensions
p = 3; mf = 3;
# generate random dimensions for system order and input vectors
nu = Int(floor(1+4*rand())); mu = Int(floor(1+4*rand()));
```

```
nd = Int(floor(1+4*rand())); md = Int(floor(1+4*rand()));

# define random Gu(s) and Gd(s) with triplex sensor redundancy
# and Gf(s) for triplex sensor faults
Gu = ones(3,1)*rss(nu,1,mu);  # enter Gu(s) in state-space form
Gd = ones(3,1)*rss(nd,1,md);  # enter Gd(s) in state-space form
Gf = eye(3);                  # enter Gf(s) for sensor faults
atol = 1.e-7;                 # tolerance for rank tests

# build model with faults sysf = [Gu Gd Gf]
sysf = fdimodset([Gu Gd Gf], c = 1:mu, d = mu.+(1:md),
                  f = (mu+md).+(1:mf))

# enter reference model
Mr = FDFilterIF(dss([ 0 1 -1; -1 0 1; 1 -1 0]); mf)

# solve an exact model-matching problem using emmsyn
Q, R, info = emmsyn(sysf,Mr; atol);

# check synthesis conditions
# Q*[Gu Gd;I 0] = 0 and Q*[Gf; 0] = Mr
Rt = fdIFeval(Q,sysf; atol) # form Q*[Gu Gd Gf;I 0 0];
@test iszero(Rt.sys[:,[Rt.controls;Rt.disturbances]]; atol) &&
       iszero(Rt.sys[:,Rt.faults]-Mr.sys; atol)
```

In what follows, we discuss the solution of the EMMP for strongly isolable systems. According to Remark 5.12, the solvability of the EMMP is automatically guaranteed in this case, regardless the choice of the reference model $M_r(\lambda)$. An important particular case in practical applications is when $M_r(\lambda)$ is diagonal, stable, proper and invertible. In this case, the solution of the EMMP allows the detection and isolation of up to m_f simultaneous faults, and thus is also a solution of a strong EFDIP (i.e., for an identity structure matrix). Fault reconstruction (or fault estimation) problems can be addressed in this way by choosing $M_r(\lambda) = I_{m_f}$. For the solution of the EMMP for strongly isolable systems, we develop a specialized synthesis procedure, which also addresses the least order synthesis aspect for a regularity-enforcing admissibility condition.

Using the factorized representation $Q(\lambda) = \overline{Q}_1(\lambda)Q_1(\lambda)$ in (5.9), the factor $\overline{Q}_1(\lambda)$ can be determined in the product form

$$\overline{Q}_1(\lambda) = \overline{Q}_2(\lambda)Q_2(\lambda),$$

where $Q_2(\lambda)$ is computed such that

$$\widetilde{G}_f(\lambda) := Q_2(\lambda)\overline{G}_f(\lambda)$$

is invertible. This regularization step is always possible, since, for a strongly isolable system, $\overline{G}_f(\lambda)$ is left invertible (see Remark 5.12). The simplest choice of $Q_2(\lambda)$ is a constant (e.g., orthogonal) projection matrix, which simply selects m_f linearly independent rows of $\overline{G}_f(\lambda)$. A more involved choice is based on an *admissibility condition*, which enforces the invertibility of $\widetilde{G}_f(\lambda)$ simultaneously with the least dynamical orders of $Q_2(\lambda)Q_1(\lambda)$ and $\widetilde{G}_f(\lambda)$. Such a choice of $Q_2(\lambda)$ is possible using minimal dynamic cover techniques (see Sect. 7.5).

The factor $\overline{Q}_2(\lambda)$ can be determined in the form

$$\overline{Q}_2(\lambda) = Q_4(\lambda)Q_3(\lambda) ,$$

where $Q_3(\lambda) = M_r(\lambda)\widetilde{G}_f^{-1}(\lambda)$ and $Q_4(\lambda) := M(\lambda)$ is chosen a diagonal, stable, proper and invertible TFM, to ensure that the resulting final filter

$$Q(\lambda) = Q_4(\lambda)Q_3(\lambda)Q_2(\lambda)Q_1(\lambda)$$

is stable and proper. The updating factor $M(\lambda)$ can be determined using stable and proper coprime factorization techniques (see Sects. 9.1.6 and 10.3.5).

The above synthesis method is sometimes called in the literature the *inversion-based method*. The **Procedure EMMS**, given below, formalizes the computational steps of the inversion-based synthesis method to solve the EMMP for strongly isolable systems.

Procedure EMMS: Exact model-matching synthesis of strong FDI filters

Inputs : $\{G_u(\lambda), G_d(\lambda), G_f(\lambda)\}$, $M_r(\lambda)$
Outputs: $Q(\lambda)$, $M(\lambda)$

(1) Compute a proper minimal basis $Q_1(\lambda)$ for the left nullspace of $G(\lambda)$ defined in (5.2); set $Q(\lambda) = Q_1(\lambda)$ and compute $R_f(\lambda) = Q_1(\lambda)\begin{bmatrix} G_f(\lambda) \\ 0 \end{bmatrix}$.

    Exit if rank $R_f(\lambda) < m_f$ (no solution).

(2) Choose $Q_2(\lambda)$ such that $Q_2(\lambda)R_f(\lambda)$ is invertible and $Q_2(\lambda)[R_f(\lambda)\ Q(\lambda)]$ has least order; update $Q(\lambda) \leftarrow Q_2(\lambda)Q(\lambda)$ and $R_f(\lambda) \leftarrow Q_2(\lambda)R_f(\lambda)$.

(3) With $Q_3(\lambda) = M_r(\lambda)R_f^{-1}(\lambda)$, update $Q(\lambda) \leftarrow Q_3(\lambda)Q(\lambda)$.

(4) Determine diagonal, stable, proper and invertible $Q_4(\lambda) := M(\lambda)$ such that $M(\lambda)Q(\lambda)$ is stable and proper; update $Q(\lambda) \leftarrow Q_4(\lambda)Q(\lambda)$.

Example 5.13 Consider a continuous-time   system   with the transfer function matrices

$$G_u(s) = \begin{bmatrix} \dfrac{s}{s^2+3s+2} & \dfrac{1}{s+2} \\ \dfrac{s}{s+1} & 0 \\ 0 & \dfrac{1}{s+2} \end{bmatrix} , \qquad G_d(s) = 0, \qquad G_f(s) = \begin{bmatrix} \dfrac{s}{s^2+3s+2} & \dfrac{1}{s+2} \\ \dfrac{s}{s+1} & 0 \\ 0 & \dfrac{1}{s+2} \end{bmatrix}$$

for which we want to solve the EMMP with the reference model

$$M_r(s) = \begin{bmatrix} 1 & 0 \\ 0 & 1 \end{bmatrix} .$$

Using **Procedure EMMS**, we choose at Step (1) the left nullspace basis $Q_1(s) = [\, I \ -G_u(s)\,]$ and initialize $Q(s) = Q_1(s)$ for which the corresponding $R_f(s)$ is simply $R_f(s) = G_f(s)$. $R_f(s)$ has full column rank (thus is left invertible), and therefore, the EMMP has a solution. Since $R_f(s)$ has zeros in the origin and at infinity, the existence condition of Lemma 9.5 for a stable solution $\overline{Q}_1(s)$ of $\overline{Q}_1(s)R_f(s) = M_r(s)$ is not fulfilled.

At Step (2), we choose $Q_2(s)$ such that $Q_2(s)[\, R_f(s) \ Q(s)\,]$ has a least order. This can be achieved with the simple choice

$$Q_2(s) = \begin{bmatrix} 0 & 1 & 0 \\ 0 & 0 & 1 \end{bmatrix}$$

and, after updating $Q(s) \leftarrow Q_2(s)Q(s)$ and $R_f(s) \leftarrow Q_2(s)R_f(s)$, we obtain

$$Q(s) = \begin{bmatrix} 0 & 1 & 0 & -\dfrac{s}{s+1} & 0 \\[2mm] 0 & 0 & 1 & 0 & -\dfrac{1}{s+2} \end{bmatrix}, \qquad R_f(s) = \begin{bmatrix} \dfrac{s}{s+1} & 0 \\[2mm] 0 & \dfrac{1}{s+2} \end{bmatrix}.$$

At Step (3), the resulting

$$Q_3(s) := M_r(s)R_f^{-1}(s) = \begin{bmatrix} \dfrac{s+1}{s} & 0 \\[2mm] 0 & s+2 \end{bmatrix}$$

is improper and unstable, and therefore, the updated $Q(s) \leftarrow \widetilde{Q}(s) := Q_3(s)Q(s)$

$$\widetilde{Q}(s) = \begin{bmatrix} 0 & \dfrac{s+1}{s} & 0 & -1 & 0 \\[2mm] 0 & 0 & s+2 & 0 & -1 \end{bmatrix}$$

is improper and unstable as well.

Finally, at Step (4), we determine a diagonal

$$Q_4(s) = M(s) = \begin{bmatrix} \dfrac{s}{s+1} & 0 \\[2mm] 0 & \dfrac{1}{s+1} \end{bmatrix},$$

which ensures the properness and stability of the solution. The final $Q(s) = Q_4(s)\widetilde{Q}(s)$ is

$$Q(s) = \begin{bmatrix} 0 & 1 & 0 & -\dfrac{s}{s+1} & 0 \\[2mm] 0 & 0 & \dfrac{s+2}{s+1} & 0 & -\dfrac{1}{s+1} \end{bmatrix}.$$

The McMillan degree of this fault detection filter is 2 and is the least achievable one among all stable and proper filters, which solve the EMMP. Note that the presence of the zero $s = 0$ in $M(s)M_r(s)$ is unavoidable for the existence of a stable solution. It follows, that while a constant fault f_2 is strongly detectable, a constant fault f_1 is only detectable during transients.

An alternative way, see Remark 5.13, to determine a least order solution of the considered EMMP is to directly solve, for the least order solution $Q_1(s)$, the linear rational matrix equation

$$Q_1(s) \begin{bmatrix} G_u(s) & G_f(s) \\ I & 0 \end{bmatrix} = \begin{bmatrix} 0 & M_r(s) \end{bmatrix}$$

and then determine $Q_2(s)$ (as above $Q_4(s)$ at Step (4) of **Procedure EMMS**) to obtain a stable and proper $Q(s) := Q_2(s)Q_1(s)$.

The script **Ex5_13c.jl** in Listing 5.6 solves the EMMP considered in this example and calls the synthesis function **emmsyn**, which implements the **Procedure EMMS**. The script also includes the verification of results by checking the synthesis conditions. A more detailed version of this script is **Ex5_13.jl** (not listed), which closely follows the computational steps of the **Procedure EMMS**. The alternative direct approach is implemented in the script **Ex5_13a.jl** (not listed).   ◊

Listing 5.6 Script **Ex5_13c.jl** to solve the EMMP of Example 5.13 using **Procedure EMMS**

```julia
using FaultDetectionTools, DescriptorSystems, Test

# Example 5.13 - Solution of an EMMP
println("Example 5.13")

# define s as an improper transfer function
s = rtf('s');
# enter Gu(s), Gf(s) and Mr(s)
Gu = [s/(s^2+3*s+2) 1/(s+2);
      s/(s+1) 0;
      0 1/(s+2)];
Mr = dss(eye(2));                          # enter Mr(s)
p, mu = size(Gu); mf = mu

# build the synthesis model with additive faults
sysf = fdimodset(dss(Gu), c = 1:mu, f = 1:mu);

# enter reference model
Mr = fdimodset(dss(eye(mf)), f = 1:mf)

atol = 1.e-7                  # tolerance for rank tests
sdeg = -1                     # set stability degree

# solve an exact model-matching problem using emmsyn
Q, R, info = emmsyn(sysf, Mr; atol, sdeg, minimal = false);

# display results
println("Q = $(dss2rm(Q.sys; atol))")
```

```
println("M = $(dss2rm(info.M; atol))")

# check synthesis conditions:
# Q*[Gu Gd;I 0] = 0 and Q*[Gf; 0] = M*Mr
Rt = fdIFeval(Q,sysf; atol) # form Q*[Gu Gd Gf;I 0 0];
@test iszero(Rt.sys[:,[Rt.controls;Rt.disturbances]]; atol) &&
      iszero(Rt.sys[:,Rt.faults]-info.M*Mr.sys; atol)
```

5.7 Solving the Approximate Model-Matching Problem

Using the factorized representation $Q(\lambda) = \overline{Q}_1(\lambda)Q_1(\lambda)$ in (5.9), with $Q_1(\lambda)$ stable and proper, allows to reformulate the *approximate model-matching problem* (AMMP), formulated in Sect. 3.5.6 for the system (3.2), in terms of the reduced system (5.11), with both $\overline{G}_f(\lambda)$ and $\overline{G}_w(\lambda)$ assumed to be stable and proper (this can be always enforced by a suitable choice of $Q_1(\lambda)$). The following corollary to Proposition 3.1 gives a sufficient condition for the solvability of the AMMP in terms of the reduced system (5.11).

Corollary 5.11 *For the system (3.2) and a given $M_r(\lambda)$, the AMMP is solvable if the EMMP is solvable for the reduced system (5.11).*

According to Remark 5.12, for a strongly isolable system (3.2), the left invertibility condition (5.52) (i.e., rank $\overline{G}_f(\lambda) = m_f$) is, therefore, a sufficient condition for the solvability of the AMMP.

To solve the AMMP for the reduced system (5.11), a standard model-matching problem can be formulated to determine the optimal stable and proper solution $\overline{Q}_1(\lambda)$ of the norm-minimization problem

$$\left\| \overline{Q}_1(\lambda)\left[\,\overline{G}_f(\lambda)\ \overline{G}_w(\lambda)\,\right] - \left[\,M_r(\lambda)\ 0\,\right] \right\| = \text{minimum}\,. \tag{5.55}$$

With $\overline{F}(\lambda) := [\,M_r(\lambda)\ 0\,]$, $\overline{G}(\lambda) := [\,\overline{G}_f(\lambda)\ \overline{G}_w(\lambda)\,]$, and the error function

$$\overline{\mathcal{E}}(\lambda) := \overline{F}(\lambda) - X(\lambda)\overline{G}(\lambda)\,, \tag{5.56}$$

a solution of the AMMP can be aimed by solving either a $\mathcal{H}_\infty$- or $\mathcal{H}_2$-*model-matching problem* (MMP) (see Sect. 9.1.10) to determine $\overline{Q}_1(\lambda)$ as the stable and proper optimal solution $X(s)$, which minimizes $\|\overline{\mathcal{E}}(\lambda)\|_\infty$ or $\|\overline{\mathcal{E}}(\lambda)\|_2$, respectively. Sufficient conditions for the solvability of the $\mathcal{H}_\infty$-MMP and $\mathcal{H}_2$-MMP are given in Lemmas 9.6 and 9.7, respectively. These sufficient conditions require that $\overline{G}(\lambda)$ has no zeros in $\partial\mathbb{C}_s$. However, these conditions are not necessary for the solvability of the AMMP, and therefore, we define the *standard case*, when $\overline{G}(\lambda)$ has no zeros in $\partial\mathbb{C}_s$, and the *nonstandard case*, when $\overline{G}(\lambda)$ has zeros in $\partial\mathbb{C}_s$.

Solution procedures for the standard case are presented in Sect. 9.1.10 and determine optimal solutions, which are stable and proper. The same procedures applied in the nonstandard case determine "optimal" solutions, which, in general, have poles in $\partial\mathbb{C}_s$, and thus are unstable or improper. If $X(\lambda)$ is such a solution, then a diagonal, stable, proper and invertible updating factor $M(\lambda)$ can be determined such that the filter $\overline{Q}_1(\lambda) := M(\lambda)X(\lambda)$ is stable and proper, and achieves the (suboptimal) performance level $\gamma_{sub} := \|M(\lambda)\overline{\mathcal{E}}(\lambda)\|$. Let $\overline{X}(\lambda)$ be an "optimal" solution (possibly unstable or improper), which minimizes the weighted error norm $\|M(\lambda)\overline{\mathcal{E}}(\lambda)\|$ and let γ_{opt} be the corresponding optimal performance level. Since $\gamma_{opt} \leq \gamma_{sub}$, the difference $\gamma_{sub} - \gamma_{opt}$ is an indicator of the achieved degree of suboptimality of the resulting filter $\overline{Q}_1(\lambda)$ for the weighted norm-minimization problem corresponding to the updated reference model $M(\lambda)M_r(\lambda)$. The choice of a diagonal $M(\lambda)$ is instrumental to preserve the zero-nonzero structure of $M_r(\lambda)$.

Example 5.14 Consider the $\mathcal{H}_\infty$-MMP in a continuous-time setting with

$$\overline{G}(s) = \left[\,\overline{G}_f(s)\,|\,\overline{G}_w(s)\,\right] := \left[\begin{array}{c|c} \dfrac{1}{s+1} & \dfrac{1}{s+2} \end{array}\right], \quad \overline{F}(s) = \left[\,M_r(s)\,|\,0\,\right] = \left[\begin{array}{c|c} \dfrac{1}{s+3} & 0 \end{array}\right].$$

This problem is nonstandard, because $\overline{G}(s)$ has a zero at infinity. Ignoring momentarily this aspect, we can formally use the solution approach in Sect. 9.1.10 relying on the quasi-co-outer–inner factorization of $\overline{G}(s)$ followed by the solution of a 2-block $\mathcal{H}_\infty$-least-distance problem. We obtain the $\mathcal{H}_\infty$-optimal solution

$$X_\infty(s) = \frac{0.041587(s+13.65)(s+2)(s+1)}{(s+3)(s+1.581)},$$

which is improper. The optimal error norm is $\gamma_{\infty,opt} := \|\overline{F}(s) - X_\infty(s)\overline{G}(s)\|_\infty = 0.1745$, thus finite. With $M(s) = \frac{1}{s+1}$, we obtain a proper candidate filter

$$\overline{Q}_1(s) = M(s)X_\infty(s) = \frac{0.041587(s+13.65)(s+2)}{(s+3)(s+1.581)},$$

for which $\gamma_{sub} := \|M(s)\overline{F}(s) - \overline{Q}_1(s)\overline{G}(s)\|_\infty = 0.1522$. The optimal solution $\overline{X}_\infty(s)$ of the $\mathcal{H}_\infty$-MMP, which minimizes $\|M(s)\overline{F}(s) - X(s)\overline{G}(s)\|_\infty$, leads to an optimal value of $\gamma_{opt} = \|M(s)\overline{F}(s) - \overline{X}_\infty(s)\overline{G}(s)\|_\infty = 0.1491$. As expected, the optimal solution $\overline{X}_\infty(s)$ is improper. Since $\gamma_{sub} - \gamma_{opt} = 0.0031$, the degree of suboptimality of the proper and stable filter $M(s)X_\infty(s)$ with respect to the optimal (but improper) solution $\overline{X}_\infty(s)$ appears to be acceptable.   ◊

Example 5.15 We can also solve the $\mathcal{H}_2$-MMP for Example 5.14. Although this problem is nonstandard, still cancellations of infinite poles and zeros make that the resulting $\mathcal{H}_2$-optimal solution is proper

$$X_2(s) = \frac{0.54572(s+2)(s+1)}{(s+3)(s+1.581)}.$$

The corresponding optimal performance is $\gamma_{opt} = \|\overline{F}(s) - X_2(s)\overline{G}(s)\|_2 = 0.2596$. Interestingly, the $\mathcal{H}_\infty$ error norm of the $\mathcal{H}_2$-optimal solution is $\|\overline{F}(s) - X_2(s)\overline{G}(s)\|_\infty = 0.1768$, which is only marginally worse than $\gamma_{\infty,opt}$, the optimal performance of the improper $\mathcal{H}_\infty$-optimal solution $X_\infty(s)$. Thus, $X_2(s)$ can be considered an acceptable $\mathcal{H}_\infty$-suboptimal solution. $\qquad\qquad\qquad\qquad\qquad\qquad\qquad\qquad\qquad\qquad\quad\diamond$

In what follows we develop a general synthesis procedure for solving AMMPs relying on the solution of $\mathcal{H}_{2/\infty}$-MMPs. We assume that the reference model $M_r(\lambda)$ has been chosen to capture a fault estimation or, equivalently, a strong fault isolation setup. Often, $M_r(\lambda)$ is chosen diagonal, and even equal to the identity matrix, when trying to solve a fault estimation problem. Therefore, $M_r(\lambda)$ will be assumed to be a stable and invertible TFM. In the case of an EMMP (when $w \equiv 0$), a necessary and sufficient condition for the existence of a proper and stable solution (possibly with an updated reference model $M(\lambda)M_r(\lambda)$, with $M(\lambda)$ a diagonal, stable and invertible factor) is that $\overline{G}_f(\lambda)$ has full column rank (i.e., left invertible) (see Corollary 5.10). For simplicity, we will assume that this condition is fulfilled and provide a synthesis procedure, which computes an optimal solution in the standard case or a suboptimal solution of a weighted problem in the nonstandard case. As it will be apparent, the final fault detection filter intrinsically results in a factored form as in (5.1), which automatically leads to a synthesis procedure relying on successive updating of partially synthesized filters.

Let $\ell \geq m_f$ be the rank of the $(p - r_d) \times (m_f + m_w)$ TFM $\overline{G}(\lambda)$. We take $\overline{Q}_1(\lambda) = \overline{Q}_2(\lambda)Q_2(\lambda)$, where $Q_2(\lambda)$ is an $\ell \times (p - r_d)$ proper TFM chosen to ensure that $Q_2(\lambda)\overline{G}(\lambda)$ has full row rank ℓ. If $\ell < p - r_d$ (i.e., $\overline{G}(\lambda)$ has not a full row rank), a possible choice of $Q_2(\lambda)$ is that one, which simultaneously minimizes the McMillan degree of $Q_2(\lambda)Q_1(\lambda)$ (see Sect. 7.5). A simpler choice with $Q_2(\lambda)$ a constant (e.g., orthogonal) matrix is also always possible. If $\ell = p - r_d$, then $Q_2(\lambda) = I_\ell$ can be chosen.

The next step is standard in solving $\mathcal{H}_{2/\infty}$-MMPs and consists in compressing the full row rank TFM $\overline{G}(\lambda)$ to a full column rank (thus invertible) TFM. For this, we compute an extended quasi-co-outer–co-inner factorization in the form

$$Q_2(\lambda)\overline{G}(\lambda) = [\, G_o(\lambda)\ 0\,] \begin{bmatrix} G_{i,1}(\lambda) \\ G_{i,2}(\lambda) \end{bmatrix} := [\, G_o(\lambda)\ 0\,]G_i(\lambda), \qquad (5.57)$$

where the quasi-co-outer part $G_o(\lambda)$ is invertible and has only zeros in $\overline{\mathbb{C}}_s$, and $G_i(\lambda)$ is a square co-inner factor (i.e., $G_i(\lambda)G^\sim(\lambda) = I$). The factor $\overline{Q}_2(\lambda)$ is determined in the product form

$$\overline{Q}_2(\lambda) = Q_5(\lambda)Q_4(\lambda)Q_3(\lambda),$$

with $Q_3(\lambda) = G_o^{-1}(\lambda)$ and $Q_4(\lambda)$, the optimal solution, which minimizes the error norm $\|\widetilde{\mathcal{E}}(\lambda)\|_{2/\infty}$, with $\widetilde{\mathcal{E}}(\lambda)$ defined as

$$\widetilde{\mathcal{E}}(\lambda) := \overline{\mathcal{E}}(\lambda)G_i^\sim(\lambda) = [\, \widetilde{F}_1(\lambda) - Q_4(\lambda)\ \widetilde{F}_2(\lambda)\,], \qquad (5.58)$$

where $\widetilde{F}_1(\lambda) := \overline{F}(\lambda)G_{i,1}^{\sim}(\lambda)$ and $\widetilde{F}_2(\lambda) := \overline{F}(\lambda)G_{i,2}^{\sim}(\lambda)$. The factor $Q_5(\lambda) := M(\lambda)$ is chosen to enforce the stability and properness of the final filter

$$Q(\lambda) = Q_5(\lambda)Q_4(\lambda)Q_3(\lambda)Q_2(\lambda)Q_1(\lambda) \,. \tag{5.59}$$

The determination of a stable and proper $Q_4(\lambda)$, which minimizes $\|\widetilde{\mathcal{E}}(\lambda)\|_{2/\infty} = \|\overline{\mathcal{E}}(\lambda)\|_{2/\infty}$ is a $\mathcal{H}_{2/\infty}$-*least-distance problem* ($\mathcal{H}_{2/\infty}$-LDP) for which solution methods are given in Sect. 9.1.10.

The overall filter $Q(\lambda)$ in (5.59) can be alternatively expressed in the form $Q(\lambda) = Q_5(\lambda)Q_4(\lambda)\overline{Q}(\lambda)$, where $\overline{Q}(\lambda) := Q_3(\lambda)Q_2(\lambda)Q_1(\lambda)$ can be interpreted as a partial synthesis. The TFMs of the internal form corresponding to this filter are

$$\begin{aligned}
[\,\overline{R}_f(\lambda)\ \overline{R}_w(\lambda)\,] &:= Q_3(\lambda)Q_2(\lambda)[\,\overline{G}_f(\lambda)\ \overline{G}_w(\lambda)\,] \\
&= [\,I_\ell\ \ 0\,]\begin{bmatrix} G_{i,1}(\lambda) \\ G_{i,2}(\lambda) \end{bmatrix} = G_{i,1}(\lambda)
\end{aligned} \tag{5.60}$$

and thus, are parts of the (stable) co-inner TFM $G_{i,1}(\lambda)$.

Generally, $\overline{Q}(\lambda)$ contains among its poles the zeros of $G_o(\lambda)$. This is also true for the product $Q_4(\lambda)\overline{Q}(\lambda)$, where $Q_4(\lambda)$ is the stable and proper solution of the $\mathcal{H}_{2/\infty}$-LDP. In the standard case (i.e., when $\overline{G}(\lambda)$ has no zeros in $\partial\mathbb{C}_s$), $G_o(\lambda)$ has only stable finite zeros and no infinite zeros, and therefore, $\overline{Q}(\lambda)$ results stable, provided $Q_2(\lambda)Q_1(\lambda)$ is stable. In this case, we take simply $Q_5(\lambda) = I$ and the updating factor $M(\lambda) = I$. In the nonstandard case (i.e., when $\overline{G}(\lambda)$ has zeros in $\partial\mathbb{C}_s$), the quasi-outer factor $G_o(\lambda)$ will have these zeros in $\partial\mathbb{C}_s$ too. Therefore, $\overline{Q}(\lambda)$ results unstable or improper, and we choose a diagonal, stable, proper and invertible $M(\lambda) := Q_5(\lambda)$, such that the final $Q(\lambda)$ is proper and stable.

The computation of suitable $M(\lambda)$ can be done using LCF-based techniques as described in Sect. 9.1.6. The choice of $M(\lambda)$ can be performed such that $\|M(\lambda)\widetilde{\mathcal{E}}(\lambda)\|_{2/\infty} \approx \|\widetilde{\mathcal{E}}(\lambda)\|_{2/\infty}$ and $M(\lambda)$ has the least possible McMillan degree. For example, to ensure properness or strict properness, $M(\lambda)$ can be chosen diagonal with the diagonal terms $M_j(\lambda)$, $j = 1, \ldots, m_f$ having the form

$$M_j(s) = \frac{1}{(\tau s + 1)^{k_j}} \quad \text{or} \quad M_j(z) = \frac{1}{z^{k_j}}\,,$$

for continuous- or discrete-time settings, respectively. Notice that both above factors have unit $\mathcal{H}_\infty$-norm.

The **Procedure AMMS**, given in what follows, formalizes the computational steps of the described synthesis method for a strongly isolable system and an invertible reference model $M_r(\lambda)$. This procedure can also be interpreted as an enhanced version of **Procedure EMMS**.

Procedure AMMS: Approximate model-matching synthesis of FDI filters

Inputs : $\{G_u(\lambda), G_d(\lambda), G_f(\lambda), G_w(\lambda)\}$, invertible $M_r(\lambda)$
Outputs: $Q(\lambda), R_f(\lambda), R_w(\lambda), M(\lambda)$

(1) Compute a minimal proper basis $Q_1(\lambda)$ for the left nullspace of $G(\lambda)$ defined
     in (5.2); set $Q(\lambda) = Q_1(\lambda)$ and compute

$$\left[\, R_f(\lambda) \,\middle|\, R_w(\lambda) \,\right] = Q_1(\lambda) \left[\begin{array}{c|c} G_f(\lambda) & G_w(\lambda) \\ 0 & 0 \end{array} \right]$$

Exit if rank $R_f(\lambda) < m_f$ (no solution).
(2) Choose $Q_2(\lambda)$ such that $Q_2(\lambda) \left[\, R_f(\lambda) \,\middle|\, R_w(\lambda) \,\right]$ has maximal full row rank
     and $Q_2(\lambda)Q(\lambda)$ has least McMillan degree; update $Q(\lambda) \leftarrow Q_2(\lambda)Q(\lambda)$,
     $R_f(\lambda) \leftarrow Q_2(\lambda)R_f(\lambda)$ and $R_w(\lambda) \leftarrow Q_2(\lambda)R_w(\lambda)$.
(3) Compute the extended quasi-co-outer–co-inner factorization

$$\left[\, R_f(\lambda) \,\middle|\, R_w(\lambda) \,\right] = [\, G_o(\lambda)\ 0\,] \left[\begin{array}{c} G_{i,1}(\lambda) \\ G_{i,2}(\lambda) \end{array} \right].$$

With $Q_3(\lambda) = G_o^{-1}(\lambda)$, update $Q(\lambda) \leftarrow Q_3(\lambda)Q(\lambda)$ and compute
$\widetilde{F}_1(\lambda) = \left[\, M_r(\lambda)\ 0\,\right] G_{i,1}^\sim(\lambda)$, $\widetilde{F}_2(\lambda) = \left[\, M_r(\lambda)\ 0\,\right] G_{i,2}^\sim(\lambda)$.
Set $\left[\, R_f(\lambda) \,\middle|\, R_w(\lambda) \,\right] = G_{i,1}(\lambda)$.
(4) Compute the solution $Q_4(\lambda)$ of the $\mathcal{H}_{2/\infty}$-LDP

$$\min_{Q_4(\lambda) \in \mathcal{H}_\infty} \left\| \left[\, \widetilde{F}_1(\lambda) - Q_4(\lambda)\ \ \widetilde{F}_2(\lambda) \,\right] \right\|_{2/\infty};$$

update $Q(\lambda) \leftarrow Q_4(\lambda)Q(\lambda)$, $R_f(\lambda) \leftarrow Q_4(\lambda)R_f(\lambda)$ and
$R_w(\lambda) \leftarrow Q_4(\lambda)R_w(\lambda)$.
(5) Determine diagonal, stable, proper and invertible $Q_5(\lambda) := M(\lambda)$ such that
     $M(\lambda)Q(\lambda)$ is stable and proper; update $Q(\lambda) \leftarrow Q_5(\lambda)Q(\lambda)$,
     $R_f(\lambda) \leftarrow Q_5(\lambda)R_f(\lambda)$ and $R_w(\lambda) \leftarrow Q_5(\lambda)R_w(\lambda)$.

Remark 5.14 The main advantage of the **Procedure AMMS** over alternative meth-
ods, as—for example, solving $\mathcal{H}_{2/\infty}$ filter synthesis problems using standard $\mathcal{H}_{2/\infty}$
optimization procedures, lies in the possibility to easily handle frequently encoun-
tered nonstandard cases (e.g., strictly proper systems). For such a case, the standard
procedures would either fail without providing any useful result, or determine unprac-
ticable solutions (e.g., with very fast dynamics or excessively large or small gains).
In contrast, the described method produces the weighting TFM $M(\lambda)$, which allows
to easily obtain a suboptimal solution of a weighted problem.   □

Remark 5.15 In the case when $M_r(\lambda)$ is an $m_f \times m_f$ invertible diagonal TFM,
the solution of the AMMP targets the solution of an AFDIP with a structure matrix
$S = I_{m_f}$. It follows, that we can apply the threshold selection approach described

in Remark 5.11 with $R_f(\lambda)$ and $R_w(\lambda)$ being $m_f \times m_f$ and, respectively, $m_f \times m_w$ TFMs. An alternative approach can be devised for the case when $M_r(\lambda)$ is a given reference model (not assumed to be structured). To account for the achieved model-matching performance, we employ instead of the residual r, the tracking error

$$\mathbf{e}(\lambda) := \mathbf{r}(\lambda) - M(\lambda)M_r(\lambda)\mathbf{f}(\lambda) = (R_f(\lambda) - M(\lambda)M_r(\lambda))\mathbf{f}(\lambda) + R_w(\lambda)\mathbf{w}(\lambda)$$

and we set the threshold $\tau_i \geq \tau_f^{(i)}$, where $\tau_f^{(i)}$ is the false alarm bound for the i-th component e_i of the tracking error defined as

$$\tau_f^{(i)} := \sup_{\substack{\|w\|_2 \leq \delta_w \\ \|f\|_2 \leq \delta_f}} \|e_i(\lambda)\|_2 .$$

As in Remark 5.11, δ_f and δ_w are the assumed bounds for the norms of the fault and noise signals, respectively. For example, τ_i can be chosen as

$$\tau_i = \| R_f^{(i)}(\lambda) - M^{(i)}(\lambda)M_r(\lambda)\|_\infty \delta_f + \| R_w^{(i)}(\lambda)\|_\infty \delta_w ,$$

where $R_f^{(i)}(\lambda)$, $M^{(i)}(\lambda)$ and $R_w^{(i)}(\lambda)$ are the i-th rows of $R_f(\lambda)$, $M(\lambda)$ and $R_w(\lambda)$, respectively. The above bound can be refined along the approach used in Remark 5.11 in the case when $M_r(\lambda)$ is a structured matrix with the corresponding structure matrix S_{M_r}. $\square$

Example 5.16 We use the LTI system of Example 2.2 to solve a robust fault detection and isolation problem for actuator faults by employing the $\mathcal{H}_\infty$-norm based version of **Procedure AMMS**. The fault system in state-space form (2.2) has a standard state-space realization with $E = I$ and

$$A = \begin{bmatrix} -0.8 & 0 & 0 \\ 0 & -0.5 & 0.6 \\ 0 & -0.6 & -0.5 \end{bmatrix} ,$$

$$B_u = \begin{bmatrix} 1 & 1 \\ 1 & 0 \\ 0 & 1 \end{bmatrix} , \quad B_d = 0, \quad B_w := \begin{bmatrix} 0 & 0 \\ 0 & 0.25 \\ 0.25 & 0 \end{bmatrix} , \quad B_f = \begin{bmatrix} 1 & 1 \\ 1 & 0 \\ 0 & 1 \end{bmatrix} ,$$

$$C = \begin{bmatrix} 0 & 1 & 1 \\ 1 & 1 & 0 \end{bmatrix} , \quad D_u = 0, \quad D_d = 0, \quad D_w = 0, \quad D_f = 0 .$$

The noise input matrix B_w accounts for the effect of parametric uncertainties in the complex conjugated eigenvalues of A and is a 0.25 times scaled version of B_w derived in Example 2.2. Let $G_u(s)$, $G_d(s) = 0$, $G_w(s)$ and $G_f(s)$ denote the TFMs defined according to (2.3). The FDI filter $Q(s)$ is aimed to provide robust fault detection and isolation of actuator faults in the presence of parametric uncertainties.

At Step (1) of **Procedure AMMS**, we choose as nullspace basis

$$Q_1(s) = [\, I \ -G_u(s)\,] = \left[\begin{array}{c|cc} sI - A & 0 & -B_u \\ \hline C & I & -D_u \end{array}\right]$$

and obtain $R_f(s) = G_f(s)$ and $R_w(s) = G_w(s)$. The solvability condition is: rank $R_f(s) = 2$, and thus fulfilled. Note that $R_f(s)$ is invertible and we can choose $Q_2(s) = I$ at Step (2).

At Step (3), the extended quasi-co-outer–co-inner factorization of $\overline{G}(s) = [\, R_f(s) \ R_w(s)\,]$ in (5.57) is computed. The state-space realization of the resulting $G_o(s)$ is obtained in the form (see dual version of Theorem 9.3)

$$G_o(s) = \left[\begin{array}{c|c} A - sI & \overline{B}_o \\ \hline C & \overline{D}_o \end{array}\right],$$

with

$$\overline{B}_o = \begin{bmatrix} -1.313 & -0.48 \\ -0.9334 & 0.3602 \\ -0.398 & -0.9538 \end{bmatrix}, \quad \overline{D}_o = 0.$$

Since $\overline{G}(s)$ has two zeros at infinity, $G_o(s)$ inherits these two zeros and has an additional stable zero at -1.7772. This stable zero is also the only pole of the first-order inner factor $G_i(s) \in \mathcal{H}(s)^{4\times 4}$. With $Q_3(s) = G_o^{-1}(s)$, the descriptor realization of the current synthesis $\overline{Q}(s) = Q_3(s)Q_2(s)Q_1(s)$ can be explicitly computed as (see (7.81) in Sect. 7.9)

$$\overline{Q}(s) = G_o^{-1}(s)Q_1(s) = \left[\begin{array}{cc|cc} A - sI & \overline{B}_o & 0 & -B_u \\ C & \overline{D}_o & I & -D_u \\ \hline 0 & -I & 0 & 0 \end{array}\right].$$

While the current filter $\overline{Q}(s)$ is improper (having two infinite poles), the updated $R_f(s)$ and $R_w(s)$ can also be expressed according to (5.60) as $[\, R_f(s) \ R_w(s)\,] \leftarrow Q_3(s)[\, R_f(s) \ R_w(s)\,] = G_{i,1}(s)$ and are, therefore, stable systems (as parts of the inner factor).

With $M_r(s) = I_2$, we compute $\widetilde{F}_1(s)$ and $\widetilde{F}_2(s)$ as

$$[\, \widetilde{F}_1(s) \ \widetilde{F}_2(s)\,] = [\, I \ 0\,][\, G_{i,1}^{\sim}(s) \ G_{i,2}^{\sim}(s)\,] = \left[\begin{array}{c|cc} \widetilde{A} - sI & \widetilde{B}_1 & \widetilde{B}_2 \\ \hline \widetilde{C} & \widetilde{D}_1 & \widetilde{D}_2 \end{array}\right],$$

where

$$\widetilde{A} = 1.7772, \qquad \widetilde{B}_1 = [\, -0.01688 \ -1.129\,], \quad \widetilde{B}_2 = [\, 4.304 \ 4.754\,],$$

$$\widetilde{C} = \begin{bmatrix} 0.04136 \\ -0.1661 \end{bmatrix}, \quad \widetilde{D}_1 = \begin{bmatrix} -0.9090 & 0.3542 \\ -0.4035 & -0.7796 \end{bmatrix}, \quad \widetilde{D}_2 = \begin{bmatrix} 0.2190 & -0.0136 \\ -0.4273 & -0.2165 \end{bmatrix}.$$

Both $\widetilde{F}_1(s)$ and $\widetilde{F}_2(s)$ are first-order systems with an unstable eigenvalue at 1.7772. At Step (4), we solve a $\mathcal{H}_\infty$-LDP and determine the optimal solution

$$Q_4(s) = \begin{bmatrix} -1.017 & 0.3501 \\ -0.448 & -0.7868 \end{bmatrix},$$

which leads to the current optimal synthesis $\widetilde{Q}(s) = Q_4(s)\overline{Q}(s)$, which is still improper. To obtain a proper and stable FDI filter $Q(s) = Q_5(s)\widetilde{Q}(s)$, we take at Step (5) $Q_5(s) = M(s) = \frac{10}{s+10}I_2$. The resulting overall filter $Q(s)$ has order three. Note that the orders of the realizations of the individual factors $Q_1(s)$, $Q_2(s)$, $Q_3(s)$, $Q_4(s)$ and $Q_5(s)$ are respectively $2, 0, 5, 0$ and 3, which sum together to 10. The corresponding (suboptimal) error norm is $\gamma_{sub} := \|M(s)\widetilde{\mathcal{E}}(s)\|_\infty = 0.4521$. The minimum error norm $\gamma_{opt} := \|\overline{X}(s)\overline{G}(s) - M(s)\overline{F}(s)\|_\infty$ corresponding to the optimal improper solution $\overline{X}(s)$ (of McMillan degree 4) is $\gamma_{opt} = 0.4502$. The relatively small difference $\gamma_{sub} - \gamma_{opt} = 0.0019$ indicates that the computed $Q(s)$ is a satisfactory suboptimal proper and stable solution of the weighted problem.

We can check the robustness of the resulting $Q(s)$ by applying this FDI filter to the original system in Example 2.2 with the parameter dependent state matrix

$$A(\rho_1, \rho_2) = \begin{bmatrix} -0.8 & 0 & 0 \\ 0 & -0.5(1+\rho_1) & 0.6(1+\rho_2) \\ 0 & -0.6(1+\rho_2) & -0.5(1+\rho_1) \end{bmatrix},$$

where ρ_1 and ρ_2 are uncertain parameters, which take values in the ranges $\rho_1 \in [-0.25, 0.25]$ and $\rho_2 \in [-0.25, 0.25]$. The simulations have been performed by determining the step responses of the internal form of the fault detection filter with uncertainties. As it can be observed from Fig. 5.2, the resulting sizes of the uncertainty bands around the nominal trajectories allow to choose appropriate detection thresholds, which permit the reliable detection and isolation of constant faults in the presence of parametric uncertainties.

The script **Ex5_16c.jl** in Listing 5.7 solves the $\mathcal{H}_\infty$-AMMP considered in this example and calls the synthesis function **ammsyn**, which implements the **Procedure AMMS**. The script also includes the verification of the computed results by checking the synthesis conditions, the computation of the nominal step responses, as well as of the lower and upper bounds of the parametric step responses (to be used for plotting). A more detailed version of this script is **Ex5_16.jl** (not listed), which closely follows the computational steps of the **Procedure AMMS**. The script **Fig5_2.jl** (not listed) can be used to generate Fig. 5.2.   ◊

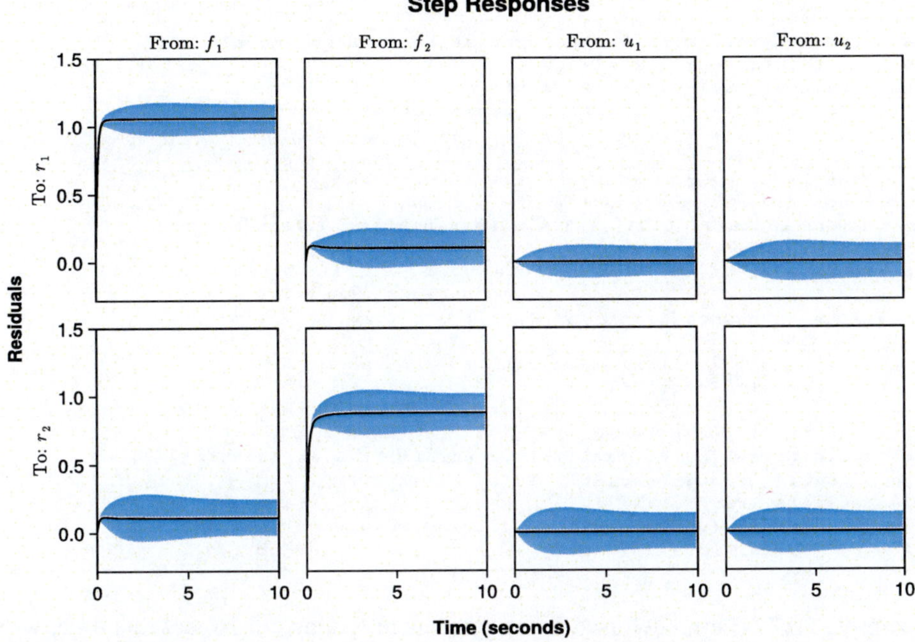

Fig. 5.2 Residual step responses with uncertainty bounds for $\mathcal{H}_\infty$-synthesis

Listing 5.7 Script **Ex5_16c.jl** to solve the AMMP of Example 5.16 using **Procedure** AMMS

```julia
using FaultDetectionTools, DescriptorSystems, Measurements, Test

# Example 5.16 - Solution of an H∞ AMMP
println("Example 5.16")

# define system with control, noise and actuator fault inputs
A = [-.8 0 0; 0 -.5 .6; 0 -.6 -.5]
Bu = [1 1;1 0;0 1]; Bw = 0.25*[0 0;0 1;1 0]; Bf = Bu
C = [0 1 1; 1 1 0]; Du = zeros(2,2); Dw = zeros(2,2)
p, mu = size(Du); mf = mu; mw = size(Dw,2)

atol = 1.e-7;                     # tolerance for rank tests

# set up synthesis model with additive actuator faults
# with Bf = Bu, Df = Du
sysf = fdimodset(dss(A,[Bu Bw],C,[Du Dw]), c = 1:mu,
                 n = mu .+ (1:mw), f = 1:mu)

# define Mr(s) = I
Mr = FDFilterIF(dss(eye(mf)); mf)

Q, R, info = ammsyn(sysf, Mr; atol, nullspace = false, sdeg = -10,
                    reltol = 5.e-4, normalize = "dcgain")

gamma_opt0 = info.gammaopt0 # optimal initial problem performance
gamma_opt = info.gammaopt   # optimal performance
gamma_sub = info.gammasub   # suboptimal performance
```

```
# check suboptimal solution
Ge = [sysf.sys[:,[sysf.faults;sysf.noise]]; zeros(mu,mf+mw)]
Me = [info.M*Mr.sys zeros(mf,mw)]
@test iszero(R.sys-Q.sys*Ge; atol) &&
      fdimmperf(R,info.M*Mr) ≈ gamma_sub &&
      glinfnorm(Me-Q.sys*Ge)[1] ≈ gamma_sub

# compute step responses with uncertainty bounds
# define uncertain parameters p = (ρ1, ρ2)
ρ1 = measurement(0, 0.25); ρ2 = measurement(0, 0.25);
# build uncertain state matrix A(p)
Ap = [-.8 0 0; 0 -.5*(1+ρ1) .6*(1+ρ2); 0 -0.6*(1+ρ2) -0.5*(1+ρ1)];
# build the uncertain system with controls and faults
uncsysf = fdimodset(dss(Ap,Bu,C,Du), c = 1:mu, f = 1:mu);

# determine the internal form with uncertainties
Rt = fdIFeval(Q, uncsysf)

# determine the step responses with uncertainties
Rtd = c2d(Rt.sys[:,[Rt.faults;Rt.controls]], 0.01,"Tustin")[1]
uncy, tout, _ = stepresp(Rtd, 10);
y = getfield.(uncy,:val)          # nominal step responses
yerr = getfield.(uncy,:err)       # error bounds
yl = y .- yerr; yu = y .+ yerr    # lower and upper bounds
```

Example 5.17 The model used is the same as in Example 5.16, but this time, we employ the $\mathcal{H}_2$-norm based version of **Procedure AMMS**. Therefore, we choose $M_r(s) = \frac{10}{s+10} I_2$, which ensures that $\|\overline{\mathcal{E}}(s)\|_2$, the $\mathcal{H}_2$-norm of the error function in (5.56), is finite. Steps (1)–(3) are the same as in Example 5.16. At Step (4), the solution $Q_4(s)$ of the $\mathcal{H}_2$-LDP is simply the stable part of $\widetilde{F}_1(s)$

$$Q_4(s) = \left[\begin{array}{cc|cc} -10-s & 0 & -9.090 & 3.582 \\ 0 & -10-s & -4.038 & -7.955 \\ \hline 1 & 0 & 0 & 0 \\ 0 & 1 & 0 & 0 \end{array}\right].$$

At Step (5), we take $Q_5(s) = M(s) = I$. The resulting FDI filter $Q(s)$ has order three. Note that the orders of the realizations of the individual factors $Q_1(s)$, $Q_2(s)$, $Q_3(s)$, $Q_4(s)$ and $Q_5(s)$ are respectively 3, 0, 5, 2, and 0, which sum together to 10. The corresponding $\mathcal{H}_2$-norm of the error is $\|\widetilde{\mathcal{E}}(s)\|_2 = 1.1172$, while the $\mathcal{H}_\infty$-norm of the error is 0.4519. It follows, that $Q(s)$ can be also interpreted as a fully satisfactory suboptimal solution of the $\mathcal{H}_\infty$-MMP. For the resulting filter, simulation results similar to those in Fig. 5.2 have been obtained, which indicates a satisfactory robustness of the FDI filter.   ◊

5.8 Notes and References

Section 5.1. The two computational paradigms, which underlie the synthesis procedures presented in this chapter, have been discussed for the first time in the authors' papers [150, 157]. The factorized form (5.1) of the resulting fault detection filters is the basis of numerically reliable integrated computational algorithms. The numerical aspects of these algorithms are presented in Chap. 7. The parametrization of fault detection filters given in Theorem 5.1 extends the product form parametrization proposed in [48] given in terms of a polynomial nullspace basis. An alternative less general parametrization, without including the disturbance inputs, is presented in [34] and [47]. The nullspace-based characterization of strong fault detectability in Proposition 5.2 generalizes the characterization proposed in [98] based on polynomial bases.

Section 5.2. The nullspace method (without using this naming), in a state-space based formulation, has been originally employed in [107] to solve the EFDIP using structured residuals and extended in [65] to descriptor systems. The least order synthesis problem has been apparently addressed for the first time in [48], where a minimal polynomial basis based solution has been proposed. The application of the polynomial basis method to systems with improper TFMs is done in [99]. A numerically reliable state-space approach to the least order synthesis relying on rational nullspace bases has been proposed in [138]. The computational details of this approach, in a state-space-based setting, are discussed in Sect. 7.4. The role of the nullspace method as a universal first computational step in all synthesis algorithms has been recognized in [157] as an important computational paradigms to address the synthesis of residual generators for a range of fault detection problems. The *sensitivity condition* (5.20) has been introduced in [51, page 353] as a criterion to be minimized for an optimal design.

Section 5.3. To solve the AFDP, $\mathcal{H}_\infty/\mathcal{H}_\infty$ optimization-based methods have been suggested by several authors, as [34, 47, 111] to cite a few of them. In this context, the H_∞-filter based solution, advocated in [40, 41], is one of several possible synthesis methods. The $\mathcal{H}_\infty/\mathcal{H}_\infty$ optimization-based problem formulation as well as similar ones (e.g., $\mathcal{H}_2/\mathcal{H}_\infty$, $\mathcal{H}_2/\mathcal{H}_2$, etc.) have a basic difficulty in enforcing the sensitivity of residual to all fault inputs. To enhance the optimization-based formulations, the $\|\cdot\|_-$ index has been introduced in [67] as a sensitivity measure covering globally all fault inputs. Based on this definition, synthesis methods to solve the AFDP have been proposed in several papers [31, 69, 81, 82, 176]. The alternative fault sensitivity measures $\|\cdot\|_{\infty-}$ and $\|\cdot\|_{2-}$ have been introduced by the author in [147], where a synthesis procedure similar to **Procedure AFD** has also been proposed. The solution of several nonstandard problems has been considered in [55]. A solution approach for the nonstandard case has been described in [31], based on a special factorization of the quasi-outer factor as a product of an outer factor and a second factor containing all zeros on the boundary of stability domain. This latter approach is implicitly contained in **Procedure AFD**, where the respective zeros are dislocated as the poles of the inverse of the quasi-outer factor using coprime factorization techniques. The extended quasi-co-outer–co-inner factorization of an arbitrary rational matrix can be computed using the dual of the algorithm of [103] for the continuous-time case and the dual of the algorithm of [100] for the discrete-time case. Specialized versions of these algorithms for proper and full column rank rational matrices are presented in Sect. 10.3.6.

Section 5.4. The solution of the EFDIP was one of the most intensively investigated problems in the fault detection literature. We only mention some of the notable works in this area by pointing out the main achievements. Historically, of fundamental importance for a range of subsequent developments was the *geometric approach* introduced by Massoumnia [85], which was the starting point of *observer-based methods*. The main limitation of this *single filter* approach is the assumed form of the fault detection filter as a full-order Luenberger observer [84], with a suitably determined output gain matrix targeting the achievement of a desired structure matrix. The strong solvability conditions can frequently not be satisfied (no single stable filter exists), even if the FDIP has a solution. The use of a *bank of filters*, as suggested in [87], therefore, appears as a natural approach to solve FDIPs for a given structure matrix. Phatak and Viswanadham proposed the use of a bank of *unknown-input observers* (UIOs) as fault detection and isolation filters [109]. Although the lack of generality of this approach is well known and ways to eliminate them have been proposed by Hou and Müler [66], the UIO-based approach preserved over the years a certain popularity (e.g., being the preferred method in [23]). The extension of the observer-based approach to the case of general proper systems has been done by Patton and Hou [107] and later extended by Hou to descriptor systems in [65]. The least order synthesis aspect, in the most general setting, has been addressed by the author in [146] and later improved in [155], where the nullspace method has been used as a first preprocessing step to reduce the complexity of the FDIP and for designing a bank of fault detection filters to provide a set of structured residuals. This improved approach underlies the **Procedure EFDI**. Similar synthesis methods can be devised using the *parity-space approach* proposed by Chow and Willsky [25], where the least order synthesis aspect has been discussed in [33]. The synthesis of FDI schemes based on structured residuals, also including the selection of structure matrices for suitable coding sets, has been discussed in several works of Gertler [51–53]. The nullspace-based algorithm for the efficient determination of the maximally achievable structure matrix has been proposed in [151]. This algorithm underlies the **Procedure GENSPEC**.

Section 5.5. The **Procedure AFDI** represents a refined version of the approach suggested in [157].

Section 5.6. The solution of the EMMP involves the solution of a linear rational matrix equation (see Sect. 9.1.9 for existence conditions and parametrization of all solutions). General computational algorithms, based on state-space representations, have been proposed by the author in [140, 141] and are discussed in details in Sect. 10.3.7. The *inversion-based method* to solve the EFDIP with the strong fault isolability requirement goes back to Massoumnia and Vander Velde [86], where only the case without disturbance inputs is addressed. For further extensions and discussions of this method, see [34, 52, 75]. A recent development, leading to the general numerically reliable computational approach in **Procedure EMMS**, has been proposed by the author in [157].

Section 5.7. The solution of the AMMP using a $\mathcal{H}_\infty$ or $\mathcal{H}_2$ optimal *controller* synthesis setup is the method of choice in some textbooks, see for example [15, 23]. Standard software tools for controller synthesis are available (e.g., the functions **hinfsyn** or **h2syn** available in MATLAB), but their general applicability to solve the (dual) *filter* synthesis problems may face difficulties. Typical bottlenecks are the assumptions on stabilizability (not fulfilled for filter synthesis for unstable plants), the lack of zeros on $\partial\mathbb{C}_s$ (typically not fulfilled if only actuator faults are considered) or the need to formulate meaningful reference models for the TFM from faults to residuals. The first two aspects can be overcome with the help of stable factorization techniques and by using more general computational frameworks (e.g., *linear matrix inequalities* (LMIs)-based formulations). However, in spite of some efforts (see for example, [97]), there are no clear guidelines for choosing reference models able to guarantee the existence of stable solutions. This is why, a new approach has been proposed by the author in [143], where the choice of a suitable $M_r(\lambda)$ is part of the solution procedure. This procedure that has been later refined in [152, 153, 156] and **Procedure AMMS** represents its final form. The main computational ingredients of this procedure are discussed in Chap. 7, in a state-space formulation based setting.

Final note: A common aspect worth to mention regarding the proposed synthesis procedures to solve the approximate synthesis problems AFDP, AFDIP and AMMP is that the main focus in developing these algorithms lies not on solving the associated optimization problems, but on obtaining "useful" solutions of these synthesis problems, in the most general setting and using reliable numerical techniques. Although the proposed solution approaches in [147, 152, 153, 156] follow the usual solution processes to determine the optimal solutions, still the resulting filters are usually not optimal in the nonstandard cases. The assessment of the "usefulness" of the resulting filters involves the evaluation of the actual signal bounds on the contribution of noise inputs in the residual signal and the determination of the minimum detectable amplitudes of fault signals. A solution can be considered as "useful" if it is possible to choose a suitable decision threshold, which allows a robust fault monitoring without false alarms and missed fault detections. For a pertinent discussion of these aspects, see [51].

Chapter 6
Synthesis of Model Detection Filters

This chapter presents general synthesis procedures of residual generators, which solve the model detection problems formulated in Chap. 4. Similar to Chap. 3, the synthesis procedures are described in terms of input–output models. The numerical aspects of equivalent state-space representation based synthesis algorithms are essentially the same as for the synthesis algorithms of fault detection and isolation filters, and the discussion of related computational techniques is also covered in Chap. 7.

6.1 Nullspace-Based Synthesis

We assume the overall residual generator filter $Q(\lambda)$ has the TFM of the form (4.3), which corresponds to a bank of N individual filters as in (4.2). Furthermore, for $i = 1, \ldots, N$, the i-th filter driven by the j-th model has the internal form in (4.4). Let $R_u^{(i,j)}(\lambda)$ and $R_d^{(i,j)}(\lambda)$ be the TFMs defined in (4.4) and (4.5). A useful parametrization of all individual filters can be obtained on the basis of the conditions $R_u^{(i,i)}(\lambda) = 0$ and $R_d^{(i,i)}(\lambda) = 0$ for $i = 1, \ldots, N$ in (4.11). For each filter with the TFM $Q^{(i)}(\lambda)$, these conditions are equivalent to

$$Q^{(i)}(\lambda) \begin{bmatrix} G_u^{(i)}(\lambda) & G_d^{(i)}(\lambda) \\ I_{m_u} & 0 \end{bmatrix} = 0 \, . \tag{6.1}$$

Therefore, $Q^{(i)}(\lambda)$ must be a left annihilator of the TFM

$$G^{(i)}(\lambda) := \begin{bmatrix} G_u^{(i)}(\lambda) & G_d^{(i)}(\lambda) \\ I_{m_u} & 0 \end{bmatrix} \, . \tag{6.2}$$

Let $r_d^{(i)}$ be the normal rank of $G_d^{(i)}(\lambda)$. It follows that there exists a maximal full row rank left annihilator $N_l^{(i)}(\lambda)$ of size $(p - r_d^{(i)}) \times (p + m_u)$ such that $N_l^{(i)}(\lambda)G^{(i)}(\lambda) = 0$. Any such $N_l^{(i)}(\lambda)$ is a *rational basis* of $\mathcal{N}_L(G^{(i)}(\lambda))$, the left (rational) nullspace of $G^{(i)}(\lambda)$. Using this fact and Theorem 5.1, we have the following straightforward parametrization of all component filters:

Theorem 6.1 *For $i = 1, \ldots, N$, let $N_l^{(i)}(\lambda)$ be a basis of $\mathcal{N}_L(G^{(i)}(\lambda))$, with $G^{(i)}(\lambda)$ defined in (6.2). Then, each filter $Q^{(i)}(\lambda)$ satisfying condition (i) of (4.11) can be expressed in the form*

$$Q^{(i)}(\lambda) = V^{(i)}(\lambda)N_l^{(i)}(\lambda), \quad i = 1, \ldots, N, \tag{6.3}$$

where $V^{(i)}(\lambda)$ is a suitable TFM.

The parametrization result of Theorem 6.1 underlies the nullspace method based synthesis procedures of model detection filters. All synthesis procedures of the model detection filters, presented in this book, rely on the initial factored forms

$$Q^{(i)}(\lambda) = \overline{Q}_1^{(i)}(\lambda)Q_1^{(i)}(\lambda), \quad i = 1, \ldots, N, \tag{6.4}$$

where each $Q_1^{(i)}(\lambda) = N_l^{(i)}(\lambda)$ is a basis of $\mathcal{N}_L(G^{(i)}(\lambda))$, while each factor $\overline{Q}_1^{(i)}(\lambda)$ has to be subsequently determined. The nullspace-based first step allows to reduce the synthesis problems of model detection filters formulated for the multiple models (4.1) to simpler problems, which allow to easily check the solvability conditions.

Using the factored form (6.4), the model detection filters in (4.2) can be rewritten in the alternative forms

$$\mathbf{r}^{(i)}(\lambda) = \overline{Q}_1^{(i)}(\lambda)Q_1^{(i)}(\lambda)\begin{bmatrix} \mathbf{y}(\lambda) \\ \mathbf{u}(\lambda) \end{bmatrix} = \overline{Q}_1^{(i)}(\lambda)\overline{\mathbf{y}}^{(i)}(\lambda), \quad i = 1, \ldots, N, \tag{6.5}$$

where

$$\overline{\mathbf{y}}^{(i)}(\lambda) := Q_1^{(i)}(\lambda)\begin{bmatrix} \mathbf{y}(\lambda) \\ \mathbf{u}(\lambda) \end{bmatrix}. \tag{6.6}$$

For $y(t) = y^{(j)}(t)$, both the residual signal $r^{(i)}(t)$ in (6.5) and $\overline{y}^{(i)}(t)$ in (6.6) depend on all system inputs $u^{(j)}(t)$, $d^{(j)}(t)$ and $w^{(j)}(t)$ via the system output $y^{(j)}(t)$. The internal form (4.4) of the i-th filter for the j-th model can be expressed as

$$\widetilde{\mathbf{r}}^{(i,j)}(\lambda) = \overline{Q}_1^{(i)}(\lambda)\widetilde{\mathbf{y}}^{(i,j)}(\lambda),$$

where

$$\widetilde{\mathbf{y}}^{(i,j)}(\lambda) := Q_1^{(i)}(\lambda)\begin{bmatrix} \mathbf{y}^{(j)}(\lambda) \\ \mathbf{u}^{(j)}(\lambda) \end{bmatrix}.$$

Using the expression of $\mathbf{y}^{(j)}(\lambda)$ from (4.1), we obtain

$$\widetilde{\mathbf{y}}^{(i,j)}(\lambda) = \overline{G}_u^{(i,j)}(\lambda)\mathbf{u}^{(j)}(\lambda) + \overline{G}_d^{(i,j)}(\lambda)\mathbf{d}^{(j)}(\lambda) + \overline{G}_w^{(i,j)}(\lambda)\mathbf{w}^{(j)}(\lambda), \tag{6.7}$$

with

$$\left[\overline{G}_u^{(i,j)}(\lambda)\left|\overline{G}_d^{(i,j)}(\lambda)\right|\overline{G}_w^{(i,j)}(\lambda)\right] := Q_1^{(i)}(\lambda)\left[\begin{array}{c|c|c} G_u^{(j)}(\lambda) & G_d^{(j)}(\lambda) & G_w^{(j)}(\lambda) \\ I_{m_u} & 0 & 0 \end{array}\right]. \tag{6.8}$$

The system (6.7) can be interpreted as the internal form of the i-th filter driven by the j-th model, corresponding to the partial synthesis $Q_1^{(i)}(\lambda)$. For $j = i$, the particular choice of $Q_1^{(i)}(\lambda)$ as a left nullspace basis of $G^{(i)}(\lambda)$ in (6.2) ensures that $\overline{G}_u^{(i,i)}(\lambda) = 0$ and $\overline{G}_d^{(i,i)}(\lambda) = 0$.

At this stage we can assume that both $Q_1^{(i)}(\lambda)$ and the TFMs (6.8) are proper and stable. This can be always achieved by using $Q_1^{(i)}(\lambda) = M^{(i)}(\lambda)N_l^{(i)}(\lambda)$ (instead $Q_1^{(i)}(\lambda) = N_l^{(i)}(\lambda)$), where $M^{(i)}(\lambda)$ is a stable and proper TFM such that

$$M^{(i)}(\lambda)\left[N_l^{(i)}(\lambda) \mid \overline{G}_u^{(1)}(\lambda) \ \overline{G}_d^{(1)}(\lambda) \ \overline{G}_w^{(1)}(\lambda) \mid \cdots \mid \overline{G}_u^{(N)}(\lambda) \ \overline{G}_d^{(N)}(\lambda) \ \overline{G}_w^{(N)}(\lambda)\right]$$

is stable and proper. Such an $M^{(i)}(\lambda)$ can be determined as the denominator matrix of a stable and proper LCF (see Sect. 9.1.6).

Relying on the parametrization result of Theorem 6.1, we have the following straightforward characterization of the model detectability of the multiple model (4.1) in terms of the N multiple models (6.7).

Proposition 6.1 *For the multiple model (4.1), let $Q_1^{(i)}(\lambda) = N_l^{(i)}(\lambda)$, $i = 1, \ldots, N$, be rational bases of $\mathcal{N}_L(G^{(i)}(\lambda))$, with $G^{(i)}(\lambda)$ defined in (6.2), and let (6.7) be the multiple model associated with the i-th residual. Then, the multiple model (4.1) with $w^{(j)} \equiv 0$ for $j = 1, \ldots, N$, is model detectable if and only if, for $i = 1, \ldots, N$*

$$\left[\overline{G}_u^{(i,j)}(\lambda) \ \overline{G}_d^{(i,j)}(\lambda)\right] \neq 0 \ \ \forall j \neq i. \tag{6.9}$$

6.2 Solving the Exact Model Detection Problem

Using Proposition 6.1, the solvability conditions of the *exact model detection problem* (EMDP) formulated in Sect. 4.4.1 for the multiple model (4.1), can be also expressed in terms of the multiple models (6.7), according to the following corollary to Theorem 4.2.

Corollary 6.1 *For the multiple model (4.1) with $w^{(j)} \equiv 0$ for $j = 1, \ldots, N$, the EMDP is solvable if and only if for the multiple model (6.7), with $w^{(j)} \equiv 0$ for $j = 1, \ldots, N$, the following conditions hold for $i = 1, \ldots, N$*

$$\left[\, \overline{G}_u^{(i,j)}(\lambda)\ \overline{G}_d^{(i,j)}(\lambda)\,\right] \neq 0 \quad \forall j \neq i. \tag{6.10}$$

The synthesis procedure of the N component filters $Q^{(i)}(\lambda)$, $i = 1, \ldots, N$, employs a common computational approach. Accordingly, the i-th filter $Q^{(i)}(\lambda)$ is determined in the factored form

$$Q^{(i)}(\lambda) = Q_3^{(i)}(\lambda) Q_2^{(i)}(\lambda) Q_1^{(i)}(\lambda) \,,$$

where: $Q_1^{(i)}(\lambda)$ is a rational basis of $\mathcal{N}_L\big(G^{(i)}(\lambda)\big)$, the left nullspace of $G^{(i)}(\lambda)$ defined in (6.2); $Q_2^{(i)}(\lambda)$ ensures that $Q_2^{(i)}(\lambda) Q_1^{(i)}(\lambda)$ has least McMillan degree; and, $Q_3^{(i)}(\lambda)$ is chosen such that $Q^{(i)}(\lambda)$ is stable and the corresponding $R^{(i,j)}(\lambda)$ defined in (4.6), for $j = 1, \ldots, N$, $j \neq i$, are stable and nonzero. Using Proposition 6.1, the existence condition of the i-th filter is satisfied if $Q_1^{(i)}(\lambda) G^{(j)}(\lambda) \neq 0$, $\forall j \neq i$.

There exists some freedom in determining model detection filters, which solve the EMDP. For example, the number of outputs of the i-th filter $Q^{(i)}(\lambda)$ can be chosen arbitrarily between 1 and $p - r_d^{(i)}$, where $r_d^{(i)} := \operatorname{rank} G_d^{(i)}(\lambda)$, provided the model detectability conditions are fulfilled. Also, least-order scalar output model detection filters can be employed to ensure that the overall bank of filters has the least achievable global order. However, filters with more outputs can occasionally provide a better sensitivity condition (see later) for model detection.

The **Procedure EMD**, given below, determines the N filters $Q^{(i)}(\lambda)$, $i = 1, \ldots,$ N, and the corresponding internal forms $R^{(i,j)}(\lambda) := \left[\, R_u^{(i,j)}(\lambda)\ \ R_d^{(i,j)}(\lambda)\,\right]$, for $i, j = 1, \ldots, N$, with the i-th filter having a maximal row dimension q_{max}.

Procedure EMD: Exact synthesis of model detection filters

Inputs : $\{G_u^{(j)}(\lambda),\, G_d^{(j)}(\lambda)\}$, for $j = 1, \ldots, N$; q_{max}
Outputs: $Q^{(i)}(\lambda)$, for $i = 1, \ldots, N$; $R^{(i,j)}(\lambda)$ for $i, j = 1, \ldots, N$

For $i = 1, \ldots, N$

(1) Compute a $(p - r_d^{(i)}) \times (p + m_u)$ minimal basis matrix $Q_1^{(i)}(\lambda)$ for the left
  nullspace of $G^{(i)}(\lambda)$ defined in (6.2), where $r_d^{(i)} := \operatorname{rank} G_d^{(i)}(\lambda)$; set
  $Q^{(i)}(\lambda) = Q_1^{(i)}(\lambda)$ and compute $R^{(i,j)}(\lambda) = Q^{(i)}(\lambda) G^{(j)}(\lambda)$ for $j = 1, \ldots, N$.
  Exit if $R^{(i,j)}(\lambda) = 0$ for any $j \in \{1, \ldots, N\}$, $j \neq i$ (no solution exists).
(2) Choose a $\min\left(q_{max},\, p - r_d^{(i)}\right) \times (p + m_u)$ rational matrix $Q_2^{(i)}(\lambda)$, such that
  $Q_2^{(i)}(\lambda) Q^{(i)}(\lambda)$ has least McMillan degree and $Q_2^{(i)}(\lambda) R^{(i,j)}(\lambda) \neq 0$ for
  $j = 1, \ldots, N$, $j \neq i$; compute $Q^{(i)}(\lambda) \leftarrow Q_2^{(i)}(\lambda) Q^{(i)}(\lambda)$ and
  $R^{(i,j)}(\lambda) \leftarrow Q_2^{(i)}(\lambda) R^{(i,j)}(\lambda)$ for $j = 1, \ldots, N$, $j \neq i$.
(3) Choose a proper and stable invertible rational matrix $Q_3^{(i)}(\lambda)$ such that
  $Q_3^{(i)}(\lambda) Q^{(i)}(\lambda)$ has a desired stable dynamics and $Q_3^{(i)}(\lambda) R^{(i,j)}(\lambda)$ for
  $j = 1, \ldots, N$, $j \neq i$ are stable; compute $Q^{(i)}(\lambda) \leftarrow Q_3^{(i)}(\lambda) Q^{(i)}(\lambda)$ and
  $R^{(i,j)}(\lambda) \leftarrow Q_3^{(i)}(\lambda) R^{(i,j)}(\lambda)$ for $j = 1, \ldots, N$, $j \neq i$.

The computational algorithms underlying **Procedure EMD** are essentially the same as those used for the synthesis of fault detection filters (see **Procedure EFD**) and rely on state-space representations as in (2.19) of the component models. These algorithms are amply described in Sects. 7.4–7.6, and therefore, we restrict our discussion to specific aspects of Steps (2) and (3). To determine filters with least dynamical orders at Step (2), a straightforward systematic approach is to build successive candidate filters $Q_2^{(i)}(\lambda)Q_1^{(i)}(\lambda)$ with increasing McMillan degrees and check the specific *admissibility condition* $Q_2(i)(\lambda)Q_1^{(i)}(\lambda)G^{(j)}(\lambda) \neq 0$ (or equivalently $Q_2^{(i)}(\lambda)\big[\,\overline{G}_u^{(i,j)}(\lambda)\ \ \overline{G}_d^{(i,j)}(\lambda)\,\big] \neq 0$) for all $j \neq i$. The least possible order of the fault detection filter $Q^{(i)}(\lambda)$ is uniquely determined by the fulfilment of the above admissibility condition. Since $Q_3^{(i)}(\lambda)$ is invertible, its choice plays no role in ensuring admissibility. However, the final orders of the individual filters can occasionally further increase at Step (3), if the cancellation of unstable poles in the component models is necessary, in accordance with the formulated requirements for the EMDP. As in the case of solving the EFDP, a least-order filter synthesis can be always achieved by a scalar output filter. Since the choice of $Q_2^{(i)}(\lambda)$ is not unique, an appropriate parametrization of $Q_2^{(i)}(\lambda)$ allows to make an optimal choice of free parameters (e.g., to achieve other desirable features; see Remark 6.1). Further aspects of selecting suitable $Q_2^{(i)}(\lambda)$, in accordance with the employed type of nullspace basis, are discussed in Sect. 5.2, in the context of solving the EFDP.

Remark 6.1 Assume that all component models in (4.1) are stable and all vectors $d^{(i)}(t)$, $i = 1, \ldots, N$, have dimension m_d. In this case, the norm of $R^{(i,j)}(\lambda)$ has a simple interpretation as a weighted distance between the i-th and j-th models. In accordance with Theorem 6.1, $Q^{(i)}(\lambda)$ can be expressed as $Q^{(i)}(\lambda) = V^{(i)}(\lambda)N_l^{(i)}(\lambda)$, with the nullspace basis $N_l^{(i)}(\lambda)$ chosen in a form similar to (5.5), as

$$N_l^{(i)}(\lambda) = N_{l,d}^{(i)}(\lambda)\,\big[\,I_p \ -G_u^{(i)}(\lambda)\,\big]\,,$$

where $N_{l,d}^{(i)}(\lambda)$ is a $(p - r_d^{(i)}) \times p$ TFM representing a basis of $\mathcal{N}_L\big(G_d^{(i)}(\lambda)\big)$. This choice leads to

$$
\begin{aligned}
R^{(i,j)}(\lambda) &= Q^{(i)}(\lambda)G^{(j)}(\lambda) \\
&= V^{(i)}(\lambda)N_{l,d}^{(i)}(\lambda)\big[\,G_u^{(j)}(\lambda) - G_u^{(i)}(\lambda)\ \ G_d^{(j)}(\lambda) - G_d^{(i)}(\lambda)\,\big].
\end{aligned}
\tag{6.11}
$$

If we define the distance between the i-th and j-th models as

$$\mathrm{dist}\,(G^{(i)}(\lambda), G^{(j)}(\lambda)) := \big\|\big[\,G_u^{(j)}(\lambda) - G_u^{(i)}(\lambda)\ \ G_d^{(j)}(\lambda) - G_d^{(i)}(\lambda)\,\big]\big\|\,,$$

then, the norm of $R^{(i,j)}(\lambda)$ can be interpreted as a weighted distance between the TFMs of the i-th and j-th models. An ideal model detection filter $Q(\lambda)$ of the form (4.3) would monotonically map the distances between two models to the corresponding norms of $R^{(i,j)}(\lambda)$, that is, if the distances of the j-th and k-th models to the i-th model satisfy

$$\text{dist}\,(G^{(i)}(\lambda),\,G^{(j)}(\lambda)) < \text{dist}\,(G^{(i)}(\lambda),\,G^{(k)}(\lambda)) \,,$$

then the weighted distances satisfy

$$\| R^{(i,j)}(\lambda)\| < \| R^{(i,k)}(\lambda)\| \,.$$

Moreover, the fulfilment of the symmetry conditions

$$\| R^{(i,j)}(\lambda)\| = \| R^{(j,i)}(\lambda)\|, \ \forall i \neq j \,,$$

is also highly desirable. A model detection filter having these properties can be employed to reliably identify the nearest model from a given set of models to the actual plant model.

Ensuring the monotonic distance mapping and symmetry properties can be seen as a global synthesis goal of model detection filters, and can be targeted in various ways, as—for example, by an optimal choice of the free parameters of the weighting functions $V^{(i)}(\lambda)N_{l,d}^{(i)}(\lambda)$, or by choosing each filter $Q^{(i)}(\lambda)$ to enforce a certain isometry (i.e., distance preserving) property (e.g., by choosing $V^{(i)}(\lambda)N_{l,d}^{(i)}(\lambda)$ a co-inner matrix). $\qquad\qquad\square$

Remark 6.2 A properly designed model detection system as in Fig. 4.1 (e.g., with the model detection filter determined using **Procedure EMD**), is always able to identify the exact matching of the current model with one (and only one) of the N component models. However, in practice, we often encounter the situation that the actual (or true) model will never match exactly any of the N component models, and therefore, the best we can aim is to correctly figure out the nearest model to the actual one. Assume that the actual model has $\widetilde{G}_u(\lambda)$ and $\widetilde{G}_d(\lambda)$, the TFMs from the control-input-to-output and disturbance-input-to-output, respectively. Therefore, $\widetilde{G}_u(\lambda)$ and $\widetilde{G}_d(\lambda)$ can be expressed in terms of their deviations to the N component models for $j = 1, \ldots, N$ as

$$\widetilde{G}_u(\lambda) = G_u^{(j)}(\lambda) + \Delta G_u^{(j)}(\lambda), \quad \widetilde{G}_d(\lambda) = G_d^{(j)}(\lambda) + \Delta G_d^{(j)}(\lambda) \,.$$

Assuming the N component models are mutually distinct, there exists for each $i = 1, \ldots, N$, a largest $\delta^{(i)} > 0$ such that the following conditions simultaneously hold:

$$\left\| \left[\, \Delta G_u^{(i)}(\lambda) \ \ \Delta G_d^{(i)}(\lambda) \,\right] \right\|_\infty \leq \delta^{(i)}, \quad \left\| \left[\, \Delta G_u^{(j)}(\lambda) \ \ \Delta G_d^{(j)}(\lambda) \,\right] \right\|_\infty > \delta^{(i)}, \forall j \neq i \,.$$

The size of $\delta^{(i)}$ defines the family of all sufficiently nearby models to the i-th model, which are distinguishable (using the $\mathcal{H}_\infty$-norm based distance) from the rest of models. In the case when the nearest model to the actual model is the i-th model (i.e., the above inequalities are fulfilled), it is highly desirable that the model detection filter ensures that the i-th evaluation signal, $\theta_i \approx \| r^{(i)}\|_2$, has the least value among the N components of θ, and thus, allow to identify the i-th model as the nearest one to the current model. The attainability of this goal usually depends on the concrete problem to be solved. With the interpretation of the norm of $R^{(i,j)}(\lambda)$ in Remark 6.1 as a weighted distance between the i-th and j-th models, a prerequisite to fulfil the above goal is the use of a model detection filter able to monotonically map the distances between the models to the corresponding norms of the internal representations (i.e., to $R^{(i,j)}(\lambda)$). $\qquad\square$

Example 6.1 To illustrate the effectiveness of the proposed nullspace-based synthesis approach of model detection filters, we consider the detection and identification of loss of efficiency of flight actuators using a model detection-based approach. The fault-free state-space model describes the continuous-time lateral dynamics of an F-16 aircraft with the matrices

$$A^{(1)} = \begin{bmatrix} -0.4492 & 0.046 & 0.0053 & -0.9926 \\ 0 & 0 & 1.0000 & 0.0067 \\ -50.8436 & 0 & -5.2184 & 0.7220 \\ 16.4148 & 0 & 0.0026 & -0.6627 \end{bmatrix}, \quad B_u^{(1)} = \begin{bmatrix} 0.0004 & 0.0011 \\ 0 & 0 \\ -1.4161 & 0.2621 \\ -0.0633 & -0.1205 \end{bmatrix},$$

$$C^{(1)} = I_4, \qquad D_u^{(1)} = 0_{4\times2}.$$

The four state variables are the sideslip angle, roll angle, roll rate and yaw rate, and the two input variables are the aileron deflection and rudder deflection. The individual failure models correspond to different levels of surface efficiency degradation. For simplicity, we build a multiple model with $N = 9$ component models on a coarse two-dimensional parameter grid for N values of the parameter vector $\rho := [\rho_1, \rho_2]^T$. For each component of ρ, the chosen three grid points are $\{0, 0.5, 1\}$. The component system matrices in (2.19) are defined for $i = 1, 2, \ldots, N$ as: $E^{(i)} = I_4$, $A^{(i)} = A^{(1)}, C^{(i)} = C^{(1)}$, and $B_u^{(i)} = B_u^{(1)} \Gamma^{(i)}$, where $\Gamma^{(i)} = \text{diag}\left(1 - \rho_1^{(i)}, 1 - \rho_2^{(i)}\right)$ and $\left(\rho_1^{(i)}, \rho_2^{(i)}\right)$ are the values of parameters (ρ_1, ρ_2) on the chosen grid:

ρ_1 :	0	0	0	0.5	0.5	0.5	1	1	1
ρ_2 :	0	0.5	1	0	0.5	1	0	0.5	1

For example, $\left(\rho_1^{(1)}, \rho_2^{(1)}\right) = (0, 0)$ corresponds to the fault-free situation, while $\left(\rho_1^{(9)}, \rho_2^{(9)}\right) = (1, 1)$ corresponds to complete failure of both control surfaces. It follows, that the TFM $G_u^{(i)}(s)$ of the i-th system can be expressed as

$$G_u^{(i)}(s) = G_u^{(1)}(s)\Gamma^{(i)}, \tag{6.12}$$

where

$$G_u^{(1)}(s) = C^{(1)}\left(sI - A^{(1)}\right)^{-1} B_u^{(1)}$$

is the TFM of the fault-free system. Note that $G_u^{(N)}(s) = 0$ describes the case of complete failure.

We applied the **Procedure EMD** to design $N = 9$ model detection filters of least dynamical order with scalar outputs. At Step (1), nullspace bases of the form

$$Q_1^{(i)}(s) = \left[\, I_4 \;\; -G_u^{(i)}(s) \,\right] = \left[\, I_4 \;\; -G_u^{(1)}(s)\Gamma^{(i)} \,\right]$$

have been chosen as initial designs. The internal forms corresponding to these designs are

$$R_1^{(i,j)}(s) := Q_1^{(i)}(s)\begin{bmatrix} G_u^{(j)}(s) \\ I_2 \end{bmatrix} = G_u^{(j)}(s) - G_u^{(i)}(s) = G_u^{(1)}(s)\left(\Gamma^{(j)} - \Gamma^{(i)}\right).$$

At this stage, the norms $\left\| R_1^{(i,j)}(s) \right\|_\infty$ monotonically map the distances between the i-th and j-th component models (similarly to the norms displayed in Fig. 6.1).

At Step (2), we target to preserve the monotonic mapping of norms (as in Fig. 6.1) after updating $Q_1^{(i)}(s)$, by choosing the updating filter $Q_2^{(i)}(s)$ such that $Q_2^{(i)}(s)Q_1^{(i)}(s)$ has the least order. For this purpose, with a suitably chosen row vector h, a linear combination of the basis vectors has been formed as $X^{(i)}(s) = hQ_1^{(i)}(s)$, and then a proper rational row vector $Y^{(i)}(s)$ has been determined such that $Q_2^{(i)}(s)Q_1^{(i)}(s) := X^{(i)}(s) + Y^{(i)}(s)Q_1^{(i)}(s)$ has least McMillan degree and $Q_2^{(i)}(s)R_1^{(i,j)}(s) \neq 0$ for all $j \neq i$. The resulting $Q_2^{(i)}(s)$ is simply $Q_2^{(i)}(s) = h + Y^{(i)}(s)$. For this computation, minimal dynamic cover techniques described in Sect. 7.5 have been used. After some trials with randomly generated h, the value

$$h = [\,0.7645 \;\; 0.8848 \;\; 0.5778 \;\; 0.9026\,]$$

led to a satisfactory dynamics of a first-order updated filter, without the need of further stabilization. Due to the particular forms of $G_u^{(i)}(s)$ in (6.12), the same $Q_2^{(i)}(s) := Q_2^{(1)}(s)$, $i = 1, \ldots, N$, can be used for all models. The resulting final filters are given by

$$Q^{(i)}(s) = Q_2^{(i)}(s)Q_1^{(i)}(s) = \left[\, Q_2^{(1)}(s) \;\; -Q_2^{(1)}(s)G_u^{(1)}(s)\Gamma^{(i)} \,\right], \tag{6.13}$$

where, for convenience, we set $Q^{(N)}(s)$ as

$$Q^{(N)}(s) = \left[\, Q_2^{(1)}(s)\; 0 \,\right],$$

with a first-order state-space realization, although $Q^{(N)}(s) = \left[\, h\; 0 \,\right]$ was also possible.

The final internal filters $R^{(i,j)}(s)$ result as

$$R^{(i,j)}(s) = Q_2^{(1)}(s)G_u^{(1)}(s)\left(\Gamma^{(j)} - \Gamma^{(i)}\right), \quad i, j = 1, \ldots, N$$

and exhibit the monotonic mapping of distances as shown in Fig. 6.1.

For practical use, the N filters $Q^{(i)}(s)$ have been scaled such that the corresponding row blocks $R^{(i,j)}(s)$ fulfil the condition $\min_{j=1:N, i \neq j} \left\| R^{(i,j)}(s) \right\|_\infty = 1$. This amounts

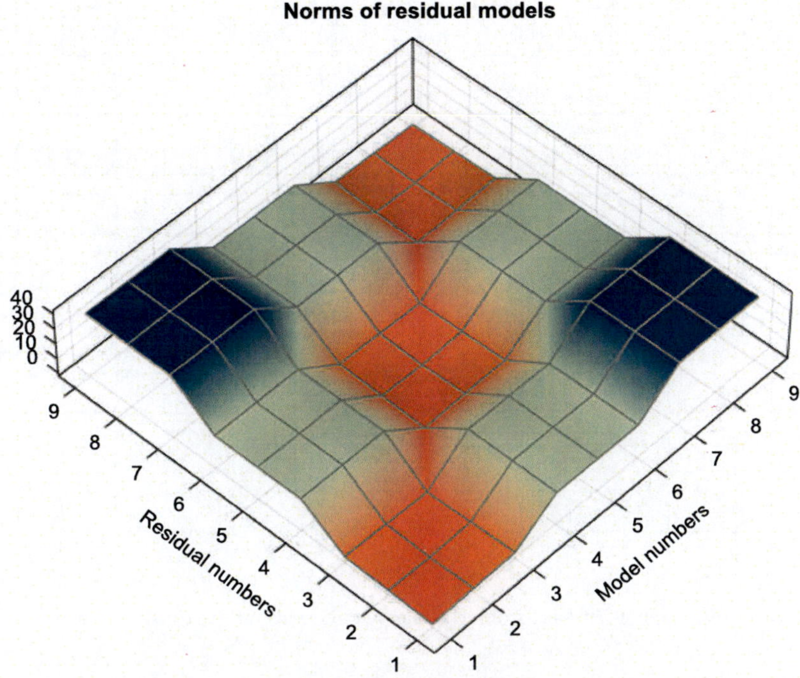

Fig. 6.1 Norms of residual models for the least-order synthesis

to replace $Q^{(i)}(s)$ by $Q^{(i)}(s)/\gamma_i$ and $R^{(i,j)}(s)$ by $R^{(i,j)}(s)/\gamma_i$, for $j = 1, \ldots, N$, where $\gamma_i = \min_{j=1:N, i \neq j} \left\| R^{(i,j)}(s) \right\|_\infty$. This scaling also enforces the symmetry conditions $\left\| R^{(i,j)}(s) \right\|_\infty = \left\| R^{(j,i)}(s) \right\|_\infty$ for all $i \neq j$.

In Fig. 6.2, the step responses from u_1 (aileron) and u_2 (rudder) are presented for the 9×9 block array, whose entries are the rescaled TFMs $R^{(i,j)}(s)$. Each column corresponds to a specific model for which the step responses of the N residuals are computed. The achieved typical structure matrix for model detection (with zeros down the diagonal) can easily be read out from this signal-based assessment.

The script **Ex6_1c.jl** in Listing 6.1 solves the EMDP considered in this example and calls the synthesis function **emdsyn**, which implements the **Procedure EMD**. The script also includes the verification of the model detectability condition. A more detailed version of this script is **Ex6_1.jl** (not listed), which closely follows the computational steps of the **Procedure EMD**. The scripts **Fig. 6_1.jl** and **Fig. 6_2.jl** (not listed) generate the plots in Fig. 6.1 and Fig. 6.2, respectively.

$\diamond$

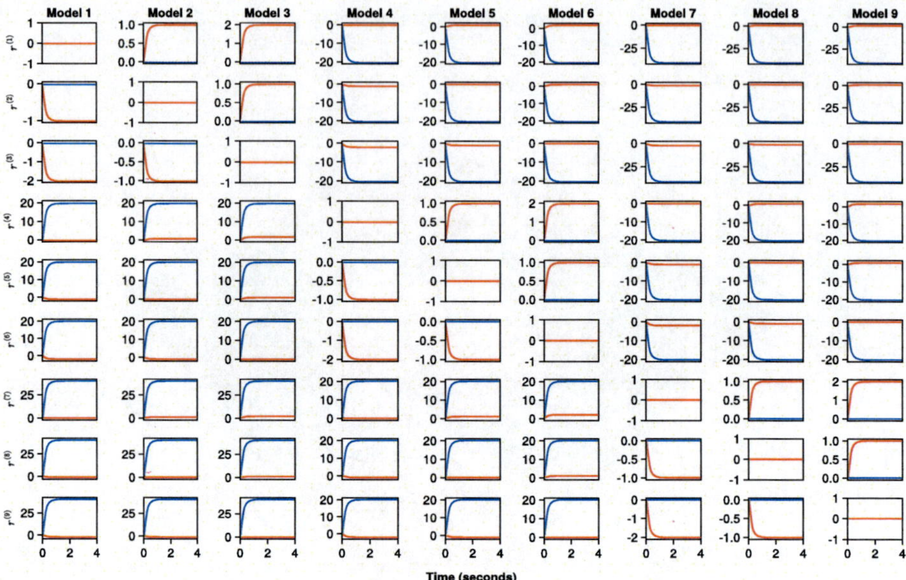

Fig. 6.2 Step responses of $R^{(i,j)}(s)$ from u_1 (blue) and u_2 (red) for least-order syntheses

Listing 6.1 Script `Ex6_1c.jl` to solve the EMDP of Example 6.1 using **Procedure EMD**

```julia
using FaultDetectionTools, DescriptorSystems, LinearAlgebra, Test

# Example 6.1 - Solution of an EMDP
println("Example 6.1")

# Lateral aircraft model without faults
A = [-.4492 0.046 .0053 -.9926;
      0 0 1 0.0067;
      -50.8436 0 -5.2184 .722;
      16.4148 0 .0026 -.6627];
Bu = [0.0004 0.0011; 0 0; -1.4161 .2621; -0.0633 -0.1205];
C = eye(4); p = size(C,1); mu = size(Bu,2);
# define the LOE faults Gamma_i
Gamma = 1 .- [ 0 0 0 .5 .5 .5 1 1 1;
               0 .5 1 0 .5 1 0 .5 1 ]';
N = size(Gamma,1);
# define multiple physical fault model Gui = Gu*Gamma_i
sysu = similar(Vector{DescriptorStateSpace},N)
for i = 1:N
    sysu[i] = dss(A,Bu*diagm(Gamma[i,:]),C,0);
end


# setup synthesis model
sysm = mdmodset(sysu, controls = 1:mu);

# call emdsyn with the options: stability degree -1,
# filter poles -1, and a common design matrix H to form
# linear combinations of the left nullspace basis vectors
H = [ 0.7645 0.8848 0.5778 0.9026 ];
```

```
Q, R, info = emdsyn(sysm, sdeg = -1, poles = [-1], HDesign = H);

distinf = info.MDperf
# check model detectability
@test all(diag(distinf).< 1.e-7) && all(distinf.+eps(2.)+I .>= 1)

println("Norms of final residual models")
display(distinf)
```

6.3 Solving the Approximate Model Detection Problem

Using Proposition 6.1, the solvability conditions of the *approximate fault detection problem* (AMDP) formulated in Sect. 4.4.2 for the multiple model (4.1), can be also expressed in terms of the multiple models (6.7), according to the following corollary to Theorem 4.3.

Corollary 6.2 *For the multiple model (4.1), the AMDP is solvable if and only if for the multiple models (6.7) the following conditions hold for $i = 1, \ldots, N$:*

$$\left[\overline{G}_u^{(i,j)}(\lambda) \ \overline{G}_d^{(i,j)}(\lambda) \right] \neq 0 \ \ \forall j \neq i . \tag{6.14}$$

We have seen in the proof of Theorem 4.3, that a solution of the AMDP can be determined by solving the related EMDP with $w^{(j)} \equiv 0$ for $j = 1, \ldots, N$, by using, for example, the **Procedure EMD**. However, potentially better solutions can be obtained by trying to maximize the gap between the requirements for high sensitivity to non-current models and strong attenuation of noise signals for the current model. An optimization-based approach, similar to that used for the solution of the AFDP, can be used to achieve this goal.

Consider the parametrization (6.4) of the i-th filter as $Q^{(i)}(\lambda) = \overline{Q}_1^{(i)}(\lambda) Q_1^{(i)}(\lambda)$. With the notation used in (6.8), we obtain from (4.5)

$$R^{(i,j)}(\lambda) = \overline{Q}_1^{(i)}(\lambda) \overline{R}^{(i,j)}(\lambda) , \tag{6.15}$$

where

$$\overline{R}^{(i,j)}(\lambda) := \left[\overline{G}_u^{(i,j)}(\lambda) \ \overline{G}_d^{(i,j)}(\lambda) \ \overline{G}_w^{(i,j)}(\lambda) \right] . \tag{6.16}$$

The above choice of $Q^{(i)}(\lambda)$ ensures that

$$\overline{R}^{(i,i)}(\lambda) = \left[0 \ 0 \ \overline{G}_w^{(i,i)}(\lambda) \right] . \tag{6.17}$$

Let $\gamma_i > 0$ be an admissible level for the effect of the noise signal $w^{(i)}(t)$ on the residual $r^{(i)}(t)$ in the case when the i-th model is the current model. In the light of (6.17), such a limitation can be imposed, for example, as a constraint of the form

$$\| R_w^{(i,i)}(\lambda) \|_{2/\infty} \leq \gamma_i , \tag{6.18}$$

where $R_w^{(i,j)}(\lambda)$ is defined in (4.5). Using (6.15)–(6.17), $R_w^{(i,i)}(\lambda)$ can expressed as $R_w^{(i,i)}(\lambda) = \overline{Q}_1^{(i)}(\lambda)\overline{G}_w^{(i,i)}(\lambda)$, and therefore, (6.18) becomes

$$\left\| \overline{Q}_1^{(i)}(\lambda)\overline{G}_w^{(i,i)}(\lambda) \right\|_{2/\infty} \leq \gamma_i . \tag{6.19}$$

For $\gamma_i > 0$, it is always possible, via a suitable scaling of the i-th filter, to use the normalized value $\gamma_i = 1$.

In the absence of noise, the influence of the j-th model on the i-th residual can be characterized by the associated gain $\left\| \begin{bmatrix} R_u^{(i,j)}(\lambda) & R_d^{(i,j)}(\lambda) \end{bmatrix} \right\|_{2/\infty}$. Therefore, as a measure of the global sensitivity of the i-th residual to the rest of $N - 1$ models different from the i-th model, the minimum values of these gains can be employed. Using the parametrization (6.4) of the i-th filter, the following sensitivity measure can be defined

$$\zeta_1^{(i)}\big(\overline{Q}_1^{(i)}(\lambda)\big) := \min_{1 \leq j \leq N, j \neq i} \left\| \overline{Q}_1^{(i)}(\lambda)\begin{bmatrix} \overline{G}_u^{(i,j)}(\lambda) & \overline{G}_d^{(i,j)}(\lambda) \end{bmatrix} \right\|_{2/\infty}, \tag{6.20}$$

where the dependence of $\zeta_1^{(i)}$ of the choice of the filter $\overline{Q}_1^{(i)}(\lambda)$ is explicitly emphasized. The requirement $\zeta_1^{(i)} > 0$ for $i = 1, \ldots, N$ can be interpreted as an alternative characterization of the *model detectability* of the N component models.

We can formulate several optimization problems (for different combinations of employed norms) to address the computation of a satisfactory (or even optimal) solution of the AMDP, having the goal of maximizing the model sensitivities (6.20) under the noise attenuation constraints (6.19). In what follows, we only discuss the $\mathcal{H}_\infty$-norm-based synthesis for which we give a detailed computational procedure. The synthesis of the i-th filter can be individually addressed by solving for each $i = 1, \ldots, N$ the following constrained optimization problem: Given $\gamma_i \geq 0$, determine $\beta_i > 0$ and a stable and proper filter $\overline{Q}_1^{(i)}(\lambda)$ such that

$$\beta_i = \max_{\overline{Q}_1^{(i)}(\lambda)} \left\{ \zeta_1^{(i)}\big(\overline{Q}_1^{(i)}(\lambda)\big) \,\Big|\, \left\| \overline{Q}_1^{(i)}(\lambda)\overline{G}_w^{(i,i)}(\lambda) \right\|_\infty \leq \gamma_i \right\}. \tag{6.21}$$

The gap $\eta_i := \beta_i/\gamma_i$ can be interpreted as a measure of the quality of i-th filter in differentiating between the i-th model and the rest of the models in the presence of noise. For $\gamma_i = 0$, the above formulation includes the exact solution (i.e., of the EMDP) and the corresponding gap is infinite.

To solve the formulated N optimization problems (6.21), we devise a synthesis procedure based on successive simplifications of the original problem by reducing it to simpler problems with the help of the factorized representations of the filters (6.4). The existence conditions of Corollary 6.2 can be immediately checked. In this context, we introduce a useful concept to simplify the presentation. A filter $Q^{(i)}(\lambda)$ is called *admissible* if the corresponding $\left[\, R_u^{(i,j)}(\lambda) \quad R_d^{(i,j)}(\lambda) \,\right]$ in (4.5) are all nonzero for $j \neq i$. Tests as those of Corollary 6.2 can be used to check admissibility. Assume that the test indicates the solvability of the AMDP.

Let q_i be the desired number of residual components for the i-th filter with output $r^{(i)}(t)$. As in the case of an EMDP, if a solution of the AMDP exists, then, in general, the use of a scalar output fault detection filter (thus, $q_i = 1$) is always possible. However, larger values of q_i can be advantageous, because may provide more free parameters, which can be appropriately tuned. In general, the choice of q_i must satisfy $q_i \leq p - r_d^{(i)}$, where $r_d^{(i)} := \operatorname{rank} G_d^{(i)}(\lambda)$. In the **Procedure AMD** to solve the AMDP, given in what follows, the choice $q_i \leq r_w^{(i)}$ is enforced, in the case when $r_w^{(i)} := \operatorname{rank} \overline{G}_w^{(i,i)}(\lambda) > 0$. This choice is only for convenience and leads to a simpler synthesis procedure.

As next step, the factor $\overline{Q}_1^{(i)}(\lambda)$ is determined in the product form $\overline{Q}_1^{(i)}(\lambda) = \overline{Q}_2^{(i)}(\lambda) Q_2^{(i)}(\lambda)$, where the $r_w^{(i)} \times (p - r_d^{(i)})$ factor $Q_2^{(i)}(\lambda)$ is determined such that $Q_2^{(i)}(\lambda) \overline{G}_w^{(i,i)}(\lambda)$ has full row rank $r_w^{(i)}$, the product $Q_2^{(i)}(\lambda) Q_1^{(i)}(\lambda)$ is admissible, and has the least possible McMillan degree. If this latter requirement is not imposed, then a simple choice is $Q_2^{(i)}(\lambda) = H^{(i)}$, where $H^{(i)}$ is an $r_w^{(i)} \times \left(p - r_d^{(i)}\right)$ full row rank constant matrix (e.g., chosen as a randomly generated matrix with orthonormal rows). This corresponds to building $Q_2^{(i)}(\lambda) Q_1^{(i)}(\lambda)$ as $r_w^{(i)}$ linear combinations of the left nullspace basis vectors contained in the rows of $Q_1^{(i)}(\lambda)$.

At this stage, the optimization problem to be solved falls into one of two categories. The *standard case* is when $Q_2^{(i)}(\lambda) \overline{G}_w^{(i,i)}(\lambda)$ has no unstable zeros on the boundary of the stability domain $\partial\mathbb{C}_s$ (i.e., the extended imaginary axis in the continuous-time case, or the unit circle centred in the origin in the discrete-time case). The *nonstandard case* corresponds to the presence of such zeros. This categorization is revealed in the next step, which also involves the computation of the respective zeros.

The quasi-co-outer–co-inner factorization of the full row rank $Q_2^{(i)}(\lambda) \overline{G}_w^{(i,i)}(\lambda)$ is

$$Q_2^{(i)}(\lambda) \overline{G}_w^{(i,i)}(\lambda) = G_{wo}^{(i)}(\lambda) G_{wi}^{(i)}(\lambda), \tag{6.22}$$

where the quasi-co-outer factor $G_{wo}^{(i)}(\lambda)$ is invertible, having only zeros in $\overline{\mathbb{C}}_s$, and $G_{wi}^{(i)}(\lambda)$ is co-inner. The factor $\overline{Q}_2^{(i)}(\lambda)$ is chosen in the product form $\overline{Q}_2^{(i)}(\lambda) = \overline{Q}_3^{(i)}(\lambda) Q_3^{(i)}(\lambda)$, with $Q_3^{(i)}(\lambda) = \left(G_{wo}^{(i)}(\lambda)\right)^{-1}$ and $\overline{Q}_3^{(i)}(\lambda)$ to be determined. Using (6.16), we define

$$\widetilde{R}^{(i,j)}(\lambda) := Q_3^{(i)}(\lambda) Q_2^{(i)}(\lambda) \overline{R}^{(i,j)}(\lambda), \tag{6.23}$$

with the component blocks defined as

$$\left[\, \widetilde{R}_u^{(i,j)}(\lambda) \mid \widetilde{R}_d^{(i,j)}(\lambda) \mid \widetilde{R}_w^{(i,j)}(\lambda) \,\right] := Q_3^{(i)}(\lambda) Q_2^{(i)}(\lambda) \left[\, \overline{R}_u^{(i,j)}(\lambda) \mid \overline{R}_d^{(i,j)}(\lambda) \mid \overline{R}_w^{(i,j)}(\lambda) \,\right].$$

This allows to express $\zeta_1^{(i)}$ in (6.20) as $\zeta_1^{(i)}\big(Q_1^{(i)}(\lambda)\big) = \zeta_3^{(i)}\big(Q_3^{(i)}(\lambda)\big)$, where

$$\zeta_3^{(i)}\big(\overline{Q}_3^{(i)}(\lambda)\big) := \min_{1 \le j \le N,\, j \ne i} \left\| \overline{Q}_3^{(i)}(\lambda)\left[\, \widetilde{R}_u^{(i,j)}(\lambda) \mid \widetilde{R}_d^{(i,j)}(\lambda) \,\right] \right\|_\infty. \tag{6.24}$$

It follows, that $\overline{Q}_3^{(i)}(\lambda)$ can be determined as the solution of

$$\beta_i = \max_{\overline{Q}_3^{(i)}(\lambda)} \left\{ \zeta_3^{(i)}\big(\overline{Q}_3^{(i)}(\lambda)\big) \,\Big|\, \big\| \overline{Q}_3^{(i)}(\lambda) \big\|_\infty \le \gamma_i \right\},$$

where we used that

$$\left\| \overline{Q}_3^{(i)}(\lambda) Q_3^{(i)}(\lambda) Q_2^{(i)}(\lambda) \overline{G}_w^{(i,i)}(\lambda) \right\|_\infty = \left\| \overline{Q}_3^{(i)}(\lambda) G_{wi}^{(i,i)}(\lambda) \right\|_\infty = \left\| \overline{Q}_3^{(i)}(\lambda) \right\|_\infty.$$

In the standard case, we can always ensure that the partial filter defined by the product of stable factors $Q_3^{(i)}(\lambda) Q_2^{(i)}(\lambda) Q_1^{(i)}(\lambda)$ is stable. However, $\widetilde{R}^{(i,j)}(\lambda)$ is generally not stable unless all component systems of the multiple model (4.1) are stable. In such a case, $\overline{Q}_3^{(i)}(\lambda)$ can be simply determined as $\overline{Q}_3^{(i)}(\lambda) = Q_4^{(i)}$, where $Q_4^{(i)}$ is a constant matrix representing the optimal solution of the simpler problem

$$\beta_i = \max_{Q_4^{(i)}} \left\{ \zeta_3^{(i)}\big(Q_4^{(i)}\big) \,\Big|\, \big\| Q_4^{(i)} \big\|_\infty \le \gamma_i \right\},$$

such that the resulting filter $Q^{(i)}(\lambda) = Q_4^{(i)} Q_3^{(i)}(\lambda) Q_2^{(i)}(\lambda) Q_1^{(i)}(\lambda)$ is admissible. For square $Q_4^{(i)}$, the choice $Q_4^{(i)} = \gamma_i I$ is the simplest optimal solution. If $\widetilde{R}^{(i,j)}(\lambda)$ is unstable or improper, the solution approach for the nonstandard case, discussed below, can be used.

The following result, given without proof, is similar to Theorem 5.2. The proof is similar to the proofs in the case of solving AFDPs in continuous- and discrete-time, see [81] and [82], respectively.

Theorem 6.2 *Using the parametrization* (6.4) *of the i-th filter and the notation in* (6.16), *let* $Q_2^{(i)}(\lambda)$ *be such that* $\left\| Q_2^{(i)}(\lambda)\left[\, \overline{G}_u^{(i,j)}(\lambda) \;\; \overline{G}_d^{(i,j)}(\lambda) \,\right] \right\|_\infty > 0$ *for all* $j \ne i$, *and, additionally,* $Q_2^{(i)}(\lambda) \overline{G}_w^{(i,i)}(\lambda)$ *has full row rank and has no zeros on the boundary of the stability domain. Then, for* $\gamma_i > 0$, *the optimal solution of the optimization problem* (6.21) *is*

$$\overline{Q}_{1,opt}^{(i)}(\lambda) := \gamma_i \left(G_{wo}^{(i)}(\lambda) \right)^{-1} Q_2^{(i)}(\lambda),$$

where $G_{wo}^{(i)}(\lambda)$ *is the co-outer factor of the co-outer–co-inner factorization* (6.22).

In the nonstandard case, both the partial filter $\widetilde{Q}^{(i)}(\lambda) := Q_3^{(i)}(\lambda) Q_2^{(i)}(\lambda) Q_1^{(i)}(\lambda)$ and the corresponding $\widetilde{R}^{(i,j)}(\lambda)$ in (6.23) for $j = 1, \ldots, N$, can result unstable or improper due the presence of poles of $Q_3^{(i)}(\lambda) = \left(G_{wo}^{(i)}(\lambda)\right)^{-1}$ in $\partial\mathbb{C}_s$ (i.e., $G_{wo}^{(i)}(\lambda)$ has zeros in $\partial\mathbb{C}_s$). In this case, $\overline{Q}_3^{(i)}(\lambda)$ is chosen in the form $\overline{Q}_3^{(i)}(\lambda) = Q_5^{(i)} Q_4^{(i)}(\lambda)$, where $Q_4^{(i)}(\lambda)$ results form a LCF with stable and proper factors

$$\left[\, \widetilde{Q}^{(i)}(\lambda)\ \widetilde{R}^{(i,1)}(\lambda) \ldots \widetilde{R}^{(i,N)}(\lambda)\,\right] = \left(Q_4^{(i)}(\lambda)\right)^{-1}\left[\, \widehat{Q}^{(i)}(\lambda)\ \widehat{R}^{(i,1)}(\lambda) \ldots \widehat{R}^{(i,N)}(\lambda)\,\right],$$

while $Q_5^{(i)}$ is a constant matrix, which solves

$$\beta_i = \max_{Q_5^{(i)}} \left\{\, \zeta_5^{(i)}\big(Q_5^{(i)}\big) \,\Big|\, \big\| Q_5^{(i)} Q_4^{(i)}(\lambda) \big\|_\infty \le \gamma_i \,\right\},$$

where

$$\zeta_5^{(i)}\big(Q_5^{(i)}\big) := \min_{1 \le j \le N, \, j \ne i} \big\| Q_5^{(i)} Q_4^{(i)}(\lambda)\big[\, \widetilde{R}_u^{(i,j)}(\lambda) \mid \widetilde{R}_d^{(i,j)}(\lambda)\,\big] \big\|_\infty.$$

The choice of a diagonal $Q_4^{(i)}(\lambda)$, with all its diagonal elements having $\mathcal{H}_\infty$-norms equal to 1, significantly simplifies the solution of the above problem. In this case, the choice $Q_5^{(i)} = \gamma_i I$ is always possible.

In the standard case, the dynamical order of the resulting filter $Q^{(i)}(\lambda)$ is the McMillan degree of $Q_3^{(i)}(\lambda)$, provided $Q_4^{(i)}(\lambda)$ is chosen a constant matrix. This order results from the conditions that $Q_2^{(i)}(\lambda)\overline{G}_w^{(i,i)}(\lambda)$ has full row rank and $Q_2^{(i)}(\lambda)Q_1^{(i)}(\lambda)$ has least order and is admissible. For each candidate $Q_2^{(i)}(\lambda)$, the corresponding optimal $Q_3^{(i)}(\lambda)$ results automatically, but the different "optimal" filters for the same level γ_i of noise attenuation performance can have significantly differing optimal performance levels β_i. Finding the best compromise between the achieved order and the achieved performance (measured via the gap β_i/γ_i), should take into account that larger orders and larger number of detector outputs q_i may potentially lead to better performance.

The **Procedure AMD**, given in what follows, allows the synthesis of least-order model detection filters, by solving the AMDP employing an $\mathcal{H}_\infty$ optimization-based approach. This procedure includes also the **Procedure EMD**, in the case when, an exact solution exists. Similar synthesis procedures, relying on alternative optimization-based formulations, can be devised by only adapting appropriately the last computational step of **Procedure AMD**.

Procedure AMD: Approximate synthesis of model detection filters

Inputs : $\{G_u^{(j)}(\lambda),\, G_d^{(j)}(\lambda),\, G_w^{(j)}(\lambda)\}$, for $j = 1, \ldots, N$; q_{max}
Outputs: $Q^{(i)}(\lambda)$, for $i = 1, \ldots, N$; $R^{(i,j)}(\lambda)$ for $i, j = 1, \ldots, N$

For $i = 1, \ldots, N$

(1) Compute a $(p - r_d^{(i)}) \times (p + m_u)$ minimal proper stable basis $Q_1^{(i)}(\lambda)$ for
the left nullspace of $G^{(i)}(\lambda)$ defined in (6.2), where $r_d^{(i)} := \operatorname{rank} G_d^{(i)}(\lambda)$;
set $Q^{(i)}(\lambda) = Q_1^{(i)}(\lambda)$, compute $\overline{G}_w^{(i,i)}(\lambda) = Q_1^{(i)}(\lambda) \begin{bmatrix} G_w^{(i)}(\lambda) \\ 0 \end{bmatrix}$, and

$$R^{(i,j)}(\lambda) = [\, R_u^{(i,j)}(\lambda) \mid R_d^{(i,j)}(\lambda) \mid R_w^{(i,j)}(\lambda)\,]$$

$$= Q_1^{(i)}(\lambda) \begin{bmatrix} G_u^{(j)}(\lambda) & G_d^{(j)}(\lambda) & G_w^{(j)}(\lambda) \\ I_{m_u} & 0 & 0 \end{bmatrix}, \quad j = 1, \ldots, N$$

Exit if $[\, R_u^{(i,j)}(\lambda)\ \ R_d^{(i,j)}(\lambda)\,] = 0$ for any $j \in \{1, \ldots, N\}$, $j \neq i$ (no solution)

(2) Compute $r_w^{(i)} = \operatorname{rank} \overline{G}_w^{(i,i)}(\lambda)$; if $r_w^{(i)} = 0$, set $q_1^{(i)} = \min(p - r_d^{(i)}, q_{max})$; else,
set $q_1^{(i)} = r_w^{(i)}$; choose a $q_1^{(i)} \times (p - r_d^{(i)})$ rational matrix $Q_2^{(i)}(\lambda)$ such that
$Q_2^{(i)}(\lambda)[\, R_u^{(i,j)}(\lambda)\ \ R_d^{(i,j)}(\lambda)\,] \neq 0$ for $j = 1, \ldots, N$, $j \neq i$, $Q_2^{(i)}(\lambda) Q^{(i)}(\lambda)$
has least McMillan degree, and, if $r_w^{(i)} > 0$, then rank $Q_2^{(i)}(\lambda) \overline{G}_w^{(i,i)}(\lambda) = r_w^{(i)}$;
compute $Q^{(i)}(\lambda) \leftarrow Q_2^{(i)}(\lambda) Q^{(i)}(\lambda)$ and $R^{(i,j)}(\lambda) \leftarrow Q_2^{(i)}(\lambda) R^{(i,j)}(\lambda)$ for
$j = 1, \ldots, N$, $j \neq i$.
(3) If $r_w^{(i)} > 0$, compute the quasi-co-outer–co-inner factorization (6.22) with
$G_{wo}^{(i)}(\lambda)$
invertible and having only zeros in $\overline{\mathbb{C}}_s$, and $G_{wi}^{(i)}(\lambda)$ co-inner;
with $Q_3^{(i)}(\lambda) = \big(G_{wo}^{(i)}(\lambda)\big)^{-1}$ compute $Q^{(i)}(\lambda) \leftarrow Q_3^{(i)}(\lambda) Q^{(i)}(\lambda)$ and
$R^{(i,j)}(\lambda) \leftarrow Q_3^{(i)}(\lambda) R^{(i,j)}(\lambda)$ for $j = 1, \ldots, N$, $j \neq i$.
(4) Choose a square rational matrix $Q_4^{(i)}(\lambda)$ such that $Q_4^{(i)}(\lambda) Q^{(i)}(\lambda)$ has a
desired stable dynamics and $Q_4^{(i)}(\lambda) R^{(i,j)}(\lambda)$ for $j = 1, \ldots, N$, $j \neq i$ are
stable; compute $Q^{(i)}(\lambda) \leftarrow Q_4^{(i)}(\lambda) Q^{(i)}(\lambda)$ and $R^{(i,j)}(\lambda) \leftarrow Q_4^{(i)}(\lambda) R^{(i,j)}(\lambda)$
for $j = 1, \ldots, N$, $j \neq i$.
(5) If $r_w^{(i)} > 0$, choose $Q_5^{(i)} \in \mathbb{R}^{\min(q_{max}, r_w^{(i)}) \times q_1^{(i)}}$ such that $\| Q_5^{(i)} Q_4^{(i)}(\lambda) \|_\infty = \gamma_i$
and $\beta_i = \min_{1 \leq j \leq N, j \neq i} \big\| Q_5^{(i)} [\, R_u^{(i,j)}(\lambda)\ \ R_d^{(i,j)}(\lambda)\,] \big\|_\infty > 0$; compute
$Q^{(i)}(\lambda) \leftarrow Q_5^{(i)} Q^{(i)}(\lambda)$ and $R^{(i,j)}(\lambda) \leftarrow Q_5^{(i)} R^{(i,j)}(\lambda)$ for $j = 1, \ldots, N$,
$j \neq i$; else, set $\beta_i = \infty$.

Remark 6.3 For the selection of the threshold τ_i for the component $r^{(i)}(t)$ of the residual vector, a similar approach to that described in Remark 5.11 can be used. The i-th residual, which results when the j-th model is the current one, is

$$\mathbf{r}^{(i)}(\lambda) = R_u^{(i,j)}(\lambda)\mathbf{u}(\lambda) + R_d^{(i,j)}(\lambda)\mathbf{d}^{(j)}(\lambda) + R_w^{(i,j)}(\lambda)\mathbf{w}^{(j)}(\lambda), \qquad (6.25)$$

where $R_u^{(i,j)}(\lambda)$, $R_d^{(i,j)}(\lambda)$, and $R_w^{(i,j)}(\lambda)$ are formed from the columns of $R^{(i,j)}(\lambda)$ corresponding to the inputs u, $d^{(j)}$ and $w^{(j)}$, respectively. To determine the false alarm bound for the i-th residual, we can use the residual, which results for the i-th filter if the i-th model is the current one. Taking into account that $R_u^{(i,i)}(\lambda) = 0$ and $R_d^{(i,i)}(\lambda) = 0$, we obtain

$$\mathbf{r}^{(i)}(\lambda) = R_w^{(i,i)}(\lambda)\mathbf{w}^{(i)}(\lambda) . \qquad (6.26)$$

If we assume, for example, a bounded energy noise input $w^{(i)}(t)$ such that $\|w^{(i)}\|_2 \leq \delta_w^{(i)}$, then the false alarm bound $\tau_f^{(i)}$ for the i-th residual vector component $r^{(i)}(t)$ can be computed as

$$\tau_f^{(i)} = \sup_{\|w^{(i)}\|_2 \leq \delta_w^{(i)}} \| R_w^{(i,i)}(\lambda)\mathbf{w}^{(i)}(\lambda)\|_2 = \| R_w^{(i,i)}(\lambda)\|_\infty \delta_w^{(i)} . \qquad (6.27)$$

The setting of the thresholds to $\tau_i = \tau_f^{(i)}$ for $i = 1, \ldots, N$ ensures no false alarms in detecting the i-th model, provided sufficient control, disturbance or noise activity is present such that

$$\|r^{(j)}\|_2 > \tau_f^{(j)}, \quad \forall j \neq i .$$

Therefore, to enhance the decision-making process it must be additionally checked that the control input u has a certain minimum energy, i.e., $\|u\|_2 > \underline{\delta}_u$, where $\underline{\delta}_u$ is the least size of the acceptable control inputs. A conservative (worst-case) estimate of $\underline{\delta}_u$ can be determined by enforcing

$$\left\| R_u^{(i,j)}(\lambda)\mathbf{u}(\lambda)\right\|_2 \geq \left\| R_d^{(i,j)}(\lambda)\mathbf{d}^{(j)}(\lambda)\right\|_2 + \left\| R_w^{(i,j)}(\lambda)\mathbf{w}^{(j)}(\lambda)\right\|_2$$

for $\left\|d^{(j)}\right\|_2 \leq \delta_d^{(j)}$ and $\left\|w^{(j)}\right\|_2 \leq \delta_w^{(j)}$, $\forall\, i, j$ with $j \neq i$. A possible choice is

$$\underline{\delta}_u = \max_{i,j; i \neq j} \frac{\left\| R_d^{(i,j)}(\lambda)\right\|_\infty \delta_d^{(j)} + \left\| R_w^{(i,j)}(\lambda)\right\|_\infty \delta_w^{(j)}}{\left\| R_u^{(i,j)}(\lambda)\right\|_\infty} .$$

$\square$

Example 6.2 This is basically the same multiple model as that used in Example 6.1, however with only two measured outputs, namely, the sideslip angle and roll angle, and additional input noise and output noise. The fault-free state-space model describes the continuous-time lateral dynamics of an F-16 aircraft with the matrices

$$
A^{(1)} = \begin{bmatrix} -0.4492 & 0.046 & 0.0053 & -0.9926 \\ 0 & 0 & 1.0000 & 0.0067 \\ -50.8436 & 0 & -5.2184 & 0.7220 \\ 16.4148 & 0 & 0.0026 & -0.6627 \end{bmatrix}, \quad B_u^{(1)} = \begin{bmatrix} 0.0004 & 0.0011 \\ 0 & 0 \\ -1.4161 & 0.2621 \\ -0.0633 & -0.1205 \end{bmatrix}, \quad B_w^{(1)} = [\, I_4 \ 0_{4\times 2}\,],
$$

$$
C^{(1)} = \begin{bmatrix} 57.2958 & 0 & 0 & 0 \\ 0 & 57.2958 & 0 & 0 \end{bmatrix}, \quad D_u^{(1)} = 0_{2\times 2}, \quad D_w^{(1)} = [\, 0_{2\times 4} \ I_2\,].
$$

The component system matrices in (2.19) are defined for $i = 1, 2, \ldots, N$ as: $E^{(i)} = I_4$, $A^{(i)} = A^{(1)}$, $C^{(i)} = C^{(1)}$, $B_w^{(i)} = B_w^{(1)}$, $D_w^{(i)} = D_w^{(1)}$, and $B_u^{(i)} = B_u^{(1)} \Gamma^{(i)}$, where $\Gamma^{(i)} = \mathrm{diag}\left(1 - \rho_1^{(i)}, 1 - \rho_2^{(i)}\right)$ and $\left(\rho_1^{(i)}, \rho_2^{(i)}\right)$ are the values of parameters (ρ_1, ρ_2) on the chosen grid points $\{0, 0.5, 1\}$ for each component of $\rho := [\rho_1, \rho_2]^T$. The values $\left(\rho_1^{(1)}, \rho_2^{(1)}\right) = (0, 0)$ correspond to the fault-free situation. The TFMs $G_u^{(i)}(s)$ and $G_w^{(i)}(s)$ of the i-th system can be expressed as

$$
G_u^{(i)}(s) = G_u^{(1)}(s)\Gamma^{(i)}, \quad G_w^{(i)}(s) = G_w^{(1)}(s), \tag{6.28}
$$

where

$$
G_u^{(1)}(s) = C^{(1)}\left(sI - A^{(1)}\right)^{-1} B_u^{(1)}, \quad G_w^{(1)}(s) = C^{(1)}\left(sI - A^{(1)}\right)^{-1} B_w^{(1)} + D_w^{(1)}.
$$

We applied the **Procedure AMD** to design $N = 9$ model detection filters of full dynamical order with two outputs. At Step (1), nullspace bases of the form

$$
Q_1^{(i)}(s) = \left[\, I_2 \ -G_u^{(i)}(s)\,\right] = \left[\, I_2 \ -G_u^{(1)}(s)\Gamma^{(i)}\,\right]
$$

have been chosen as initial designs. The internal forms corresponding to these designs are

$$
R_{u,1}^{(i,j)}(s) := Q_1^{(i)}(s)\begin{bmatrix} G_u^{(j)}(s) \\ I_2 \end{bmatrix} = G_u^{(1)}(s)\left(\Gamma^{(j)} - \Gamma^{(i)}\right),
$$

$$
R_{w,1}^{(i,j)}(s) := Q_1^{(i)}(s)\begin{bmatrix} G_w^{(j)}(s) \\ 0 \end{bmatrix} = G_w^{(1)}(s).
$$

At Step (2), the choice $Q_2^{(i)}(s) = I$ ensures that $Q_2^{(i)}(s)G_w^{(1)}(s)$ has full row rank and no zeros. Therefore, the co-outer–co-inner factorization (6.22) of $Q_2^{(i)}(s)G_w^{(1)}(s)$ computed at Step (3) allows to obtain the optimal solution for $\gamma_i = 1$ (see Theorem 6.2) as

$$
Q^{(i)}(s) = \left(G_{wo}^{(i)}(s)\right)^{-1} Q_1^{(i)}(s).
$$

The final internal forms of the filters, $R^{(i,j)}(s) = \left[\, R_u^{(i,j)}(s) \ R_w^{(i,j)}(s)\,\right]$, result for $i, j = 1, \ldots, N$ with

$$R_u^{(i,j)}(s) = \big(G_{wo}^{(i)}(s)\big)^{-1} G_u^{(1)}(s)\big(\Gamma^{(j)} - \Gamma^{(i)}\big), \qquad R_w^{(i,j)}(s) = G_{wi}^{(i)}(s),$$

and therefore, $R_u^{(i,j)}(s)$ preserves the monotonic mapping of distances between the i-th and j-th models. The performance of each filter $Q^{(i)}(s)$ is given by the resulting gap $\eta_i = \beta_i/\gamma_i (= \beta_i)$, where $\beta_i = \min_{j=1:N, i \neq j} \big\| R_u^{(i,j)}(s) \big\|_\infty$. For the resulting design, we have $\eta_i = 0.0525$, for $i = 1, \ldots, N$.

Each of the filters $Q^{(i)}(s)$ has McMillan degree 4, and therefore, the overall filter $Q(s)$ has the same complexity as a filter based on a bank of Kalman filters. A Kalman filter based approach is well-suited in the case when the input and measurement noise are Gaussian white noise processes. Assuming the input noise has a covariance of $\Sigma_x = 0.01^2 I_4$ and the measurement noise has a covariance of $\Sigma_y = 0.2^2 I_2$, then N Kalman filter-based residual generators $\widetilde{Q}^{(i)}(s)$, with state-space realizations of the form

$$\dot{x}_e^{(i)}(t) = (A^{(i)} - K^{(i)} C^{(i)}) x_e^{(i)}(t) + K^{(i)} y(t) + B^{(i)} u(t),$$
$$r^{(i)}(t) = C^{(i)} x_e^{(i)}(t) - y(t)$$

can be determined, where the optimal gains $K^{(i)}$ result by solving suitable algebraic Riccati equations. The achieved gaps for these filters are $\widetilde{\eta}_i = 0.0152$, and therefore, below the values achieved in the optimal $\mathcal{H}_\infty$ synthesis. An optimal synthesis with second-order scalar output residual generators achieves a gap of 0.0323.

In Fig. 6.3, the time responses of the residual evaluation signals $\theta_i(t)$ are presented, where $\theta_i(t)$ are computed using a Narendra-type evaluation filter (3.40) with input

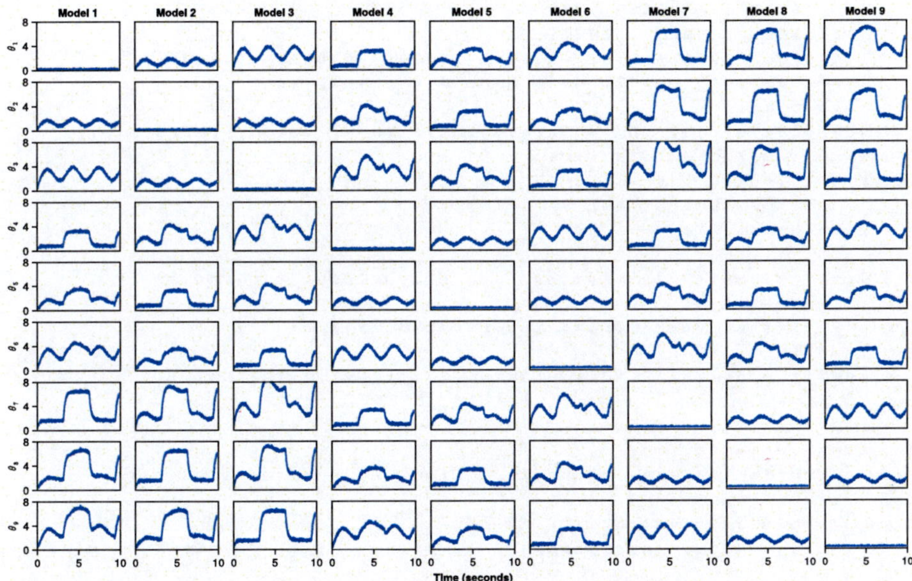

Fig. 6.3 Time responses of evaluation signals for the optimal syntheses

$\|r^{(i)}(t)\|_2^2$ and parameters $\alpha = 0.9$, $\beta = 0.1$, $\gamma = 10$ (see Sect. 3.6). The control inputs have been chosen as follows: $u_1(t)$ is a step of amplitude 0.3 added to a square wave of period 2π, and $u_2(t)$ is a step of amplitude 1.5 added to a sinus function of unity amplitude and period π. The noise inputs are zero mean white noise of amplitude 0.01 for the input noise and 0.03 for the measurement noise. Each column corresponds to a specific model for which the time responses of the N residual evaluation signals are computed. The achieved typical structure matrix for model detection (with zeros down the diagonal) can easily be read out from this signal-based assessment, even in the presence of noise.

The script **Ex6_2c.jl** in Listing 6.2 solves the AMDP considered in this example and calls the synthesis function **amdsyn**, which implements the **Procedure AMD**. The script also includes the verification of the model detectability condition. A more detailed version of this script is **Ex6_2.jl** (not listed), which closely follows the computational steps of the **Procedure AMD**. The script **Fig. 6_3.jl** (not listed) generates the plots in Fig. 6.3.   ◊

Listing 6.2 Script **Ex6_2c.jl** to solve the AMDP of Example 6.2 using **Procedure AMD**

```julia
using FaultDetectionTools, DescriptorSystems, LinearAlgebra, Test

# Example 6.2 - Solution of an AMDP
println("Example 6.2")

# Lateral aircraft model without faults
A = [-.4492 0.046 .0053 -.9926;
      0 0 1 0.0067;
      -50.8436 0 -5.2184 .722;
      16.4148 0 .0026 -.6627];
Bu = [0.0004 0.0011; 0 0; -1.4161 .2621; -0.0633 -0.1205];
n, mu = size(Bu); p = 2; mw = n+p; m = mu+mw;
Bw = eye(n,mw);
C = 180/pi*eye(p,n); Du = zeros(p,mu); Dw = [zeros(p,n) eye(p)];
# define the LOE faults Gamma_i
Gamma = 1 .- [ 0 0 0 .5 .5 .5 1 1 1;
               0 .5 1 0 .5 1 0 .5 1 ]';
N = size(Gamma,1);

# define the multiple physical fault model Gui = Gu*Gamma_i
# with noise inputs
sysuw = similar(Vector{DescriptorStateSpace},N)
for i = 1:N
  sysuw[i] = dss(A,[Bu*diagm(Gamma[i,:]) Bw],C,[Du Dw]);
end

# setup synthesis model
sysm = mdmodset(sysuw, controls = 1:mu, noise = mu+1:mu+mw);

# use nonminimal design with amdsyn
Q, R, info = amdsyn(sysm,sdeg = -1, poles = [-1], minimal = false);

distinf = info.MDperf
# check model detectability
@test all(diag(distinf) .< 1.e-7) && all(distinf.+eps(2.)+I .>= 1)
```

```
println("Norms of final residual models")
display(distinf)

println("Resulting gaps")
display(info.MDgap)
```

6.4 Notes and References

Section 6.1. The nullspace-based computational paradigm, which underlies the synthesis procedures presented in this chapter, has been discussed for the first time in the author's papers [150, 157] in the context of solving fault detection and isolation problems. The resulting factorized form of the component filters is similar to that for fault detection filters (see (5.1)) and is the basis of numerically reliable integrated computational algorithms. Specific numerical aspects of these algorithms are presented in Chap. 7. The parameterization of component filters given in Theorem 6.1 is similar to that used for solving FDI synthesis problems stated in Theorem 5.1. The nullspace-based characterization of model detectability in Proposition 6.1 can be interpreted as an extension of a special version of Theorem 3.5 for a particular structure matrix S.

Section 6.2. The nullspace-based synthesis method to solve the EMDP using least-order component filters has been proposed in [148]. The multiple model used in Example 6.1 has been used in [73] to address a fault tolerant control problem using interacting multiple-model Kalman filters. A solution with $N = 25$ models, allowing a more accurate identification of the degree of loss of efficiency, has been presented in [148].

Section 6.3. The solution method of the AMDP using an optimization-based method, summarized in **Procedure AMD**, represents a straightforward adaptation of the synthesis method for solving the AFDIP given in **Procedure AFDI**. The Kalman filter-based multiple-model approaches have been investigated by Wilsky in [174], where the Baram's proximity measure, introduced in [4], has been used to define the distance between two stochastic models. This measure is also the basis for discriminating among stochastic models in recently proposed methods for robust multiple-model adaptive control [44].

Chapter 7
Computational Issues

This chapter discusses the main computational issues underlying the synthesis procedures of fault detection filters presented in Chap. 5. While all synthesis procedures have been developed in terms of input–output system representations, the underlying numerical algorithms exclusively rely on state-space representation-based computational methods. The preference for state-space models is justified by discussing the suitability of different (input–output, polynomial, state-space) system representations for the development of reliable numerical algorithms. In this context, an important aspect discussed is the improvement of the conditioning of the state-space models using coordinate transformations.

All discussed computational procedures rely on the numerical algorithms described in detail in Chap. 10 for which quality assessments regarding their generality, numerical reliability and computational efficiency have been already established in the literature. Since the development of quality numerical algorithms was seldom considered in the fault detection literature, we discuss in a separate section the main attributes, which characterize a satisfactory numerical algorithm.

The core of our presentation is the discussion in depth of several basic computational paradigms, which are employed at typical computational steps of the synthesis procedures. These paradigms have been already mentioned in the introduction to Part II and, for the sake of completeness, are recalled here once again: (a) the product form representation of the synthesized filters, which leads to updating-based filter synthesis techniques; (b) the use of the nullspace method as a first synthesis step to reduce all synthesis problems to simpler forms, which allow to easily check solvability conditions; (c) the use of minimum dynamic cover algorithms, which allows addressing least-order synthesis problems; and (d) the use of coprime factorization techniques, which allows to conveniently enforce a desired filter dynamics. Without entering into algorithmic details, our discussion of the computational issues of the synthesis procedures primarily focuses on three aspects: (1) emphasizing the structural features present in the computed intermediary results (e.g., particular shapes of the matrices of the resulting state-space representations), (2) exploiting the

structural features achieved at the termination of a computational step at the subsequent steps and (3) developing, if possible, explicit updating formulas for the
state-space representations of partial filter syntheses.

The synthesis procedures of model detection filters, presented in Chap. 6, employ
similar computational paradigms, and therefore, the involved computational issues
are practically the same.

7.1 Developing Satisfactory Numerical Algorithms

The development of computational methods for solving the synthesis problems of
fault detection filters was a constant activity, which complemented most theoretical
works. Unfortunately, there are many signs of a general lack of numerical aware-
ness in the fault diagnosis community. For example, many of the proposed methods
employ highly questionable numerical techniques, as polynomial manipulations,
operations involving matrix products and powers, or even the computation of highly
sensitive canonical forms. Most of the proposed computational methods suffer from
the lack of guaranteed numerical reliability, and therefore, may produce inaccurate
results even for well-conditioned computational problems. Hence, such methods are
generally unsuited for solving large order problems.

Despite many algorithmic developments, it is rather surprising that, with a few
notable exceptions, the vast literature on fault detection until around 2000 contains
almost no results on the development of reliable numerical methods along the well-
established criteria for satisfactory algorithms in the field of numerical linear algebra.
Because of the lack of generality or the lack of numerical reliability, most of popular
synthesis techniques of fault detection filters (e.g., parity-space methods, geometric
methods, unknown-input-observer-based methods) cannot be considered as satisfactory numerical approaches. To remedy this situation, a new generation of numerically
reliable computational algorithms has been developed by the author during the last
decade. The new algorithms are able to solve various synthesis problems of fault
detection filter in the most general setting, without employing any technical assumptions.

In what follows, we shortly review the general principles that lead to the development of satisfactory numerical algorithms. The term *satisfactory numerical algorithm*
has been coined in the field of numerical linear algebra to designate an algorithm,
which is suitable to be implemented as quality numerical software, as exists nowadays for most standard linear algebra problems such as the solution of linear algebraic
equations, computation of eigenvalues and eigenvectors, etc. The standard requirements for a satisfactory algorithm, in the order of their importance, are generality,
reliability, stability, accuracy, and efficiency. As it will be shown, according to these
requirements, very few of the existing synthesis algorithms of fault detection filters
are completely satisfactory.

Generality means that the synthesis method is applicable to a wide class of
LTI systems, which fulfil the existence conditions of various synthesis problems.

Therefore, we only consider algorithms, which are able to compute a solution, whenever a solution exists. To ensure the highest level of generality, all proposed algorithms in this book are applicable to both standard and generalized systems, in both continuous- and discrete-time settings. Although the requirement for generality appears as a legitimate desideratum for any synthesis algorithm, many algorithms in the field of fault detection do not fulfil this elementary requirement. For example, synthesis methods based on unknown-input observers, and also the equivalent geometric synthesis methods, are not applicable for systems having unstable zeros in the disturbance channel (i.e., when $G_d(\lambda)$ is not minimum phase).

To define terms like reliability, stability and accuracy, we consider an algorithm as an abstract function f, which computes $y = f(u)$, where f acts on the data $u \in \mathcal{U} \subset \mathbb{R}^m$ to produce the "true" result $y \in \mathbb{R}^p$. Let u^* be an approximation of u (i.e., assume that u^* is "near" to u). If $f(u^*)$ is "near" to $f(u)$, the computational problem f is said to be *well conditioned* in u. If $f(u^*)$ and $f(u)$ may differ greatly, the problem is said to be *ill-conditioned*. The concept of "near" can be made precise by using appropriate norms. The perturbation analysis is the main mathematical tool to analyze the conditioning of problems and to assess the potential loss of accuracy (see below). Although the problem conditioning is independent of any specific algorithm used to solve the problem, it is important that the employed algorithms do not increase the problem sensitivity to small variations in the data. Consequently, any method based on employing ill-conditioned coordinate transformations (e.g., to highly sensitive "canonical" forms) can not be considered satisfactory.

The computation of $y = f(u)$ is done using finite-precision arithmetic, typically using double-precision floating-point computations involving 16 decimal accurate digits. Since the performed elementary computations such as, additions and multiplications, are always rounded to 16 digits, so-called roundoff errors occur at each performed floating-point operation. The cumulated effect of all roundoff errors makes that the finite-precision computational algorithm to evaluate $f(u)$ will in general produce $\overline{y} = f^*(u)$, where f^* denotes the function, which corresponds to the floating-point-based evaluation of f (sometimes denoted by $f^*(\cdot) = fl(f(\cdot))$). The algorithm f^* is said *numerically (backward) stable* if, for all $u \in \mathcal{U}$, there exists $u^* \in \mathcal{U}$ "near" u such that $f^*(u) = f(u^*)$, that is, the computed solution is the *exact* solution of a slightly perturbed problem.

The concept of *accuracy* is related to the resulting error $y - \overline{y}$ between the exact and computed solutions. If a stable algorithm is used to solve a well-conditioned problem, the computed solution $f^*(u)$ must be near to the exact solution $f(u)$. However, when a stable algorithm is used to solve an ill-conditioned problem, the resulting error may be large, since there is no guarantee that $f^*(u)$ and $f(u)$ are near from one another. The accuracy of an algorithm also refers to the error introduced by truncating infinite series or terminating iterations.

Proving the numerical stability of an algorithm allows to have guarantees that reasonably accurate results can be expected for well-conditioned computational problems. For complex algorithms, the proof of numerical stability can be very tedious, and even not possible, although all involved computational steps can be provably numerically stable. This is a typical case for complex synthesis procedures as those presented in this book, where the best we can achieve is to exclusively build on numerically stable computational methods. Fortunately, for many algorithms, it is relatively easy to recognize that they are numerically unstable. For example, the parity-space-based synthesis methods frequently perform multiplications with non-orthogonal matrices and also raising of matrices to powers. Such computations are, in general, numerically unstable, and, therefore, these synthesis methods can not be considered satisfactory. Interestingly, an algorithm can be unstable and still be reliable if the instability can be detected. An algorithm is said to be *reliable* if it gives some warning whenever it introduces excessive errors. A well-known example is the solution of systems of algebraic equations using Gaussian elimination with either partial or complete pivoting, where the growth of matrix elements during the elimination can be easily detected. In practice, however, such growth is rare, and therefore this algorithm is the method of choice when solving linear systems.

Efficiency is measured by the amount of computer time required to solve a particular problem. We are primarily concerned with the order of magnitude of the involved computational effort, in terms of the number of performed *floating-point operations* (flops), where 1 flop accounts for either one multiplication or one addition. For algorithms based on matrix computations, $\mathcal{O}(n^3)$ flops is the usual order of acceptable magnitude, where n is the largest dimension of the problem (e.g., the order of state-space realization). For example, a method, which would require $\mathcal{O}(n^4)$ flops, would be considered grossly inefficient unless the asymptotic estimate of the coefficient of n^4 is particularly small.

When developing numerical algorithms for solving synthesis problems, we rely on a vast collection of proven numerically stable or numerically reliable algorithms. It is generally not possible to show that complete synthesis algorithms are numerically stable, but it is always possible to show that all synthesis steps in these algorithms can be performed using numerically stable or numerically reliable methods. To assess this property, we rely on well-established techniques, which promote or even guarantee numerical stability, as—for example, the use of orthogonal transformations, while completely avoiding "dubious" computations, as—for example, forming matrix powers or performing reductions to canonical forms.

7.2 Modelling Issues

In this section, we discuss some modelling-related aspects, which are important when solving numerical problems. Two of the discussed aspects are of particular importance: the choice of the best-suited system representation for numerical computations and the conditioning of the chosen representation and its improvement. For simplicity, we consider only continuous-time representations, but all discussed aspects equally apply to discrete-time representations.

7.2.1 *System Representations*

In this section, we discuss the three main system representations of linear time-invariant systems: the input–output representation based on transfer function matrices, the polynomial-fraction models relying on polynomial matrices and the generalized state-space representations based on descriptor system representations. The basic data in the first two representations are polynomials with real coefficients, while state-space models use real matrices. The main aim of our discussion is to compare the three model classes with respect to their suitability for developing reliable numerical algorithms for the synthesis of fault detection filters. As it will be apparent, the main concern against using polynomial-based representations is the potential ill-conditioning of polynomial roots with respect to small variations in the coefficients. A second aspect is the lack of numerically reliable algorithms for polynomial or rational matrix manipulations to cover all computational aspects of solving the synthesis problems. In contrast, state-space representations are better suited for handling relatively large order models (e.g., up to a few hundred state variables) and, for addressing the solution of the synthesis problems in this book, a huge arsenal of numerically reliable algorithms and associated software is available.

Transfer-function-based representation of systems are widely used in the control and fault detection-related literature to describe input–output representations of the form

$$\mathbf{y}(s) = G(s)\mathbf{u}(s),\tag{7.1}$$

where $\mathbf{y}(s)$ and $\mathbf{u}(s)$ are the Laplace-transformed output $y(t) \in \mathbb{R}^p$ and input $u(t) \in \mathbb{R}^m$, respectively, and the *transfer function matrix* (TFM) $G(s)$ is $p \times m$ rational matrix. The (i, j)-th element of $G(s)$ has the form $g_{ij}(s) = \alpha(s)/\beta(s)$, where $\alpha(s)$ and $\beta(s)$ are polynomials with real coefficients. The complexity of TFM-based models is characterized by the number of outputs p and the number of inputs m, and additionally by the McMillan degree n, which represents the number of poles (finite and infinite) counting multiplicities. TFM-based models often result from physical model building or system identification. Typical values of n are not larger than 50–100. TFM-based models are well suited to describe elementary coupling operations of systems such as series, parallel or feedback couplings, to perform frequency-domain analysis (Bode, Nichols, etc) and to build discretized sampled-data representation. Input–output representations based on TFMs can be easily converted to alternative representations, as polynomial or state-space models. Due to their compactness, we used TFM-based models to illustrate some of the synthesis procedures presented in this book. However, as we will see, these models are not the best choice to perform numerical computations, especially for large values of n.

Polynomial models, in their simplest form, can be expressed as

$$M(s)\mathbf{y}(s) = N(s)\mathbf{u}(s),\tag{7.2}$$

where $M(s)$ and $N(s)$ are $p \times p$ and $p \times m$ polynomial matrices, respectively, with $M(s)$ nonsingular. The matrices of the two models (7.1) and (7.2) are related as

$$G(s) = M^{-1}(s)N(s),$$

which is the reason why these models are also called polynomial-fraction models. Polynomial models frequently arise from the first principle modelling of subsystems, as—for example, second-order linear differential equations describing the dynamics of mechanical systems.

The third representation we consider is the state-space model in the generalized form

$$\begin{aligned} E\dot{x}(t) &= Ax(t) + Bu(t), \\ y(t) &= Cx(t) + Du(t), \end{aligned} \tag{7.3}$$

where $x(t) \in \mathbb{R}^n$ is the system state variable, E, A, B, C and D are real matrices, and the pencil $A - \lambda E$ is regular. The state-space model of the form (7.3), with E possibly singular, is also called a *descriptor system model*, while for $E = I_n$, it is called a *standard state-space model*. The corresponding TFM of the input–output model (7.1) is given by

$$G(s) = C(sE - A)^{-1}B + D.$$

Models in the form (7.3) often result from the linearization of nonlinear physical plant models or from the spatial discretization of linear partial differential equation models. In the latter case, n may range from a few hundreds to several ten-thousands. The main appeal of state-space models is that they are generally better suited for numerical computations than their rational function or polynomial-fraction-based counterparts.

It is well known that polynomials with multiple roots are very sensitive to small variations in the coefficients. For example, the polynomial $s^2 - \varepsilon$ (a perturbation of size ε of s^2) has roots at $\pm\sqrt{\varepsilon}$, which is much bigger than ε, when ε is small. However, this large sensitivity may be present even in the case of polynomials with well-separated roots, if the order of polynomials is sufficiently large. This will be illustrated by the following example.

Example 7.1 The simple transfer function

$$g(s) = \frac{1}{(s+1)(s+2)\cdots(s+25)} = \frac{1}{s^{25} + 325s^{24} + \cdots + 25!}$$

has the exact poles $\{-1, -2, \ldots, -25\}$. The denominator is a modification of the famous Wilkinson polynomial (originally of order 20 and with positive roots), which has been used in many works to illustrate the pitfalls of algorithms for computing eigenvalues of a matrix by computing the roots of its characteristic polynomial. The Julia code in Listing 7.1 explicitly constructs the transfer function $g(s)$ and computes

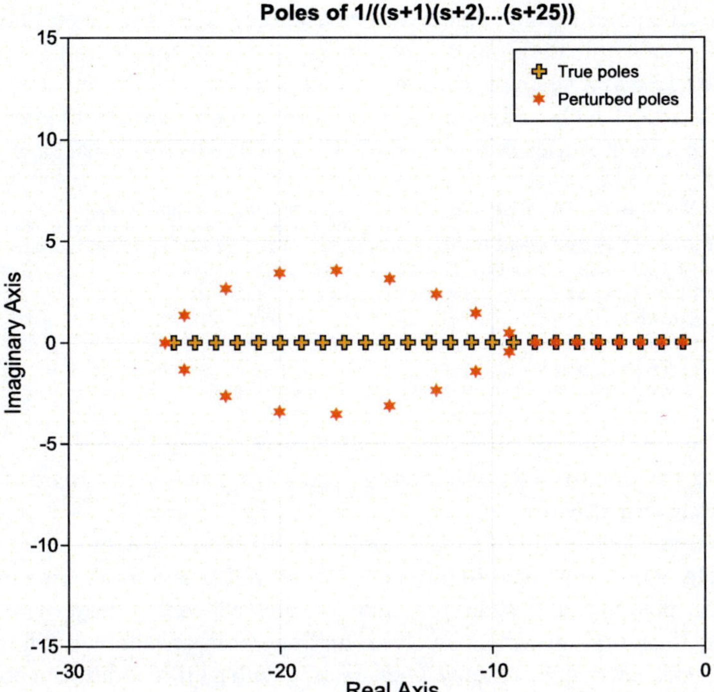

Fig. 7.1 Example for high sensitivity of polynomial poles

its poles. As can be observed from Fig. 7.1, some inaccurate poles with significant imaginary parts result (e.g., $-19.8511 \pm 3.2657i$ instead -19 and -20).

Listing 7.1 Script **Ex7_1.jl** to compute the results of Example 7.1

```
using DescriptorSystems , Polynomials
pexact = Vector(-25.:1.:-1.)
g = rtf(1,Polynomial(fromroots(pexact),:s))
poles = gpole(g)
println("Exact poles   Computed poles")
display([pexact poles])
```

The main reason for these inaccurately computed poles is the high sensitivity of the polynomial roots to small variations in the coefficients. In this case, inherent truncations take place in representing the large integer coefficients due to the finite representation with 16 accurate digits of double-precision floating-point numbers. For example, the constant term in the denominator $25! \approx 1.55 \cdot 10^{25}$ has 25 decimal digits, and thus can not be exactly represented with 16 digits precision. While the relative error in representing $25!$ is of the order of the machine-precision $\varepsilon_M \approx 10^{-16}$, the absolute error is of the order 10^9!

Interestingly, Julia allows to compute the poles to high accuracy using an extended precision representations of floating-point numbers. For example, the following code determines the poles to more than 20-digits accuracy, using the **BigFloat** representation of floating-point numbers in conjunction with the **GenericLinearAlgebra** package, which supports the computation of eigenvalues of matrices with **BigFloat** (i.e., with arbitrary precision) entries or with extended precision entries as provided by the **DoubleFloats** package.

Listing 7.2 Script **Ex7_1a.jl** to compute the exact results of Example 7.1

```
using DescriptorSystems, Polynomials, GenericLinearAlgebra
pexact = BigFloat.(Vector(-25.:1.:-1.))
g = rtf(1,Polynomial(fromroots(pexact),:s))
poles = gpole(g)
println("Exact poles  Computed poles")
display([pexact poles])
```

◊

Essentially the same results are obtained using the alternative extended precision package **GenericSchur**.

The extreme sensitivity of roots of polynomials with respect to small variations in the coefficients, illustrated in Example 7.1, is well known in the literature and is inherent for polynomial-based representations above a certain degree (say $n > 10$). Therefore, all algorithms, which involve rounding errors, are doomed to fail by giving results of extremely poor accuracy when dealing with an ill-conditioned polynomial. This potential loss of accuracy is one of the main reasons why polynomial-based system representations with rational or polynomial matrices are generally not suited for numerical computations.

It was frequently argued that polynomial-description-based algorithms are more efficient in terms of the involved computational effort due to an intrinsically more compact system representation. For a single-input single-output system of degree n, there are $2n$ coefficients to represent the polynomial model data, while for a state-space model there are $n^2 + 2n$ model parameters. This corresponds roughly to $\frac{n}{2}$ times less parameters in the polynomial representation, and this advantage also holds for multi-input multi-output models (albeit is less significant for larger number of inputs and outputs). At the algorithmic level, typically the polynomial algorithms have a complexity of n time smaller than equivalent state-space-based algorithms. In spite of this advantage in efficiency, it is frequently the case that algorithms for polynomial models suffer from numerical instabilities, whereas there exist satisfactory numerical algorithms for most problems formulated in state space. The main cause usually lies in the classes of allowed transformation performed in these algorithms. While for state-space representations usually well-conditioned orthogonal transformations can be routinely used for most of the required problem transformations, in the case of polynomial models unimodular transformations have to be employed. The usual pivot selection step merely involves checking for nonzero elements, and therefore usually there is no way to enforce the numerical stability of computations by selecting the "largest" pivots, as done in reducing real matrices using stabilized transformations. Therefore, the second major reason for the unsuitability of polynomial-based models is the lack of satisfactory numerical algorithms for most of the basic computations required for the solution of the main computational problems.

7.2.2 Model Conditioning

Occasionally, numerical algorithms may fail just because of poor problem formulations involving ill-conditioned models. This is why, we shortly discuss in this section the main aspects related to conditioning of state-space models. To simplify the discussions, we confine to LTI continuous-time standard state-space models of the form

$$\dot{x}(t) = Ax(t) + Bu(t), \\ y(t) = Cx(t) + Du(t),$$

(7.4)

which are specified by the quadruple of matrices (A, B, C, D). Two sources of possible ill-conditioning are discussed more in detail: poorly scaled models and poorly balanced models. To improve the model conditioning in each case, suitable coordinate transformations can be used.

Poorly scaled models usually exhibit a large spreading of numerical values in the elements of the system matrices. If the system matrices have elements with both very small and very large magnitudes, then numerical methods can simply fail because considering small, but physically nonzero elements, as zero values. Other times, small elements merely represent roundoff numerical errors and their handling as zeros is perfectly legitimate. In the first case, rescaling the system matrices can drastically improve the overall model conditioning. For example, by employing diagonal scaling matrices T_x, T_y and T_u, it is possible to replace in many problem formulations, the initial (ill-conditioned) model (A, B, C, D) by

$$(\widetilde{A}, \widetilde{B}, \widetilde{C}, \widetilde{D}) := (T_x^{-1} A T_x, T_x^{-1} B T_u, T_y C T_x, T_y D T_u).$$

The general aims for choosing the scaling matrices are to equilibrate the system matrices by compressing the numerical range of their elements and to normalize the dynamic ranges of input and output variables. It is important also to introduce no roundoff errors if possible, thus the diagonal elements of scaling matrices can be enforced to be powers of 2, the basis of numerical representation of floating-point numbers. Since automatic scaling is in some software a default option, it is important to mention some caveats regarding possible negative effects of scaling such as, increasing the sensitivity of (some) system poles and zeros, worsening the controllability or observability degrees of the system, or amplifying of roundoff errors in the data. This latter aspect is particularly important when employing structure exploiting and structure preserving algorithms, where roundoff errors may blur the fixed zero entries in the system matrices. In such cases, it is advisable to set explicitly to zero small elements resulting due to roundoff errors (if these can be easily recognized).

A second source of ill-conditioning is present in so-called unbalanced models, where the degree of controllability and degree of observability are significantly differing. Measures of the degrees of controllability and observability for a stable system can be defined in terms of the controllability gramian P and observability gramian R, respectively. The two symmetric and positive semi-definite gramians satisfy the respective Lyapunov equations

$$AP + PA^T + BB^T = 0, \qquad A^T R + RA + C^T C = 0.$$

The controllability and observability properties of the system (7.4) can be (equivalently) characterized by the positive definiteness of the controllability and observability gramians, $P > 0$ and $R > 0$, respectively. The *degree of controllability* can be defined as the least eigenvalue of P, while the *degree of observability* as the least eigenvalue of R. If any of these eigenvalues is zero, the system is not minimal (i.e., uncontrollable, or unobservable, or both). For a *balanced system*, the two positive definite gramians are equal and diagonal, $P = R = \Sigma$, with the diagonal elements of Σ ordered with decreasing magnitudes. While the eigenvalues of Σ (also called *Hankel singular values*) are invariant to a system coordinate transformation of the form $x = T\widetilde{x}$, with T nonsingular, the two gramians are generally not. To balance a system, a coordinate transformation $x = T\widetilde{x}$ is performed with a nonsingular transformation matrix T, to obtain the transformed system

$$(\widetilde{A}, \widetilde{B}, \widetilde{C}, \widetilde{D}) := (T^{-1}AT, T^{-1}B, CT, D). \tag{7.5}$$

The system transformation (7.5) is called a *similarity transformation* and the two systems (A, B, C, D) and $(\widetilde{A}, \widetilde{B}, \widetilde{C}, \widetilde{D})$ have the same TFM $G(s)$. From the singular value decomposition $RP = U\Sigma^2 V^T$, where U and V are orthogonal matrices and Σ diagonal, we can determine the balancing transformation matrix as $T = PV\Sigma^{-1}$ (also $T^{-1} = \Sigma^{-1}U^T R$). The main aim of system balancing is to determine statespace representations with minimum sensitivity to small variations in the matrix elements. An important application area of system balancing is in model reduction. Unfortunately, balancing often relies on ill-conditioned transformations if the underlying system is nearly non-minimal (the least eigenvalue of Σ is very small) or nearly unstable. For non-minimal stable systems in descriptor form, a general approach to compute balanced minimal realizations is described in Sect. 10.4.4 (see also **Procedure GBALMR**).

We discuss shortly the effects of a similarity transformation with a transformation matrix T on small errors in the system matrices. Assume A is perturbed with a small error δA. The effect of this perturbation via the similarity transformation with T can be expressed as

$$\widetilde{A} + \delta\widetilde{A} := T^{-1}(A + \delta A)T,$$

where

$$\delta\widetilde{A} = T^{-1}\delta AT.$$

It follows that

$$\|\delta\widetilde{A}\| \leq \kappa(T)\|\delta A\|,$$

where

$$\kappa(T) := \|T^{-1}\|\|T\| \geq 1.$$

is the *condition number* of T with respect to matrix inversion. The condition number $\kappa(T)$ measures the worst possible amplification of perturbations in A, which can be induced in the transformed $\widetilde{A}$ by the transformation matrix T. Evidently, a large value of $\kappa(T)$ indicates a large potential loss of accuracy and, therefore, T is called an ill-conditioned transformation. Interestingly, model conditioning transformations are frequently ill-conditioned, but the resulting scaled or balanced models have usually better numerical properties for subsequent computations. If T is an orthogonal transformation matrix satisfying $T^T T = I$, then $\kappa(T) = 1$ (the least achievable value), and therefore T is a perfectly conditioned transformation matrix, which ensures that $\|\delta\widetilde{A}\| = \|\delta A\|$. This desirable property justifies the use of orthogonal similarity transformations as the principal class of transformations for which the numerical stability can be often proved.

In some works on fault diagnosis, transformations to so-called canonical forms are used in the proposed synthesis algorithms. In general, canonical forms such as the Jordan form of the state matrix A, or controllability (observability) related forms of the matrix pair (A, B) $((A, C))$, or the Kronecker-form of a matrix pencil $A - \lambda E$ are notorious examples, whose computation may involve ill-conditioned transformations leading to extremely sensitive condensed forms. The dangers of such an approach are illustrated in the following simple example to compute the controllability companion form of a state-space model.

Example 7.2 Consider a state-space realization of order $n = N$ of $g(s)$ in Example 7.1 with

$$A = \begin{bmatrix} -1 & 1 & 0 & \cdots & 0 & 0 \\ 0 & -2 & 1 & \cdots & 0 & 0 \\ \vdots & \vdots & \vdots & \ddots & \ddots & \vdots \\ 0 & 0 & 0 & \cdots & -N+1 & 1 \\ 0 & 0 & 0 & \cdots & 0 & -N \end{bmatrix}, \quad B = \begin{bmatrix} 0 \\ 0 \\ \vdots \\ 0 \\ 1 \end{bmatrix},$$

$$C = \begin{bmatrix} 1 & 0 & 0 & \cdots & 0 & 0 \end{bmatrix}, \qquad D = \begin{bmatrix} 0 \end{bmatrix}.$$

The transformation matrix to the controllable companion form of the pair (A, B) can be explicitly built as

$$T = \begin{bmatrix} B & AB & \cdots & A^{N-1}B \end{bmatrix}.$$

If P is the $N \times N$ permutation matrix

$$
P = \begin{bmatrix} 0 & 0 & \cdots & 0 & 1 \\ 0 & 0 & \cdots & 1 & 0 \\ \vdots & \vdots & & \vdots & \vdots \\ 0 & 1 & \cdots & 0 & 0 \\ 1 & 0 & \cdots & 0 & 0 \end{bmatrix},
$$

then it is easy to see that PT is an upper triangular matrix with unit elements on the diagonal. Thus, T^{-1} can be easily computed as $T^{-1} = (PT)^{-1}P$. If we denote

$$
(s+1)(s+2)\cdots(s+N) = s^N + \beta_1 s^{N-1} + \ldots + \beta_{N-1}s + \beta_N,
$$

then the transformed state-space matrices are

$$
\widetilde{A} = T^{-1}AT = \begin{bmatrix} 0 & 0 & \cdots & 0 & -\beta_N \\ 1 & 0 & \cdots & 0 & -\beta_{N-1} \\ 0 & 1 & \cdots & 0 & -\beta_{N-2} \\ \vdots & \vdots & \ddots & \vdots & \vdots \\ 0 & 0 & \cdots & 1 & -\beta_1 \end{bmatrix}, \quad \widetilde{B} = T^{-1}B = \begin{bmatrix} 1 \\ 0 \\ \vdots \\ 0 \end{bmatrix},
$$

$$
\widetilde{C} = CT = \begin{bmatrix} 0 & 0 & \cdots & 0 & 1 \end{bmatrix}.
$$

Up to $N = 13$, the computed $\widetilde{A}$ is exact, although for $N = 13$ the condition number $\kappa(T)$ is of order 10^{23}. However, starting with $N = 14$, the difference between the above theoretical form and the numerically computed form is nonzero. For $N = 25$, this difference is extremely large (of order 10^{26}), which is partly the consequence of the large condition number $\kappa(T)$ of order 10^{59}. $\Diamond$

Finally, we discuss the conversion from a descriptor system representation of the form

$$
\begin{aligned}
E\dot{x}(t) &= Ax(t) + Bu(t), \\
y(t) &= Cx(t) + Du(t),
\end{aligned} \tag{7.6}
$$

with $x(t) \in \mathbb{R}^n$, to a standard state-space system of the form

$$
\begin{aligned}
\dot{\widetilde{x}}(t) &= \widetilde{A}\widetilde{x}(t) + \widetilde{B}u(t), \\
y(t) &= \widetilde{C}\widetilde{x}(t) + \widetilde{D}u(t),
\end{aligned} \tag{7.7}
$$

with $\widetilde{x}(t) \in \mathbb{R}^{\tilde{n}}$ and $\tilde{n} \leq n$. This conversion is necessary, for example, after the final synthesis step, to obtain the final fault detection filter in a standard state-space form, which is better suited for real-time processing. However, we cautiously recommend to avoid such conversions at early steps of the synthesis procedures, unless we can guarantee that no significant loss of accuracy takes place due to ill-conditioned transformations.

The conversion to a standard state-space form can be performed only if the associated TFM $G(s) = C(sE - A)^{-1}B + D$ is proper. For simplicity, we consider only the case when the descriptor realization is irreducible. Therefore, we assume that the regular pencil $A - sE$ has r finite eigenvalues and $n - r$ simple eigenvalues at infinity, where r is the rank of E. When E is nonsingular, we can simply choose $\widetilde{x}(t) = x(t)$ and

$$\widetilde{A} = E^{-1}A, \quad \widetilde{B} = E^{-1}B, \quad \widetilde{C} = C, \quad \widetilde{D} = D,$$

or alternatively choose $\widetilde{x}(t) = Ex(t)$ and

$$\widetilde{A} = AE^{-1}, \quad \widetilde{B} = B, \quad \widetilde{C} = CE^{-1}, \quad \widetilde{D} = D.$$

In these conversion formulas, the inverse of E is explicitly involved, and therefore, severe loss of accuracy can occur if the condition number $\kappa(E)$ is large. A somewhat better choice is to use the *singular value decomposition* (SVD) $E = U\Sigma V^T$, with U and V orthogonal matrices and Σ a diagonal matrix whose diagonal elements are the decreasingly ordered singular values $\sigma_1 \geq \sigma_2 \geq \cdots \geq \sigma_n > 0$. We can choose $\widetilde{x}(t) = \Sigma^{\frac{1}{2}} V^T x(t)$ and

$$\widetilde{A} = \Sigma^{-\frac{1}{2}} U^T A V \Sigma^{-\frac{1}{2}}, \quad \widetilde{B} = \Sigma^{-\frac{1}{2}} U^T B, \quad \widetilde{C} = C V \Sigma^{-\frac{1}{2}}, \quad \widetilde{D} = D.$$

From the SVD of E, we can easily compute the condition number $\kappa(E) = \sigma_1/\sigma_n$, and thus have a rough estimation of potential loss of accuracy induced by using the above transformation.

More involved transformation is necessary when E is singular, with rank $E = r < n$. In this case, we employ the singular value decomposition of E in the form

$$E = U\Sigma V^T := \begin{bmatrix} U_1 & U_2 \end{bmatrix} \begin{bmatrix} \widetilde{\Sigma} & 0 \\ 0 & 0 \end{bmatrix} \begin{bmatrix} V_1 & V_2 \end{bmatrix}^T,$$

where $\widetilde{\Sigma}$ is a nonsingular diagonal matrix of order $\tilde{n} := r$ with the nonzero singular values of E on the diagonal, and U and V are compatibly partitioned orthogonal matrices. If we apply a system similarity transformation with the transformation matrices

$$\widetilde{U} = \mathrm{diag}\left(\widetilde{\Sigma}^{-\frac{1}{2}}, I_{n-r}\right) U^T, \quad \widetilde{V} = V \, \mathrm{diag}\left(\widetilde{\Sigma}^{-\frac{1}{2}}, I_{n-r}\right)$$

we obtain

$$\widetilde{U}(A - sE)\widetilde{V} = \begin{bmatrix} A_{11} - sI_r & A_{12} \\ A_{21} & A_{22} \end{bmatrix}, \quad \widetilde{U}B = \begin{bmatrix} B_1 \\ B_2 \end{bmatrix}, \quad C\widetilde{V} = \begin{bmatrix} C_1 & C_2 \end{bmatrix},$$

where A_{22} is nonsingular, due to the assumption of only simple infinite eigenvalues of the regular pencil $A - sE$. The above transformed matrices correspond to the coordinate transformation $\overline{x} = \widetilde{V}^{-1} x(t)$ and lead to the partitioned system representation

$$\dot{\overline{x}}_1(t) = A_{11}\overline{x}_1(t) + A_{12}\overline{x}_2(t) + B_1u(t),$$
$$0 = A_{21}\overline{x}_1(t) + A_{22}\overline{x}_2(t) + B_2u(t),$$
$$y(t) = C_1\overline{x}_1(t) + C_2\overline{x}_2(t) + Du(t),$$

where $\overline{x}(t) = \begin{bmatrix} \overline{x}_1(t) \\ \overline{x}_2(t) \end{bmatrix}$ is partitioned such that $\overline{x}_1(t) \in \mathbb{R}^r$ and $\overline{x}_2(t) \in \mathbb{R}^{n-r}$. We can solve the second (algebraic) equation for $\overline{x}_2(t)$ to obtain

$$\overline{x}_2(t) = -A_{22}^{-1}A_{21}\overline{x}_1(t) - A_{22}^{-1}B_2u(t)$$

and arrive to a standard system representation with $\widetilde{x}(t) = \overline{x}_1(t)$ and the corresponding matrices

$$\widetilde{A} = A_{11} - A_{12}A_{22}^{-1}A_{21}, \quad \widetilde{B} = B_1 - A_{12}A_{22}^{-1}B_2,$$
$$\widetilde{C} = C_1 - C_2A_{22}^{-1}A_{21}, \quad \widetilde{D} = D - C_2A_{22}^{-1}B_2.$$

In this case, if any of the condition numbers $\kappa(\widetilde{\Sigma})$ or $\kappa(A_{22})$ is large, potential accuracy losses can be induced by the conversion to a standard state-space form.

This discussion emphasizes that the synthesis algorithms must be able to cover both standard and descriptor system representations, because any conversion to a standard state-space form may induce severe accuracy losses in the whole chain of subsequent computations due to the use of possibly ill-conditioned transformations. This is one of the reasons why an extensive research effort took place over several decades to develop algorithms for descriptor system representations. These algorithms work directly on the original descriptor representation, without the need for conversion to a standard system form (even if this is feasible). However, if all involved transformations are well conditioned, such a conversion is numerically harmless, and therefore, can be safely performed. Such a conversion can occasionally reduce the computational effort, because the algorithms for descriptor systems are usually more involved than their counterparts for standard systems.

7.3 Basic Procedural Framework

In the rest of this chapter, we discuss the computational issues related to the synthesis procedures of fault detection filters presented in Chap. 5. These discussions also cover the main numerical issues associated with the synthesis procedures of model detection filters presented in Chap. 6. It was shown in the previous section that polynomial-based system representations are generally not suited for numerical computations due to potential high sensitivity of such models, which precludes any guarantee for numerical reliability. Therefore, for the development of numerically reliable computational synthesis algorithms, which are suitable for robust software implementations, we focus exclusively on algorithms based on the manipulation of state-space representations. For all computational subproblems formulated in terms of TFMs in the synthesis procedures presented in Chap. 5, we indicate the state-space-based computational methods, which are best suited for solving these problems. The choice of suitable algorithms is guided by the requirements formulated in Sect. 7.1 for satisfactory numerical algorithms.

The synthesis procedures of the FDI filters are formulated in Chap. 5 in terms of three mathematical objects, which are recalled below:

– the system model with additive faults has the input–output representation

$$\mathbf{y}(\lambda) = G_u(\lambda)\mathbf{u}(\lambda) + G_d(\lambda)\mathbf{d}(\lambda) + G_f(\lambda)\mathbf{f}(\lambda) + G_w(\lambda)\mathbf{w}(\lambda) \qquad (7.8)$$

and represents the main input data object to all synthesis procedures;
– the fault detection filter in the input–output implementation form

$$\mathbf{r}(\lambda) = Q(\lambda)\begin{bmatrix} \mathbf{y}(\lambda) \\ \mathbf{u}(\lambda) \end{bmatrix} = Q_y(\lambda)\mathbf{y}(\lambda) + Q_u(\lambda)\mathbf{u}(\lambda) \qquad (7.9)$$

is the main output object of all synthesis procedures;
– the fault detection filter in the input–output internal form

$$\mathbf{r}(\lambda) = R_u(\lambda)\mathbf{u}(\lambda) + R_d(\lambda)\mathbf{d}(\lambda) + R_f(\lambda)\mathbf{f}(\lambda) + R_w(\lambda)\mathbf{w}(\lambda) \qquad (7.10)$$

is the second main output object of most of synthesis procedures.

Recall that the implementation and internal forms are related as follows

$$\left[\, R_u(\lambda)\,\middle|\,R_d(\lambda)\,\middle|\,R_f(\lambda)\,\middle|\,R_w(\lambda)\,\right] := Q(\lambda)\begin{bmatrix} G_u(\lambda) & \middle|G_d(\lambda) & \middle|G_f(\lambda) & \middle|G_w(\lambda) \\ I_{m_u} & 0 & 0 & 0 \end{bmatrix}. \qquad (7.11)$$

In all synthesis procedures, the nullspace method is used as the first synthesis step to enforce $R_u(\lambda) = 0$ and $R_d(\lambda) = 0$. Also, for all exact synthesis methods $w \equiv 0$ is assumed, and therefore no $R_w(\lambda)$ term exists.

For a typical synthesis procedure with K synthesis steps, we can express the resulting filter $Q(\lambda)$ in a factored form

$$Q(\lambda) = Q_K(\lambda)\cdots Q_2(\lambda)Q_1(\lambda)\,, \qquad (7.12)$$

where $Q_i(\lambda)$, $i = 1, \ldots, K$ are the factors computed at the i-th synthesis step. The computational steps usually involve the updating of both the implementation form and the nonzero parts of the internal forms of the fault detection filter. All synthesis procedures presented in Chap. 5 have the following simple logical structure:

> (1) Compute $Q_1(\lambda)$, a left nullspace basis of $\begin{bmatrix} G_u(\lambda)\, G_d(\lambda) \\ I_{m_u} \quad 0 \end{bmatrix}$;
>
>   set $Q(\lambda) = Q_1(\lambda)$ and $R(\lambda) = Q_1(\lambda)\begin{bmatrix} G_f(\lambda) & G_w(\lambda) \\ 0 & 0 \end{bmatrix}$;
>
>   **Exit** if solvability conditions are not fulfilled.
>   **For** $i = 2, \ldots, K$
> (i) Choose $Q_i(\lambda)$; update $\begin{bmatrix} Q(\lambda) & R(\lambda) \end{bmatrix} \leftarrow Q_i(\lambda)\begin{bmatrix} Q(\lambda) & R(\lambda) \end{bmatrix}$

The use of the nullspace method as the first synthesis is instrumental in simplifying the formulations of all synthesis problems. This approach emerged in the last decade as an important computational paradigm for all synthesis algorithms presented in this book. The main appeals of the nullspace-based fault detection filter synthesis are: generality, being applicable to both standard and singular (or non-proper) systems; numerical reliability, by relying on numerically sound computational techniques; and flexibility, by leading to simplified problem formulations, which allow to easily check solvability conditions and address the least- order synthesis problems.

The choice of each $Q_i(\lambda)$ at subsequent steps is based on the current partial synthesis $Q(\lambda)$ and $R(\lambda)$ computed at the previous iteration step. In this chapter, we discuss in detail both the determination of $Q_i(\lambda)$ at each computational step as well as the derivation of suitable updating formulas for $Q(\lambda)$ and $R(\lambda)$, which exploit all possible pole–zero cancellations. In this context, a second paradigm emerged when developing generally applicable, numerically reliable and computationally efficient synthesis methods. The so-called integrated synthesis algorithms rely on the successive updating of partial syntheses, where each step addresses specific synthesis requirements. Since each partial synthesis may represent a valid fault detection filter, the updating-based approach has a built-in flexibility in devising several particular synthesis approaches. However, the main strength of the integrated algorithms lies in their ability to exploit at each updating step all available structural information at the previous step, which globally leads to very efficient structure exploiting computations to perform the necessary updating.

To develop numerically reliable computational procedures, we exclusively rely on state-space representations of the different synthesis objects. The state-space representation of the system model with faults (see also (2.2)) is

$$\begin{bmatrix} G_u(\lambda) & G_d(\lambda) & G_f(\lambda) & G_w(\lambda) \end{bmatrix} := \left[\begin{array}{c|cccc} A - \lambda E & B_u & B_d & B_f & B_w \\ \hline C & D_u & D_d & D_f & D_w \end{array}\right], \qquad (7.13)$$

where we can generally assume that the state-space realization (7.13) is minimal. A particular feature of the realization (7.13) is that the individual realizations of $G_u(\lambda)$, $G_d(\lambda)$, $G_f(\lambda)$, and $G_w(\lambda)$ share the common observable pair $(A - \lambda E, C)$. Although for most practical problems the original system model corresponds to a causal physical system for which we can always enforce $E = I$, we prefer to use the generalized system representation in (7.13) in conjunction with computational techniques for descriptor systems to prevent possible loss of accuracy due to ill-conditioned inversion of E and, simultaneously, to preserve the most general problem formulation for arbitrary E. As it will be apparent in the next section, the computational aspects are practically the same for systems in both standard and descriptor forms.

For the intermediary (or even final) forms of the fault detection filter $Q(\lambda)$ (implementation form) and $R(\lambda)$ (internal form) it is possible to enforce at the end of each computational step state-space representations of the form

$$\left[\, Q(\lambda) \; R(\lambda) \,\right] = \left[\begin{array}{c|cc} A_Q - \lambda E_Q & B_Q & B_R \\ \hline C_Q & D_Q & D_R \end{array}\right],\tag{7.14}$$

where the descriptor pair $(A_Q - \lambda E_Q, C_Q)$ is observable. This is done either explicitly, via explicit updating formulas or, implicitly, by using minimal realization techniques to enforce all possible pole–zero cancellations.

In the following sections, we describe the computational aspects of the main steps of the synthesis procedures presented in Chaps. 5 and 6. The emphasis of the presentation is to give an overview of the basic computations performed at each synthesis step, but without entering into algorithmic details. However, for the interested readers, these details are partly explained in the Chap. 10, and further, in the references provided in the Sects. 7.11 and 10.6. Some special algorithms for descriptor systems, which are instrumental for solving the computational problems for the synthesis of fault detection filters, are described in Sect. 10.4. Available software implementations are mentioned in Sect. 10.5.

7.4 Nullspace-Based Reduction

In this section we discuss the numerical computations performed at Step (1) of all synthesis procedures presented in this book. In particular, this is the main step of **Procedure EFD**, where the resulting (stable) filter represents a solution of the EFDP, provided the fault detectability conditions are fulfilled. This step typically involves two main computations. The first one is the computation of $N_l(\lambda)$, a proper rational left nullspace basis of the $(p + m_u) \times (m_u + m_d)$ TFM

$$G(\lambda) = \left[\begin{array}{cc} G_u(\lambda) & G_d(\lambda) \\ I_{m_u} & 0 \end{array}\right].\tag{7.15}$$

This serves to set the first factor of $Q(\lambda)$ to $Q_1(\lambda) := N_l(\lambda)$ and to initialize the filter synthesis by setting the TFM of the implementation form of the filter to $Q(\lambda) = Q_1(\lambda)$. The second computation is the determination of the nonzero TFMs of the reduced proper system (5.11)

$$\left[\, \overline{G}_f(\lambda) \; \overline{G}_w(\lambda) \,\right] := N_l(\lambda)F(\lambda),\tag{7.16}$$

where

$$F(\lambda) := \begin{bmatrix} G_f(\lambda) & G_w(\lambda) \\ 0 & 0 \end{bmatrix}. \tag{7.17}$$

This serves to initialize the TFM of the internal form as $R(\lambda) = \begin{bmatrix} \overline{G}_f(\lambda) & \overline{G}_w(\lambda) \end{bmatrix}$. For both computations, we rely on the numerically stable algorithm described in Sect. 10.3.2 to compute proper nullspace basis. Additionally, we discuss the checking of solvability conditions for several problems using the resulting state-space representation of the TFM $\overline{G}_f(\lambda)$.

In what follows, we assume that $p > r_d := \operatorname{rank} G_d(\lambda)$, which guarantees the existence of a nonempty left nullspace basis with $p - r_d$ rational basis vectors. Using the realization (7.13), state-space realizations of $G(\lambda)$ and $F(\lambda)$ are

$$G(\lambda) = \left[\begin{array}{c|cc} A - \lambda E & B_u & B_d \\ \hline C & D_u & D_d \\ 0 & I_{m_u} & 0 \end{array} \right], \qquad F(\lambda) = \left[\begin{array}{c|cc} A - \lambda E & B_f & B_w \\ \hline C & D_f & D_w \\ 0 & 0 & 0 \end{array} \right]. \tag{7.18}$$

The computational method of left nullspace bases exploits the simple fact that $N_l(\lambda)$ is a left nullspace basis of $G(\lambda)$ if and only if, for a suitable $M_l(\lambda)$,

$$Y_l(\lambda) := \begin{bmatrix} M_l(\lambda) & N_l(\lambda) \end{bmatrix} \tag{7.19}$$

is a left nullspace basis of the associated system matrix

$$S(\lambda) = \begin{bmatrix} A - \lambda E & B_u & B_d \\ C & D_u & D_d \\ 0 & I_{m_u} & 0 \end{bmatrix}. \tag{7.20}$$

Thus, to compute a proper rational left nullspace basis $N_l(\lambda)$ of $G(\lambda)$ we can determine first a proper rational left nullspace basis $Y_l(\lambda)$ of $S(\lambda)$ and then, $N_l(\lambda)$ simply results as

$$N_l(\lambda) = Y_l(\lambda) \begin{bmatrix} 0 \\ I_{p+m_u} \end{bmatrix}. \tag{7.21}$$

Since $Y_l(\lambda)$, of the form (7.19), is a left nullspace basis of $S(\lambda)$ in (7.20), it is easy to show that

$$Y_l(\lambda) \left[\begin{array}{c|c|c} A - \lambda E & B_f & B_w \\ C & D_f & D_w \\ 0 & 0 & 0 \end{array} \right] = \begin{bmatrix} 0 & \overline{G}_f(\lambda) & \overline{G}_w(\lambda) \end{bmatrix},$$

and, therefore, $N_l(\lambda) F(\lambda)$ in (7.16) results as

$$\begin{bmatrix} \overline{G}_f(\lambda) & \overline{G}_w(\lambda) \end{bmatrix} = Y_l(\lambda) \begin{bmatrix} B_f & B_w \\ D_f & D_w \\ 0 & 0 \end{bmatrix}. \tag{7.22}$$

We recall shortly the computation of $Y_l(\lambda)$, using the approach presented in Sect. 10.3.2. Let U and V be orthogonal matrices such that the transformed pencil $\widetilde{S}(\lambda) := U S(\lambda) V$ is in the Kronecker-like staircase form (see Sect. 10.1.6)

$$\widetilde{S}(\lambda) = \left[\begin{array}{c|c} A_r - \lambda E_r & A_{r,l} - \lambda E_{r,l} \\ \hline 0 & A_l - \lambda E_l \\ \hline 0 & C_l \end{array}\right], \tag{7.23}$$

where the descriptor pair $(A_l - \lambda E_l, C_l)$ is observable, E_l is nonsingular, and $A_r - \lambda E_r$ has full row rank excepting possibly a finite set of values of λ (i.e., the invariant zeros of $S(\lambda)$). The proper rational left nullspace basis $Y_l(\lambda)$ of $S(\lambda)$ can be determined as

$$Y_l(\lambda) = \left[\begin{array}{c|c|c} 0 & C_l(\lambda E_l - A_l)^{-1} & I_{p-r_d} \end{array}\right] U. \tag{7.24}$$

We compute now

$$U \left[\begin{array}{cc|c|c} 0 & 0 & B_f & B_w \\ I_p & 0 & D_f & D_w \\ 0 & I_{m_u} & 0 & 0 \end{array}\right] = \left[\begin{array}{c|c|c} * & * & * \\ \hline B_l & \overline{B}_f & \overline{B}_w \\ \hline D_l & \overline{D}_f & \overline{D}_w \end{array}\right], \tag{7.25}$$

where the row partitioning of the right hand side corresponds to the row partitioning of $\widetilde{S}(\lambda)$ in (7.23). With $Y_l(\lambda)$ in the form (7.24) and by using (7.25), we obtain from (7.21) and (7.22)

$$[\, N_l(\lambda)\ \overline{G}_f(\lambda)\ \overline{G}_w(\lambda)\,] = \left[\begin{array}{c|ccc} A_l - \lambda E_l & B_l & \overline{B}_f & \overline{B}_w \\ \hline C_l & D_l & \overline{D}_f & \overline{D}_w \end{array}\right]. \tag{7.26}$$

The descriptor representation (7.26) has been obtained by performing exclusively orthogonal transformations on the system matrices. We can prove that all computed matrices are exact for a slightly perturbed system matrix pencil. It follows that the approach to compute the matrices of the realization (7.26) is *numerically backward stable*.

According to Proposition 10.2, the realization $(A_l - \lambda E_l, B_l, C_l, D_l)$ of $N_l(\lambda)$ resulted in (7.26) represents a minimal proper rational left nullspace basis provided the realization (7.15) is controllable. In this case, according to Proposition 10.3, the realization of $N_l(\lambda)$ is also maximally controllable. Since, in general, $N_l(\lambda)$ has no infinite zeros, D_l has full row rank. However, in the case, when the realization (7.18) of $G(\lambda)$ is not controllable, then the descriptor realization $(A_l - \lambda E_l, B_l, C_l, D_l)$ does not represent, in general, a minimal proper left nullspace basis, because, it can be uncontrollable, or it can be controllable, but not maximally controllable. In both cases, a lower-order basis exists.

Remark 7.1 The realization (7.13) can be always assumed minimal, in which case the realization (7.18) of $G(\lambda)$ is observable, but, in general, may be uncontrollable. For example, uncontrollable generalized eigenvalues of the pair (A, E) may exist, if some poles of $[\, G_w(\lambda)\ G_f(\lambda)\,]$, are not simultaneously poles of $[\, G_u(\lambda)\ G_d(\lambda)\,]$. Fortunately, the minimality of the realization of $G(\lambda)$ is typically fulfilled if only additive actuator and sensor faults are considered and $w \equiv 0$. In this case, B_f has partly the same columns as B_u (in the case of actuator faults) or zero columns (in the case of sensor faults). Controllability is also guaranteed if the noise input w accounts exclusively for the effects of parametric uncertainties and the nominal model is controllable (see Sect. 2.2.1). $\qquad\square$

Remark 7.2 We can always determine a stable nullspace basis, using an output injection matrix K such that the pair $(A_l + K C_l, E_l)$ has stable generalized eigenvalues, or, alternatively, the spectrum $\Lambda(A_l + K C_l, E_l) = \Gamma$, where Γ is any symmetric set of n_l complex values in $\mathbb{C}_s$. Following the approach in Sect. 10.3.2, we perform an additional similarity transformation on $\widetilde{S}(\lambda)$ in (7.23), with the transformation matrix

$$\widehat{U} = \begin{bmatrix} I & 0 & 0 \\ 0 & I & K \\ 0 & 0 & I \end{bmatrix}, \tag{7.27}$$

to obtain $\widehat{S}(\lambda) := \widehat{U}\widetilde{S}(\lambda)$ as

$$\widehat{S}(\lambda) = \left[\begin{array}{c|c} A_r - \lambda E_r & A_{r,l} - \lambda E_{r,l} \\ \hline 0 & A_l + K C_l - \lambda E_l \\ \hline 0 & C_l \end{array}\right]. \tag{7.28}$$

After computing

$$\widehat{U}U \left[\begin{array}{cc|c|c} 0 & 0 & B_f & B_w \\ I_p & 0 & D_f & D_w \\ 0 & I_{m_u} & 0 & 0 \end{array}\right] = \left[\begin{array}{c|c|c} * & * & * \\ \hline B_l + K D_l & \overline{B}_f + K\overline{D}_f & \overline{B}_w + K\overline{D}_w \\ \hline D_l & \overline{D}_f & \overline{D}_w \end{array}\right], \tag{7.29}$$

we can form the realization for an alternative basis $\widetilde{N}_l(\lambda)$ and the corresponding reduced system $[\, \widetilde{G}_f(\lambda)\ \widetilde{G}_w(\lambda)\,]$ in the form

$$\left[\, \widetilde{N}_l(\lambda)\ \widetilde{G}_f(\lambda)\ \widetilde{G}_w(\lambda)\,\right] = \left[\begin{array}{c|ccc} A_l + K C_l - \lambda E_l & B_l + K D_l & \overline{B}_f + K\overline{D}_f & \overline{B}_w + K\overline{D}_w \\ \hline C_l & D_l & \overline{D}_f & \overline{D}_w \end{array}\right]. \tag{7.30}$$

It is easy to check that, with

$$\widetilde{M}_l(\lambda) = \left[\begin{array}{c|c} A_l + K C_l - \lambda E_l & K \\ \hline C_l & I \end{array}\right], \tag{7.31}$$

we implicitly determined the stable LCF

$$\left[\, N_l(\lambda)\ \overline{G}_f(\lambda)\ \overline{G}_w(\lambda)\,\right] = \tilde{M}_l^{-1}(\lambda)\left[\, \tilde{N}_l(\lambda)\ \tilde{G}_f(\lambda)\ \tilde{G}_w(\lambda)\,\right].\qquad\square$$

An important property of the resulting realizations of $N_l(\lambda)$, $\overline{G}_f(\lambda)$ and $\overline{G}_w(\lambda)$ in (7.26) is that they share the same observable pair $(A_l - \lambda E_l, C_l)$. In general, all three individual realizations may be uncontrollable (thus not minimal). However, we can easily show the following result.

Proposition 7.1 *If the realization (7.13) of* $\left[\, G_u(\lambda)\ G_d(\lambda)\ G_f(\lambda)\ G_w(\lambda)\,\right]$ *is irreducible, then the realization (7.26) is minimal.*

Proof The pair $(A_l - \lambda E_l, C_l)$ is observable by construction. To prove the controllability of the pair $(A_l - \lambda E_l, [\, B_l\ \overline{B}_f\ \overline{B}_w\,])$ we apply the same technique as in the proof of Proposition 10.2. Since the realization (7.13) of $\left[\, G_u(\lambda)\ G_d(\lambda)\ G_f(\lambda)\ G_w(\lambda)\,\right]$ is controllable, the realization

$$\left[\begin{array}{cccc} G_u(\lambda)\ G_d(\lambda)\ G_f(\lambda)\ G_w(\lambda) \\ I_{m_u} \qquad 0 \qquad 0 \qquad 0 \end{array}\right] = \left[\begin{array}{c|cccc} A - \lambda E & B_u & B_d & B_f & B_w \\ \hline C & D_u & D_d & D_f & D_w \\ 0 & I_{m_u} & 0 & 0 & 0 \end{array}\right]$$

is controllable as well. Due to the controllability of the pair $(A - \lambda E, [\, B_u\ B_d\ B_f\ B_w\,])$, the pencil $[\, A - \lambda E\ B_u\ B_d\ B_f\ B_w\,]$ has full row rank, and thus the reduced pencil

$$U\left[\begin{array}{cc|c|c|cc} A - \lambda E & B_u\ B_d & B_f & B_w & 0 & 0 \\ C & D_u\ D_d & D_f & D_w & I_p & 0 \\ 0 & I_{m_u}\ 0 & 0 & 0 & 0 & I_{m_u} \end{array}\right]\left[\begin{array}{c|c} V & 0 \\ \hline 0 & I \end{array}\right] = \left[\begin{array}{cc|c|c|c} A_r - \lambda E_r & A_{r,l} - \lambda E_{r,l} & * & * & * \\ 0 & A_l - \lambda E_l & \overline{B}_f & \overline{B}_w & B_l \\ 0 & C_l & \overline{D}_f & \overline{D}_w & D_l \end{array}\right]$$

has full row rank as well. It follows that $\left[\, A_l - \lambda E_l\ B_l\ \overline{B}_f\ \overline{B}_w\,\right]$ has full row rank and thus the pair $(A_l - \lambda E_l, [\, B_l\ \overline{B}_f\ \overline{B}_w\,])$ is controllable. The minimality is implied by irreducibility, because E_l is invertible. $\blacksquare$

Remark 7.3 If $m_u = m_d = 0$ and $[\, G_f(\lambda)\ G_w(\lambda)\,]$ is not proper, then realizations of the form (7.26) can be obtained by computing $[\, N_l(\lambda)\ \overline{G}_f(\lambda)\ \overline{G}_w(\lambda)\,]$, a proper left nullspace basis satisfying

$$[\, N_l(\lambda)\ \overline{G}_f(\lambda)\ \overline{G}_w(\lambda)\,]\left[\begin{array}{cc} G_f(\lambda) & G_w(\lambda) \\ -I_{m_f} & 0 \\ 0 & -I_{m_w} \end{array}\right] = 0\,.$$

The described nullspace computation approach leads to a state-space realization of the three TFMs $N_l(\lambda)$, $\overline{G}_f(\lambda)$ and $\overline{G}_w(\lambda)$ as in (7.26), which share the observable pair $(A_l - \lambda E_l, C_l)$. $\square$

Proposition 7.1 and Remark 7.3 show that the initial synthesis problems formulated for the system $\left[\, G_u(\lambda)\ G_d(\lambda)\ G_f(\lambda)\ G_w(\lambda)\,\right]$ with minimal realization (7.13) have been reduced to simpler problems formulated for a reduced proper system $[\,0\ 0\ \overline{G}_f(\lambda)\ \overline{G}_w(\lambda)\,]$, without control and disturbance inputs, such that the compound TFM

$$[\,Q(\lambda)\ R_f(\lambda)\ R_w(\lambda)\,] := [\,N_l(\lambda)\ \overline{G}_f(\lambda)\ \overline{G}_w(\lambda)\,]\,, \tag{7.32}$$

representing the initial synthesis at Step (1) of the synthesis procedures, has a minimal realization given by (7.26). Moreover, the stability of the initial synthesis can be enforced, as discussed in Remark 7.2, by using a suitable output injection matrix K such that the pair $(A_l + KC_l, E_l)$ has only stable generalized eigenvalues. Accordingly, the alternative initial synthesis

$$[\,Q(\lambda)\ R_f(\lambda)\ R_w(\lambda)\,] := [\,\widetilde{N}_l(\lambda)\ \widetilde{G}_f(\lambda)\ \widetilde{G}_w(\lambda)\,] \tag{7.33}$$

can be chosen, whose minimal realization is given in (7.30). An important aspect to mention is that, similarly to the original realization (7.13), both realizations (7.26) and (7.30), share the same observable pairs $(A_l - \lambda E_l, C_l)$ and $(A_l + KC_l - \lambda E_l, C_l)$, respectively. Any further updating of the initial synthesis can be done by preserving these properties.

Updating of the initial synthesis takes place in **Procedure EFDI** to solve the EFDIP, in **Procedure AFDI** to solve the AFDIP, and in **Procedure GENSPEC** to compute the maximally achievable fault specifications. The updating techniques employed in these procedures, can be conveniently described in terms of two rational matrices: $G(\lambda)$, the rational matrix for which a left nullspace basis has to be determined, and $F(\lambda)$, the rational matrix, which has to be multiplied from left with the computed basis. The key property of the state-space realizations of $G(\lambda)$ and $F(\lambda)$, which is instrumental to perform the necessary updating is that they share the same state, descriptor, and output matrices. Note that, for the initial synthesis, the choices in (7.15) for $G(\lambda)$ and (7.17) for $F(\lambda)$, with the corresponding state-space realizations in (7.18), have been used.

The resulting realizations of $R_f(\lambda)$ (i.e., either $\overline{G}_f(\lambda)$ in (7.26) or $\widetilde{G}_f(\lambda)$ in (7.30)) allow to check various solvability conditions. The following result is the state-space version of Corollary 5.2 to characterize the solvability of the EFDP.

Corollary 7.1 *For the system (7.13) with $w \equiv 0$ the EFDP is solvable if and only if*

$$\left[\frac{\overline{B}_{f_j}}{\overline{D}_{f_j}}\right] \neq 0, \quad j = 1, \ldots m_f, \tag{7.34}$$

where $\overline{B}_{f_j}$ and $\overline{D}_{f_j}$ are, respectively, the j-th columns of $\overline{B}_f$ and $\overline{D}_f$.

Proof Since the pair $(A_l - \lambda E_l, C_l)$ is observable, the fault input observability conditions $\overline{G}_{f_j}(\lambda) \neq 0$, for $j = 1, \ldots, m_f$, of Corollary 5.2 are equivalent to the conditions (7.34). ∎

For the solvability of the EFDP, by imposing the strong detectability condition with respect to a set of frequencies Ω, we have the following state-space version of Corollary 5.3.

Corollary 7.2 *Let Ω be the set of frequencies, which characterize the persistent fault signals, and assume that the resulting descriptor realization in (7.26) is such that $\Lambda(A_l, E_l) \cap \Omega = \emptyset$. For the system (7.13) with $w \equiv 0$ the EFDP is solvable with the strong detectability condition with respect to Ω if and only if for all $\lambda_z \in \Omega$*

$$\operatorname{rank} \begin{bmatrix} A_l - \lambda_z E_l & \overline{B}_{f_j} \\ C_l & \overline{D}_{f_j} \end{bmatrix} > n_l, \quad j = 1, \dots m_f. \tag{7.35}$$

Proof The conditions (7.35) are equivalent with the strong detectability requirement of Corollary 5.3 for $\overline{G}_{f_j}(\lambda)$ to have no zero in Ω, for $j = 1, \dots, m_f$. ∎

These corollaries can be extended in a straightforward way to cover the fault isolability conditions for the solvability of EFDIP. For the solvability of the EFDIP with strong isolability condition we have the following state-space version of Corollary 5.6.

Corollary 7.3 *For the system (7.13) with $w \equiv 0$ the EFDIP with strong isolability is solvable if and only if*

$$\operatorname{rank} \begin{bmatrix} A_l - \lambda E_l & \overline{B}_f \\ C_l & \overline{D}_f \end{bmatrix} = n_l + m_f. \tag{7.36}$$

Proof The strong isolability condition for the solvability of the EFDIP is equivalent with the left invertibility condition (3.21) for the reduced model $\overline{G}_f(\lambda)$. However, this is equivalent to the normal rank condition (7.36). ∎

This corollary serves also to check the solvability conditions for the AMMP.

Example 7.3 Consider the continuous-time system with the TFMs

$$G_u(s) = \begin{bmatrix} \dfrac{s+1}{s+2} \\ \dfrac{s+2}{s+3} \end{bmatrix}, \quad G_d(s) = \begin{bmatrix} \dfrac{s-1}{s+1} \\ 0 \end{bmatrix}, \quad G_f(s) = \begin{bmatrix} \dfrac{s+1}{s-2} & 0 \\ \dfrac{s+2}{s-3} & 1 \end{bmatrix}, \quad G_w(s) = 0.$$

The compound TFM $[\, G_u(s)\ G_d(s)\ G_f(s) \,]$ has the standard state-space realization with matrices $E = I_5$ and

$$A = \begin{bmatrix} -2 & 0 & 0 & 0 & 0 \\ 0 & -3 & 0 & 0 & 0 \\ 0 & 0 & -1 & 0 & 0 \\ 0 & 0 & 0 & 2 & 0 \\ 0 & 0 & 0 & 0 & 3 \end{bmatrix}, \quad [\, B_u \mid B_d \mid B_f \,] = \begin{bmatrix} 1 & 0 & 0 & 0 \\ 1 & 0 & 0 & 0 \\ 0 & 2 & 0 & 0 \\ 0 & 0 & 2 & 0 \\ 0 & 0 & 2 & 0 \end{bmatrix},$$

$$C = \begin{bmatrix} -1 & 0 & -1 & \frac{3}{2} & 0 \\ 0 & -1 & 0 & 0 & \frac{5}{2} \end{bmatrix}, \quad [\, D_u \mid D_d \mid D_f \,] = \begin{bmatrix} 1 & 1 & 1 & 0 \\ 1 & 0 & 1 & 1 \end{bmatrix}.$$

It is easy to observe that the pair $(A, [\, B_u \; B_d \,])$ is not controllable, and therefore the realization of $G(s)$ in (7.18) is not controllable as well.

Using the nullspace computation approach, we obtain for $N_l(s)$ and $\overline{G}_f(s)$ the matrices of the descriptor realizations

$$A_l - sE_l = \begin{bmatrix} 1.5522 & -1.0552 \\ -2.0011 & -2.5848 \end{bmatrix} - s \begin{bmatrix} -0.8442 & -0.1362 \\ 0 & -0.9672 \end{bmatrix},$$

$$[\, B_l \mid \overline{B}_f \,] = \begin{bmatrix} 0 & 0.3666 & 0.3666 \mid -0.5133 & 0.3666 \\ 0 & -0.1796 & -0.1796 \mid -1.9757 & -0.1796 \end{bmatrix},$$

$$C_l = [\, 0 \quad 1.904 \,],$$

$$[\, D_l \mid \overline{D}_f \,] = [\, 0 \quad 0.7071 \quad -0.7071 \mid 0.7071 \quad 0.7071 \,].$$

The generalized eigenvalues of the pair (A_l, E_l) are -2.5 and 3. The pair $(A_l - sE_l, B_l)$ is not controllable, and even not stabilizable, because the unstable eigenvalue 3 is not controllable. The corresponding TFMs, scaled with $\sqrt{2}$, are

$$\sqrt{2}N_l(s) = \begin{bmatrix} 0 & \dfrac{s+3}{s+2.5} & -\dfrac{s+2}{s+2.5} \end{bmatrix}, \quad \sqrt{2}\overline{G}_f(s) = \begin{bmatrix} \dfrac{(s+3)(s+2)}{(s-3)(s+2.5)} & \dfrac{s+3}{s+2.5} \end{bmatrix}.$$

Since $\overline{G}_f(s)$ is unstable, we can choose the output injection matrix

$$K = \begin{bmatrix} 0.8503 \\ 3.0480 \end{bmatrix},$$

such that the generalized eigenvalues of the pair $(A_l + KC_l, E_l)$ become -2.5 and -3. With $\widetilde{M}_l(s)$ defined in (7.31), we obtain the updated basis $\widetilde{N}_l(s) = \widetilde{M}_l(s)N_l(s)$ and the corresponding $\widetilde{G}_f(s) = \widetilde{M}_l(s)\overline{G}_f(s)$, whose realizations are given in (7.30). The resulting realization of $\widetilde{N}_l(s)$ is now controllable (due to output injection) and $Q(s) := \sqrt{2}\widetilde{N}_l(s)$ is a least McMillan degree solution, of order two, of the EFDP. The TFM $Q(s)$ of the implementation form and the corresponding TFM of the internal form $R_f(s)$ are

$$Q(s) = \begin{bmatrix} 0 & \dfrac{s-3}{s+2.5} & -\dfrac{(s+2)(s-3)}{(s+3)(s+2.5)} \end{bmatrix}, \quad R_f(s) = \begin{bmatrix} \dfrac{s+2}{s+2.5} & \dfrac{s-3}{s+2.5} \end{bmatrix}.$$

The denominator factor $\widetilde{M}_l(s)$ employed for the stabilization of $\overline{G}_f(s)$ is

$$\widetilde{M}_l(s) = \frac{s-3}{s+3}.$$

Note that this factor needs not be computed explicitly.

It is worth mentioning that the resulting filter $Q(s)$ is a proper nullspace basis of $G(s)$, but is not a minimal proper nullspace basis (it contains the non-constant factor $\widetilde{M}_l(s)$). Thus, this example illustrates the case when the peculiarities of the fault dynamics of $G_f(s)$ (e.g., unstable poles, which are not poles of the underlying system) impose the use of a higher order fault detection filter than the order of a minimal nullspace basis.

The script **Ex7_3.jl** in Listing 7.3 computes the results obtained in this example. The script **Ex7_3c.jl** (not listed) is its compact version, which calls several functions of the **FaultDetectionTools** package. $\Diamond$

Listing 7.3 Script **Ex7_3.jl** to compute the results of Example 7.3

```julia
using DescriptorSystems, LinearAlgebra

# Example 7.3 - Nullspace-based synthesis
# Uses only DescriptorSystems.jl
println("Example 7.3")

A = diagm([ -2, -3, -1, 2, 3]);
Bu = [1 1 0 0 0]'; Bd = [0 0 2 0 0]';
Bf = [0 0 0 2 2;0 0 0 0 0]';
C = [-1 0 -1 1.5 0; 0 -1 0 0 2.5];
Du = [ 1 1]'; Dd = [1 0]'; Df = [1 0; 1 1];
p = 2; mu = 1; md = 1; mf = 2;          # enter dimensions
sys = dss(A,[Bu Bd Bf],C,[Du Dd Df]); # define system

# compute [Q Rf], where Q is a left nullspace basis of
# [Gu Gd;I 0] and Rf = Q*[Gf;0];
Q_Rf = glnull([sys;eye(mu,mu+md+mf)],mf)[1];

# stabilize using left coprime factorization
Q_Rf, M = glcf(Q_Rf, sdeg = -3, smarg = -2);

# compute Q and Rf; scale to meet example
Q = -sqrt(2)*Q_Rf[:,1:p+mu]; Rf = -sqrt(2)*Q_Rf[:,p+mu+1:end];

# display results
println("Q = $(dss2rm(Q))")
println("Rf = $(dss2rm(Rf))")
println("M = $(dss2rm(M))")
```

Remark 7.4 We can easily extend the nullspace method to systems with parametric faults, described by equivalent linear models with additive faults of the form

$$E\lambda x(t) = Ax(t) + B_u u(t) + B_d d(t) + B_f(t)f(t)\,,$$
$$y(t) = Cx(t) + D_u u(t) + D_d d(t) + D_f(t)f(t)\,, \tag{7.37}$$

where the fault input channel contains time-varying matrices with special structures (see (2.17) in Sect. 2.2.2). If we denote $\widetilde{f}_1(t) := B_f(t)f(t) \in \mathbb{R}^n$, $\widetilde{f}_2(t) := D_f(t)f(t) \in \mathbb{R}^p$, and $\widetilde{f}(t) = [\,\widetilde{f}_1^T(t)\ \widetilde{f}_2^T(t)\,]^T$, then the time-varying system (7.37) can be equivalently expressed in the form

$$E\lambda x(t) = Ax(t) + B_u u(t) + B_d d(t) + [\, I_n \; 0\,]\widetilde{f}(t)\,,$$
$$y(t) = Cx(t) + D_u u(t) + D_d d(t) + [\, 0 \; I_p\,]\widetilde{f}(t)\,. \tag{7.38}$$

For this LTI system, we can compute, similarly as in (7.25),

$$U \begin{bmatrix} 0 & 0 & I_n & 0 \\ I_p & 0 & 0 & I_p \\ 0 & I_{m_u} & 0 & 0 \end{bmatrix} = \begin{bmatrix} * & * \\ \hline B_l & \overline{B}_f \\ \hline D_l & \overline{D}_f \end{bmatrix} \tag{7.39}$$

from which we obtain the descriptor system realizations of the proper left nullspace basis $(A_l - \lambda E_l, B_l, C_l, D_l)$ and of the reduced proper system $(A_l - \lambda E_l, \overline{B}_f, C_l, \overline{D}_f)$, with outputs $\overline{y}(t)$, as defined in (5.11), and inputs $\widetilde{f}(t)$. According to Remark 7.2, we can arbitrarily assign the dynamics of both the nullspace basis as well as of the reduced system, by using an additional LTI prefilter of the form (7.31). Recall that any stable nullspace basis can serve as a fault detection filter, with outputs $r(t)$ and inputs $y(t)$ and $u(t)$, provided the fault detectability conditions for the reduced system are fulfilled (jointly with the stability requirement). This comes down to check (e.g., via simulations or by exploiting the special structures of matrices $B_f(t)$ and $D_f(t)$, see Sect. 2.2.2) that the residual signal $r(t)$ is sensitive to each parametric fault $f_i(t)$, for $i = 1, \ldots, k$. It follows, that the nullspace method allows to address fault detection problems for parametric faults in a simple way, involving only multiplications of (structured) time-varying matrices with constant matrices. $\square$

7.5 Least-Order Synthesis

The synthesis of fault detection filters of least McMillan degree underlies an important computational paradigm, typically employed at Step (2) of several of the presented synthesis procedures. This paradigm concerns with the updating of the proper left nullspace basis $Q(\lambda) = N_l(\lambda)$, computed at Step (1), by determining a factor $Q_2(\lambda)$ such that the product $Q_2(\lambda)N_l(\lambda)$ has the least possible McMillan degree under the constraint that certain *admissibility conditions* are simultaneously fulfilled. A basic admissibility condition is the (problem dependent) solvability condition, which must be always fulfilled by the updated reduced system with the TFMs $Q_2(\lambda)\overline{G}_w(\lambda)$ and $Q_2(\lambda)\overline{G}_f(\lambda)$. For example, for the solvability of the EFDP and AFDP, all columns of $Q_2(\lambda)\overline{G}_f(\lambda)$ must be nonzero (see Corollaries 5.2 and 5.4), while for the solvability of the EMMP and AMMP with enforced strong isolability, $Q_2(\lambda)\overline{G}_f(\lambda)$ must be left invertible (i.e., must have full column rank) (see Corollaries 5.10 and 5.11). Certain regularization conditions are additionally imposed, as—for example, $Q_2(\lambda)\overline{G}_w(\lambda)$ to have full row rank when solving the AFDP, $Q_2(\lambda)\overline{G}_f(\lambda)$ to be invertible when solving the EMMP with strong isolability, or $[\, Q_2(\lambda)\overline{G}_f(\lambda) \;\; Q_2(\lambda)\overline{G}_w(\lambda)\,]$ to have full row rank when solving the AMMP. These additional conditions are enforced to ease the solution of some synthesis problems, however, they are not necessary for the solvability of the respective problems.

We assume that the initial nullspace-based synthesis computed at Step (1) of all synthesis procedures is $[\,Q(\lambda)\ R_f(\lambda)\ R_w(\lambda)\,] := [\,N_l(\lambda)\ \overline{G}_f(\lambda)\ \overline{G}_w(\lambda)\,]$ in (7.32) and has the state-space realization (7.26), with D_l of full row rank. To simplify the presentation, we denote $W(\lambda) := [\,Q(\lambda)\ R_f(\lambda)\ R_w(\lambda)\,]$ the $(p - r_d) \times (p + m_u + m_f + m_w)$ proper rational matrix, where $r_d = \operatorname{rank} G_d(\lambda)$. For the $q \times (p - r_d)$ TFM $Q_2(\lambda)$ we assume the state-space realization

$$Q_2(\lambda) = \left[\begin{array}{c|c} U_l(A_l + KC_l - \lambda E_l)V_l & U_l K \\ \hline HC_l V_l & H \end{array}\right], \tag{7.40}$$

where $H \in \mathbb{R}^{q \times (p - r_d)}$ is a full row rank matrix (also called *design matrix*), to be chosen, and $K \in \mathbb{R}^{n_l \times (p - r_d)}$ is an output injection matrix, to be determined together with the two nonsingular transformation matrices U_l and V_l. It is straightforward to check that the state-space realization of $Q_2(\lambda)W(\lambda)$ is

$$\left[\begin{array}{c|ccc} U_l(A_l + KC_l - \lambda E_l)V_l & U_l(B_l + KD_l) & U_l(\overline{B}_f + K\overline{D}_f) & U_l(\overline{B}_w + K\overline{D}_w) \\ \hline HC_l V_l & HD_l & H\overline{D}_f & H\overline{D}_w \end{array}\right]. \tag{7.41}$$

For H fixed, the main computation is to determine the output injection matrix K, jointly with the transformation matrices U_l and V_l, such that $Q_2(\lambda)Q(\lambda)$, whose state-space realization is contained in (7.41), has the least possible McMillan degree for which the solvability conditions, as those in Corollaries 7.1, 7.2 or 7.3, are fulfilled. For example, with H chosen as a $p - r_d$ dimensional row vector (i.e., $q = 1$), we want to determine the least order fault detection filter with scalar output, which solves the EFDP in **Procedure EFD**. Or, we choose H such that $Q_2(\lambda)R_w(\lambda)$ in **Procedure AFD**, $Q_2(\lambda)R_f(\lambda)$ in **Procedure EMMS**, or $Q_2(\lambda)[\,R_f(\lambda)\ R_w(\lambda)\,]$ in **Procedure AMMS** are all full row rank TFMs.

We discuss first how to determine, for a given H, the output injection matrix K such that $Q_2(\lambda)Q(\lambda)$ has the least possible McMillan degree. For this purpose, we will attempt, with a suitable choice of K, to achieve the least possible McMillan degree of $Q_2(\lambda)Q(\lambda)$ (and also of $Q_2(\lambda)W(\lambda)$) by making the pair $(A_l + KC_l - \lambda E_l, HC_l)$ maximally unobservable. Minimal dynamic cover techniques can be employed to perform this computation (see Sects. 10.4.2 and 10.4.3 for the definitions of Type I and Type II dynamic covers and for numerical algorithms for their computations).

The minimum dynamic cover algorithm described in **Procedure GRMCOVER1** in Sect. 10.4.2 relies on a preliminary orthogonal similarity transformation performed on the state-space realization of $[\,W^T(\lambda)H^T\ W^T(\lambda)\,]$, with $W(\lambda)$ given in (7.26). In a first stage, the controllable descriptor pair $(A_l^T - \lambda E_l^T, [\,C_l^T H^T\ C_l^T\,])$ is reduced to a special controllability staircase form by applying **Procedure GSCSF** presented in Sect. 10.4.1. Then, with additional block permutations and non-orthogonal block row and block column transformations, the system matrices are transformed into a *special form*, which allows to cancel, using a suitable output injection, the maximum number of unobservable eigenvalues. For the so-called Type I dynamic covers, two nonsingular transformation matrices U_l and V_l result such that

$$
U_l(A_l - \lambda E_l)V_l = \left[\begin{array}{c|c} \widehat{A}_{11} - \lambda\widehat{E}_{11} & \widehat{A}_{12} - \lambda\widehat{E}_{12} \\ \hline \widehat{A}_{21} & \widehat{A}_{22} - \lambda\widehat{E}_{22} \end{array}\right],
$$

$$
U_l[\,B_l\ \overline{B}_f\ \overline{B}_w\,] = \left[\begin{array}{ccc} \widehat{B}_1 & \widehat{B}_{f,1} & \widehat{B}_{w,1} \\ \hline \widehat{B}_2 & \widehat{B}_{f,2} & \widehat{B}_{w,2} \end{array}\right], \tag{7.42}
$$

$$
\left[\frac{HC_l}{C_l}\right]V_l = \left[\begin{array}{c|c} 0 & \widehat{C}_{22} \\ \hline \widehat{C}_{11} & \widehat{C}_{12} \end{array}\right],
$$

where the pairs $\left(\widehat{A}_{11} - \lambda\widehat{E}_{11}, \widehat{C}_{11}\right)$ and $\left(\widehat{A}_{22} - \lambda\widehat{E}_{22}, \widehat{C}_{22}\right)$ are observable, and the submatrices $\widehat{C}_{11}$ and $\widehat{A}_{21}$ have the particular structure

$$
\left[\begin{array}{c} \widehat{A}_{21} \\ \widehat{C}_{11} \end{array}\right] = \left[\begin{array}{cc} 0 & A_{21} \\ 0 & C_{11} \end{array}\right],
$$

with C_{11} having full column rank. By taking

$$
K = U_l^{-1}\left[\begin{array}{c} 0 \\ K_2 \end{array}\right],
$$

with K_2 satisfying

$$
K_2C_{11} + A_{21} = 0, \tag{7.43}
$$

we annihilate $\widehat{A}_{21} + K_2\widehat{C}_{11}$, the (2,1)-block of $U_l(A_l + KC_l)V_l$, and thus make all eigenvalues of $\widehat{A}_{11} - \lambda\widehat{E}_{11}$ unobservable. The resulting realization of $Q_2(\lambda)W(\lambda)$ has a maximum number of unobservable eigenvalues and allows to determine from (7.41) an observable realization, by removing the unobservable part. The resulting observable state-space realization of $Q_2(\lambda)W(\lambda)$ can be explicitly written down as

$$
Q_2(\lambda)W(\lambda) = \left[\begin{array}{c|ccc} \widehat{A}_{22} + K_2\widehat{C}_{12} - \lambda\widehat{E}_{22} & \widehat{B}_2 + K_2D_l & \widehat{B}_{f,2} + K_2\overline{D}_f & \widehat{B}_{w,2} + K_2\overline{D}_w \\ \hline \widehat{C}_{22} & HD_l & H\overline{D}_f & H\overline{D}_w \end{array}\right]. \tag{7.44}
$$

Since both H and D_l have full row rank, it follows that HD_l has also full row rank.

Using similar arguments as in the proof of Proposition 7.1, we can easily prove the following result.

Proposition 7.2 *If the realization (7.26) of* $[\,Q(\lambda)\ R_f(\lambda)\ R_w(\lambda)\,]$ *in (7.32) is minimal, then the realization (7.44) is minimal.*

The algorithm underlying **Procedure GRMCOVER1** to compute the output injection matrix K is *not* numerically stable, because it involves similarity transformations with non-orthogonal transformation matrices U_l and V_l in (7.42) and the computation of the solution K_2 of the matrix equation (7.43). Still, this algorithm can be considered a *numerically reliable* algorithm, because the potential loss of numerical stability can be easily detected, either by detecting large norms of the employed transformation matrices U_l and V_l, or a large norm of the resulting K_2. In both cases, a possible remedy is to employ a different choice of H or to increase the targeted order.

The resulting realization (7.44) can be used to check the admissibility conditions for different synthesis problems. To simplify the notations, we denote the resulting state-space realization of $Q_2(\lambda)W(\lambda)$, of order $\tilde{n}_l$, as

$$Q_2(\lambda)\left[\, Q(\lambda)\ R_f(\lambda)\ R_w(\lambda)\,\right] = \left[\begin{array}{c|ccc} \widetilde{A}_l - \lambda\widetilde{E}_l & \widetilde{B}_l & \widetilde{B}_f & \widetilde{B}_w \\ \hline \widetilde{C}_l & \widetilde{D}_l & \widetilde{D}_f & \widetilde{D}_w \end{array}\right]. \tag{7.45}$$

The selected $H \in \mathbb{R}^{q \times (p-r_d)}$ is a valid choice when solving the EFDP, provided the following solvability conditions of the EFDP (see Corollary 7.1) are fulfilled

$$\left[\begin{array}{c} \widetilde{B}_{f_j} \\ \widetilde{D}_{f_j} \end{array}\right] \neq 0, \quad j = 1, \ldots, m_f, \tag{7.46}$$

where $\widetilde{B}_{f_j}$ and $\widetilde{D}_{f_j}$ are the j-th columns of $\widetilde{B}_f$ and $\widetilde{D}_{f_j}$, respectively. When solving the AFDP, besides checking the conditions (7.46), the full row rank condition for $Q_2(\lambda)R_w(\lambda)$ must be also checked. Using the resulting realization in (7.45), this involves to check that

$$\mathrm{rank}\left[\begin{array}{cc} \widetilde{A}_l - \lambda_c\widetilde{E}_l & \widetilde{B}_w \\ \widetilde{C}_l & \widetilde{D}_w \end{array}\right] = \tilde{n}_l + q, \tag{7.47}$$

where λ_c is any real value (e.g., randomly chosen), which is not a pole or zero of $Q_2(\lambda)R_w(\lambda)$. To fulfil (7.47), the choice of the number of filter outputs, q, must satisfy $q \leq r_w \leq p - r_d$, where $r_w := \mathrm{rank}\,\overline{G}_w(\lambda)$ (in the case when $r_w < p - r_d$, see Remark 5.10 for choosing q satisfying $r_w < q \leq p - r_d$). Similar tests have to be performed to check the admissibility conditions for the EMMP or AMMP. If the admissibility test fails, a different choice of H is necessary. In what follows, we discuss possible choices of H, which lead to the least McMillan degree of $Q_2(\lambda)Q(\lambda)$ and satisfy the admissibility conditions.

We assume that a minimal proper rational left nullspace basis $N_l(\lambda)$ with a minimal (i.e., controllable and observable) descriptor realization $(A_l - \lambda E_l, B_l, C_l, D_l)$ has been determined using the approach described in Sect. 10.3.2. Accordingly, we can assume that the observable descriptor pair $(A_l - \lambda E_l, C_l)$ is in the observability staircase form

$$\left[\frac{A_l - \lambda E_l}{C_l} \right] = \left[\begin{array}{c|c|c|c} A_{\ell,\ell+1} & A_{\ell,\ell} - \lambda E_{\ell,\ell} & \cdots & A_{\ell,1} - \lambda E_{\ell,1} \\ \hline & A_{\ell-1,\ell} & \ddots & \vdots \\ \hline & & \ddots & A_{1,1} - \lambda E_{1,1} \\ \hline & & & A_{0,1} \end{array} \right], \qquad (7.48)$$

where $A_{i-1,i} \in \mathbb{R}^{\mu_{i-1} \times \mu_i}$ are full column rank upper triangular matrices, for $i = 1, \ldots, \ell + 1$, with $\mu_{\ell+1} := 0$. The matrices $A_{i-1,i}$ have the form

$$A_{i-1,i} = \left[\begin{array}{c} R_{i-1,i} \\ 0 \end{array} \right], \qquad (7.49)$$

where $R_{i-1,i}$ is an invertible upper triangular matrix of order μ_i. The left (or row) Kronecker indices of the system pencil $S(\lambda)$ in (7.20) are the column numbers of the $L_{\eta_i}^T(\lambda)$ blocks of the Kronecker canonical form (see Lemma 9.9) and result as follows: there are $\mu_{i-1} - \mu_i$ Kronecker blocks $L_{\eta_i}^T(\lambda)$ of size $i \times (i-1)$, for $i = 1, \ldots, \ell + 1$. Notice that $\mu_{i-1} - \mu_i$ is also the row dimension of the zero block of $A_{i-1,i}$. From standard linear algebra results it follows that the number of linearly independent basis vectors (i.e., the number of rows of $N_l(\lambda)$) is $p - r_d$, where $r_d = \text{rank } G_d(\lambda)$. This number is equal to the total number of Kronecker indices, thus $\sum_{i=1}^{\ell+1}(\mu_{i-1} - \mu_i) = \mu_0$. According to Proposition 10.1, $\mu_{i-1} - \mu_i$ represents the number of polynomial vectors of degree $i - 1$ in a minimal polynomial basis, and therefore, also the number of rational vectors of McMillan degree $i - 1$ in a simple proper basis.

Let n_j, $j = 1, \ldots, p - r_d$, be the (decreasingly ordered) *row minimal indices* (or left Kronecker indices, see Sect. 9.1.3), representing the degrees of the polynomial basis vectors in a minimal polynomial basis (or the McMillan degrees of the rational basis vectors in a simple minimal rational basis). The order of the descriptor realization of $N_l(\lambda)$ is $\nu_l = \sum_{i=1}^{\ell} \mu_i$ and is equal to the degree $\sum_{j=1}^{p-r_d} n_j$ of a minimal polynomial basis (also the sum of McMillan degrees of the rational basis vectors in a simple proper basis). For a desired number of filter outputs q, it is possible to choose a suitable matrix $H \in \mathbb{R}^{q \times (p-r_d)}$, which leads to a least McMillan degree of the resulting $Q_2(\lambda)N_l(\lambda)$, such that the (problem dependent) admissibility conditions are fulfilled. The possible values of the least McMillan degree of $Q_2(\lambda)N_l(\lambda)$, are among the possible dimensions of the controllability subspaces of the dual standard pair $\left(E_l^{-T} A_l^T, E_l^{-T} C_l^T \right)$ containing span $\left(E_l^{-T} C_l^T H^T \right)$, and according to [172, Theorem 1], are directly determined by the minimal indices n_j, $i = 1, \ldots, p - r_d$. For example, if k is an index such that $1 \le k \le p - r_d - 1$, and $n_{k+1} > \sum_{i=1}^{k} n_i + 1$, then there exists no controllability subspace of dimension l, with $\sum_{i=1}^{k} n_i < l < n_{k+1}$. (This result is the statement of Corollary 1 in [172].) In what follows, we discuss how to choose H, for obtaining fault detection filters of McMillan degree smaller than ν_l.

For simplicity, we only discuss the determination of the least order solution of the EFDP for which we consider two possible approaches. (Both approaches can be easily extended to all other least order synthesis problems). The first approach is direct (i.e., non-iterative) and is based on the selection of a suitable H by exploiting the binary information contained in the structure matrix corresponding to an alternative simple proper rational basis of the left nullspace of $G(\lambda)$ in (7.15). According to Proposition 10.4, for the choice $H^{(j)} = e_j^T$, where e_j is the j-th column of the identity matrix of order $p - r_d$, there exists an output injection matrix K_j, such that the McMillan degree of the resulting proper rational vector $v_j(\lambda) := Q_2^{(j)}(\lambda)N_l(\lambda)$ is the j-th minimal index n_j, where

$$Q_2^{(j)}(\lambda) := H^{(j)}C_l(\lambda E_l - A_l - K_jC_l)^{-1}K_j + H^{(j)}D_l .$$

The row vectors $v_j(\lambda)$, $j = 1, \ldots, p - r_d$, form a simple proper rational basis of the left nullspace of $G(\lambda)$ in (7.15). Let $s^{(j)}$ be the resulting $1 \times m_f$ binary structure matrix (see Sect. 3.4) of $Q_2^{(j)}(\lambda)\overline{G}_f(\lambda)$ and let $S^{(i)}$ be the binary matrix formed by stacking the i row vectors $s^{(j)}$ for $j = p - r_d - i + 1, \ldots, p - r_d$. Let i be the least value such that $S^{(i)}$ contains at least one nonzero element in all its columns. It follows, that the filter formed by stacking the i vectors $v_j(\lambda)$, for $j = p - r_d - i + 1, \ldots, p - r_d$, is admissible for the solution of the EFDP. In the case $q = 1$, it follows from Corollary 10.1, that there exists a linear combination (with rational coefficients)

$$v(\lambda) = \sum_{j=1}^{i} \phi^{(j)}(\lambda)v_{p-r_d-i+j}(\lambda)$$

of the i rational basis vectors with McMillan degrees up to n_i, such that $v(\lambda)$ has McMillan degree n_i and the filter $Q(\lambda) = v(\lambda)$ is admissible.

The above analysis allows an easy selection of a suitable H leading to reduced order filters. In the case $q = 1$, instead of determining explicitly the linear combination using rational coefficients, such a linear combination can be directly computed with the choice

$$H = h^{(i)} := [\, 0 \cdots 0 \ h_i \cdots h_1 \,], \tag{7.50}$$

with $h_j \neq 0$ for $j = 1, \ldots, i$, which ensures that $Q_2(\lambda)N_l(\lambda)$ has McMillan degree at most n_i and the admissibility conditions in (7.46) are fulfilled. In the case $q > 1$, H can be chosen in the form

$$H = \begin{bmatrix} 0 \ H^{(i)} \end{bmatrix}, \tag{7.51}$$

where $H^{(i)}$ is a $q \times \max(q, i)$ full row rank matrix, which builds q linear combinations of the basis vectors up to McMillan degrees $n_{\max(q,i)}$. A practical approach, to be used in both cases, is to generate the nonzero elements of H as random numbers.

The use of the direct approach for the selection of H requires the computation of (at least) i basis vectors of a simple proper rational left nullspace basis, using the minimum dynamic-covers-based technique described in Sect. 10.3.2 (see Proposition 10.4). Additionally, for the selected H, the same technique is used to determine the output injection gain matrix K leading to the least McMillan degree of $Q_2(\lambda)N_l(\lambda)$. For these computations, the **Procedure GRMCOVER1** can be employed, which can also detect potential accuracy losses due to the usage of ill-conditioned transformations when determining the vectors of a simple proper rational basis. To avoid the use of ill-conditioned transformations, another approach, presented below, may be better suited.

The second approach for the selection of H is based on a systematic search by using successive candidates for H as in (7.51) (or in (7.50) for $q = 1$) with increasing number i, of nonzero trailing columns, and checking, for each i, the admissibility conditions. The corresponding observable realizations of $Q_2(\lambda)N_l(\lambda)$, $Q_2(\lambda)\overline{G}_f(\lambda)$, and $Q_2(\lambda)\overline{G}_w(\lambda)$, obtained in (7.44) or in (7.45), allow to check the admissibility conditions as those in (7.46) (or those in (7.47)). The successively determined partial filters $Q_2(\lambda)N_l(\lambda)$ have non-decreasing orders and therefore, the first admissible filter represents a satisfactory least McMillan degree synthesis. To speed up the selection in the case $q = 1$, the choices of $i = \mu_0 - \mu_j$, $j = 1, \ldots, \ell$, nonzero components of $h^{(i)}$ in (7.50) ensures a tentative order n_i, by building a linear combination of all $\mu_0 - \mu_i$ basis vectors of orders less than or equal to n_i. In this way, repeated checks for the same order are avoided and the search is terminated in at most ℓ steps.

If $p - r_d > 1$, the resulting admissible value of H, in the form (7.51), is not unique. From a numerical point of view, it is desirable to use an "optimal" choice of H, which increases the overall reliability of computations, as—for example, that one, which minimizes the condition numbers of the transformation matrices U_l and V_l employed in (7.42), or, the norm of the employed injection matrix K_2 in (7.43). From the point of view of the performance of the resulting fault detection filters, the minimization of the sensitivity conditions, introduced in Remark 5.6, can equally represent valuable goals for an optimal tuning of the nonzero elements of H.

Example 7.4 Consider the continuous-time system with the standard state-space realization with matrices $E = I_5$ and

$$
A = \begin{bmatrix}
0 & 0 & 1.1320 & 0 & -1.0000 \\
0 & -0.0538 & -0.1712 & 0 & 0.0705 \\
0 & 0 & 0 & 1.0000 & 0 \\
0 & 0.0485 & 0 & -0.8556 & -1.0130 \\
0 & -0.2909 & 0 & 1.0532 & -0.6859
\end{bmatrix},
$$

$$
\begin{bmatrix} B_u & | & B_d & | & B_f \end{bmatrix} = \begin{bmatrix}
0 & 0 & 0 & 0 & 0 & 0 & 0 & 0 & 0 \\
-0.1200 & 1 & 0 & 0 & 0 & 0 & 0 & -0.1200 & 1 \\
0 & 0 & 0 & 0 & 0 & 0 & 0 & 0 & 0 \\
4.4190 & 0 & -1.6650 & -1.6650 & 0 & 0 & 0 & 4.4190 & 0 \\
1.5750 & 0 & -0.0732 & -0.0732 & 0 & 0 & 0 & 1.5750 & 0
\end{bmatrix},
$$

$$
C = \begin{bmatrix}
1 & 0 & 0 & 0 & 0 \\
0 & 1 & 0 & 0 & 0 \\
0 & 0 & 1 & 0 & 0
\end{bmatrix},
$$

$$[\; D_u \;|\; D_d \;|\; D_f \;] = \begin{bmatrix} 0 & 0 & 0 & 0 & 1 & 0 & 0 & 0 & 0 \\ 0 & 0 & 0 & 0 & 0 & 1 & 0 & 0 & 0 \\ 0 & 0 & 0 & 0 & 0 & 0 & 1 & 0 & 0 \end{bmatrix}.$$

It is easy to observe that the pair $(A, [\, B_u \; B_d \,])$ is controllable, and therefore the realization of $G(s)$ in (7.18) is controllable as well.

The resulting rational left nullspace basis $N_l(s)$ of $G(s)$ has two vectors, which together form a minimal proper rational basis of order 3, while a minimal polynomial basis has two vectors, one of degree one and a second of degree two. A first-order synthesis, obtained with $h^{(1)} = [\, 0 \; 1\,]$ and imposing a pole at -1, is

$$Q^{(1)}(s) = \frac{1}{s+1}[\; -0.07033\,s \quad -0.9975\,s - 0.05367 \quad -0.09117 \quad -0.1197 \;\; 0.9975 \;\; 0\;].$$

A second-order synthesis, obtained with $h^{(2)} = [\, 1 \; 1\,]$ and imposing a stability degree of -1, is

$$Q^{(2)}(s) = \frac{1}{s^2 + 2s + 1.159}[\; 0.9262\,s^2 + 0.5839\,s \quad -1.068\,s^2 - 0.8869\,s - 0.3381 \quad \cdots$$

$$0.04403\,s^2 - 0.1388\,s - 0.8029 \quad -0.1281\,s + 1.283 \quad 1.068\,s + 0.8294 \;\; 0\;].$$

For the resulting

$$R_f^{(1)}(s) := Q^{(1)}(s)\begin{bmatrix} G_f(s) \\ 0 \end{bmatrix}, \qquad R_f^{(2)}(s) := Q^{(2)}(s)\begin{bmatrix} G_f(s) \\ 0 \end{bmatrix},$$

we can check the admissibility conditions for an arbitrary value of s, say $s = i = \sqrt{-1}$. The resulting absolute values of the gains $R_f^{(1)}(i)$ are

$$|R_f^{(1)}(i)| = [\, 0.0497 \;\; 0.7064 \;\; 0.0645 \;\; 0.0846 \;\; 0.7054\,],$$

with the ratio of the maximum and minimum gains of 14.2049. The resulting absolute values of the gains $R_f^{(2)}(i)$ are

$$|R_f^{(2)}(i)| = [\, 0.5457 \;\; 0.5724 \;\; 0.4278 \;\; 0.6428 \;\; 0.6739\,],$$

with the ratio of the maximum and minimum gains of 1.5754. Therefore, the second-order filter provides a more uniform detection performance at frequencies nearby $s = i$.

The script **Ex7_4.jl** in Listing 7.4 computes the results obtained in this example. The script **Ex7_4c.jl** (not listed) is its compact version, which calls several functions of the **FaultDetectionTools** package. $\diamond$

Listing 7.4 Script `Ex7_4.jl` to compute the results of Example 7.4

```julia
using DescriptorSystems, Test

# Example 7.4 - Nullspace-based synthesis
# Uses only DescriptorSystems.jl
println("Example 7.4")

# define the state-space realizations of Gu, Gd and Gf
n = 5; p = 3; mu = 3; md = 1; mf = 5;        # enter dimensions
# define matrices of the state-space realization
A = [ 0 0 1.132 0 -1;
0 -0.0538 -0.1712 0 0.0705;
0 0 0 1 0;
0 0.0485 0 -0.8556 -1.013;
0 -0.2909 0 1.0532 -0.6859];
Bu = [0 0 0;-0.12 1 0;0 0 0;4.419 0 -1.665;1.575 0 -0.0732];
Bd = Bu[:,mu]; Bf = [zeros(n,p) Bu[:,1:mu-1]];
C=eye(p,n); Du=zeros(p,mu); Dd=zeros(p,md); Df=eye(p,mf);
sys = dss(A,[Bu Bd Bf],C,[Du Dd Df]);           # define system

# compute [Q Rf], where Q is a left nullspace basis of
# [Gu Gd;I 0] and Rf = Q*[Gf;0];
Q_Rf, info = glnull([sys;eye(mu,mu+md+mf)],mf);
info.degs

# determine 1-st and 2-nd order scalar output designs
Q1_Rf1 = glmcover1([0 1;eye(2)]*Q_Rf,1)[1];      # [Q1 Rf1]
Q2_Rf2 = glmcover1([-1 1;eye(2)]*Q_Rf,1)[1];     # [Q2 Rf2]

# compute stable left coprime factorizations
Q1_Rf1 = glcf(Q1_Rf1, sdeg=-1, smarg = -1)[1];
# choose poles to meet example
Q2_Rf2 = glcf(Q2_Rf2, evals = [-1. + 0.3992im, -1. - 0.3992im],
              sdeg = -1, smarg = -1)[1];

# check admissibility
# compute Q1 and Rf1;
Q1 = Q1_Rf1[:,1:p+mu]; Rf1 = Q1_Rf1[:,p+mu+1:end];
# compute reciprocal sensitivity condition
g1 = abs.(evalfr(Rf1,im)); rscond1 = maximum(g1)/minimum(g1)
# compute Q2 and Rf2;
Q2 = Q2_Rf2[:,1:p+mu]; Rf2 = Q2_Rf2[:,p+mu+1:end];
# compute reciprocal sensitivity condition
g2 = abs.(evalfr(Rf2,im)); rscond2 = maximum(g2)/minimum(g2)
println("|Rf1(im)| = $g1")
println("rscond1 = $rscond1 ")
println("|Rf2(im)| = $g2")
println("rscond2 = $rscond2")
# check finite reciprocal sensitivity conditions
@test !isinf(rscond1) && !isinf(rscond2)

# display results
println("Q1 = $(dss2rm(Q1))")
println("Q2 = $(dss2rm(Q2))")
```

7.6 Coprime Factorization Techniques

A computational paradigm used in all synthesis algorithms is the adjustment of the filter dynamics by premultiplying the current filter $Q(\lambda)$, and, if appropriate, the TFMs $R_f(\lambda)$ and $R_w(\lambda)$ of its internal form, with a square $M(\lambda)$, such that the updated filters $\widehat{Q}(\lambda) := M(\lambda)Q(\lambda)$, as well as $\widehat{R}_f(\lambda) := M(\lambda)R_f(\lambda)$ and $\widehat{R}_w(\lambda) := M(\lambda)R_w(\lambda)$, are physically realizable (i.e., proper) and have only poles in a "good" domain $\mathbb{C}_g$ of the complex plane $\mathbb{C}$. These computations can be implicitly performed by determining a left fractional factorization (see Sect. 9.1.6) of the compound TFM $[\, Q(\lambda)\ R_f(\lambda)\ R_w(\lambda)\,]$ in the form

$$[\, Q(\lambda)\ R_f(\lambda)\ R_w(\lambda)\,] = M^{-1}(\lambda)\big[\,\widehat{Q}(\lambda)\ \widehat{R}_f(\lambda)\ \widehat{R}_w(\lambda)\,\big], \tag{7.52}$$

such that both the numerator factor $\big[\,\widehat{Q}(\lambda)\ \widehat{R}_f(\lambda)\ \widehat{R}_w(\lambda)\,\big]$ and the denominator factor $M(\lambda)$ are proper and have poles only in $\mathbb{C}_g$. Let us assume that $Q(\lambda)$ and the TFMs $R_f(\lambda)$ and $R_w(\lambda)$ of its internal form have the joint descriptor realization

$$[\, Q(\lambda)\ R_f(\lambda)\ R_w(\lambda)\,] = \left[\begin{array}{c|ccc} \widetilde{A}_l - \lambda\widetilde{E}_l & \widetilde{B}_l & \widetilde{B}_f & \widetilde{B}_w \\ \hline \widetilde{C}_l & \widetilde{D}_l & \widetilde{D}_f & \widetilde{D}_w \end{array}\right], \tag{7.53}$$

and, therefore, share the descriptor pair $(\widetilde{A}_l - \lambda\widetilde{E}_l, \widetilde{C}_l)$. As we have seen in the previous sections, such state-space realizations are instrumental in developing filter updating formulas for both the implementation and internal forms of the fault detection filters. The LCF-based filter updating techniques, discussed in this section, fully support the updating-based synthesis methods presented in this book. Consequently, the resulting factors $\big[\,\widehat{Q}(\lambda)\ \widehat{R}_f(\lambda)\ \widehat{R}_w(\lambda)\,\big]$ and $M(\lambda)$ are determined with descriptor realizations of the form (see also Sect. 9.2.6)

$$\big[\,\widehat{Q}(\lambda)\ \widehat{R}_f(\lambda)\ \widehat{R}_w(\lambda)\,\big] = \left[\begin{array}{c|ccc} \widehat{A}_l - \lambda\widehat{E}_l & \widehat{B}_l & \widehat{B}_f & \widehat{B}_w \\ \hline \widehat{C}_l & \widehat{D}_l & \widehat{D}_f & \widehat{D}_w \end{array}\right], \tag{7.54}$$

$$M(\lambda) = \left[\begin{array}{c|c} \widehat{A}_l - \lambda\widehat{E}_l & \widehat{B}_M \\ \hline \widehat{C}_l & \widehat{D}_M \end{array}\right], \tag{7.55}$$

which, once again, share the same descriptor pair $(\widehat{A}_l - \lambda\widehat{E}_l, \widehat{C}_l)$.

In the synthesis procedures presented in Chap. 5 we encounter two distinct cases, where the factorization (7.52) is necessary. The *first* case consists in determining $M(\lambda)$ as the denominator factor of a stable and proper LCF of a generalized (possibly improper) system (see Sect. 9.2.6). This case is encountered—for example, at Step (3) of the **Procedure EFD**, where the proper TFMs $Q(\lambda)$, $R_f(\lambda)$ and $R_w(\lambda)$ resulted at Step (2) have descriptor realizations of the form (7.53) (see also (7.45)), with $\widetilde{E}_l$ invertible, and where the pencil $\widetilde{A}_l - \lambda\widetilde{E}_l$ may have "bad" eigenvalues, which are either unstable or exhibit unsatisfactory dynamics for the filter $Q(\lambda)$. A similar computation is performed at Step (4) of the **Procedure AFD**, where a proper and stable LCF of a generalized system is computed, which is possibly improper, or unstable, or both. Such a computation can be also encountered—for example, at Step (1) of all synthesis algorithms, when the original system (2.1) is not proper and has no control and disturbance inputs, or it may be necessary (see Remark 5.2) to obtain a proper and stable rational left nullspace basis, from an ad-hoc choice as in (5.5).

To determine the LCF in (7.52), the **Procedure GRCF**, presented in Sect. 10.3.5, can be used to compute the RCF of the transposed (dual) compound TFM

$$[\, Q(\lambda)\ R_f(\lambda)\ R_w(\lambda)\,]^T = \big[\, \widehat{Q}(\lambda)\ \widehat{R}_f(\lambda)\ \widehat{R}_w(\lambda)\,\big]^T \big(M^T(\lambda)\big)^{-1}.$$

When applying **Procedure GRCF** to the dual TFM $[\, Q(\lambda)\ R_f(\lambda)\ R_w(\lambda)\,]^T$, the updating operation

$$[\, Q(\lambda)\ R_f(\lambda)\ R_w(\lambda)\,]^T \leftarrow [\, Q(\lambda)\ R_f(\lambda)\ R_w(\lambda)\,]^T M^T(\lambda) = \big[\, \widehat{Q}(\lambda)\ \widehat{R}_f(\lambda)\ \widehat{R}_w(\lambda)\,\big]^T$$

is implicitly performed, and the resulting realizations of $\widehat{Q}(\lambda)$, $\widehat{R}_f(\lambda)$, and $\widehat{R}_w(\lambda)$ in (7.54) share the descriptor pair $(\widehat{A}_l - \lambda \widehat{E}_l, \widehat{C}_l)$.

The RCF algorithm underlying **Procedure GRCF** can be interpreted as a recursive (partial) pole assignment method, which successively moves the "bad" observable eigenvalues of $\widetilde{A}_l - \lambda \widetilde{E}_l$ into the "good" region $\mathbb{C}_g$. A useful feature of this factorization algorithm is that "bad" unobservable eigenvalues of $\widetilde{A}_l - \lambda \widetilde{E}_l$ are automatically detected and, therefore, are removed from the resulting final descriptor representation (7.54). This feature automatically ensures that the order of the resulting descriptor realization (7.54) is always equal to or less than the order of the initial descriptor realization (7.53).

The *second* case consists in computing a diagonal $M(\lambda)$ as the denominator factor of a stable and proper fractional representation of a possibly improper fault detection filter $Q(\lambda)$. This computation is performed at Step (3) of **Procedure EMM**, at Step (4) of **Procedure EMMS**, or at Step (5) of **Procedure AMMS**. In all these procedures, the resulting $M(\lambda)$ serves for updating an initial reference model $M_r(\lambda)$ and has the form

$$M(\lambda) = \mathrm{diag}\left(M^{(1)}(\lambda), \ldots, M^{(m_f)}(\lambda)\right), \tag{7.56}$$

where $M^{(i)}(\lambda)$, for $i = 1, \ldots, m_f$, is a scalar proper and stable TFM. To determine each $M^{(i)}(\lambda)$, the approach to determine proper and stable LCFs, described previously, can be applied to the i-th row $Q^{(i)}(\lambda)$ of $Q(\lambda)$, and, if appropriate, jointly to the i-th rows $R_f^{(i)}(\lambda)$ and $R_w^{(i)}(\lambda)$ of $R_f(\lambda)$ and $R_w(\lambda)$, respectively. Assume that $[\, Q(\lambda)\ R_f(\lambda)\ R_w(\lambda)\,]$ has the realization in (7.53) and we denote $\widetilde{C}_l^{(i)}$, $\widetilde{D}_l^{(i)}$, $\widetilde{D}_f^{(i)}$, and $\widetilde{D}_w^{(i)}$, the i-th rows of the matrices $\widetilde{C}_l$, $\widetilde{D}_l$, $\widetilde{D}_f$, and $\widetilde{D}_w$, respectively. The descriptor realization of $\big[\, Q^{(i)}(\lambda)\ R_f^{(i)}(\lambda)\ R_w^{(i)}(\lambda)\,\big]$ is

$$\left[\, Q^{(i)}(\lambda)\ R_f^{(i)}(\lambda)\ R_w^{(i)}(\lambda)\,\right] = \left[\begin{array}{c|ccc} \widetilde{A}_l - \lambda\widetilde{E}_l & \widetilde{B}_l & \widetilde{B}_f & \widetilde{B}_w \\ \hline \widetilde{C}_l^{(i)} & \widetilde{D}_l^{(i)} & \widetilde{D}_f^{(i)} & \widetilde{D}_w^{(i)} \end{array}\right], \tag{7.57}$$

and is, in general, unobservable (and even undetectable). The factor $M^{(i)}(\lambda)$ results from the stable and proper LCF

$$[\, Q^{(i)}(\lambda)\ R_f^{(i)}(\lambda)\ R_w^{(i)}(\lambda)\,] = \left(M^{(i)}(\lambda)\right)^{-1} [\, \widehat{Q}^{(i)}(\lambda)\ \widehat{R}_f^{(i)}(\lambda)\ \widehat{R}_w^{(i)}(\lambda)\,], \tag{7.58}$$

where $\widehat{Q}^{(i)}(\lambda)$, $\widehat{R}_f^{(i)}(\lambda)$, and $\widehat{R}_w^{(i)}(\lambda)$ are the i-th rows of the resulting $\widehat{Q}(\lambda)$, $\widehat{R}_f(\lambda)$, and $\widehat{R}_w(\lambda)$ in (7.54). The descriptor system realizations of the factors $\widehat{Q}^{(i)}(\lambda)$, $\widehat{R}_f^{(i)}(\lambda)$, $\widehat{R}_w^{(i)}(\lambda)$ and $M^{(i)}(\lambda)$ result in the form

$$\left[\, \widehat{Q}^{(i)}(\lambda)\ \widehat{R}_f^{(i)}(\lambda)\ \widehat{R}_w^{(i)}(\lambda)\,\right] = \left[\begin{array}{c|ccc} \widehat{A}_l^{(i)} - \lambda\widehat{E}_l^{(i)} & \widehat{B}_l^{(i)} & \widehat{B}_f^{(i)} & \widehat{B}_w^{(i)} \\ \hline \widehat{C}_l^{(i)} & \widehat{D}_l^{(i)} & \widehat{D}_f^{(i)} & \widehat{D}_w^{(i)} \end{array}\right], \tag{7.59}$$

$$M^{(i)}(\lambda) = \left[\begin{array}{c|c} \widehat{A}_l^{(i)} - \lambda\widehat{E}_l^{(i)} & \widehat{B}_M^{(i)} \\ \hline \widehat{C}_l^{(i)} & \widehat{D}_M^{(i)} \end{array}\right]. \tag{7.60}$$

In general, both state-space realizations in (7.59) and (7.60) are non-minimal, because the descriptor pair $(\widehat{A}_l^{(i)} - \lambda\widehat{E}_l^{(i)}, \widehat{C}_l^{(i)})$ may be unobservable (albeit detectable, being stable). However, controllability is generically preserved, provided the original realization (7.53) (thus also (7.57)) is controllable.

To compute the LCF in (7.58), the **Procedure GRCF** can be employed, by computing the RCF of the appropriate dual system. A main advantage of using this procedure is its general applicability, regardless the underlying descriptor realization is finite detectable or not, or if it is infinite-observable or not (undetectable eigenvalues are simply removed from the resulting factors). The dimension of the state-space realization of the resulting descriptor system (7.59) can be evaluated in terms of the eigenvalues of $\widetilde{A}_l - \lambda\widetilde{E}_l$, and is given by the sum of four numbers: (1) the number of stable eigenvalues (which are kept in the resulting $\widehat{A}_l^{(i)} - \lambda\widehat{E}_l^{(i)}$); (2) the number of observable unstable eigenvalues (which are moved to the stable domain); (3) the number of observable nonsimple infinite eigenvalues (which are moved to finite values in the stable domain); and (4) the number of simple infinite eigenvalues (which are the same as of $\widetilde{A}_l - \lambda\widetilde{E}_l$). Therefore, to determine an irreducible realization, it is usually sufficient to only remove the unobservable stable eigenvalues of $\widehat{A}_l^{(i)} - \lambda\widehat{E}_l^{(i)}$ (which are the same as the unobservable stable eigenvalues of $\widetilde{A}_l - \lambda\widetilde{E}_l$). The overall descriptor realization in (7.54) is obtained by simply stacking the m_f computed rows of the numerator factors (see Sect. 9.2.3 for the formulas for building column concatenation of descriptor systems). The resulting final descriptor realization (7.54) has block diagonal matrices $\widehat{A}_l$, $\widehat{E}_l$, and $\widehat{C}_l$, and is, in general, non-minimal. Irreducible and minimal realizations of descriptor systems can be computed using the algorithms described in Sect. 10.3.1.

The above considerations apply also to the realization of $M^{(i)}(\lambda)$ in (7.60). However, a useful feature of using **Procedure GRCF** is the possibility to obtain a minimal descriptor system realization

$$M^{(i)}(\lambda) = \left[\begin{array}{c|c} \widehat{A}_{l,o}^{(i)} - \lambda \widehat{E}_{l,o}^{(i)} & \widehat{B}_{M,o}^{(i)} \\ \hline \widehat{C}_{l,o}^{(i)} & \widehat{D}_{M,o}^{(i)} \end{array} \right], \tag{7.61}$$

which can be simply read out from the resulting realization of $M^{(i)}(\lambda)$ in (7.60) (see Sect. 10.3.5 for details). The resulting realization of $M(\lambda)$ in (7.56) can be obtained by employing the diagonal stacking formulas of descriptor systems (see Sect. 9.2.3).

7.7 Outer–Inner Factorizations

An important computation in solving several approximate synthesis problems involves the quasi-co-outer–co-inner factorization of a full row rank TFM. We recall (see Sect. 9.1.8), that a quasi-co-outer TFM has full column rank (i.e., is injective) and has only zeros in the closed stability domain $\overline{\mathbb{C}}_s$. A particular aspect enforced in all approximate synthesis procedures is that the quasi-co-outer factor has also full row rank, and therefore, is invertible. The main advantage of this feature is the easiness of performing the subsequent filter updating operations, which involve the inverses of the quasi-co-outer factors. In some cases, we need to determine the extended quasi-outer–inner factorization, where a square co-inner factor results.

The quasi-outer–co-inner factorization of the $q \times m_w$ TFM $R_w(\lambda)$, of full row rank, is computed at Step (3) of the **Procedure AFD** in the form

$$R_w(\lambda) = G_o(\lambda)G_i(\lambda), \tag{7.62}$$

where the resulting $q \times q$ quasi-outer factor $G_o(\lambda)$ is invertible and the $q \times m_w$ co-inner TFM $G_i(\lambda)$ has full row rank. The extended quasi-outer–co-inner factorization of the compound $q \times (m_f + m_w)$ TFM $\left[\, R_f(\lambda) \,\middle|\, R_w(\lambda) \,\right]$, of full row rank $q \geq m_f$, is computed at Step (3) of **Procedure AMMS** in the form

$$\left[\, R_f(\lambda) \,\middle|\, R_w(\lambda) \,\right] = [\, G_o(\lambda)\ 0\,]G_i(\lambda), \tag{7.63}$$

where the $q \times q$ quasi-outer factor $G_o(\lambda)$ is invertible and $G_i(\lambda)$ is an $(m_f + m_w) \times (m_f + m_w)$ square co-inner factor. Subsequently, with $Q_3(\lambda) = G_o^{-1}(\lambda)$, the following updating operations are performed

$$[\, Q(\lambda)\ R_f(\lambda)\ R_w(\lambda)\,] \leftarrow [\, \overline{Q}(\lambda)\ \overline{R}_f(\lambda)\ \overline{R}_w(\lambda)\,] := Q_3(\lambda)[\, Q(\lambda)\ R_f(\lambda)\ R_w(\lambda)\,]. \tag{7.64}$$

In this section we discuss the computation of the above factorizations and show how the subsequent updating can be efficiently performed.

We assume that at the end of Step (2) of either **Procedure AFD** or **Procedure AMMS**, the proper TFMs $Q(\lambda)$, $R_f(\lambda)$ and $R_w(\lambda)$ have the observable state-space realizations in (7.45), which, for convenience, are reproduced below

$$
\left[\, Q(\lambda)\ R_f(\lambda)\ R_w(\lambda)\,\right] = \left[\begin{array}{c|ccc} \widetilde{A}_l - \lambda \widetilde{E}_l & \widetilde{B}_l & \widetilde{B}_f & \widetilde{B}_w \\ \hline \widetilde{C}_l & \widetilde{D}_l & \widetilde{D}_f & \widetilde{D}_w \end{array}\right],
\tag{7.65}
$$

where $\widetilde{E}_l$ is invertible. Furthermore, we can always assume that $\widetilde{D}_l$ has full row rank. To compute the factorizations in (7.62) or (7.63), we apply the inner-outer factorization method presented in Sect. 10.3.6 to the transposed TFMs $R_w^T(\lambda)$ and $[\, R_f(\lambda)\ R_w(\lambda)\,]^T$, respectively, to obtain the quasi-outer factor $G_o^T(\lambda)$ and inner factor $G_i^T(\lambda)$. The resulting descriptor system realization of the quasi-co-outer factor $G_o(\lambda)$ has the form

$$
G_o(\lambda) = \left[\begin{array}{c|c} \widetilde{A}_l - \lambda \widetilde{E}_l & \widetilde{B}_o \\ \hline \widetilde{C}_l & \widetilde{D}_o \end{array}\right],
\tag{7.66}
$$

where $\widetilde{B}_o$ and $\widetilde{D}_o$ are matrices with q columns. For the co-inner factors $G_i(\lambda)$, in each case, explicit minimal order descriptor realizations are obtained from either Proposition 10.6, in the continuous-time case, or Proposition 10.7, in the discrete-time case.

To perform the updating operations in (7.64), we can alternatively solve the linear rational system of equations

$$
G_o(\lambda)\left[\, \overline{Q}(\lambda)\ \overline{R}_f(\lambda)\ \overline{R}_w(\lambda)\,\right] = \left[\, Q(\lambda)\ R_f(\lambda)\ R_w(\lambda)\,\right].
\tag{7.67}
$$

Observe that $G_o(\lambda)$, $Q(\lambda)$, $R_f(\lambda)$ and $R_w(\lambda)$ have descriptor realizations, which share the same observable pair $(\widetilde{A}_l - \lambda \widetilde{E}_l, \widetilde{C}_l)$. This allows to compute the solution as (see also Sect. 7.9 for more details)

$$
\left[\, \overline{Q}(\lambda)\ \overline{R}_f(\lambda)\ \overline{R}_w(\lambda)\,\right] = \left[\, 0\ I_q\,\right] Y(\lambda),
$$

where $Y(\lambda)$ is the rational solution of the linear polynomial equation

$$
\left[\begin{array}{cc} \widetilde{A}_l - \lambda \widetilde{E}_l & \widetilde{B}_o \\ \widetilde{C}_l & \widetilde{D}_o \end{array}\right] Y(\lambda) = \left[\begin{array}{ccc} \widetilde{B}_l & \widetilde{B}_f & \widetilde{B}_w \\ \widetilde{D}_l & \widetilde{D}_f & \widetilde{D}_w \end{array}\right].
\tag{7.68}
$$

Since the system matrix

$$
S_o(\lambda) = \left[\begin{array}{cc} \widetilde{A}_l - \lambda \widetilde{E}_l & \widetilde{B}_o \\ \widetilde{C}_l & \widetilde{D}_o \end{array}\right]
\tag{7.69}
$$

of the descriptor system realization (7.66) is invertible, we obtain

$$
\left[\, \overline{Q}(\lambda)\ \overline{R}_f(\lambda)\ \overline{R}_w(\lambda)\,\right] = \left[\, 0\ I_q\,\right] S_o^{-1}(\lambda) \left[\begin{array}{ccc} \widetilde{B}_l & \widetilde{B}_f & \widetilde{B}_w \\ \widetilde{D}_l & \widetilde{D}_f & \widetilde{D}_w \end{array}\right],
$$

which leads to the explicit descriptor realizations of the updated terms

$$\left[\,\overline{Q}(\lambda)\ \overline{R}_f(\lambda)\ \overline{R}_w(\lambda)\,\right] = \left[\begin{array}{cc|ccc} \widetilde{A}_l - \lambda \widetilde{E}_l & \widetilde{B}_o & \widetilde{B}_l & \widetilde{B}_f & \widetilde{B}_w \\ \hline \widetilde{C}_l & \widetilde{D}_o & \widetilde{D}_l & \widetilde{D}_f & \widetilde{D}_w \\ \hline 0 & -I_q & 0 & 0 & 0 \end{array}\right]. \tag{7.70}$$

In obtaining the realization (7.70), implicit pole–zero cancellations take place, leading to the cancellation of all eigenvalues of $\widetilde{A}_l - \lambda \widetilde{E}_l$. This feature allows to perform the updating operation in (7.64) using an explicit updating formula, based on the realization (7.70), and was the main reason for enforcing the full row rank property of $R_w(\lambda)$, in **Procedure AFD**, or of $[\,R_f(\lambda)\ R_w(\lambda)\,]$, in **Procedure AMMS**.

Regarding the structural properties of the resulting realization (7.70), we have the following result.

Proposition 7.3 *If the realization (7.65) of $\left[\,Q(\lambda)\ R_f(\lambda)\ R_w(\lambda)\,\right]$ is irreducible and $\widetilde{D}_l$ has full row rank, then the realization (7.70) is irreducible.*

Proof Since the realization (7.65) is observable, $\left[\begin{array}{c} \widetilde{A}_l - \lambda \widetilde{E}_l \\ \widetilde{C}_l \end{array}\right]$ has full column rank for all $\lambda \in \mathbb{C}$. The observability of the realization (7.70), then follows from the full column rank property of

$$\left[\begin{array}{cc} \widetilde{A}_l - \lambda \widetilde{E}_l & \widetilde{B}_o \\ \widetilde{C}_l & \widetilde{D}_o \\ 0 & -I_q \end{array}\right].$$

To prove the controllability of the realization (7.70), we have to show that for all $\lambda \in \mathbb{C}$, the matrix

$$\widetilde{S}(\lambda) := \left[\begin{array}{ccccc} \widetilde{A}_l - \lambda \widetilde{E}_l & \widetilde{B}_o & \widetilde{B}_l & \widetilde{B}_f & \widetilde{B}_w \\ \widetilde{C}_l & \widetilde{D}_o & \widetilde{D}_l & \widetilde{D}_f & \widetilde{D}_w \end{array}\right]$$

has full row rank. This matrix is the system matrix of an irreducible proper descriptor realization of the compound TFM $\left[\,G_o(\lambda)\ Q(\lambda)\ R_f(\lambda)\ R_w(\lambda)\,\right]$, which has no zeros (finite or infinite) because $Q(\infty) = \widetilde{D}_l$ has full row rank. It follows that $\widetilde{S}(\lambda)$ has no invariant zeros as well, and therefore has full row rank for all $\lambda \in \mathbb{C}$. ∎

Even if the realization (7.70) is irreducible, it may still be non-minimal due to the presence of simple infinite eigenvalues. For example, if $G_o(\lambda)$ has no infinite zeros (e.g., when solving the standard cases of the AFDP or AMMP), then $\widetilde{D}_o$ is invertible and the descriptor realization (7.70) is proper. In this case, the simple infinite eigenvalues can be eliminated to obtain a reduced order descriptor realization. For example, a reduced order realization of $\overline{Q}(\lambda)$ results as

$$\overline{Q}(\lambda) = \left[\begin{array}{c|c} \widetilde{A}_l - \widetilde{B}_o \widetilde{D}_o^{-1} \widetilde{C}_l - \lambda \widetilde{E}_l & \widetilde{B}_l - \widetilde{B}_o \widetilde{D}_o^{-1} \widetilde{D}_l \\ \hline \widetilde{D}_o^{-1} \widetilde{C}_l & \widetilde{D}_o^{-1} \widetilde{D}_l \end{array}\right].$$

Similar reduced order realizations can be obtained for $\overline{R}_f(\lambda)$ and $\overline{R}_w(\lambda)$ too.

When solving the standard cases of the AFDP or AMMP, the resulting realization (7.70) has only stable poles, which are the finite zeros of $G_o(\lambda)$. Looking to the details of the inner-outer factorization procedure (see Sect. 10.3.6), these zeros originate partly from the performed column compression and partly from the unstable zeros of the underlying TFMs to be factorized (i.e., $R_f(\lambda)$ or $[\,R_f(\lambda)\ R_w(\lambda)\,]$), which are moved into stable positions. These stable zeros are also the (only) poles of the inner factor $G_i(\lambda)$. If the TFMs to be factorized have zeros also in $\partial\mathbb{C}_s$ (e.g., in the nonstandard cases), then these zeros become also zeros of $G_o(\lambda)$, but not poles of $G_i(\lambda)$. Therefore, in this case, the resulting realizations of $\overline{R}_w(\lambda)$ or of $[\,\overline{R}_f(\lambda)\ \overline{R}_w(\lambda)\,]$ are uncontrollable.

For the quasi-co-outer–co-inner factorization of $R_w(\lambda)$ in (7.62), used in **Procedure AFD**, a proper minimal state-space realization of $\overline{R}_w(\lambda)$ can be explicitly determined, by observing that $\overline{R}_w(\lambda) = G_o^{-1}(\lambda)R_w(\lambda) = G_i(\lambda)$. Therefore, when using the inner-outer factorization approach presented in Sect. 10.3.6, the resulting minimal proper state-space realization of the co-inner factor $G_i(\lambda)$ is obtained from either Proposition 10.6, in the continuous-time case, or Proposition 10.7, in the discrete-time case. Similarly, for the quasi-co-outer–co-inner factorization of $[\,R_f(\lambda)\ R_w(\lambda)\,]$ in (7.63), used in **Procedure AMMS**, an explicit proper minimal state-space realization of $[\,\overline{R}_f(\lambda)\ \overline{R}_f w(\lambda)\,]$ can be determined from the resulting descriptor realization of the $(m_f + m_w) \times (m_f + m_w)$ co-inner factor $G_i(\lambda)$. Consider the co-inner factor $G_i(\lambda)$, partitioned row-wise to correspond to the column structure of $[\,G_o(\lambda)\ 0\,]$ in (7.63), in the form

$$G_i(\lambda) = \begin{bmatrix} G_{i,1}(\lambda) \\ G_{i,2}(\lambda) \end{bmatrix}, \tag{7.71}$$

where $G_{i,1}(\lambda)$ and $G_{i,2}(\lambda)$ have q and $m_f + m_w - q$ rows, respectively. It follows

$$[\,\overline{R}_f(\lambda)\ \overline{R}_w(\lambda)\,] = G_o^{-1}(\lambda)[\,R_f(\lambda)\ R_w(\lambda)\,] = [\,I_q\ 0\,]G_i(\lambda) = G_{i,1}(\lambda)\,.$$

Remark 7.5 At Step (3) of the **Procedure AMMS**, we need additionally to compute the TFMs $\widetilde{F}_1(\lambda)$ and $\widetilde{F}_2(\lambda)$, which define the least distance problem to be solved at the next step. For a given $m_f \times m_f$ reference model with TFM $M_r(\lambda)$ and with the square co-inner factor $G_i(\lambda)$ partitioned as in (7.71), we need to evaluate

$$[\,\widetilde{F}_1(\lambda)\ \widetilde{F}_2(\lambda)\,] := [\,M_r(\lambda)\ 0\,]\,G_i^\sim(\lambda) = [\,M_r(\lambda)\ 0\,][\,G_{i,1}^\sim(\lambda)\ G_{i,2}^\sim(\lambda)\,]\,,$$

Assume that the conjugate TFM $G_i^\sim(\lambda)$ has a descriptor system realization of the form

$$G_i^\sim(\lambda) = [\,G_{i,1}^\sim(\lambda)\ G_{i,2}^\sim(\lambda)\,] = \left[\begin{array}{c|cc} A_i - \lambda E_i & B_{i,1} & B_{i,2} \\ \hline C_i & D_{i,1} & D_{i,2} \end{array}\right], \tag{7.72}$$

where all generalized eigenvalues of the pair $(A_i,\ E_i)$ are unstable (see Sect. 9.2.3 for formulas to build realizations of conjugated TFMs). The use of a descriptor system realization (instead of a standard state-space realization) is necessary to cover all possible cases. While for a continuous-time system with TFM $G_i(s)$, the resulting E_i is always invertible, this is not generally true for a discrete-time system with TFM $G_i(z)$ for which E_i can be singular if $G_i(z)$ has poles in the origin. In such a case, the pair (A_i, E_i) has infinite (unstable) generalized eigenvalues. Let the quadruple (A_r, B_r, C_r, D_r) describe the state-space realization of $[\, M_r(\lambda)\ 0\,]$. Then, using (7.72), the state-space realization of $[\, \widetilde{F}_1(\lambda)\ \widetilde{F}_2(\lambda)\,]$ has the form

$$[\, \widetilde{F}_1(\lambda)\ \widetilde{F}_2(\lambda)\,] = \left[\begin{array}{cc|cc} A_r - \lambda I & B_r C_i & B_r D_{i,1} & B_r D_{i,2} \\ 0 & A_i - \lambda E_i & B_{i,1} & B_{i,2} \\ \hline C_r & D_r C_i & D_r D_{i,1} & D_r D_{i,2} \end{array}\right], \tag{7.73}$$

where A_r has only stable eigenvalues, while the pair (A_i, E_i) has only unstable generalized eigenvalues. $\qquad\square$

7.8 Spectral Factorizations

Let $G(\lambda)$ be a proper TFM. The first problem we discuss in this section is the computation, for a stable TFM $G(\lambda)$, of a minimum-phase spectral factor $G_o(\lambda)$, which solves the left spectral factorization problem

$$\gamma^2 I + G(\lambda)G^{\sim}(\lambda) = G_o(\lambda)G_o^{\sim}(\lambda), \tag{7.74}$$

where $\gamma > 0$. This computation appears in addressing the solution of the nonstandard AFDP using the regularization approach described in Remark 5.8 (see also [55]). The second problem we consider is the computation, for a TFM $G(\lambda)$ without poles on the boundary of the stability domain $\partial\mathbb{C}_s$, of a stable and minimum-phase spectral factor $G_o(\lambda)$, which solves the left spectral factorization problem

$$\gamma^2 I - G(\lambda)G^{\sim}(\lambda) = G_o(\lambda)G_o^{\sim}(\lambda), \tag{7.75}$$

where $\gamma > \|G(\lambda)\|_\infty$. This computation is repeatedly performed at Step (4) of **Procedure AMMS**, when solving the 2-block $\mathcal{H}_\infty$-least distance problem, discussed in Sect. 9.1.10, using the approach described in Sect. 7.10.

For the solution of the considered spectral factorization problems, the numerically reliable computation of the solutions of the generalized continuous- and discrete-time algebraic Riccati equation (GCARE, GDARE) is of paramount importance. Numerically reliable methods for the solution of GCARE and GDARE are discussed in Sect. 10.2.2.

Assume $G(\lambda)$ has a descriptor system realization $(A - \lambda E, B, C, D)$, with invertible E. The spectral factorization problem (7.74) can be equivalently written in the form

$$\gamma^2 I + G(\lambda)G^{\sim}(\lambda) = \left[\,\gamma I\ G(\lambda)\,\right]\left[\begin{array}{c} \gamma I \\ G^{\sim}(\lambda) \end{array}\right],$$

which is a standard *minimum-phase left spectral factorization problem* for the full row rank $\widetilde{G}(\lambda) := \left[\,\gamma I\ G(\lambda)\,\right]$ (see Sect. 9.1.8). Since $G_o(\lambda)$ in (7.74) is the co-outer factor of the co-outer–co-inner factorization of $\widetilde{G}(\lambda)$, we can use, for the dual descriptor system realization of $\widetilde{G}^T(\lambda)$, the results of Theorem 9.3, in the continuous-time case, and of Theorem 9.4, in the discrete-time case. The co-outer factor is obtained, in both cases, in the form

$$G_o(\lambda) = \left[\begin{array}{c|c} A - \lambda E & -K_s R^{1/2} \\ \hline C & R^{1/2} \end{array} \right],$$

where K_s is the stabilizing output injection gain, obtained from the stabilizing solution of the appropriate continuous- or discrete-time Riccati equation, with R defined accordingly.

For the continuous-time case, $R = \gamma^2 I + DD^T$ and K_s is determined as

$$K_s = -(EY_s C^T + BD^T)R^{-1},$$

where Y_s is the stabilizing solution the dual (filter) GCARE

$$AYE^T + EYA^T - (EYC^T + BD^T)R^{-1}(CYE^T + DB^T) + BB^T = 0.$$

In the discrete-time case, $R = R_D + CY_s C^T$, with $R_D := \gamma^2 I + DD^T$, and K_s is determined as

$$K_s = -(AY_s C^T + BD^T)R^{-1},$$

where Y_s is the stabilizing solution of the dual (filter) GDARE

$$AYA^T - EYE^T - (AYC^T + BD^T)(R_D + CYC^T)^{-1}(CYA^T + DB^T) + BB^T = 0.$$

For the solution of the second left spectral factorization problem (7.75) we assume that $G(\lambda)$ has no poles on the boundary of stability domain $\partial\mathbb{C}_s$. Furthermore, in the continuous-time case, we assume $G(\lambda)$ is proper. Under these conditions, the computation of the spectral factor $G_o(\lambda)$ can be done in two successive steps. In the first step, we compute a right coprime factorization $G(\lambda) = N(\lambda)M^{-1}(\lambda)$, where the denominator factor $M(\lambda)$ is inner. It follows that

$$\gamma^2 I - G(\lambda)G^\sim(\lambda) = \gamma^2 I - N(\lambda)N^\sim(\lambda),$$

where $N(\lambda)$ is proper and has only poles in $\mathbb{C}_s$. Assume $G(\lambda)$ has a descriptor system realization $(A - \lambda E, B, C, D)$. In the continuous-time case, we assume E is invertible, while in the discrete-time case, E may be singular, in which case, the pair (A, E) has (unstable) infinite eigenvalues. The computation of the RCF with inner denominator can be done using the **Procedure GRCFID** described in Sect. 10.3.5, which determines $N(\lambda)$ and $M(\lambda)$ in the form

$$N(\lambda) = \left[\begin{array}{c|c} \widetilde{A} - \lambda\widetilde{E} & \widetilde{B} \\ \hline \widetilde{C}_N & \widetilde{D}_N \end{array}\right], \quad M(\lambda) = \left[\begin{array}{c|c} \widetilde{A} - \lambda\widetilde{E} & \widetilde{B} \\ \hline \widetilde{C}_M & \widetilde{D}_M \end{array}\right]. \tag{7.76}$$

This procedure can be interpreted as a recursive partial pole assignment method, which uses the generalized real Schur form of the pair (A, E) for an initial separation of the stable and unstable eigenvalues and determines the resulting pair $(\widetilde{A}, \widetilde{E})$ also in a generalized real Schur form, where the stable eigenvalues of the pair (A, E) are preserved, while the controllable unstable generalized eigenvalues are successively moved to stable locations symmetrically placed with respect to the stability domain boundary $\partial\mathbb{C}_s$. The factorization method is applicable regardless the original descriptor system realization is stabilizable or not, because all unstable uncontrollable eigenvalues are automatically removed from the resulting factors.

In the second step, we determine the stable and minimum-phase left spectral factor $G_o(\lambda)$, which satisfies

$$\gamma^2 I - N(\lambda)N^{\sim}(\lambda) = G_o(\lambda)G_o^{\sim}(\lambda).$$

The descriptor system realization of $G_o(\lambda)$ is obtained in the form

$$G_o(\lambda) = \left[\begin{array}{c|c} \widetilde{A} - \lambda\widetilde{E} & -K_s R^{1/2} \\ \hline \widetilde{C} & R^{1/2} \end{array}\right],$$

where K_s is the stabilizing output injection gain, obtained from the stabilizing solution of the appropriate continuous- or discrete-time Riccati equation, with R defined accordingly.

In the continuous-time case, using the results of Lemma 9.14, we have

$$\begin{aligned} R &= \gamma^2 I - \widetilde{D}\widetilde{D}^T, \\ K_s &= (\widetilde{E}Y_s\widetilde{C}^T + \widetilde{B}\widetilde{D}^T)R^{-1} \end{aligned}$$

and Y_s is the stabilizing solution of the GCARE

$$\widetilde{A}Y\widetilde{E}^T + \widetilde{E}Y\widetilde{A}^T + (\widetilde{E}Y\widetilde{C}^T + \widetilde{B}\widetilde{D}^T)R^{-1}(\widetilde{C}Y\widetilde{E}^T + \widetilde{D}\widetilde{B}^T) + \widetilde{B}\widetilde{B}^T = 0.$$

In the discrete-time case, using the results of Lemma 9.15, we have

$$\begin{aligned} R_D &= \gamma^2 I - \widetilde{D}\widetilde{D}^T, \\ R &= R_D - \widetilde{C}Y_s\widetilde{C}^T, \\ K_s &= (\widetilde{B}\widetilde{D}^T + \widetilde{A}Y_s\widetilde{C}^T)R^{-1} \end{aligned}$$

and Y_s is the stabilizing solution of the GDARE

$$\widetilde{A}Y\widetilde{A}^T - \widetilde{E}Y\widetilde{E}^T - (\widetilde{A}Y\widetilde{C}^T + \widetilde{B}\widetilde{D}^T)(-R_D + \widetilde{C}Y\widetilde{C}^T)^{-1}(\widetilde{C}Y\widetilde{A}^T + \widetilde{D}\widetilde{B}^T) + \widetilde{B}\widetilde{B}^T = 0.$$

7.9 Linear Rational Equations

In this section we discuss the computational aspects of solving linear rational equations encountered in the synthesis algorithms presented in Chap. 5. First, we consider an important particular case of solving the equation

$$G(\lambda)X(\lambda) = F(\lambda),\tag{7.77}$$

with $G(\lambda)$ a $p \times p$ invertible rational matrix and $F(\lambda)$ an $p \times q$ (arbitrary) rational matrix. Such a computation is encountered in performing the filter updating operations at Step (3) of the **Procedure AFD**, at Step (3) of **Procedure EMMS** and at Step (3) of the **Procedure AMMS**. In the **Procedures AFD** and AMMS, $G(\lambda) = G_o(\lambda)$, where $G_o(\lambda)$ is the invertible quasi-outer factor resulting from particular quasi-co-outer–co-inner factorizations. The corresponding updating operations have been already discussed in Sect. 7.7. In **Procedure EMMS**, a similar computation is performed with $G(\lambda) = R_f(\lambda)$, where $R_f(\lambda)$ is an invertible TFM representing the current TFM from faults to residual.

A straightforward way to determine a descriptor realization of $X(\lambda)$ starting with existing descriptor realizations of $G(\lambda)$ and $F(\lambda)$ is to form first $G^{-1}(\lambda)$ explicitly and then to compute a minimal realization of $G^{-1}(\lambda)F(\lambda)$. Fortunately, we can exploit a nice common feature of all the above computations, where the descriptor realizations of $G(\lambda)$ and $F(\lambda)$ have the forms

$$G(\lambda) = \left[\begin{array}{c|c} A - \lambda E & B_G \\ \hline C & D_G \end{array}\right], \qquad F(\lambda) = \left[\begin{array}{c|c} A - \lambda E & B_F \\ \hline C & D_F \end{array}\right],\tag{7.78}$$

which share the observable descriptor system pair $(A - \lambda E, C)$. It is easy to observe that any solution of (7.77) is also part of the solution of the linear polynomial equation

$$\left[\begin{array}{cc} A - \lambda E & B_G \\ C & D_G \end{array}\right] Y(\lambda) = \left[\begin{array}{c} B_F \\ D_F \end{array}\right],\tag{7.79}$$

where $Y(\lambda) = \left[\begin{array}{c} W(\lambda) \\ X(\lambda) \end{array}\right]$. Therefore, alternative to solving (7.77) using explicit inversion of $G(\lambda)$, we can solve instead (7.79) for $Y(\lambda)$ and compute $X(\lambda)$ as

$$X(\lambda) = \left[\begin{array}{cc} 0 & I_p \end{array}\right] Y(\lambda).\tag{7.80}$$

It is straightforward to see that a descriptor system realization of $X(\lambda)$ can be explicitly obtained as

$$X(\lambda) = \left[\begin{array}{cc|c} A - \lambda E & B_G & B_F \\ C & D_G & D_F \\ \hline 0 & -I_p & 0 \end{array}\right].\tag{7.81}$$

Thus, all updating computations can be performed explicitly without any numerical computations. For example, in the **Procedure AFD** the linear rational equation (7.67) is solved via the equivalent linear pencil-based equation (7.68) leading to the explicit realization (7.70). Entirely similar realizations can be employed also in **Procedure AMMS**.

An important case is when the Eq. (7.77) has a non-unique solution and the non-uniqueness can be exploited to determine a special particular solution, for example, one having the least McMillan degree. Assume that $G(\lambda)$ and $F(\lambda)$ are $p \times m$ and $p \times q$ rational matrices, respectively, and $X(\lambda)$ is a (non-unique) $m \times q$ solution. The solvability condition of equation (7.77) results from Lemma 9.4 (applied to the dual problem $G^T(\lambda)X^T(\lambda) = F^T(\lambda)$) as

$$\operatorname{rank} G(\lambda) = \operatorname{rank}[\, G(\lambda) \ F(\lambda) \,]$$

and assume that it is fulfilled. The general solution of (7.77) can be expressed as

$$X(\lambda) = X_0(\lambda) + X_r(\lambda)Y(\lambda),$$

where $X_0(\lambda)$ is any particular solution of (7.77) and $X_r(\lambda)$ is a rational basis matrix for the right nullspace of $G(\lambda)$. By choosing $Y(\lambda)$ appropriately, we aim to achieve that $X(\lambda)$ has the least possible McMillan degree. The numerical solution of this problem is addressed in detail in Sect. 10.3.7, where a general numerical approach is provided, which relies on combining orthogonal pencil reduction techniques and minimal dynamic-covers-based order reduction. The computational algorithm can be equally employed to solve the dual equation $X(\lambda)G(\lambda) = F(\lambda)$, by solving for $Y(\lambda) = X^T(\lambda)$ the equation $G^T(\lambda)Y(\lambda) = F^T(\lambda)$.

The solution of the dual equation $X(\lambda)G(\lambda) = F(\lambda)$ is encountered at Step (2) of the **Procedure EMM**, when solving the EMMP formulated in Sect. 3.5.5. In this case, the rational matrix $G(\lambda)$ is built from the reduced system (5.11) as $G(\lambda) = \overline{G}_f(\lambda)$, while $F(\lambda)$ is set as $F(\lambda) = M_r(\lambda)$, where $M_r(\lambda)$ is the TFM of a desired reference model. In this case, the solution $X(\lambda)$ represents one of the factors in the product form representation of the FDI filter. However, according to Remark 5.13, the EMMP can be also solved directly to determine a solution $X(\lambda)$, which represents a preliminary or even the final synthesis of a FDI filter. In this case, $G(\lambda)$ and $F(\lambda)$ stand for

$$G(\lambda) = \begin{bmatrix} G_u(\lambda) \ G_d(\lambda) \ G_f(\lambda) \\ I_{m_u} \qquad 0 \qquad 0 \end{bmatrix}, \quad F(\lambda) = \begin{bmatrix} 0 \ 0 \ M_r(\lambda) \end{bmatrix}.$$

The general solution of the dual equation $X(\lambda)G(\lambda) = F(\lambda)$ can be expressed as

$$X(\lambda) = X_0(\lambda) + Y(\lambda)X_l(\lambda),$$

where $X_0(\lambda)$ is any particular solution and $X_l(\lambda)$ is a rational basis matrix of the left nullspace of $G(\lambda)$. By choosing $Y(\lambda)$ appropriately, we aim to achieve that $X(\lambda)$ has the least possible McMillan degree.

7.10 Solution of Least-Distance Problems

In this section, we discuss the computational aspects involved by the solution of the $\mathcal{H}_\infty$ least distance problem ($\mathcal{H}_\infty$-LDP)

$$\min_{Y(\lambda)\in\mathcal{H}_\infty} \left\| \left[\; \widetilde{F}_1(\lambda) - Y(\lambda) \;\; \widetilde{F}_2(\lambda) \;\right] \right\|_\infty , \tag{7.82}$$

where $\widetilde{F}_1(\lambda)$ and $\widetilde{F}_2(\lambda)$ are rational matrices without poles in $\partial\mathbb{C}_s$, the boundary of the stability domain. In the continuous-time setting we additionally assume that $\widetilde{F}_1(s)$ and $\widetilde{F}_2(s)$ are proper TFMs. If $\widetilde{F}_2(\lambda) = 0$, then we have an 1-*block* $\mathcal{H}_\infty$-LDP, while if $\widetilde{F}_2(\lambda) \neq 0$ then we have a 2-*block* $\mathcal{H}_\infty$-LDP. The solution of a $\mathcal{H}_\infty$-LDP is performed at Step (4) of **Procedure AMMS** presented in Sect. 5.7.

Conceptual algorithms for the solution of the 1-*block* and 2-*block* $\mathcal{H}_\infty$-LDP are described in Sect. 9.1.10, in the context of solving the $\mathcal{H}_\infty$ *model-matching problem* ($\mathcal{H}_\infty$-MMP). In this section, we discuss the underlying computational algorithms in terms of descriptor system representations. Assume $\left[\; \widetilde{F}_1(\lambda)\;\; \widetilde{F}_2(\lambda) \;\right]$ has the descriptor representation

$$\left[\; \widetilde{F}_1(\lambda)\;\; \widetilde{F}_2(\lambda) \;\right] = \left[\begin{array}{c|cc} \widetilde{A} - \lambda\widetilde{E} & \widetilde{B}_1 & \widetilde{B}_2 \\ \hline \widetilde{C} & \widetilde{D}_1 & \widetilde{D}_2 \end{array}\right] . \tag{7.83}$$

In the continuous-time case, we assume $\widetilde{E}$ invertible.

When solving the AMMP using **Procedure AMMS**, $\left[\; \widetilde{F}_1(\lambda)\;\; \widetilde{F}_2(\lambda) \;\right]$ is computed as

$$\left[\; \widetilde{F}_1(\lambda)\;\; \widetilde{F}_2(\lambda) \;\right] = \left[\; M_r(\lambda)\; 0 \;\right] G_i^\sim(\lambda),$$

where $M_r(\lambda)$ is the TFM of a desired reference model (assumed proper and stable) and $G_i(\lambda)$ is a square inner factor (i.e., $G_i^{-1}(\lambda) = G_i^\sim(\lambda)$ is anti-stable). In Remark 7.5, the descriptor system realization (7.83) is constructed as

$$\left[\begin{array}{c|cc} \widetilde{A} - \lambda\widetilde{E} & \widetilde{B}_1 & \widetilde{B}_2 \\ \hline \widetilde{C} & \widetilde{D}_1 & \widetilde{D}_2 \end{array}\right] := \left[\begin{array}{cc|cc} A_r - \lambda I & B_r C_i & B_r D_{i,1} & B_r D_{i,2} \\ 0 & A_i - \lambda E_i & B_{i,1} & B_{i,2} \\ \hline C_r & D_r C_i & D_r D_{i,1} & D_r D_{i,2} \end{array}\right] , \tag{7.84}$$

where the quadruple (A_r, B_r, C_r, D_r) is a standard state-space realization of the stable TFM $[\; M_r(\lambda)\; 0\;]$ and the anti-stable TFM $G_i^\sim(\lambda)$ has the descriptor realization (7.72). It follows that A_r has only stable eigenvalues, while the pair (A_i, E_i) has only unstable generalized eigenvalues. This structure can be exploited to simplify the computations when solving the $\mathcal{H}_\infty$-LDP.

Solution of the 1-block $\mathcal{H}_\infty$-LDP. The stable optimal solution $Y(\lambda)$ of the 1-block problem can be computed by solving an optimal Nehari problem. Let $L_s(\lambda)$ be the stable part and let $L_u(\lambda)$ be the unstable part of the additive decomposition

$$L_s(\lambda) + L_u(\lambda) = \widetilde{F}_1(\lambda) . \tag{7.85}$$

Then, for the optimal solution $Y(\lambda)$ we have successively

$$\|\widetilde{F}_1(\lambda) - Y(\lambda)\|_\infty = \|L_u(\lambda) - \widetilde{Y}(\lambda)\|_\infty = \|L_u^{\sim}(\lambda)\|_H \,, \qquad (7.86)$$

where $\widetilde{Y}(\lambda)$ is the optimal Nehari solution and $Y(\lambda) = \widetilde{Y}(\lambda) + L_s(\lambda)$.

For the computation of the additive spectral decomposition (7.85), the **Procedure GSDEC**, described in Sect. 9.2.5, can be used. Basically, using suitable invertible matrices U and Z, the transformed pencil $U(\widetilde{A} - \lambda\widetilde{E})Z$ results in a block diagonal form with two diagonal blocks, where one block contains the stable eigenvalues and the second block contains the unstable eigenvalues. By performing a system similarity transformation using the matrices U and Z, we obtain the system representation with partitioned matrices of the form

$$\left[\begin{array}{c|c} U\widetilde{A}Z - \lambda U\widetilde{E}Z & U\widetilde{B}_1 \\ \hline \widetilde{C}Z & D \end{array} \right] = \left[\begin{array}{cc|c} A_s - \lambda E_s & 0 & B_{1,s} \\ 0 & A_u - \lambda E_u & B_{1,u} \\ \hline C_s & C_u & D \end{array} \right], \qquad (7.87)$$

where $\Lambda(A_s - \lambda E_s) \subset \mathbb{C}_s$ and $\Lambda(A_u - \lambda E_u) \subset \mathbb{C}_u$. It follows that

$$L_s(\lambda) = \left[\begin{array}{c|c} A_s - \lambda E_s & B_{1,s} \\ \hline C_s & D \end{array} \right], \quad L_u(\lambda) = \left[\begin{array}{c|c} A_u - \lambda E_u & B_{1,u} \\ \hline C_u & 0 \end{array} \right].$$

To determine U and Z using the approach described in Sect. 9.2.5, the two main computations are the reduction of the matrix pair $(\widetilde{A}, \widetilde{E})$ to an ordered *generalized real Schur form* (GRSF) using the QZ-algorithm (see Sect. 10.1.4) and the solution of a generalized Sylvester system of equations (see Sect. 10.2.1).

For the computation of the solution $Y(\lambda)$ of the optimal Nehari problem (7.86), the **Procedure GNEHARI**, presented in Sect. 10.4.5, can be employed. If the resulting optimal solution is $\widehat{Y}(\lambda) = (\widehat{A} - \lambda\widehat{E}, \widehat{B}, \widehat{C}, \widehat{D})$, then the overall solution $Y(\lambda) = L_s(\lambda) + \widehat{Y}(\lambda)$ has the descriptor system realization

$$Y(\lambda) = \left[\begin{array}{cc|c} A_s - \lambda E_s & 0 & B_{1,s} \\ 0 & \widehat{A} - \lambda\widehat{E} & \widehat{B} \\ \hline C_s & \widehat{C} & D + \widehat{D} \end{array} \right].$$

When solving the AMMP using **Procedure AMMS**, the structure of the realization (7.84) can be exploited to simplify the computation of the additive spectral separation (7.85). Since A_r has only stable eigenvalues and all generalized eigenvalues of the pair (A_i, E_i) are unstable, the initial descriptor system realization (7.84) is already with separated stable-unstable spectrum. Therefore, by using orthogonal transformations, the matrix A_r can be separately reduced to a RSF, while the matrix pair (A_i, E_i) can be separately reduced to a GRSF. Therefore, we can assume that the initial realizations of $M_r(\lambda)$ and $G_i^{\sim}(\lambda)$ are such that A_r is in a RSF, while the pair (A_i, E_i) is in a GRSF.

To achieve the block diagonal form in (7.87) of matrices $U\widetilde{A}Z$ and $U\widetilde{E}Z$, we can use the particular transformation matrices

$$U = \begin{bmatrix} I & X \\ 0 & I \end{bmatrix}, \qquad Z = \begin{bmatrix} I & -XE_i \\ 0 & I \end{bmatrix},$$

where X satisfies the generalized Sylvester equation

$$XA_i - A_r X E_i = -B_r C_i \, .$$

This equation can be solved using algorithms discussed in Sect. 10.2.1. In this way, instead of solving a system of two Sylvester equations, we have to only solve a single generalized Sylvester equation. The resulting elements of the additive decomposition (7.85) are

$$L_s(\lambda) = \left[\begin{array}{c|c} A_r - \lambda I & B_r D_{i,1} + X B_{i,1} \\ \hline C_r & D_r D_{i,1} \end{array} \right], \quad L_u(\lambda) = \left[\begin{array}{c|c} A_i - \lambda E_i & B_{i,1} \\ \hline D_r C_i - C_r X E_i & 0 \end{array} \right].$$

If the resulting optimal solution is $\widetilde{Y}(\lambda) = (\widehat{A} - \lambda\widehat{E}, \widehat{B}, \widehat{C}, \widehat{D})$, then the overall solution $Y(\lambda) = L_s(\lambda) + \widetilde{Y}(\lambda)$ has the descriptor system realization

$$Y(\lambda) = \left[\begin{array}{cc|c} A_r - \lambda I & 0 & B_r D_{i,1} + X B_{i,1} \\ 0 & \widehat{A} - \lambda\widehat{E} & \widehat{B} \\ \hline C_r & \widehat{C} & D_r D_{i,1} + \widehat{D} \end{array} \right].$$

Solution of the 2-block $\mathcal{H}_\infty$-LDP. Following the conceptual procedure described in Sect. 9.1.10, a stable optimal solution $Y(\lambda)$ of the 2-block least distance problem can be approximately determined as the solution of the suboptimal 2-block least distance problem

$$\| [\, \widetilde{F}_1(\lambda) - Y(\lambda) \;\; \widetilde{F}_2(\lambda) \,] \|_\infty < \gamma, \tag{7.88}$$

where $\gamma_{opt} < \gamma \leq \gamma_{opt} + \varepsilon$, with ε an arbitrary user specified (accuracy) tolerance for the least achievable value γ_{opt} of γ. The standard solution approach is a bisection-based γ-iteration method, where the solution of the 2-block problem is approximated by successively computed γ-suboptimal solutions of appropriately defined 1-block problems.

To start the γ-iteration, we have to determine γ_l and γ_u, the lower and the upper bounds for γ_{opt}, respectively. Such bounds can be computed, for example, as

$$\gamma_l = \left\| \widetilde{F}_2(\lambda) \right\|_\infty, \quad \gamma_u = \left\| [\, \widetilde{F}_1(\lambda) \;\; \widetilde{F}_2(\lambda) \,] \right\|_\infty.$$

To compute these $\mathcal{L}_\infty$-norms, efficient algorithms can be employed based on extensions of the method of [21]. Note that for computing γ_u we can exploit that $\gamma_u = \|M_r(\lambda)\|_\infty$ as a consequence of the all-pass property of $G_i^\sim(\lambda)$.

The determined bounds γ_l and γ_u on γ_{opt} serve for the initialization of the γ-iteration, which successively updates the lower and upper bounds until their distance is less than or equal to a given threshold (i.e., $\gamma_u - \gamma_l \leq \varepsilon$). The main computations to be performed in one iteration are:

(1) Set $\gamma = (\gamma_l + \gamma_u)/2$) and compute, using the method described in Sect. 7.8, the left spectral factorization

$$\gamma^2 I - \widetilde{F}_2(\lambda)\widetilde{F}_2^\sim(\lambda) = V(\lambda)V^\sim(\lambda), \tag{7.89}$$

where $V(\lambda)$ is biproper, stable and minimum-phase.

(2) Compute, using the **Procedure GSDEC**, the additive decomposition

$$L_s(\lambda) + L_u(\lambda) = V^{-1}(\lambda)\widetilde{F}_1(\lambda), \tag{7.90}$$

where $L_s(\lambda)$ is the stable part and $L_u(\lambda)$ is the unstable part.

(3) Compute $\gamma_H := \|L_u^\sim(\lambda)\|_H$ using **Procedure GBALMR** in Sect. 10.4.4; if $\gamma_H < 1$, then set $\gamma_u := \gamma$ ($\gamma > \gamma_{opt}$); else, set $\gamma_l := \gamma$ ($\gamma < \gamma_{opt}$).

These steps are repeated until $\gamma_u - \gamma_l \leq \varepsilon$, where ε is a given threshold. The suboptimal 2-block problem (7.88) has been reduced to the suboptimal 1-block problem

$$\left\|V^{-1}(\lambda)\left(\widetilde{F}_1(\lambda) - Y(\lambda)\right)\right\|_\infty \leq 1. \tag{7.91}$$

The stable solution of (7.91) can be computed as

$$Y(\lambda) = V(\lambda)(L_s(\lambda) + Y_s(\lambda)), \tag{7.92}$$

where $Y_s(\lambda)$ is the stable solution of the optimal Nehari problem

$$\|L_u(\lambda) - Y_s(\lambda)\|_\infty = \|L_u^\sim(\lambda)\|_H \tag{7.93}$$

and can be computed using the **Procedure GNEHARI** presented in Sect. 10.4.5.

When solving the AMMP using **Procedure AMMS**, the structure of the realization (7.84) can be exploited to simplify the solution of the *2-block* $\mathcal{H}_\infty$-LDP. The computation of the spectral factorization (7.89), using the method described in Sect. 7.8, involves two steps. First, we compute a RCF of $\widetilde{F}_2(\lambda)$ with inner denominator such that $\widetilde{F}_2(\lambda) = \widetilde{N}_2(\lambda)\widetilde{M}_2^{-1}(\lambda)$, where $\widetilde{M}_2(\lambda)$ is inner. It follows that $\widetilde{F}_2(\lambda)\widetilde{F}_2^\sim(\lambda) = \widetilde{N}_2(\lambda)\widetilde{N}_2^\sim(\lambda)$. This computation needs to be performed only once, before starting the γ-iteration, and suitable algorithms are described in Sect. 10.3.5 (see **Procedure GRCFID**) for both continuous- and discrete-time cases. The resulting stable factor $\widetilde{N}_2(\lambda)$ has a descriptor system realization of the form

$$
\widetilde{N}_2(\lambda) =
\left[
\begin{array}{cc|c}
A_r - \lambda I & \overline{A}_{ri} - \lambda \overline{E}_{ri} & B_r D_{i,2} \\
0 & \overline{A}_i - \lambda \overline{E}_i & B_{i,2} \\
\hline
C_r & \overline{C}_i & D
\end{array}
\right],
$$

where $\overline{E}_i$ is nonsingular and $\Lambda(\overline{A}_i - \lambda \overline{E}_i) \subset \mathbb{C}_s$. The second step of the method presented in Sect. 7.8 determines the biproper, stable and minimum-phase factor $V(\lambda)$, which solves the left spectral factorization problem

$$
\gamma^2 I - \widetilde{N}_2(\lambda) \widetilde{N}_2^{\sim}(\lambda) = V(\lambda) V^{\sim}(\lambda)
$$

in the form

$$
V(\lambda) =
\left[
\begin{array}{cc|c}
A_r - \lambda I & \overline{A}_{ri} - \lambda \overline{E}_{ri} & \overline{B}_{ri} \\
0 & \overline{A}_i - \lambda \overline{E}_i & \overline{B}_i \\
\hline
C_r & \overline{C}_i & \overline{D}_{ri}
\end{array}
\right],
$$

where $\overline{D}_{ri}$ is invertible. Recall that this computation involves the solution of a GCARE or GDARE at each iteration. An explicit realization of the inverse can be computed in the form

$$
V^{-1}(\lambda) =
\left[
\begin{array}{ccc|c}
A_r - \lambda I & \overline{A}_{ri} - \lambda \overline{E}_{ri} & \overline{B}_{ri} & 0 \\
0 & \overline{A}_i - \lambda \overline{E}_i & \overline{B}_i & 0 \\
C_r & \overline{C}_i & \overline{D}_{ri} & -I \\
\hline
0 & 0 & I & 0
\end{array}
\right],
$$

where recall that the inverse $V^{-1}(\lambda)$ is proper and stable. The non-dynamic part can be eliminated without accuracy loss provided $\overline{D}_{ri}$ is well conditioned (e.g., $\kappa(\overline{D}_{ri}) < 10^4$) to obtain a descriptor realization without non-dynamic modes.

To obtain a realization of $V^{-1}(\lambda) \widetilde{F}_1(\lambda) = V^{-1}(\lambda) \left[M_r(\lambda)\ 0 \right] G_{i,1}^{\sim}(\lambda)$, we compute first a minimal realization of $V^{-1}(\lambda) \left[M_r(\lambda)\ 0 \right]$ as a descriptor system of the form $(\overline{A} - \lambda \overline{E}, \overline{B}, \overline{C}, \overline{D})$, with $\overline{E}$ invertible. For this purpose, we can employ the numerically stable **Procedure GIR** (see Sect. 10.3.1) to compute irreducible realizations of descriptor systems, or alternatively, the balancing-technique-based **Procedure GBALMR** (see Sect. 10.4.4) to compute minimal realizations of stable descriptor systems. Using the realization (7.72) of $G_{i,1}^{\sim}(\lambda)$, we obtain

$$
V^{-1}(\lambda) \widetilde{F}_1(\lambda) =
\left[
\begin{array}{cc|c}
\overline{A} - \lambda \overline{E} & \overline{B} C_i & \overline{B} D_{i,1} \\
0 & A_i - \lambda E_i & B_{i,1} \\
\hline
\overline{C} & \overline{D} C_i & \overline{D} D_{i,1}
\end{array}
\right],
$$

where the pair $(\overline{A}, \overline{E})$ has only stable eigenvalues, while the pair (A_i, E_i) has only unstable eigenvalues. This form is structurally similar with the realization of $\widetilde{F}_1(\lambda)$ in (7.84), and therefore simplifications arise in the subsequent computations.

To compute the spectral separation (7.90), we perform a similarity transformation, which annihilates the off-diagonal terms. Using the transformation matrices

$$
U = \begin{bmatrix} I & X \\ 0 & I \end{bmatrix}, \qquad Z = \begin{bmatrix} I & Y \\ 0 & I \end{bmatrix},
$$

we achieve

$$
U \begin{bmatrix} \overline{A} - \lambda\overline{E} & \overline{B}C_i \\ 0 & A_i - \lambda E_i \end{bmatrix} Z = \begin{bmatrix} \overline{A} - \lambda\overline{E} & 0 \\ 0 & A_i - \lambda E_i \end{bmatrix},
$$

provided X and Y satisfy the system of two Sylvester equations

$$
\begin{aligned}
X A_i + \overline{A} Y &= -\overline{B} C_i, \\
X E_i + \overline{E} Y &= 0.
\end{aligned}
$$

After applying the transformations to the input and output matrices we obtain

$$
U \begin{bmatrix} \overline{B} D_{i,1} \\ B_{i,1} \end{bmatrix} = \begin{bmatrix} \overline{B} D_{i,1} + X B_{i,1} \\ B_{i,1} \end{bmatrix}, \qquad \begin{bmatrix} \overline{C} & \overline{D} C_i \end{bmatrix} Z = \begin{bmatrix} \overline{C} & \overline{C} Y + \overline{D} C_i \end{bmatrix}.
$$

The stable and unstable terms are given by

$$
L_s(\lambda) = \left[\begin{array}{c|c} \overline{A} - \lambda\overline{E} & \overline{B} D_{i,1} + X B_{i,1} \\ \hline \overline{C} & \overline{D} D_{i,1} \end{array} \right], \qquad L_u(\lambda) = \left[\begin{array}{c|c} A_i - \lambda E_i & B_{i,1} \\ \hline \overline{C} Y + \overline{D} C_i & 0 \end{array} \right].
$$

In the light of the cancellation theory for continuous-time two-block problems of [80], pole–zero cancellations occur when forming $Y(s)$ in (7.92). In accordance with this theory, the expected order of $Y(s)$ is $\tilde{n} - 1$, where $\tilde{n}$ is the order of the realization (7.83). It is conjectured that similar cancellations will occur also for discrete-time systems, where a cancellation theory for two-block problems is still missing. Although we were not able to derive an explicit minimal state-space realization of $Y(\lambda)$, we can safely employ minimal realization procedures as **Procedure GIR** to compute irreducible descriptor realizations (see Sect. 10.3.1), or **Procedure GBALMR**, which exploits that the resulting $Y(\lambda)$ is stable (see Sect. 10.4.4).

7.11 Notes and References

Section 7.1. The requirements for satisfactory algorithms in the field of numerical linear algebra have been already formulated in [92]. Two classical textbooks, which cover most of the standard linear algebra algorithms (e.g., the solution of linear systems of equations, linear least-squares problems, standard and generalized eigenvalue problems), are [58, 118].

Section 7.2. The ill-conditioning of polynomial-based representations was a constant discussion subject in the control literature to justify the advantage of using state-space realization-based models for numerical computations. For an authoritative discussion, see [122]. More details on these issues are provided in Chap. 6 of [125]. The ill-conditioned polynomial $f(s) = (s - 1)(s - 2) \cdots (s - 20)$, with distinct roots, is known in the numerical literature as the *Wilkinson polynomial* and the sensitivity of its roots has been analyzed by Wilkinson in [173]. The ill-conditioned transfer function in Example 7.1 has been coined by Daniel Kressner (private communication).

Section 7.3. The general procedural framework for the synthesis of fault detection filters has been formally introduced by the author in [157] and relies on the concept of integrated synthesis algorithms proposed for solving $\mathcal{H}_2$- and $\mathcal{H}_\infty$-optimal FDI synthesis problems [152, 153, 156].

Section 7.4. The first use of the nullspace method to solve the EFDIP has been proposed in [107] using a state-space-based approach. The term *nullspace method* has been formally introduced in [48], where a polynomial basis-based approach was used to design minimal order residual generators for fault detection. This approach has been later extended in [138] to solve the EFDP by using rational bases instead polynomial ones. The computation of these bases can be done by using numerically stable algorithms based on the reduction of an extended system pencil to a Kronecker-like form using orthogonal similarity transformations. The solution of the EFDIP by using a bank of least order fault detection filters synthesized using the rational nullspace method has been addressed in [146], where easy to check solvability conditions have been established in terms of explicit realization of the internal form of the fault detection filter. The presentation of this section basically relies on [155], where the use of the nullspace method, as a preliminary reduction step, is the basis for an efficient solution of the EFDIP. This emerged later as an important computational paradigm to solve all fault detection problems formulated in this book. It was shown in [150] that the nullspace method provides a unifying synthesis paradigm of fault detection filters for most of existing approaches, which can be interpreted as special cases of this method. The computation of Kronecker-like forms of linear matrix pencils is discussed in Sect. 10.1.6 and suitable numerically stable algorithms have been proposed, for example, in [8, 9, 27, 101, 121, 134]. Alternative synthesis methods of fault detection filters for systems with parametric faults have been proposed—for example, in [29, 51].

Section 7.5. The presentation of this section is based on [155] and strongly relies on the structural details of matrices of the descriptor realization of the nullspace basis. The order reduction is achieved using Type I minimal dynamic covers [74]. For their computation, reliable numerical algorithms have been proposed in [139] for the standard system case and in [142] for the descriptor system case.

Section 7.6. The computation of proper coprime factorizations can be obtained using the nullspace-based technique proposed in [155], or the stabilization and pole-assignment-based techniques proposed in [135]. The algorithms for the computation of stable coprime factorizations of proper systems rely on algorithms based on updating generalized Schur forms, which have been also proposed in [135]. An extension of these techniques to compute proper and stable coprime factorizations is the basis of **Procedure GRCF** presented in Sect. 10.3.5.

Section 7.7. For the computation of the quasi-co-outer–co-inner factorizations, we can employ the dual of the general algorithm of [103] for the continuous-time case and the dual of the algorithm of [100] for the discrete-time case. Simplified versions of these algorithms for the case of proper and full row rank TFMs are presented in Sect. 10.3.6.

Section 7.8. The methods to compute minimum-phase spectral factorizations, and stable and minimum-phase spectral factorizations represent straightforward extensions, to proper descriptor systems, of the standard solution methods described in [177]. The new, recursive pole-assignment-based algorithm proposed in Sect. 10.3.5, which underlies **Procedure GRCFID** for the computation of right coprime factorizations with inner denominators, is generally applicable regardless of the properties of the employed descriptor system realization. Alternative direct methods, applicable to stabilizable or detectable realizations, have been proposed in [102] for the continuous-time case and in [100] for the discrete-time case.

Section 7.9. This section partly relies on the algorithm proposed in [157, Appendix C] to solve rational systems of equations of the form $X(\lambda)G(\lambda) = F(\lambda)$, which can serve to obtain particular solutions of the EMMP in **Procedure EMM**. This algorithm can be seen as the first step of a general algorithm proposed in [140] to solve the dual rational systems of equations of the form $G(\lambda)X(\lambda) = F(\lambda)$, which also includes the computation of a least order solution. The complete algorithm to solve this equation is presented in Sect. 10.3.7 and forms the basis of a complete synthesis procedure to solve the EMMP proposed in [141].

Section 7.10. The computational approach to solve the $\mathcal{H}_\infty$-LDP represents a specialization of more general procedures described in [46, 115].

Chapter 8
Case Studies

This chapter is intended to illustrate the formulation of typical FDD problems for monitoring faults in complex technological systems and the application of the linear synthesis techniques, described in this book, to address the challenges of practical applications. Common to many industrial setups is that the underlying plant dynamics are nonlinear and depend on parameters whose values may vary during plant operations. Furthermore, various uncertainties may exist, which either manifest as exogenous (unknown) disturbances or are inherent inaccuracies in the plant parameters. For synthesis purposes, often only a set of linearized models is available, which covers the relevant plant operating conditions and main parameter variations. The aim of the synthesis is to design a robust FDD system, whose fault monitoring performance is satisfactory in the presence of variabilities induced by the nonlinear plant behaviour, parameter variations and various uncertainties.

Linear synthesis techniques can be used in many ways to address the challenges of solving robust FDD problems, as—for example, in assessing solvability, analysing performance limitations or determining meaningful design specifications (e.g., reference models). Moreover, linear synthesis approaches often underlie the gains-scheduling-based synthesis methodologies, intended to cope with the robustness requirement. The use of linear fault detection filters in conjunction with signal-processing techniques for the online identification of various types of faults often enhances the robustness of the fault detection and provides useful information for control reconfiguration purposes.

Two case studies are presented in this chapter to illustrate the above aspects. The chosen examples are related to flight control applications and address the monitoring of flight actuator faults and air data sensor faults. Our choice is partly a recognition of the leading role of the aerospace industry as one of the main drivers of the FDD research activities. On the other side, the choice also reflects the author's involvements in several research projects focussing on fault monitoring aspects of flight control, where some of the synthesis techniques described in this book have been assessed using realistic transport aircraft models. The considered examples can be seen representative to illustrate the applicability, but also the limitations, of linear synthesis techniques in addressing challenging industrial fault monitoring problems.

8.1   Monitoring Flight Actuator Faults

The monitoring of primary actuator failures of an aircraft is of paramount importance for the aircraft safe operation, prevention of excessive structural loads and reduction of the environmental footprint. For civil aircraft several fault scenarios are of particular interest. For example, the ability to detect single primary actuator faults is of critical importance, because it is part of the aircraft certification requirements. Hence, a minimal request from an active fault accommodation perspective is the requirement, for any modern aircraft design, that no single failure can lead to a catastrophic consequence. Simultaneous actuator faults can also occur—for example, in the case of more severe surface damages. The detection and isolation of simultaneous faults is therefore a standard requirement for any FDD system for monitoring flight actuator faults.

There are several classes of flight actuator faults such as jamming, runaway, oscillatory failure or loss of efficiency, whose early detection and timely handling contribute to the efficient operation of aircraft, avoid excessive fuel consumption (with all associated negative environmental effects) and increase the aircraft operational autonomy. These faults can be often considered as additive faults for which the synthesis methods of FDI filters developed in this book are applicable. In practice, the detection and isolation of one or several faults is usually followed by fault identification (i.e., determination of fault characteristics as size and shape, or even fault estimation), to allow performing appropriate reconfiguration actions in order to minimize the effects of malfunctioning and ensure acceptable performance.

Two basic approaches to monitoring flight actuator faults are, in principle, possible. The global (or system level) monitoring uses the complete available information on *all* actuator inputs and *all* measured outputs to perform the detection and isolation of actuator failures. The main advantage of the global approach is that, virtually, all categories of actuator faults (even multiplicative faults) can be detected. The disadvantage of the global approach is the inherent complexity of the global FDI filters and the challenges associated with providing guarantees for the robustness of fault diagnosis. The main advantage of local (or component level) monitoring is that the basic fault detection functionality automatically provides the fault isolation capability too. The robustness aspects can be tackled using straightforward gain-scheduling schemes, by employing low complexity (e.g., first- or second-order) fault detection filters. However, local monitoring may have difficulties in detecting parametric faults, as the loss of efficiency due to control surface damages and icing, or actuator disconnection due to a broken rod. The monitoring of these categories of faults may require the use of alternative approaches (e.g., based on model detection techniques), which are suitable for the detection of parametric faults.

In Sect. 8.1.1, we apply the global approach for the synthesis of a least-order LTI FDI filter for the nominal case. For a chosen set of fault signatures, a bank of low-order filters is synthesized, leading to a least-order global filter. Since the robustness of the nominal filter in the presence of mass variations is not satisfactory for practical usage, we address in Sect. 8.1.2 the same problem, using additionally measurements of the control surface positions associated with all monitored actuators. The resulting bank of first-order filters is a solution of a strong FDI problem, which can isolate an arbitrary number of faults occurring simultaneously. The remarkable robustness of this synthesis to variations of aircraft parameters and the low complexity (first order) of the resulting component filters are strong indications to employ a local (actuator level) fault monitoring approach instead of a global one. Specific aspects of the usage of the local monitoring of flight actuator faults in an industrially relevant FDD system are discussed in Sect. 8.1.3. The models used for global monitoring are described in Sect. 8.1.4 and consist of a collection of full-sized linearized civil aircraft models augmented with simple (first-order) actuator dynamics. The models describe the aircraft dynamics during cruise and each component model corresponds to a specific value of the aircraft mass.

8.1.1 Nominal Synthesis

In this section, we present a global (or system level) monitoring approach of primary actuator faults of a civil aircraft having 11 primary control surface actuators (4 elevators, 1 stabilizer, 4 ailerons, 2 rudders). To study the model-based FDI aspects, we employ a LTI model, which describes a normal cruise flight in the presence of wind disturbances. The main goal of our study is to emphasize the intrinsic difficulties of using a global monitoring approach for a reliable detection and isolation of actuator faults. Concretely, we will show for the considered nominal case, that although we can assess the feasibility of a global FDI system capable to localize individual or, partly, even simultaneous actuator faults, still the reliable (i.e., robust) monitoring of actuator failures using a global approach is a nonrealistic task, in the presence of various model and environmental uncertainties (e.g., variations of operating conditions and mass, wind disturbances). Alternative approaches, able to overcome these limitations, are discussed in the next sections.

To address the robustness aspects, a multiple model consisting of $N = 11$ linearized aircraft models, including the actuator dynamics and the additive actuator faults, has been developed in Sect. 8.1.4. The ith model has the state-space form

$$\begin{aligned}
\dot{x}^{(i)}(t) &= A^{(i)}x^{(i)}(t) + B_u^{(i)}u(t) + B_d^{(i)}d(t) + B_f^{(i)}f(t), \\
y^{(i)}(t) &= C^{(i)}x^{(i)}(t) \qquad\qquad\quad + D_d^{(i)}d(t),
\end{aligned} \tag{8.1}$$

where the dimensions of the vectors $x^{(i)}(t)$, $y^{(i)}(t)$, $u(t)$, $d(t)$ and $f(t)$ are respectively, $n = 32$, $p = 10$, $m_u = 22$, $m_d = 3$ and $m_f = 8$. The significance of the

Fig. 8.1 A closed-loop fault diagnosis setup

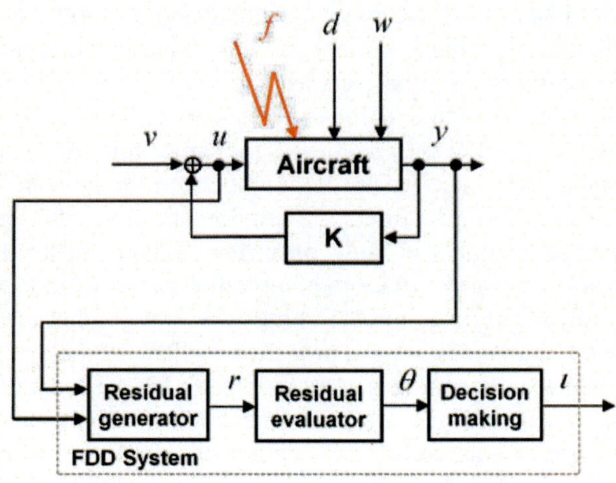

components of these vectors and the numerical values of the matrices of the nominal model (for $i = 7$) are given in Sect. 8.1.4. Each of the N component model (8.1) corresponds to a certain value of the aircraft mass and all other parametric or operational point variabilities are ignored in this study. The chosen mass values m_i, for $i = 1, \ldots, N$, cover the whole range of mass variations, from a minimum value $m_{\min}$ to a maximum value $m_{\max}$. The i-th mass value is given by $m_i = m_{\min} + \Delta m_i (m_{\max} - m_{\min})$, where Δm_i is the percentage of the mass variation from the total mass variation $m_{\max} - m_{\min}$.

The resulting linearized models (8.1) are not minimal. Besides an uncontrollable zero eigenvalue in the aircraft model, there are 10 unobservable eigenvalues, all equal to -5. This lack of observability originates from the fact that the actuators of spoilers are coupled to the aircraft surfaces via a summation of their effects, thus of the 12 eigenvalues (poles) introduced by the spoiler actuators, 10 are not observable. For our studies, we employed minimal order realizations of order $n = 21$, which have been determined using standard minimal realization tools.

All component models being unstable, the fault monitoring can be performed only in a closed-loop setting, because, even for the nominal case, slight deviations of model parameters may induce unstable behaviour in the FDD system dynamics. For evaluation purposes we employed the setting in Fig. 8.1, where the feedback block **K** is simply a constant output feedback gain K, chosen such that, with the control input $u = v + K y^{(i)}$, each of the resulting closed-loop system

$$\dot{x}_{cl}^{(i)}(t) = (A^{(i)} + B_u^{(i)} K C^{(i)}) x_{cl}^{(i)}(t) + B_u^{(i)} v(t) + (B_d^{(i)} + B_u^{(i)} K D_d^{(i)}) d(t) + B_f^{(i)} f(t),$$
$$y^{(i)}(t) = \qquad C^{(i)} x_{cl}^{(i)}(t) + \qquad\qquad D_d^{(i)} d(t), +$$
$$u(t) = \qquad K C^{(i)} x_{cl}^{(i)}(t) + \quad v(t) + \qquad\qquad K D_d^{(i)} d(t) +$$

$$\tag{8.2}$$

is robustly stable (i.e., $\Lambda(A^{(i)} + B_u^{(i)} K C^{(i)}) \subset \mathbb{C}_s$ for $i = 1, \ldots, N$). The chosen feedback gain K has no relation to any particular flight control law design and it has been determined to solely provide a stable closed-loop environment, which allows the assessment of the robustness of the fault monitoring system in presence of parametric variations. Note that without ensuring closed-loop stability, any fault detection filter will lead to an unstable FDD system due to the presence of inherent parametric uncertainties in the employed linearized synthesis models.

In Listing 8.1, we present the Julia code used for the setup of the augmented LTI aircraft models and for the determination of the stability degree of the closed-loop systems. The multiple-model, which serves for synthesis purposes, is loaded from the file **cs1data.jld2** as a vector of systems **SYSACM** together with the robustly stabilizing output feedback gain K and actuators model **ACT**. Three input groups, **'c'** for control inputs, **'d'** for disturbance inputs and **'f'** for fault inputs, have been defined for the resulting minimal realization **sysactf** of the augmented LTI aircraft models with faults. The robust stability of the closed-loop models is checked by verifying that the maximum real part of the closed-loop eigenvalues **sdeg_cloop** $= -0.02243$ is negative.

Listing 8.1 Part 1 of script **CS1_1.jl**: Model setup

```julia
using FaultDetectionTools, DescriptorSystems, LinearAlgebra, Test
using JLD2

# CS1_1 - Case-study example: Monitoring flight actuator faults
# No measurements of control surface angles are used

# Part 1 - Model setup
# load matrices of the aircraft multiple-model SYSACM,
# actuator model ACT and output-feedback gain K
cd(joinpath(pkgdir(FaultDetectionTools), "Examples"))
SYSACM, K, ACT = load("cs1data.jld2", "SYSACM", "K", "ACT");

# set dimensions
N = size(SYSACM,1)          # number of models
p, m = size(SYSACM[1].D)    # output and input dimensions
atol = 1.e-6                # absolute tolerance
nom = 7                     # index of nominal system
T = Float64                 # type of floating-point data

# set primary actuator indices for
# [ ailerons, elevators, stabilizer, ruder ]
act_prim = [ [1,2,15,16]; [17,19]; 18; 20 ];
mu = size(ACT,1); md = m-mu; mf = length(act_prim);
indu = 1:mu        # system control input indicess
indy = 1:p         # system measured output indices

# build minimal realizations of AC-models with actuators
sysact = similar(Vector{DescriptorStateSpace},N)
for i = 1:N
    sysact[i] = gir(SYSACM[i] * append(ACT,dss(eye(md)))); atol)
end

# build synthesis models with actuator faults
sysactf = fdimodset(sysact; c = Vector(1:mu),
        d = mu .+ Vector(1:md), f = act_prim)

# determine closed-loop stability margin
```

```julia
sdeg_cloop = -5
for i = 1:N
    syst = feedback(sysact[i],K,indu,indy; negative = false)
    global sdeg_cloop = max(sdeg_cloop,
                            maximum(real(eigvals(syst.A))))
end
@test sdeg_cloop < 0
println("sdeg_cloop = $sdeg_cloop")
```

For the synthesis of the FDD system for monitoring the primary actuator faults we solve an EFDIP for the nominal case by using the **Procedure EFDI** described in Sect. 5.4. To set up the fault signatures, which allow the isolation of the actuator faults, we computed first the maximally achievable structure matrix for the augmented aircraft model with additive faults. For this purpose, we employed the function **fdigenspec**, which implements **Procedure GENSPEC** described in Sect. 5.4. In Listing 8.2 we present the Julia code used for the determination of the achievable structure matrix S containing the weak fault specifications and the structure matrix S_{strong} containing the strong fault specifications. This latter matrix served for the selection of the desired specifications contained in S_{FDI}.

The first call of **fdigenspec** determines S, the maximally achievable (weak) specifications for the nominal system. A threshold for nonzero frequency-response gains is used in **fdigenspec** to generate the achievable specifications. This threshold is set to 10^{-5} via the keyword argument **FDTol**. The resulting structure matrix S is a 57×8 binary matrix, containing the 57 achievable (weak) specifications. In a second call of **fdigenspec**, we also set the keyword arguments **FDfreq** to 0, **FDGainTol** to 0.001 and **sdeg** to -0.05. The keyword argument **FDfreq** specifies the frequencies values Ω to be used for checking strong fault detectability according to Corollary 5.3 (i.e., $\Omega = \{0\}$ for constant faults), while **FDGainTol** is the internally used threshold for the frequency-response gains computed for each frequency in Ω (i.e., the DC-gains for $\Omega = \{0\}$). The keyword argument **sdeg** is used to specify a desired stability degree for the real parts of the poles of the internally computed nullspace bases. The resulting structure matrix S_{strong} contains 55 strongly achievable specifications and it is checked that they are included among the weak specifications. From the resulting S_{strong} we selected the following structure matrix S_{FDI} with 6 specifications, each of them with at least three zero elements

$$S_{FDI} = \begin{bmatrix} 0 & 1 & 1 & 1 & 1 & 0 & 1 & 0 \\ 1 & 0 & 1 & 1 & 0 & 1 & 1 & 0 \\ 1 & 1 & 0 & 1 & 1 & 0 & 1 & 0 \\ 1 & 1 & 1 & 0 & 1 & 1 & 0 & 0 \\ 1 & 1 & 1 & 1 & 0 & 0 & 0 & 0 \\ 0 & 0 & 0 & 0 & 0 & 0 & 0 & 1 \end{bmatrix}. \tag{8.3}$$

Listing 8.2 Part 2 of script `CS1_1.jl`: Setup of synthesis specifications

```
# Part 2 - Setup of the synthesis specifications

# compute the achievable weak specifications
FDtol = 1.e-5        # weak fault detection threshold
S = fdigenspec(sysactf[nom]; atol, FDtol);

# compute achievable strong specifications for constant faults
FDGainTol = 0.001   # strong fault detection threshold
FDfreq = 0              # frequency strong detectability checks
S_strong = fdigenspec(sysactf[nom]; atol, FDtol, FDGainTol,
                   FDfreq, sdeg = -0.05)

# check inclusion of strong specification
@test issubset(binmat2dec(S_strong), binmat2dec(S))

# define SFDI, the signatures for isolation of single faults
SFDI = [0 1 1 1 1 0 1 0
        1 0 1 1 0 1 1 0
        1 1 0 1 1 0 1 0
        1 1 1 0 1 1 0 0
        1 1 1 1 0 0 0 0
        0 0 0 0 0 0 0 1] .> 0;
```

The selected structure matrix S_{FDI} in (8.3) ensures the strong isolation of ruder faults (independently of all other faults) and the weak isolation of the rest of faults occurring one at a time. Moreover, it ensures the detection of all combinations of a ruder fault with another (arbitrary) second fault, as shown in Table 8.1.

Using the structure matrix S_{FDI} in (8.3), we applied the **Procedure EFDI** for the synthesis of a bank of $n_b = 6$ fault detection filters $Q^{(j)}(s)$, $j = 1, \ldots, n_b$, with scalar outputs. Listing 8.3 illustrates the call of the function **efdisyn**, which implements **Procedure EFDI**.

Table 8.1 Additional admissible signatures for detection of simultaneous faults

$f_1 \& f_8$	$f_2 \& f_8$	$f_3 \& f_8$	$f_4 \& f_8$	$f_5 \& f_8$	$f_6 \& f_8$	$f_7 \& f_8$
0	1	1	1	1	0	1
1	0	1	1	0	1	1
1	1	0	1	1	0	1
1	1	1	0	1	1	0
1	1	1	1	0	0	0
1	1	1	1	1	1	1

Listing 8.3 Part 3 of script `CS1_1.jl`: Synthesis using **Procedure EFDI**

```
# Part 3 - Synthesis using Procedure EFDI
# perform least-order synthesis of component filters
Q, Rf, info = efdisyn(sysactf[nom], SFDI; atol, FDGainTol, FDfreq,
                  rdim = 1, sdeg = -5, smarg = -0.2)
println("orders = $(order.(Q.sys))")
```

The results computed by **efdisyn** for the nominal model **sysactf[nom]** and the selected structure matrix **SFDI** are the bank of six scalar output fault detection filters $Q^{(j)}(s)$, for $j = 1, \ldots, 6$, and the corresponding six internal forms $\widetilde{R}_f^{(j)}(s)$, for $j = 1, \ldots, 6$, which are stored in the filter objects **Q** and **Rf**, respectively. The j-th filter $Q^{(j)}(s)$ achieves the j-th specification contained in the j-th row of the structure matrix S_{FDI} and has the least possible McMillan degree. The choice of scalar output filters is specified by setting the keyword argument **rdim** to 1. The stability margin of -0.2 for the real parts of the poles of the filters $Q^{(j)}(s)$ is specified via the keyword argument **smarg**, and the desired stability degree of -5 for all poles having real parts greater than the value of **smarg** is specified via the keyword argument **sdeg**. The strong detectability of constant faults is enforced by specifying the frequency value 0 using the keyword argument **FDfreq**. The resulting overall filter $Q(s)$ has a global order 31, where the six scalar output fault detection filters $Q^{(j)}(s)$ for $j = 1, \ldots, 6$ have the individual orders $\{6, 6, 6, 5, 4, 4\}$, respectively. Without employing the least-order synthesis option in **Procedure EFD** (at Step (2) of **Procedure EFDI**), the resulting order of the overall filter is 49 and the individual filters have orders $\{9, 9, 9, 8, 7, 7\}$. The resulting **info** contains additional synthesis related information.

In Listing 8.4 we present the Julia code used for the assessment of the synthesis results. The overall filters, $Q(s)$ and $\widetilde{R}_f(s)$, are formed by stacking the six computed scalar output filters $Q^{(j)}(s)$ and $\widetilde{R}_f^{(j)}(s)$, respectively. The first check is to verify the nominal synthesis conditions

$$Q(s) \begin{bmatrix} G_u^{(7)}(s) \ G_d^{(7)}(s) \\ I & 0 \end{bmatrix} = 0, \qquad Q(s) \begin{bmatrix} G_f^{(7)}(s) \\ 0 \end{bmatrix} = \widetilde{R}_f(s). \qquad (8.4)$$

The second check is to verify that the achieved structure matrix $S_{\widetilde{R}_f}$ (see Sect. 3.4) is equal to S_{FDI}. This comes down to verify that the DC-gain $\widetilde{R}_f(0)$ and S_{FDI} have the same zero–nonzero pattern.

To overcome stability related problems in studying the robustness of the obtained synthesis, we use the closed-loop setting in Fig. 8.1. With a robust output feedback $u = Ky^{(i)} + v$, the resulting i-th extended closed-loop system with inputs $\{v, d, f\}$ and outputs $\{y^{(i)}, u\}$ is given in (8.2). Let $\overline{G}_e^{(i)}(s)$ be the TFM of this system and define the partitioned internal form of the filter acting on the i-th model as

$$\overline{R}^{(i)}(s) = \left[\overline{R}_v^{(i)}(s) \middle| \overline{R}_d^{(i)}(s) \middle| \overline{R}_f^{(i)}(s) \right] := Q(s)\overline{G}_e^{(i)}(s). \qquad (8.5)$$

Table 8.2 Robustness analysis results for the nominal synthesis

Model #	$\left\|\overline{R}_u^{(i)}(s)\right\|_\infty$	$\left\|\overline{R}_d^{(i)}(s)\right\|_\infty$	$\left\|\overline{R}_f^{(i)}(s) - \widetilde{R}_f(s)\right\|_\infty$
1	136.0382	32.1952	135.9077
2	109.1135	25.4839	109.0137
3	84.1191	19.4112	84.0321
4	60.8408	13.8988	60.7787
5	39.1823	8.8672	39.1025
6	18.9390	4.2518	18.9147
7	0.0000	0.0000	0.0000
8	17.7513	3.9317	17.7023
9	34.4163	7.5797	34.3598
10	50.0823	10.9744	50.0064
11	64.8628	14.1415	64.7504

Since $Q(s)$ and $\overline{G}_e^{(i)}(s)$, for $i = 1, \ldots, N$, are stable, the resulting $\overline{R}^{(i)}(s)$ are stable as well. For the robustness of the nominal synthesis, we need to check that $\left\|\overline{R}_u^{(i)}(s)\right\|_\infty$, $\left\|\overline{R}_d^{(i)}(s)\right\|_\infty$ and $\left\|\overline{R}_f^{(i)}(s) - \widetilde{R}_f(s)\right\|_\infty$ are reasonably small for $i = 1, \ldots, N$.

The robustness analysis results, in terms of infinity norms, are summarized in Table 8.2. The large gains for some of mass values indicate that the nominal synthesis is not robust with respect to mass variations.

In Fig. 8.2, we present the step responses from the fault inputs of the 31st order internal form of the (nominal) filter $\widetilde{R}_f^{(7)}(s)$ $\left(= \overline{R}_f^{(7)}(s)\right)$ from which the achieved fault signature can be easily read out. Observe the perfect cancellations of fault signals in the residuals, which match the zero elements in the corresponding specifications. In Fig. 8.3, we present the N step responses of $\overline{R}_f^{(i)}(s)$ for $i = 1, \ldots, N$. It can be easily observed that the nominal filter is not robust with respect to large mass variations. For example, step signals in the stabilizer input, which must be rejected in the residuals, have comparable amplitudes with the signals to be detected.

Listing 8.4 Part 4 of script `CS1_1.jl`: Assessment of synthesis results

```
# Part 4 - Assessment of synthesis results
# form extended open-loop systems [ Gu Gd Gf; I 0 0]
syse = similar(Vector{DescriptorStateSpace},N)
for i = 1:N
    syse[i] = [sysactf[i].sys;eye(mu,mu+md+mf)]
end


# build overall detector filter and its internal form
Qtot = vcat(Q.sys...)
Rftilde = vcat(Rf.sys...)


# check open-loop synthesis conditions:
# (1) Q*[Gu Gd;I 0] = 0
# (2) Q*[Gf; 0] = Rftilde
# (3) achieved structure matrix = SFDI
```

```julia
@test iszero(Qtot*syse[nom][:,1:mu+md]; atol) &&
      iszero(Qtot*syse[nom][:,mu+md+1:end]-Rftilde; atol) &&
      isequal(fditspec_(Rftilde; FDfreq, atol)[:,:,1],SFDI)

# form extended closed-loop systems with the output feedback
# u = K*y+v and evaluate the corresponding internal forms

sysefb = similar(Vector{DescriptorStateSpace},N)
Rtot = similar(Vector{DescriptorStateSpace},N)
for i = 1:N
    sysefb[i] = feedback(syse[i],K,indu,indy; negative = false)
    Rtot[i] = Qtot*sysefb[i]
end

# check robustness by computing the H-inf norms
NormRu = similar(Vector{T},N)
NormRd = similar(Vector{T},N)
NormRfmRfnom = similar(Vector{T},N)
indc = sysactf[1].controls      # indices of control inputs
indd = sysactf[1].disturbances  # indices of disturbances inputs
indf = sysactf[1].faults        # indices of fault inputs
for i = 1:N
    NormRu[i] = glinfnorm(Rtot[i][:,indc])[1]
    NormRd[i] = glinfnorm(Rtot[i][:,indd])[1]
    NormRfmRfnom[i] = glinfnorm(Rtot[i][:,indf]-Rftilde)[1]
end

println(" Robustness analysis results")
println(" NormRu          NormRd          NormRfmRfnom")
display([NormRu NormRd NormRfmRfnom])
@test NormRu[nom] < 1.e-6 &&  NormRd[nom] < 1.e-6 &&
      NormRfmRfnom[nom] < 1.e-6
```

Several remarks regarding the nominal synthesis are appropriate at this point.

1. For a nominal linearized aircraft model, augmented with first-order actuator models, an FDD system for the detection and isolation of additive primary actuator faults has been successfully designed by using an exact FDI synthesis methodology. The synthesis has been performed using **Procedure EFDI** described in Sect. 5.4 and resulted in a bank of six scalar output fault detection filters, which globally produce a structured residual set with a predefined structure matrix. The resulting FDD system is able to isolate single faults, as well as two simultaneous faults, provided one of them is the ruder actuator fault. The resulting overall fault detection filter has the least possible order for the imposed fault signatures. The nominal synthesis results are highly accurate and can serve, for example, as reference model for more involved synthesis methodologies, which target robustness of the FDD system with respect to parameter variations.

2. The structure matrix S_{FDI} in (8.3) does *not* have the very desirable property that no column of S_{FDI} can be obtained from any other column by changing a single element from zero to one or viceversa. A structure matrix with this property is called *bidirectional strongly isolating of degree 1* in [51] and can be constructed by selecting all specifications with three zero elements, one of which being in the last position, and including additionally the specification in the last row of S_{FDI}

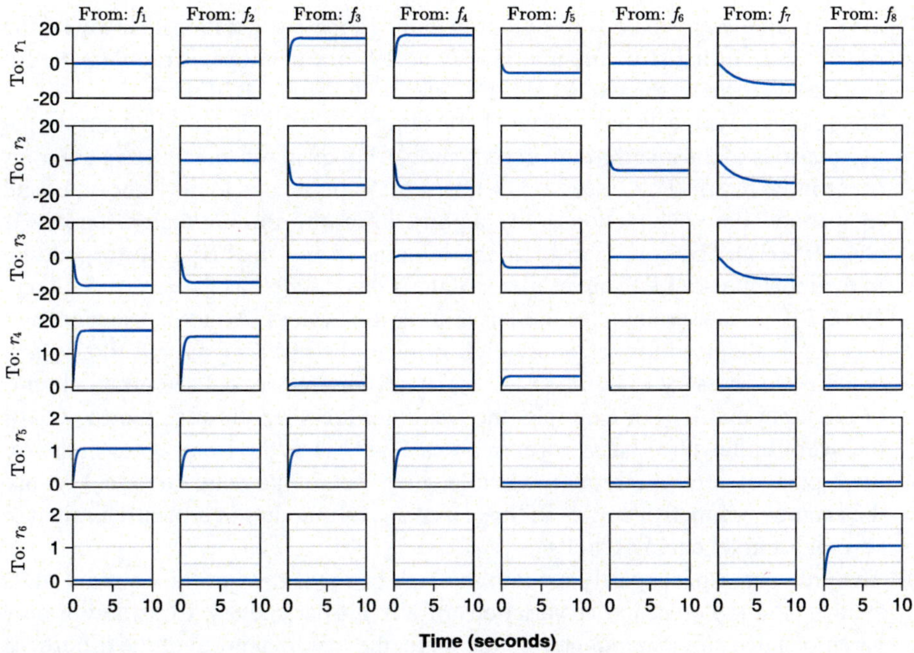

Fig. 8.2 Fault input step responses for nominal synthesis

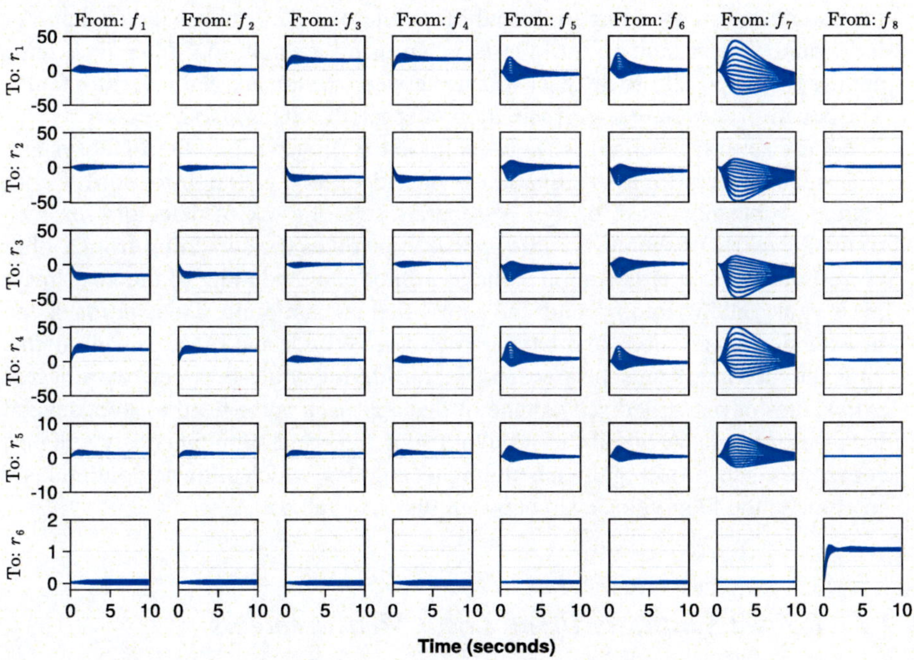

Fig. 8.3 Parametric fault input step responses for nominal synthesis

in (8.3) (all elements are zero except that one in the last position). However, the use of the resulting structure matrix, having 19 specifications, is problematic for practical use due to the expected high order of the global FDI filter.

3. The nominal synthesis has several challenging aspects, which originate from the peculiarities of the underlying aircraft models. For example, the first two columns of the input matrix $B_{ac,u}^{(7)}$ (see Sect. 8.1.4), which correspond to the right-outer and right-inner aileron deflections, only slightly differ (the norm of their difference is about 0.0037). The same is valid for the columns 15 and 16 of $B_{ac,u}^{(7)}$ corresponding to the left-inner and left-outer aileron deflections, as well as for columns 17 and 19 of $B_{ac,u}^{(7)}$ corresponding to the left and right elevators. A first consequence of this fact is the need for relatively high gains in the resulting overall filter $Q(s)$, which are necessary to compensate the small differences between the columns of the input matrix. For example, the resulting direct feedthrough matrix D_Q of the overall filter $Q(s)$ has a norm of about 3240! Thus the isolation of faults for these pairs of control effectors is challenging even in the nominal case, because the accuracy of matrix entries in linearized aircraft models is typically limited to several accurate decimal digits.

4. The previous aspects partly explain the lack of robustness of the nominal FDD synthesis in the presence of variations of aircraft parameters. Considering only variations in a single parameter, as the mass, the maximal norm of the difference between the first columns of $B_{ac,u}^{(i)}$ and first column of $B_{ac,u}^{(7)}$ is as large as 0.01. This difference is significantly larger than the above difference of 0.0037 between the columns corresponding to two adjacent ailerons. This makes the isolation of individual aileron faults a futile task, when using a single FDI filter. The same applies for the possibility to discriminate between the left and right elevator faults. The situation is even worse, if isolation of spoiler faults is also necessary.

5. There are several potential ways to increase the robustness of the FDD system. If a reliable aircraft mass estimation is provided, then a straightforward way to increase robustness is to design N separate sets of bank of detectors for each of the N values of the mass. This approach involves the switching from a filter corresponding to a larger mass value to a filter corresponding to the next lower mass value, taking into account the actual fuel consumption. This approach can be extended to variations of other parameters as altitude and speed as well. In this case, the robustness must be assessed by considering points between two adjacent grid values of mass. A disadvantage of this approach is the need to store several high-order filters. A possibility to avoid switching and simultaneously enhance robustness is to employ gain-scheduling techniques, which allow to continuously interpolate the filter gains even between two grid values.

8.1.2 *Robust Synthesis Using Local Measurements*

In this section, we explore another way to improve the robustness of the FDD system by using additional measurements of the primary control surface positions. These measurements are usually available for large civil transport aircraft and therefore can be used for fault monitoring purposes. The resulting augmented aircraft model has the form (8.1) with the system matrices defined in Sect. 8.1.4. The dimensions of the vectors $x^{(i)}(t)$, $y^{(i)}(t)$, $u(t)$, $d(t)$ and $f(t)$ are respectively, $n = 32$, $p = 18$, $m_u = 22$, $m_d = 3$ and $m_f = 8$. The $N = 11$ linearized models correspond to an equidistant grid of mass values from the minimum value m_{min} to the maximum value m_{max}.

In Listing 8.1, we present the Julia code used for the setup of the augmented LTI aircraft models with additional measurements and for the determination of the stability degree of the closed-loop systems. Three input groups, **'controls'**, **'disturbances'** and **'faults'**, are defined for the resulting minimal realization **sysactf** of the augmented LTI aircraft models with faults.

Listing 8.5 Part 1 of script **CS1_2.jl**: Model setup

```julia
using FaultDetectionTools, DescriptorSystems, LinearAlgebra, Test
using JLD2

# CS1_2 - Case-study example: Monitoring flight actuator faults
# Local measurements of control surface angles are used
println("Case study CS1_2")

# Part 1 - Model setup
# load matrices of the aircraft multiple-model SYSACM,
# actuator model ACT and output-feedback gain K
cd(joinpath(pkgdir(FaultDetectionTools), "Examples"))
SYSACM, K, ACT = load("cs1data.jld2", "SYSACM", "K", "ACT")

# set dimensions
N = size(SYSACM,1)          # number of models
p, m = size(SYSACM[1].D)    # output and input dimensions
atol = 1.e-6                # absolute tolerance
nom = 7                     # index of nominal system

# set primary actuator indices for
# [ ailerons, elevators, stabilizer, ruder ]
act_prim = [ [1,2,15,16]; [17,19]; 18; 20 ];
mu = size(ACT,1); md = m-mu; mf = length(act_prim);
indu = 1:mu        # system control input indicess
indy = mf.+(1:p)   # system measured output indices

# form augmented aircraft model with extended measurement set
ee = eye(m); p1 = p+mf;
sysact = similar(Vector{DescriptorStateSpace},N)
for i = 1:N
    sysact[i] = gir([ee[act_prim,:]; SYSACM[i]] *
                    append(ACT,dss(eye(md))); atol)
end

# build synthesis models with actuator faults
sysactf = fdimodset(sysact; c = Vector(1:mu),
            d = mu .+ Vector(1:md), f = act_prim)
```

```
# determine closed-loop stability margin
sdeg_cloop = -5
for i = 1:N
    syst = feedback(sysact[i],K,indu,indy;negative = false)
    global sdeg_cloop = max(sdeg_cloop,
                            maximum(real(eigvals(syst.A))))
end

@test sdeg_cloop < 0
println("sdeg_cloop = $sdeg_cloop")
```

For the synthesis of the FDD system for monitoring the primary actuator faults we solve an EFDIP for the nominal case by using the **Procedure EFDI** described in Sect. 5.4. For the setup of the fault signatures, which allow the isolation of the actuator faults, we computed the achievable strong specifications in the structure matrix S_{strong} for the augmented aircraft with additional surface position measurements. The achievable strong specifications include the entire set of $255 (= 2^8 - 1)$ possible specifications. Therefore, we chose the structure matrix to be achieved $S_{FDI} = I_8$, which allows the simultaneous isolation of an arbitrary number of up to 8 primary actuator faults. In Listing 8.2, we present the Julia code used for the determination of the achievable structure matrix S_{strong} containing the strong fault specifications.

Listing 8.6 Part 2 of script **CS1_2.jl**: Setup of synthesis specifications

```
# Part 2 - Setup of synthesis specifications
# compute achievable strong specifications for constant faults
FDtol = 1.e-5        # weak fault detection threshold
FDGainTol = 0.001    # strong fault detection threshold
FDfreq = 0           # frequency strong detectability checks
S_strong = fdigenspec(sysactf[nom]; atol, FDtol, FDGainTol,
                FDfreq, sdeg = -0.05)

# check resulting specifications
@test size(S_strong,1) == 2^mf-1

# define SFDI, the signatures for isolation of single faults
SFDI = eye(mf) .> 0;
```

Using the structure matrix $S_{FDI} = I_8$, we performed the **Procedure EFDI** for the synthesis of a bank of $n_b = 8$ fault detection filters $Q^{(j)}(s)$, $j = 1, \ldots, n_b$, with scalar outputs. The Listing 8.3 illustrates the call of the function **efdisyn**, which implements **Procedure EFDI**. The results computed by **efdisyn** are the bank of eight scalar output fault detection filters $Q^{(j)}(s)$, for $j = 1, \ldots, 8$, and the corresponding eight internal form filters $\widetilde{R}_f^{(j)}(s)$, for $j = 1, \ldots, 8$, which are stored in the filter objects **Q** and **Rf**, respectively. The j-th filter $Q^{(j)}(s)$ achieves the j-th specification contained in the j-th row of the structure matrix S_{FDI} and has the least possible McMillan degree of one. The stability margin for the eigenvalues of the filters $Q^{(j)}(s)$, for $j = 1, \ldots, 8$, has been set to -1 via the keyword argument **smarg**, and all eigenvalues having real parts less than this value have been assigned to -5 using the keyword argument **sdeg**. The resulting overall filter $Q(s)$ obtained by stacking the 8 scalar output filters $Q^{(j)}(s)$, for $j = 1, \ldots, 8$, has a global order 8.

Listing 8.7 Part 3 of script **CS1_2.jl**: Synthesis using **Procedure EFDI**

```julia
# Part 3 - Synthesis using Procedure EFDI
# set options for least-order synthesis with efdisyn
Q, Rf, info = efdisyn(sysactf[nom], SFDI; atol, FDGainTol, FDfreq,
                      rdim = 1, sdeg = -5, smarg = -1)
println("orders = $(order.(Q.sys))")
```

The assessment of the synthesis results can be performed using the same Julia code as that presented in Listing 8.4. The overall filters $Q(s)$ and $\widetilde{R}_f(s)$ are formed by stacking the eight scalar output filters $Q^{(j)}(s)$ and $\widetilde{R}_f^{(j)}(s)$, respectively. The first check is to verify the nominal synthesis conditions (8.4). The second check is to verify that the achieved structure matrix $S_{\widetilde{R}_f}$ (see Sect. 3.4) is equal to S_{FDI}. This comes down to simply verify that the DC-gain $\widetilde{R}_f(0)$ and S_{FDI} have the same zero–nonzero pattern (i.e., $\widetilde{R}_f(0)$ must be diagonal and nonsingular).

For the robustness of the nominal synthesis, we formed the internal representations $\overline{R}^{(i)}(s)$ in (8.5) for $i = 1, \ldots, N$, on the basis of the N closed-loop systems and computed $\left\|\overline{R}_u^{(i)}(s)\right\|_\infty$, $\left\|\overline{R}_d^{(i)}(s)\right\|_\infty$ and $\left\|\overline{R}_f^{(i)}(s) - \widetilde{R}_f(s)\right\|_\infty$, for $i = 1, \ldots, N$. The robustness analysis results in terms of infinity norms are summarized in Table 8.3. The very small norms in Table 8.3 indicate that the nominal synthesis is completely satisfactory from the point of view of the robustness requirements. We cautiously remark that these results correspond to a particular choice of a first-order filter, and other choices may lead to different robustness performance.

In Fig. 8.4 we present the N step responses of $\overline{R}_f^{(i)}(s)$ for $i = 1, \ldots, N$. It can be easily observed that the nominal filter is very robust with respect to mass variations. Practically, there are no differences between the nominal and non-nominal step responses.

In summarizing the achieved synthesis results, we have to emphasize the remarkable achievement to arrive to an overall FDI filter of lowest possible order 8, which is able to isolate up to 8 simultaneous primary actuator faults, and provides robust fault

Table 8.3 Robustness analysis results for the nominal synthesis with position measurements

Model #	$\left\|\overline{R}_u^{(i)}(s)\right\|_\infty$	$\left\|\overline{R}_d^{(i)}(s)\right\|_\infty$	$\left\|\overline{R}_f^{(i)}(s) - \widetilde{R}_f(s)\right\|_\infty$
1	1.5e-5	2.8e-6	1.5e-5
2	1.3e-5	2.5e-6	1.3e-5
3	1.1e-5	2.1e-6	1.1e-5
4	9.1e-6	1.7e-6	9.1e-6
5	6.4e-6	1.2e-6	6.4e-6
6	3.4e-6	6.3e-7	3.4e-6
7	8.6e-14	6.6e-15	4.2e-14
8	3.7e-6	6.9e-7	3.7e-6
9	7.8e-6	1.5e-6	7.8e-6
10	1.2e-5	2.2e-6	1.2e-5
11	1.7e-5	3.2e-6	1.7e-5

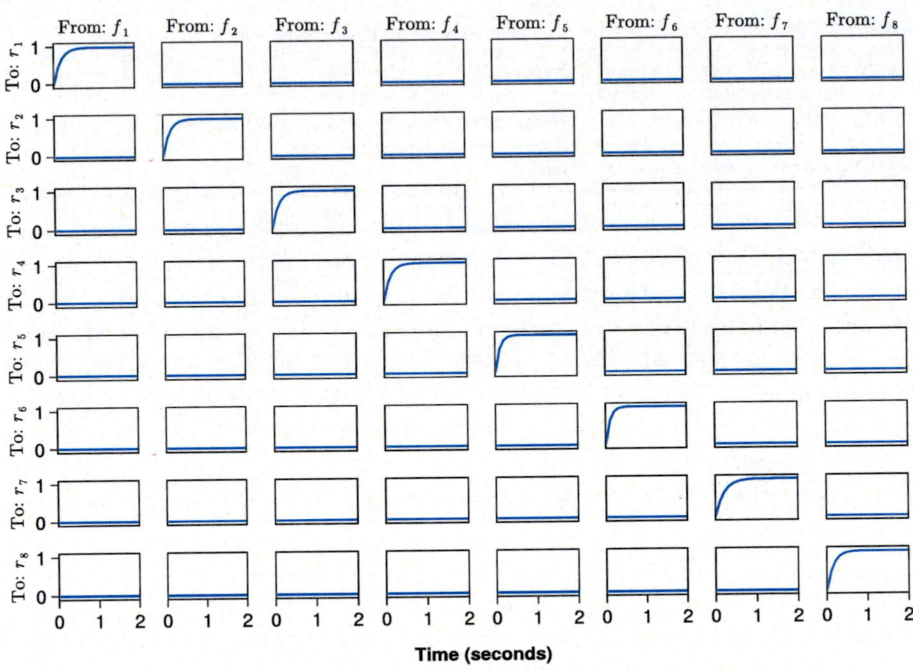

Fig. 8.4 Parametric fault input step responses for nominal synthesis with position measurements

monitoring performance over the whole range of mass variations. A clear sign for good robustness is that all gains in the filter matrices have moderate sizes. We expect that similar performance robustness is also provided in the presence of variations of other key parameters as the position of the centre of gravity, altitude and speed. For computations, we employed the Julia function **efdisyn**, which implements the **Procedure EFDI**. We claim that none of the alternative methods proposed in the fault detection literature and none of existing alternative software tools are able to reproduce our synthesis results.

All individual filters are first-order systems and this order is equal to the order of the corresponding monitored actuator. This fact is a strong indication to consider an alternative approach for monitoring flight actuator faults, namely monitoring at a component level. This approach is discussed in the next section, where additional robustness aspects are also addressed.

8.1.3 Local Monitoring of Actuator Faults—Industrial Aspects

The results of the previous section strongly suggest the use of the alternative local monitoring of flight actuator faults (see Fig. 8.4), where each flight actuator has its own dedicated FDD system relying on local measurements of surface position (or, equivalently, of actuator rod position) and of the commanded surface position as control input. For an industrial usage of model-based fault monitoring on civil aircraft, this could be an appealing option, which allows to gradually move from a pure signal-processing-based monitoring to the potentially more powerful model-based approaches. In fact, as it will be apparent from our presentation, a combination of model-based and signal-processing-based approaches seems to be the best solution, guaranteeing timely and robust fault detection, and additional provision for fault identification, which is necessary when control system reconfiguration has to be performed.

The local FDD system attached to each actuator has two basic functions: (1) the robust detection of any actuator fault, using a model-based fault detection filter, and (2) the identification of fault characteristics by using signal-processing techniques. In what follows, we succinctly discuss these two steps.

Detection of Flight Actuator Faults

For fault detection purposes, the flight actuator dynamics are frequently modelled as first- or second-order linear systems (see Example 2.1). A first-order flight actuator model, which relates the commanded surface position u_c and the corresponding aircraft surface position u_{ac}, can be described by the state-space model

$$\dot{x}_a(t) = -kx_a(t) + ku_c(t),$$
$$u_{ac}(t) = x_a(t). \tag{8.6}$$

Assuming a constant gain k, we can equivalently describe the actuator dynamics by an input-output model via the transfer function

$$G_u(s) = \frac{k}{s+k}. \tag{8.7}$$

A constant gain k is always an approximation and its value usually represents an average gain over all flight conditions. The range of typical values of k for primary actuators is between 0.5 (e.g., for a stabilizer actuator) and 50 (e.g., for ruder and aileron actuators), with somewhat lower values for elevator actuators. To setup an actuator model with additive input faults of the form

$$\mathbf{u}_{ac}(s) = G_u(s)\mathbf{u}_c(s) + G_f(s)\mathbf{f}(s),$$

we can choose $G_f(s) = G_u(s)$, by assuming control input located faults (the choice $G_f(s) = 1$ is appropriate for measurement located faults).

A simple fault detection filter is $Q(s) = \begin{bmatrix} 1 & -G_u(s) \end{bmatrix}$, which corresponds to an estimator-based fault detection scheme. However, the dynamics of the filter can be arbitrarily assigned by choosing a more general filter

$$Q(s) = M(s)\left[1 - G_u(s)\right],\tag{8.8}$$

where $M(s)$ is a stable transfer function with arbitrary poles and a zero in $-k$. For example, $M(s)$ can be chosen as

$$M(s) = \frac{a}{k_0}\frac{s+k}{s+a},\tag{8.9}$$

which has an arbitrary pole at $-a$ and a DC-gain equal to a/k_0. Usually $k_0 = k$ is chosen, but k_0 can be any nonzero constant gain. A state-space realization of the filter (8.8) is

$$\begin{aligned}\dot{x}_Q(t) &= -ax_Q(t) + \tfrac{a(k-a)}{k_0}u_{ac}(t) - \tfrac{ak}{k_0}u_c(t),\\ r(t) &= x_Q(t) + \tfrac{a}{k_0}u_{ac}(t).\end{aligned}\tag{8.10}$$

For the filter (8.8), with $M(s)$ given by (8.9), the resulting transfer function from the fault to residual (i.e., the internal form of the filter) is

$$R_f(s) = M(s)G_f(s) = \frac{k}{k_0}\frac{a}{s+a},$$

with the corresponding state-space realization

$$\begin{aligned}\dot{\widetilde{x}}_Q(t) &= -a\widetilde{x}_Q(t) + \tfrac{ak}{k_0}f(t),\\ r(t) &= \widetilde{x}_Q(t),\end{aligned}\tag{8.11}$$

where $\widetilde{x}_Q(t) := x_Q(t) + x_a(t)a/k_0$. Thus, the residual signal provides a filtered estimation of the fault, allowing to easily reconstruct the actuator input fault signal f (e.g., for further use in fault identification).

The simple fault detection filter (8.8) is primarily intended for the detection of several classes of additive faults as jamming (also called lock-in-place failure), runaway (also called hard-over failure), or oscillatory failure cases. Other types of actuator faults discussed in Example 2.5 as actuator disconnection fault (also known as free-play or float-type failure), stall load fault, or loss-of-effectiveness fault belong to the category of parametric faults because basically they involve changes in the actuator gain k. These faults are more difficult to be reliably detected using methods based on the additive fault assumption, because their effects can be only sensed during short-time transitory dynamics. For such type of faults, model detection techniques, as described in Chap. 6, are more suitable.

The use of the simple LTI fault detection filters like in (8.10) for monitoring (additive type) actuator faults on a civil aircraft may lead to difficulties, because the use of a constant gain k in the actuator model (8.6) basically ignores the interactions between the actuator dynamics and the aerodynamic forces acting on the control surfaces. In reality, the "effective" actuator gain exhibits a complex dependency $k(u_{ac}, \dot{u}_{ac}, \eta)$ of the position u_{ac} of the attached control surface, the direction of surface movement $\dot{u}_{ac}$, as well as of other flight and aircraft parameters grouped in a parameter vector η (e.g., speed, altitude, mass, etc.). For example, the "effective" gain is larger for small surface deflections than for large deflections, while for a given surface position, gain variations also occur due to the direction of surface movements. Upward movements increase the aerodynamic forces and thus automatically decrease the "effective" gain, while downward movements diminish the aerodynamics forces and thus increase the actuator gains. Gain increases make the actuator more agile, while gain decreases make the actuator more sluggish. Ignoring these physical aspects makes the FDD system more susceptible to false alarms due to the variations of the flight parameters and presence of model uncertainties.

Assuming the explicit dependencies of the gain $k(u_{ac}, \dot{u}_{ac}, \eta)$ are known, a gain-scheduling fault detection filter can be simply constructed, by using $k(u_{ac}, \dot{u}_{ac}, \eta)$ instead its assumed constant value in the realization of the filter (8.10). The development of high fidelity actuator models with properly modelled nonlinear gains is a nontrivial task to be performed by the aircraft or actuator manufacturers. The resulting expressions of the gain are usually too complicated to serve for gain-scheduling purposes. Therefore, simpler approximations of gains are usually employed for real-time implementations of the filters. By taking into account the above physical considerations, any flight actuator gain $k(u_{ac}, \dot{u}_{ac}, \eta)$ can be approximated by a gain $\hat{k}(u_{ac}, \dot{u}_{ac}, \eta)$ of the form

$$\hat{k}(u_{ac}, \dot{u}_{ac}, \eta) = C_0(\eta) + C_1(\eta)\,\mathrm{sgn}(\dot{u}_{ac})(u_{ac} + C_2(\eta)), \qquad (8.12)$$

where, for a fixed value of parameter η, $C_0(\eta)$ can be interpreted as the nominal gain; $C_1(\eta)$ describes the influence of the surface deflection u_{ac} on the gain, while the factor $\mathrm{sgn}(\dot{u}_{ac})$ allows to distinguish between upward and downward movements of the control surface; and $C_2(\eta)$ can be interpreted as a position offset. The chosen functional dependence on u_{ac} and $\mathrm{sgn}(\dot{u}_{ac})$ reflects the actual physical behaviour of the actuator for different control surface positions and signs of deflection rate. The terms $C_i(\eta)$ for $i = 0, 1, 2$ can be approximated using data fitting techniques. Often simple affine approximations are sufficient to provide good matching with the original gains.

The use of the gain-scheduling-based fault detection filter (8.10), with a simpler gain $\hat{k}(u_{ac}, \dot{u}_{ac}, \eta)$ as in (8.12) replacing k, is a prerequisite for a robust fault detection performance (i.e., lack of false alarms, lack of missed detection, satisfactory detection times). To achieve this goal, we have to additionally choose appropriate thresholds as well as the parameters of evaluation filters to be used in the fault evaluation and decision-making blocks of the FDD system (see Fig. 3.1). The low real-time computational burden associated with the on-board implementation of the FDD system is guaranteed by the use of a low complexity (e.g., first-order) fault detection filter for each monitored flight actuator.

Identification of Flight Actuator Faults

The detection of an actuator fault triggers the second phase of the fault diagnosis, which involves the execution of the fault identification algorithms. These algorithms are intended to discover the nature of the fault (e.g., jamming, runaway, or oscillatory failure) and to extract key characteristics, which are essential for a control reconfiguration action. By confirming the occurrence of a specific fault, fault identification additionally enhances the robustness of the fault detection. Special signal-processing algorithms are employed to determine the nature of the occurred fault. In what follows we consider three categories of additive faults, namely jamming, runaway and oscillatory failure, and discuss shortly their identification.

The jamming of an aircraft control surface creates a dissymmetry in the aircraft configuration, which must be compensated by appropriate deflections of other control surfaces. Therefore, the jamming leads to the degradation of the aircraft performance due to the increased drag, which depends on the amplitude and localization of the failure. To confirm the jamming of a control surface in a fixed position, it is sufficient to check that the variance of n surface position measurements $u_{ac}(t_i)$, for $i = 1, \ldots, n$ is zero (or below a small threshold), where $t_i = t_0 + iT$, with t_0 the starting time (e.g., the detection time) and T a given sampling interval. A related fault is the surface (or rod displacement) sensor bias, which can be confirmed by checking that the variance of the generated residual values $r(t_i)$ is zero. Another special case is the jamming in null position, which can be identified only if sufficient control activity is provided (i.e., during a certain manoeuvre). This fault is confirmed if both the mean and variance of $u_{ac}(t_i)$ are zero, but the variance of $u_c(t_i)$ is nonzero. The real-time computation of variance of a signal involves also the computation of the mean value, which provides the jamming deflection. To avoid the storage of a many samples of measurements, the use of recursive algorithms to evaluate the mean and covariance is recommended. A suitable algorithm for this purpose is based on the following recursion to evaluate the mean E (u_{ac}) and variance Var (u_{ac}) in n steps

$$
\mu_i = \mu_{i-1} + (u_{ac}(t_i) - \mu_{i-1})/i,
$$
$$
\sigma_i = \sigma_{i-1} + (u_{ac}(t_i) - \mu_{i-1})(u_{ac}(t_i) - \mu_i),
$$

for $i = 2, \ldots, n$, with the initializations $\mu_1 = u_{ac}(t_1)$ and $\sigma_1 = 0$. After $n - 1$ steps, we obtain E $(u_{ac}) = \mu_n$ and Var $(u_{ac}) = \sigma_n/(n - 1)$. The quantities μ_i and $\sigma_i/(i - 1)$ are the mean value and variance after i steps.

An actuator runaway takes place when a large, not commanded, surface deflection occurs and the surface tends to lock in its extreme position. In the case of a runaway, excessive structural loads can be expected, and therefore it must be very quickly detected and identified by the FDD system, before its full development. The fast identification of runaway can be done by checking that the mean value of the absolute surface deflection variation rates $|\dot{u}_{ac}(t_i)|$ is greater than a certain allowed maximum slew rate $\dot{u}_{\max}$. The mean value E $(|\dot{u}_{ac}|)$, for $i > 1$ samples, can be simply estimated as E $(|\dot{u}_{ac}|) \approx |u_{ac}(t_i) - u_{ac}(t_1)|/(t_i - t_1)$.

The so-called "oscillatory failure" (e.g., of a rudder) is a result of limit-cycle oscillations, which take place in the actuator positioning loop. Two types of oscillatory failure cases (OFCs) are usually considered. A liquid failure is an additive oscillatory fault signal inside the actuator positioning control loop. A solid failure involves an oscillatory signal, which completely replaces a normal signal in the actuator positioning loop. The early detection of an OFC (even with small amplitude oscillations) in a physically relevant frequency domain is important to prevent excessive structural loads of the aircraft. For the identification of the OFC, the most reliable method is based on determining the power spectrum of the residual signal using the discrete Fourier transform (DFT). The use of DFT allows a satisfactory accurate evaluation of the oscillation frequency, together with a strong statistical guarantee of the presence of the oscillations in a signal. For a monitored frequency ω, the DFT for n values $r(t_j)$, $j = 1, \ldots, n$, of the residual signal, is computed as

$$X(\omega) = \sum_{j=1}^{n} r(t_j) e^{-i\omega t_j} .$$

An oscillation of frequency nearby to ω is confirmed if the power spectrum $|X(\omega)|$ is greater than a certain threshold value τ_ω. For N monitored frequencies, the corresponding N power spectra must be computed. The fast Fourier transform (FFT) is usually used for this purpose and determines n values of the power spectrum for n frequency values. However, the use of FFT requires the storage of n values of the residual, and since usually $n \gg N$, the FFT-based DFT computation is possibly not the most efficient way to evaluate the power spectra only for a few frequency values. An alternative, computationally more economical approach relies on a recursive computation of the power spectrum using the straightforward recursion

$$Y_k(\omega) = Y_{k-1}(\omega) + r(t_k) e^{-i\omega t_k},$$

for $k = 2, \ldots, n$, where $Y_k(\omega) := \sum_{j=1}^{k} r(t_j) e^{-i\omega t_j}$ is the partial sum over k samples. Evidently, $X(\omega) = Y_n(\omega)$. However, for the confirmation of an oscillation with frequency ω we can use the alternative detection condition $|Y_k(\omega)| \geq \tau_\omega$, which, in the case of presence of oscillations, is usually fulfilled for a value $k \ll n$. This ensures a fast detection of the presence of oscillations nearby ω.

It can be argued that the described fault identification methods of several classes of additive flight actuator faults can be employed even without previously performing the model-based fault detection step. This would eliminate the potentially expensive modelling step to develop suitable LPV-models of flight actuators for gain-scheduling purposes. Although this seems reasonable to reduce the design costs of civil aircraft, still the two step approach, fault detection–fault identification, has certain advantages, which should prevail for its industrial usage. First, the overall reliability of fault diagnosis is clearly superior, since the fault identification can be seen as a supplementary check of the appearance of a certain type of fault. Being in many cases insensitive to modelling related uncertainties, fault identification can significantly enhance the robustness of FDD performance, by fully avoiding false alarms and missed detections. A second advantage is the smaller overall on-board computational burden. When using solely fault identification to detect faults, this requires running continuously, in parallel, several fault identification algorithms for each actuator. In contrast, when using the two step approach, the execution of computations for fault identification is only triggered by the detection of a fault for a single actuator. Therefore, for most of the time, the only real-time computational effort is that for running the gain-scheduling-based first-order fault detection filters for the monitored actuators. The associated computational burden is substantially lower than running many fault identification algorithms, all the time, in parallel, for the same actuators.

8.1.4 Linearized State-Space Models with Additive Actuator Faults

Each of the linearized models (8.1) with additive actuator faults has been obtained by augmenting a linearized aircraft model, corresponding to a standard cruise situation (i.e., in straight and level flight), with simple (first-order) actuator models. As usual in the aeronautics industry, linearized aircraft models are obtained by the linearization of nonlinear aircraft models in specific flight conditions (i.e., speed V and altitude h) and for specific values of certain aircraft parameters (e.g., aircraft mass m and relative position of its centre of gravity X_{cg}). Therefore, the numerical values of the entries of the system matrices of any linearized aircraft model depend on such parameters. The employed nonlinear model describes a generic two engine civil aircraft with 8 primary control surfaces (2 elevators, 1 stabilizer, 1 ruder, 4 ailerons) and 12 spoilers (as secondary control surfaces). The linearized aircraft models have the multiple-model form (see Sect. 2.2.3)

$$
\begin{aligned}
\dot{x}_{ac}^{(i)}(t) &= A_{ac}^{(i)} x_{ac}^{(i)}(t) + B_{ac,u}^{(i)} u_{ac}(t) + B_{ac,d}^{(i)} d(t), \\
y_{ac}^{(i)}(t) &= C_{ac}^{(i)} x_{ac}^{(i)}(t) \qquad\qquad + D_{ac,d}^{(i)} d(t),
\end{aligned}
\tag{8.13}
$$

where the dimensions of vectors $x_{ac}^{(i)}(t)$, $y_{ac}^{(i)}(t)$, $u_{ac}(t)$ and $d(t)$ are respectively, $n_{ac} = 10$, $p_{ac} = 10$, $m_u = 22$ and $m_d = 3$. These variables approximate small deviations of the system variables of the aircraft nonlinear model from their equilibrium (or trim) values. There are $N = 11$ models of the form (8.13), for $i = 1, \ldots, N$, which correspond to N values $m_i, i = 1, \ldots, N$, of the aircraft mass. The chosen values of mass cover the whole range of variation of the mass $[m_{\min}, m_{\max}]$, where $m_{\min}$ and $m_{\max}$ are the smallest and largest values of mass, respectively. The employed grid of mass values is defined as $m_i = m_{\min} + \Delta m_i (m_{\max} - m_{\min})$ for $i = 1, \ldots, N$, where Δm_i is the percentage of mass variation from its whole range. The chosen nominal model corresponds to $\Delta m_7 = 60\%$ of variation. For all linearized models, the rest of parameters are constant and equal to their nominal values.

The system variables in the state-space model (8.13) are defined as follows:

$$
y_{ac}^{(i)} = \begin{pmatrix} \textit{roll angle} \\ \textit{pitch angle} \\ \textit{yaw angle} \\ \textit{angle of attack} \\ \textit{angle of sideslip} \\ \textit{flight path angle} \\ \textit{roll rate} \\ \textit{pitch rate} \\ \textit{yaw rate} \\ \textit{calibrated airspeed} \end{pmatrix}, \quad
x_{ac}^{(i)} = \begin{pmatrix} \textit{first component of quaternion} \\ \textit{second component of quaternion} \\ \textit{third component of quaternion} \\ \textit{fourth component of quaternion} \\ \textit{ground speed X body axis} \\ \textit{ground speed Y body axis} \\ \textit{ground speed Z body axis} \\ \textit{roll rate} \\ \textit{pitch rate} \\ \textit{yaw rate} \end{pmatrix},
$$

$$
u_{ac} = \begin{pmatrix} \textit{right outer aileron deflection} \\ \textit{right inner aileron deflection} \\ \textit{spoiler}_1 \textit{ deflection} \\ \vdots \\ \textit{spoiler}_{12} \textit{ deflection} \\ \textit{left inner aileron deflection} \\ \textit{left outer aileron deflection} \\ \textit{right elevator deflection} \\ \textit{stabilizer trim angle} \\ \textit{left elevator deflection} \\ \textit{rudder deflection} \\ \textit{left engine thrust} \\ \textit{right engine thrust} \end{pmatrix}, \quad
d = \begin{pmatrix} \textit{wind speed X axis} \\ \textit{wind speed Y axis} \\ \textit{wind speed Z axis} \end{pmatrix}.
$$

The numerical values of system matrices for the nominal model are given at the end of this section. A particular feature of the employed linearized aircraft model (8.13) is that it is unstable. For example, the eigenvalues of the state matrix $A_{ac}^{(7)}$ are

$$
\Lambda\left(A_{ac}^{(7)}\right) = \begin{bmatrix} -0.6646 +1.1951\mathrm{i} \\ -0.6646 -1.1951\mathrm{i} \\ -0.0016 +0.0600\mathrm{i} \\ -0.0016 -0.0600\mathrm{i} \\ -1.6550 \\ 0.0186 +0.8768\mathrm{i} \\ 0.0186 -0.8768\mathrm{i} \\ 0.0094 \\ 0 \\ 0 \end{bmatrix}.
$$

Moreover, one of the two structurally fixed eigenvalues in the origin is uncontrollable for all system pairs $\left(A_{ac}^{(i)}, [\, B_{ac,u}^{(i)} \;\; B_{ac,d}^{(i)} \,]\right)$.

The actuator and engine models are approximated by first-order LTI systems with the following transfer functions: $10/(s + 10)$ for each of the two elevators, $0.5/(s + 0.5)$ for the stabilizer, $6.67/(s + 6.67)$ for each of the four ailerons and the ruder, $2.5/(s + 5)$ for each of the twelve spoilers and $0.667/(s + 0.667)$ for each of the two engines. The system formed by the actuators, augmented with additive faults for the primary actuators, has a state-space realization of the form

$$\begin{aligned}\dot{x}_a(t) &= A_a x_a(t) + B_{a,u} u(t) + B_{a,f} f(t), \\ u_{ac}(t) &= C_a x_a(t),\end{aligned}$$

$$(8.14)$$

where $x_a(t)$ is the state vector of dimension 22 and $u(t)$ contains the 20 deflection demands and the 2 thrust demands. The elements of the vector f are the following 8 additive faults of the primary actuators

$$f = \begin{pmatrix} right\text{-}outer\ aileron\ actuator\ fault \\ right\text{-}inner\ aileron\ actuator\ fault \\ left\text{-}inner\ aileron\ actuator\ fault \\ left\text{-}outer\ aileron\ actuator\ fault \\ right\ elevator\ actuator\ fault \\ left\ elevator\ actuator\ fault \\ stabilizer\ actuator\ fault \\ ruder\ actuator\ fault \end{pmatrix}.$$

If we denote with B_{a,u_j} the j-th column of $B_{a,u}$, then the fault input matrix $B_{a,f}$ is defined as

$$B_{a,f} = \begin{bmatrix} B_{a,u_1} & B_{a,u_2} & B_{a,u_{15}} & B_{a,u_{16}} & B_{a,u_{17}} & B_{a,u_{19}} & B_{a,u_{18}} & B_{a,u_{20}} \end{bmatrix}.$$

The matrices A_a, $B_{a,u}$ and C_a are diagonal and defined as

$$\begin{aligned} A_a &= \mathrm{diag}(-6.667 \cdot I_2, -5 \cdot I_{12}, -6.667 \cdot I_2, -10, -0.5, -10, -6.667, -0.667 \cdot I_2), \\ B_{a,u} &= \mathrm{diag}(6.667 \cdot I_2, 5 \cdot I_{12}, 6.667 \cdot I_2, 10, 0.5, 10, 6.667, 0.667 \cdot I_2), \\ C_a &= \mathrm{diag}(I_2, 0.5 \cdot I_{12}, I_8). \end{aligned}$$

The complete aircraft model (8.1) is obtained by series coupling of the actuator model (8.14) and aircraft model (8.13). The state and output vectors are defined as

$$x^{(i)}(t) = \begin{bmatrix} x_{ac}^{(i)}(t) \\ x_a(t) \end{bmatrix}, \qquad y^{(i)}(t) = \begin{bmatrix} y_{ac}^{(i)}(t) \\ \Pi u_{ac}(t) \end{bmatrix},$$

where Π is an actuator output selection matrix. The system matrices (8.1) are determined as

$$A^{(i)} = \begin{bmatrix} A_{ac}^{(i)} & B_{ac,u}^{(i)} C_a \\ 0 & A_a \end{bmatrix}, \quad B_u^{(i)} = \begin{bmatrix} 0 \\ B_{a,u} \end{bmatrix}, \quad B_d^{(i)} = \begin{bmatrix} B_{ac,d}^{(i)} \\ 0 \end{bmatrix}, \quad B_f = \begin{bmatrix} 0 \\ B_{a,f} \end{bmatrix},$$

$$C^{(i)} = \begin{bmatrix} C_{ac}^{(i)} & 0 \\ 0 & \Pi C_a \end{bmatrix}, \qquad\qquad\qquad D_d^{(i)} = \begin{bmatrix} D_{ac,d} \\ 0 \end{bmatrix}.$$

Each model in (8.1) has a state vector of dimension $n = 32$, $m_u = 22$ control inputs, $m_3 = 3$ disturbance inputs, $m_f = 8$ fault inputs and the number of measured variables is either $p = 10$ or $p = 18$. The latter case is when all control surfaces corresponding to the monitored primary actuators are provided with angle (or equivalent road position) sensors. The case without angle sensors corresponds to formally choosing Π an 0×22 matrix (with empty rows), while in the case when surface angle sensors are available Π is formed from 8 stacked rows e_i^T of the identity matrix I_{22} to select the primary actuator outputs, where $i = 1, 2, 15, 16, 17, 19, 18, 20$.

The matrices of the nominal state-space model (8.13), for $i = 7$, have the following numerical values:

$$
A_{ac}^{(7)} =
\begin{bmatrix}
0 & 0 & 0 & 0 & 0 & 0 & 0 \\
0 & 0 & 0 & 0 & 0 & 0 & 0 \\
0 & 0 & 0 & 0 & 0 & 0 & 0 \\
0 & 0 & 0 & 0 & 0 & 0 & 0 \\
-0.4447 & 0 & -19.62 & 0 & -0.003 & 0 & 0.061 \\
0 & 19.62 & 0 & 0.4447 & 0 & -0.062 & 0 \\
19.62 & 0 & -0.4447 & 0 & -0.0777 & 0 & -0.8004 \\
0 & 0 & 0 & 0 & 0 & -0.0239 & 0 \\
0 & 0 & 0 & 0 & 0.0001 & 0 & -0.0073 \\
0 & 0 & 0 & 0 & 0 & 0.0019 & 0
\end{bmatrix}
\cdots
$$

$$
\begin{bmatrix}
0 & -0.0113 & 0 \\
0.5 & 0 & 0.0113 \\
0 & 0.5 & 0 \\
-0.0113 & 0 & 0.5 \\
0 & -9.0618 & 0 \\
8.5315 & 0 & -199.3932 \\
0 & 197.8868 & 0 \\
-1.5599 & 0 & 0.3470 \\
0 & -0.5290 & 0 \\
-0.0934 & 0 & 0.0136
\end{bmatrix},
$$

$$
B_{ac,d}^{(7)} =
\begin{bmatrix}
0 & 0 & 0 \\
0 & 0 & 0 \\
0 & 0 & 0 \\
0 & 0 & 0 \\
0.0001 & 0 & -0.0314 \\
0 & 0.0319 & 0 \\
0.0586 & 0 & 0.4097 \\
0 & 0.0123 & 0 \\
0.0001 & 0 & 0.0038 \\
0 & -0.0010 & 0
\end{bmatrix},
\qquad
D_{ac,d}^{(7)} =
\begin{bmatrix}
0 & 0 & 0 \\
0 & 0 & 0 \\
0 & 0 & 0 \\
0 & 0 & -0.1467 \\
0 & -0.1467 & 0 \\
0 & 0 & 0 \\
0 & 0 & 0 \\
0 & 0 & 0 \\
0 & 0 & 0 \\
-0.7321 & 0 & 0
\end{bmatrix},
$$

$$
B_{ac,u}^{(7)} =
\begin{bmatrix}
0 & 0 & 0 & 0 & 0 & 0 & 0 \\
0 & 0 & 0 & 0 & 0 & 0 & 0 \\
0 & 0 & 0 & 0 & 0 & 0 & 0 \\
0 & 0 & 0 & 0 & 0 & 0 & 0 \\
-0.0009 & -0.0009 & -0.0006 & -0.0006 & -0.0006 & -0.0006 & -0.0006 & \cdots \\
0 & 0 & 0 & 0 & 0 & 0 & 0 \\
0.0552 & 0.0552 & 0.0071 & 0.0071 & 0.0071 & 0.0071 & 0.0071 \\
-0.0219 & -0.0183 & 0.0143 & 0.0127 & 0.0114 & 0.0095 & 0.0079 \\
-0.0059 & -0.0051 & 0.0001 & 0.0001 & 0.0001 & 0.0001 & 0.0001 \\
-0.0007 & -0.0005 & 0.0007 & 0.0007 & 0.0006 & 0.0005 & 0.0004
\end{bmatrix}
$$

$$
\begin{array}{cccccccc}
0 & 0 & 0 & 0 & 0 & 0 & 0 & 0 \\
0 & 0 & 0 & 0 & 0 & 0 & 0 & 0 \\
0 & 0 & 0 & 0 & 0 & 0 & 0 & 0 \\
0 & 0 & 0 & 0 & 0 & 0 & 0 & 0 \\
-0.0006 & -0.0006 & -0.0006 & -0.0006 & -0.0006 & -0.0012 & -0.0012 & -0.0009 \cdots \\
0 & 0 & 0 & 0 & 0 & 0 & 0 & 0 \\
0.0071 & 0.0071 & 0.0071 & 0.0071 & 0.0071 & 0.0142 & 0.0142 & 0.0552 \\
0.0053 & -0.0053 & -0.0079 & -0.0095 & -0.0114 & -0.0254 & -0.0286 & 0.0183 \\
0.0001 & 0.0001 & 0.0001 & 0.0001 & 0.0001 & 0.0002 & 0.0002 & -0.0051 \\
0.0003 & -0.0003 & -0.0004 & -0.0005 & -0.0006 & -0.0013 & -0.0015 & 0.0005
\end{array}
$$

$$
\begin{bmatrix}
0 & 0 & 0 & 0 & 0 & 0 & 0 \\
0 & 0 & 0 & 0 & 0 & 0 & 0 \\
0 & 0 & 0 & 0 & 0 & 0 & 0 \\
0 & 0 & 0 & 0 & 0 & 0 & 0 \\
-0.0009 & 0.0021 & 0.0094 & 0.0021 & 0 & 0.1712 & 0.1712 \\
0 & 0 & 0 & 0 & 0.0769 & 0 & 0 \\
0.0552 & -0.1360 & -0.6063 & -0.1360 & 0 & 0 & 0 \\
0.0219 & -0.0077 & 0 & 0.0077 & 0.0051 & 0.0007 & -0.0007 \\
-0.0059 & -0.0280 & -0.1250 & -0.0280 & 0 & 0.0015 & 0.0015 \\
0.0007 & -0.0002 & 0 & 0.0002 & -0.0079 & 0.0079 & -0.0079
\end{bmatrix},
$$

$$
C_{ac}^{(7)} = \begin{bmatrix}
0 & 2.5971 & 0 & 114.5916 & 0 & 0 \\
2.5998 & 0 & 114.7095 & 0 & 0 & 0 \\
0 & 114.5916 & 0 & 2.5971 & 0 & 0 \\
0 & 0 & 0 & 0 & -0.0129 & 0 \\
0 & 0 & 0 & 0 & 0 & 0.2850 \\
-2.5971 & 0 & 114.5916 & 0 & 0.0129 & 0 \\
0 & 0 & 0 & 0 & 0 & 0 \\
0 & 0 & 0 & 0 & 0 & 0 \\
0 & 0 & 0 & 0 & 0 & 0 \\
0 & 0 & 0 & 0 & 1.4210 & 0
\end{bmatrix} \cdots
$$

$$
\begin{bmatrix}
0 & 0 & 0 & 0 \\
0 & 0 & 0 & 0 \\
0 & 0 & 0 & 0 \\
0.2847 & 0 & 0 & 0 \\
0 & 0 & 0 & 0 \\
-0.2849 & 0 & 0 & 0 \\
0 & 57.2958 & 0 & 0 \\
0 & 0 & 57.2958 & 0 \\
0 & 0 & 0 & 57.2958 \\
0.0644 & 0 & 0 & 0
\end{bmatrix} .
$$

8.2 Monitoring Air Data Sensor Faults

The monitoring of the accuracy of air data sensor measurements, as the *angle of attack* (AoA) and *calibrated air speed* (VCAS), is of paramount importance for the satisfactory operation of civil aircraft. For example, the measurements of AoA are used to implement several protection control laws as pitch axis and AoA protections for the longitudinal control of the aircraft. These control laws are activated when the measured AoA exceeds a certain critical value, to prevent potentially dangerous aircraft stall situations. The measurements of VCAS are used in gain-scheduling-based longitudinal control laws, where VCAS is one of the main scheduling variables. Therefore, inaccurate measurements of VCAS may negatively influence the longitudinal control performance of the aircraft. Since the loss of accuracy in the measurements of air data sensors for the AoA and VCAS may degrade the overall aircraft control performance, the timely and reliable detection and isolation of air data sensor malfunctions (faults) is an important task to be fulfilled for the satisfactory operation of any aircraft.

The monitoring of air data sensor faults for AoA and VCAS is the second case study considered in this book. Specifically, we consider the synthesis of a robust FDI filter, which, besides the detection and isolation of AoA and VCAS sensor faults, additionally provides estimations of these additive faults. Provided these estimations are sufficiently accurate, they can be used for the reconstruction of the correct measurements, and thus for building virtual, model-based air data sensors. The reliability of any fault diagnosis system depends on its ability to avoid false alarms and missed detections in the presence of variabilities due to changes of operational conditions of the aircraft, parameter variations and inherent uncertainties. In this case study, we only consider a standard cruise situation of an aircraft flying at a constant altitude and with a constant speed (i.e., in straight and level flight). The main parametric variability during cruise is the variation of the aircraft mass m due to fuel consumption. The mass variation influences the aircraft dynamics behaviour, and, therefore, it is an important parameter when addressing robustness aspects of the FDD system for monitoring air data sensor faults. A collection of $N = 11$ linearized models described in Sect. 8.2.4 serves for the synthesis of a robust FDD system. This case study illustrates the use of linear synthesis approaches as basis of synthesis methodologies that are able to fully address robustness aspects.

Typical air data sensor malfunctions, which can be assimilated with additive fault signals are bias, drift, frozen value, random or even oscillatory values. Such faults can be caused by (temporary) atmospheric influences as icing, or simply by dirt (e.g., due to dust or sand), which obturates the sensor's orifices. The basic difficulty in monitoring air data sensor faults is the need to discriminate the (assumed) additive faults from the additive effects of wind disturbances. For example, a VCAS sensor measures the sum of the airspeed and wind speed, and therefore any additive fault can be also interpreted as a change in the wind input characteristics. Therefore, the only possibility to decouple the effects of the wind from an additive sensor fault is to account for the global effects of the wind on the aircraft dynamics, while sensor faults, being strictly localized, intervene only in the air data sensor outputs. Besides these physical constraints, the fault detection problem is even more challenging, because the decoupling of wind effects in the residual signals must be done in presence of varying aircraft operating conditions and parameters. The current industrial practice is to employ redundant measurements (e.g., triplex sensor redundancy) and to eliminate the wind effects (assuming they are the same on all sensors) by forming residual signals as pair-wise differences between the measured signals. An almost zero residual component jointly with two nonzero residual components is a strong indication for a single sensor fault, which can be thus easily located. Although, this voting-based approach for the detection of single sensor faults is perfectly robust with respect to all potential variabilities (e.g., unknown wind inputs and parameter uncertainties), its main limitation arises due to the ambiguity of interpreting the values of residual components when two or three sensors fail simultaneously.

In Sect. 8.2.1, we consider the synthesis of least-order robust LTI FDI filters in presence of mass variations. The usage of a single LTI FDI filter, to guarantee the robustness of the FDI performance for the whole range of mass variations, imposes some limitations of the amplitudes of input signals. This is why, in Sect. 8.2.2 we consider the mass as a scheduling variable to be used for the synthesis of least-order robust LPV FDI filters. Specific aspects of the usage of the synthesized FDI filters in an industrially relevant FDD system for monitoring air data sensors are discussed in Sect. 8.2.3. The employed synthesis models consist of a collection of reduced order linearized civil aircraft models, which describe the aircraft longitudinal dynamics during cruise. Each component model corresponds to a specific value of aircraft mass. The underlying models are described in Sect. 8.2.4.

8.2.1 Robust LTI FDI Filter Synthesis

In this section, we consider the synthesis of a constant (LTI) filter for the detection and isolation of AoA and VCAS sensor faults for a civil aircraft. We assume the aircraft is in a cruise condition with constant altitude and speed, and the only parametric variability is the mass, which varies due to fuel consumption. We aim to synthesize a LTI FDI filter, whose fault detection and isolation performance is robust in presence of mass variations.

For synthesis purposes we use the multiple linearized models with additive sensor faults described in Sect. 8.2.4. Recall that the $N = 11$ linearized longitudinal aircraft models including the additive sensor faults have the state-space forms

$$\dot{x}^{(i)}(t) = A^{(i)}x^{(i)}(t) + B_u^{(i)}u(t) + B_d^{(i)}d(t),$$
$$y^{(i)}(t) = C^{(i)}x^{(i)}(t) + D_u^{(i)}u(t) + D_d^{(i)}d(t) + D_f f(t), \qquad (8.15)$$

where the dimensions of the vectors $x^{(i)}(t)$, $y^{(i)}(t)$, $u(t)$, $d(t)$ and $f(t)$ are respectively, $n = 4$, $p = 8$, $m_u = 3$, $m_d = 2$ and $m_f = 2$. The significance of the components of these vectors and the numerical values of the considered nominal model matrices (for $i = 7$) are given in Sect. 8.2.4. The employed models (8.15) are minimal and stable, therefore no stabilization is required to study the performance of the FDD system. The considered mass variations cover the whole range of mass values from a minimal value $m_{\min}$ to a maximum value $m_{\max}$. The i-th value of mass is given by $m_i = m_{\min} + \Delta m_i(m_{\max} - m_{\min})$, where Δm_i is the percentage of the mass variation from the total mass variation $m_{\max} - m_{\min}$.

In Listing 8.1 we present the Julia code used for the setup of the LTI aircraft models, with additive AoA and VCAS sensor faults, to be used to solve an EFDIP. Three input groups, `'c'` for control inputs, `'d'` for disturbance inputs and `'f'` for fault inputs, have been defined for the resulting LTI aircraft models with sensor faults **syssenf**.

Listing 8.8 Part 1 of script **CS2_1.jl**: Model setup

```julia
using FaultDetectionTools, DescriptorSystems, LinearAlgebra,
      JLD2, Optim, Test


# CS2_1  - Case-study example: Monitoring air data sensor faults
# Robust least order LTI synthesis
println("Case study CS2_1")


# Part 1 - Model setup
# load matrices of the aircraft multiple-model SYSACSM


cd(joinpath(pkgdir(FaultDetectionTools), "Examples"))
```

```julia
SYSACSM = load("cs2data.jld2", "SYSACSM")

# set dimensions
N = size(SYSACSM,1)            # number of models
p, m = size(SYSACSM[1].D)      # output and input dimensions
n = order(SYSACSM[1])          # state dimension
atol = 1.e-6                   # absolute tolerance
nom = 7;                       # index of nominal system
T = Float64                    # type of floating-point numbers

# set sensor indices for [AoA VCAS] and set dimensions of inputs
sen = [ 2, 4 ]
md = 2; mu = m-md; mf = length(sen)

# build minimal realizations of AC-models with sensor faults
idm = eye(p);
syssen = similar(Vector{DescriptorStateSpace},N)
for i = 1:N
    syssen[i] = [ SYSACSM[i] idm[:,sen]]
end

# build synthesis models with sensor faults
syssenf = fdimodset(syssen; c = Vector(1:mu),
                    d = mu.+Vector(1:md), f = (mu+md).+(1:mf))
```

For the synthesis of the FDD system for the isolation of air data sensor faults, we can solve an EFDIP for each of the N aircraft models by using the **Procedure EFDI** described in Sect. 5.4. To choose the desired fault signatures, we computed the strongly achievable specifications in the case of constant faults as

$$S_{strong} = \begin{bmatrix} 1 & 1 \\ 0 & 1 \\ 1 & 0 \end{bmatrix},$$

which shows that both AoA sensor faults and VCAS sensor faults can be strongly isolated. Therefore, the specification $S_{FDI} = I_2$ can be targeted in the synthesis.

Listing 8.2 presents the Julia code used for the determination of the achievable structure matrix S_{strong} containing the strong fault specifications. This matrix is the basis for the selection of the specifications contained in S_{FDI}. For the setting of the keyword arguments see the explanations for Listing 8.2.

Listing 8.9 Part 2 of script **CS2_1.jl**: Setup of synthesis specifications

```julia
# Part 2 - Setup of the synthesis specifications
# compute achievable strong specifications for constant faults
FDtol = 1.e-5       # weak fault detection threshold
FDGainTol = 0.0001  # strong fault detection threshold
FDfreq = 0          # frequency for strong detectability checks
S_strong = fdigenspec(syssenf[nom]; atol, FDtol, FDGainTol,
                FDfreq, sdeg = -0.05);

# check number of maximum possible specifications
@test size(S_strong,1) == 2^mf-1

# define SFDI, the signatures for isolation of single faults
SFDI = eye(mf) .> 0;
```

Using the structure matrix $S_{FDI} = I_2$, the **Procedure EFDI** has been performed for each of the component systems. Specifically, for the i-th system, we determined (a bank of) two fault detection filters $Q^{(i,1)}(s)$ and $Q^{(i,2)}(s)$, with scalar outputs, such that the i-th FDI filter is given by

$$Q^{(i)}(s) = \begin{bmatrix} Q^{(i,1)}(s) \\ Q^{(i,2)}(s) \end{bmatrix},$$

where $Q^{(i,1)}(s)$ is the filter for the isolation of AoA sensor faults (thus achieving the specification $[\,1\ 0\,]$), while $Q^{(i,2)}(s)$ is the filter for the isolation of VCAS sensor faults (thus achieving the specification $[\,0\ 1\,]$). It follows, that each of the N FDI filters $Q^{(i)}(s)$ achieves the strong isolation of AoA and VCAS sensor faults. For practical use (e.g., for a switching-based gain-scheduling approach with respect to estimated mass variations), this solution requires the storage of N LTI filters, which can also be used for interpolation purposes, in the case when the estimated mass lies between two grid points. The synthesized individual filters $Q^{(i)}(s)$, for $i = 1, \ldots, N$ can be seen as a reference synthesis providing the best achievable performance with respect to the chosen mass variation grid points.

Each $Q^{(i)}(s)$ of the designed N filters satisfies the two synthesis conditions

$$Q^{(i)}(s) \begin{bmatrix} G_u^{(i)}(s)\ G_d^{(i)}(s) \\ I_{m_u} \qquad 0 \end{bmatrix} = 0, \qquad Q^{(i)}(s) \begin{bmatrix} G_f^{(i)}(s) \\ 0 \end{bmatrix} = I_{m_f}, \qquad (8.16)$$

where the second condition is a fault estimation condition, which can be achieved by a suitable scaling of the component filters $Q^{(i,1)}(s)$ and $Q^{(i,2)}(s)$. The internal forms of the achieved filters for all $N = 11$ models are

$$R^{(i)}(s) = \left[\, R_u^{(i)}(s) \,\middle|\, R_d^{(i)}(s) \,\middle|\, R_f^{(i)}(s) \,\right] := Q^{(i)}(s) G_e^{(i)}(s), \qquad (8.17)$$

where $G_e^{(i)}(s)$ is the i-th extended system

$$G_e^{(i)}(s) = \left[\, \begin{array}{c|c|c} G_u^{(i)}(s) & G_d^{(i)}(s) & G_f^{(i)}(s) \\ I & 0 & 0 \end{array} \,\right], \qquad (8.18)$$

with inputs $\{u, d, f\}$ and outputs $\{y^{(i)}, u\}$.

Listing 8.3 illustrates the usage of the function **efdisyn**, which implements **Procedure EFDI** to solve the EFDIP. With the option for minimum degree scalar output design, enforced with the keyword argument **rdim** set to 1, all N resulting FDI filters are pure gains (i.e., without dynamics), thus in the state-space realization $(A_{Q^{(i)}}, B_{Q^{(i)}}, C_{Q^{(i)}}, D_{Q^{(i)}})$ of each $Q^{(i)}(s)$ only the direct feedthrough matrix $D_{Q^{(i)}}$ is non-empty. For example, for the nominal synthesis the resulted $Q^{(7)}(s)$ is

$$D_{Q^{(7)}} = \begin{bmatrix} -0.780 & 1 & 0.0040 & 0 & 0 & 0 & -44.77 & 0.0777 & 0.0107 & 0.0475 & 0.0107 \\ 27.31 & 0 & 0.4970 & 1 & 0 & 0 & 1561 & 119.7 & 1.3258 & 5.9119 & 1.3258 \end{bmatrix}.$$

Inspecting the numerical values of all $D_{Q^{(i)}}$, we observe that for the isolation of AoA sensor faults, no measurements from the VCAS sensor are used, because all $(2, 2)$ elements of $D_{Q^{(i)}}, i = 1, \ldots, N$ are zero. Similarly, for the isolation of VCAS sensor faults, no measurements from the AoA sensor are used, because all $(1, 4)$ elements of $D_{Q^{(i)}}$. Interestingly, no measurements from the measured ground speed components V_x and V_z are necessary to implement these filters, since the corresponding columns are zero too. The relatively large gains in the $(2, 7)$ elements of $D_{Q^{(i)}}$ is a clear indication of the already mentioned difficulties regarding the isolation of VCAS sensor faults. We note in passing, that if the minimum synthesis option is not used by setting the keyword argument **minimal = false** and leaving the keyword argument **rdim** unset, each of the resulting filter $Q^{(i)}(s)$ would have a state-space realization of order 8! This clearly illustrates the effectiveness of the employed synthesis techniques in obtaining low complexity FDI filters.

Remark 8.1 Incidentally, the computed zeroth-order solutions $Q^{(i)}(s)$, satisfying conditions (8.16), can be alternatively computed by solving, for each $D_{Q^{(i)}}$, the linear equation

$$D_{Q^{(i)}} \begin{bmatrix} C^{(i)} & D_u^{(i)} & D_d^{(i)} & D_f \\ 0 & I_{m_u} & 0 & 0 \end{bmatrix} = \begin{bmatrix} 0 & 0 & 0 & I_{m_f} \end{bmatrix},$$

where the $(p + m_u) \times (n + m_u + m_d + m_f)$ coefficient matrix, multiplying $D_{Q^{(i)}}$, is square and nonsingular. $\square$

Listing 8.10 Part 3 of script **CS2_1.jl**: Multiple filter synthesis using **Procedure EFDI**

```
# Part 3 - Multiple filter synthesis using Procedure EFDI

Q = similar(Vector{DescriptorStateSpace},N)
for i = 1:N
    Qi = efdisyn(syssenf[i], SFDI;   rdim = 1, FDfreq, FDGainTol,
             atol, sdeg = -5, smarg = -0.05)[1]
    Q[i] = vcat(Qi.sys...)
end
@test all(order.(Q) .== 0)
```

Listing 8.4 presents the Julia code used for the assessment of the synthesis results. The step responses of $R^{(i)}(s)$ defined in (8.17) are shown in Fig. 8.5, and clearly illustrate the fulfilment of conditions (8.16). The $\mathcal{H}_\infty$-norms of the error systems

$$E_{Q^{(i)}}^{(i)}(s) := Q^{(i)}(s)G_e^{(i)}(s) - \begin{bmatrix} 0 & 0 & I_2 \end{bmatrix} \tag{8.19}$$

are of the order $O(10^{-9})$ and, therefore, practically zero.

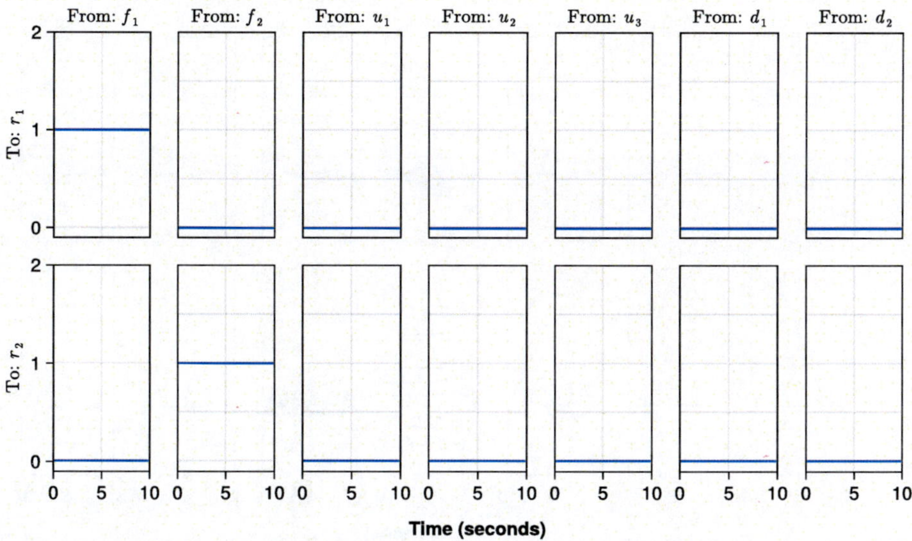

Fig. 8.5 Step responses of individual filter syntheses

Listing 8.11 Part 4 of script **CS2_1.jl**: Assessment of synthesis results

```
# Part 4 - Assesment of multiple filter synthesis results

# evaluate R[i] = Q[i]*[ Gu[i] Gd[i] Gf[i]; I 0 0]
# and determine Nref[i] = ||R[i]-[0 0 I]||∞ for i = 1, ..., N
syse = similar(Vector{DescriptorStateSpace},N)
Nref = similar(Vector{T},N)
for i = 1:N
    syse[i] = [syssenf[i].sys;eye(mu,mu+md+mf)]
    rezi = Q[i]*syse[i] - [zeros(mf,mu+md) eye(2)]
    Nref[i] = glinfnorm(rezi)[1]
end
```

A second analysis has been performed by simply taking a unique filter, which corresponds to the chosen nominal system, thus by setting $Q^{(i)}(s) = Q^{(7)}(s)$ for $i = 1, \ldots, N$. Listing 8.2 presents the Julia code used for assessment. The corresponding step responses of $R^{(i)}(s)$ defined in (8.17) are presented in Fig. 8.6 and show the lack of robustness of the nominal synthesis, which could therefore become a source of false alarms. This can be also seen by the large norms in Table 8.4 of

$$E_{Q^{(7)}}^{(i)}(s) := Q^{(7)}(s)G_e^{(i)}(s) - \begin{bmatrix} 0 & 0 & I_2 \end{bmatrix} \qquad (8.20)$$

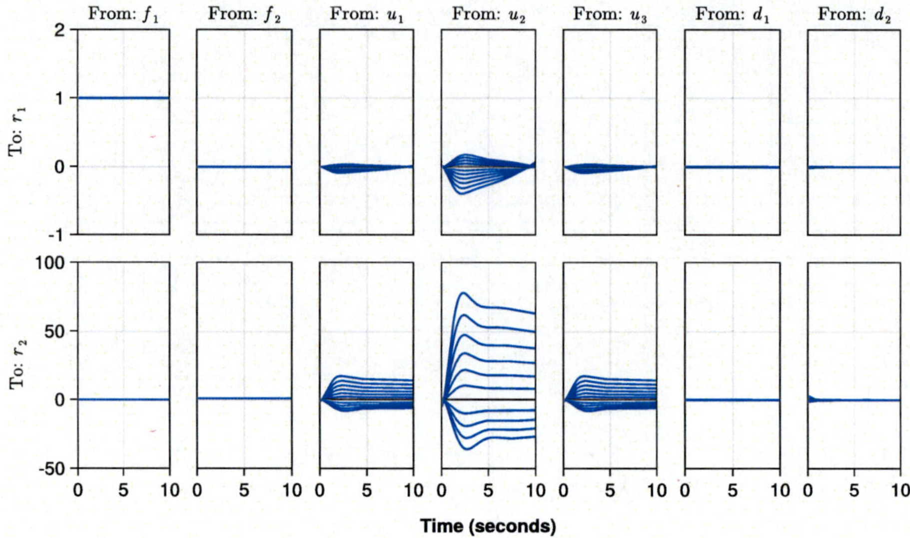

Fig. 8.6 Parametric step responses for the nominal filter synthesis

Table 8.4 Robustness analysis results for constant approximations

Model #	$\left\| E^{(i)}_{Q^{(i)}}(s) \right\|_\infty$	$\left\| E^{(i)}_{Q^{(7)}}(s) \right\|_\infty$	$\left\| E^{(i)}_{Q_0}(s) \right\|_\infty$	$\left\| E^{(i)}_{\overline{Q}_{opt}}(s) \right\|_\infty$
1	0	643.5602	504.8205	0.7730
2	0	530.5627	381.1323	0.7618
3	0	421.0268	262.3373	0.7542
4	0	313.3132	146.8795	0.7573
5	0	206.9658	34.7817	0.7692
6	0	102.2794	75.7578	0.7730
7	0	0.0000	180.9971	0.7680
8	0	98.9002	281.4598	0.7580
9	0	193.3942	376.5448	0.7489
10	0	282.5274	464.7577	0.7477
11	0	365.5045	545.6800	0.7727

Listing 8.12 Part 5 of script **CS2_1.jl**: Assessment of synthesis results for nominal synthesis

```
# Part 5 - Assesment of synthesis results for nominal synthesis

# evaluate R[i] = Q[nom]*[ Gu[i] Gd[i] Gf[i]; I 0 0]
# and determine Nnom[i] = ||R[i]-[0 0 I]||∞ for i = 1, ..., N
Nnom = similar(Vector{T},N)
for i = 1:N
    rezi = Q[nom]*syse[i] - [zeros(mf,mu+md) eye(2)]
    Nnom[i] = glinfnorm(rezi)[1]
end
```

A better constant approximation $Q_0(s)$ could be expected by approximating the elements of the direct feedthrough gains $D_{Q^{(i)}}$ with their constant mean value

$$D_{Q_0} = \sum_{i=1}^{N} D_{Q^{(i)}}/N. \tag{8.21}$$

Listing 8.3 presents the Julia code used for the determination of D_{Q_0} and the resulting filter assessment. The resulting feedthrough matrix D_{Q_0} is

$$D_{Q_0} = \begin{bmatrix} -0.7281 & 1 & 0.0049 & 0 & 0 & 0 & -41.77 & 0.2178 & 0.0132 & 0.0589 & 0.0132 \\ 27.43 & 0 & 0.4992 & 1 & 0 & 0 & 1568.4 & 110.8 & 1.3316 & 5.9380 & 1.3316 \end{bmatrix}$$

and no notable improvements with respect to the nominal filter can be seen when comparing the maximum norms of the errors $E_{Q^{(7)}}^{(i)}(s)$ and $E_{Q_0}^{(i)}(s)$ in Table 8.4. The corresponding step responses of $R^{(i)}(s)$ defined in (8.17) with $Q^{(i)}(s) = Q_0(s)$ for $i = 1, \ldots, N$ are also very similar to those in Fig. 8.6 and show the lack of robustness of this synthesis.

Listing 8.13 Part 6 of script **CS2_1.jl**: Assessment of synthesis results for mean value gains

```julia
# Part 6 - Assesment of synthesis results for mean value gains

# compute mean values of gains
D0 = similar(Matrix{T},mf,p+mu)
for i = 1:mf
    for j = 1:p+mu
        D0[i,j] = sum([Q[k].D[i,j] for k in 1:N])/N
    end
end

# evaluate R[i] := D0*[ Gu[i] Gd[i] Gf[i]; I 0 0]
# and determine N0[i] = ||R[i]-[0 0 I]||∞ for i = 1, ..., N
N0 = similar(Vector{T},N)
for i = 1:N
    rezi = D0*syse[i] - [zeros(mf,mu+md) eye(2)]
    N0[i] = glinfnorm(rezi)[1]
end
```

A more uniform performance in terms of error norms can be potentially achieved by solving for a constant filter $\overline{Q}(s) = D_{\overline{Q}}$ the min-max synthesis problem

$$D_{\overline{Q}_{opt}} := \arg\min_{\overline{Q}} \max_{i=1:N} \left\| E_{\overline{Q}}^{(i)}(s) \right\|_\infty, \tag{8.22}$$

where $E_{\overline{Q}}^{(i)}(s) := D_{\overline{Q}}G_e^{(i)}(s) - \begin{bmatrix} 0 & 0 & I_2 \end{bmatrix}$. This is a non-smooth convex optimization problem, which can be solved using adequate optimization tools, as the non-smooth optimization-based tuning tool **systune**, available in the *Control System Toolbox* of MATLAB [88]. Since no such tool is available among the optimization tools of Julia, a possible approach to solve this optimization problem is to use a global search-based tool (e.g., as available in the **Evolutionary.jl** package) to determine a good initialization for a local search-based tool (e.g., as available in the **Optim.jl** package).

Listing 8.5 illustrates the use of the gradient-free Nelder-Mead local search-based method, available in the **Optim.jl** package, to determine the optimal constant gain FDI filter, for the given multiple model (8.15) of the linearized longitudinal aircraft dynamics. For the initialization of the local search, we used the optimal solution $D_{\overline{Q}_0}$ computed with **systune** (see [161, p. 233]), rounded to two exact decimal figures. Taking into account that the (1, 2)th, (1, 4)th, (2, 2)th and (2, 4)th entries of $D_{\overline{Q}}$ must have the constant values 1, 0, 0 and 1, respectively, we only performed the optimization for the rest of entries. The optimization criterion $J(D_{\overline{Q}}) := \max_{i=1:N} \left\| E_{\overline{Q}}^{(i)}(s) \right\|_\infty$ has been implemented in the internal function **maxerror**.

Listing 8.14 Part 7 of script **CS2_1.jl**: Optimal constant filter synthesis using multiple models

```julia
# Part 7 - Multiple model synthesis of a constant gain

# optimal solution computed with MATLAB's SYSTUNE
# rounded to two exact decimal figures
Dopt0 = [
-0.79 1 -0.03 0 0.01 -0.25 4.2 0.39 -0.15 -0.08 0.03
 0.06 0 -0.02 1 -1.42 -0.01 4.6 0.27 -0.03 -0.10 -0.02
]

# filter gains set to false are fixed
mask = trues(mf,mu+p)
mask[1,2] = false; mask[1,4] = false
mask[2,2] = false; mask[2,4] = false

# define optimization function
function maxerror(x)
    err = 0.0
    D = similar(Matrix{T},mf,mu+p)
    D[mask] = x
    D[.!mask] = Dres
    for i in 1:N
        rezi = D*syse[i] - [zeros(mf,mu+md) eye(2)]
        err = max(err, glinfnorm(rezi)[1])
    end
    return err
end

# initialize with Dopt0
Dinit = Dopt0
Dres = Dinit[.!mask]

# use the gradient-free Nelder-Mead method
res = optimize(maxerror, Dinit[mask], NelderMead(),
               Optim.Options(iterations = 20000))

println("Ninit = $(maxerror(Dinit[mask]))")
println("Nfinal = $(maxerror(Optim.minimizer(res)))")

# form optimal solution
Dopt = similar(Matrix{T},mf,mu+p)
Dopt[mask] = Optim.minimizer(res)
```

```
Dopt[.!mask] = Dres

# evaluate R[i] := Dopt*[ Gu[i] Gd[i] Gf[i]; I 0 0]
# and determine ||R[i]-[0 0 I]||∞ for i = 1, ..., N
Nopt = similar(Vector{T},N)
for i = 1:N
    rezi = Dopt*syse[i] - [zeros(mf,mu+md) eye(2)]
    Nopt[i] = glinfnorm(rezi)[1]
end
```

The resulting constant optimal filter has the feedthrough gain

$$
D_{\overline{Q}_{opt}} = \begin{bmatrix} -0.8021 & 1.0 & -0.0361 & 0.0 & 0.0075 & -0.2491 \\ 0.0594 & 0.0 & -0.0134 & 1.0 & -1.4178 & -0.0064 \end{bmatrix} \cdots
$$

$$
\begin{bmatrix} 4.1252 & 0.4744 & -0.1437 & -0.0755 & 0.0475 \\ 4.6900 & 0.2977 & -0.0274 & -0.1023 & -0.0187 \end{bmatrix}.
$$

The criterion values for the initial and final gains are $J(D_{\overline{Q}_0}) = 22.14659$ and $J(D_{\overline{Q}_{opt}}) = 0.7731$, respectively. The corresponding norms of the errors $E_{\overline{Q}_{opt}}^{(i)}(s)$, given in Table 8.4, indicate a substantial improvement of performance robustness with respect to the previous LTI syntheses.

The step responses for the obtained optimal constant filter $\overline{Q}_{opt}(s)$ are shown in Fig. 8.7, where a much improved robustness is apparent. However, while the robustness of isolation, and even of estimation, of the AoA faults seems to be satisfactory for moderate size deflections of the control surfaces, the isolation of VCAS sensor faults in the presence of large wind amplitudes in the (longitudinal) x-axis could impose strong limitations on the minimum amplitude of detectable faults. An additional aspect worth mentioning is that the measurements of VCAS are not used for the isolation of AoA sensor faults (i.e., the $(1, 4)$th entry of $D_{\overline{Q}_{opt}}$ is zero), and vice versa, the measurements of AoA sensors are not used for the isolation of VCAS sensor faults (i.e., the $(2, 2)$th entry of $D_{\overline{Q}_{opt}}$ is zero). These desirable features have been enforced, by exempting these entries from the optimization.

Several concluding remarks regarding the synthesis of a LTI FDI filter are appropriate at this point.

1. The performed analyses provide strong indications that a zeroth-order synthesis can be employed to solve the FDIP for monitoring AoA and VCAS sensor faults. More insight into the synthesis procedure can be obtained by using alternatively a simple-basis-based approach for the synthesis of individual filters. This can be achieved by setting the keyword argument **simple = true** in Listing 8.3 and keeping the keyword argument **rdim** unset. For example, the resulting $Q^{(i,1)}$ consists of 5 basis vectors of degrees $\{1, 1, 1, 1, 0\}$ and $Q^{(i,2)}$ has a completely similar structure. It is easy to choose first-order designs for both $Q^{(i,1)}$ and $Q^{(i,2)}$ by simply selecting suitable basis vectors. The overall design $Q^{(i)}$ has thus order two and its dynamics can be set arbitrarily fast. It is interesting to note that although the individual second-order designs provide almost the same FDI performance

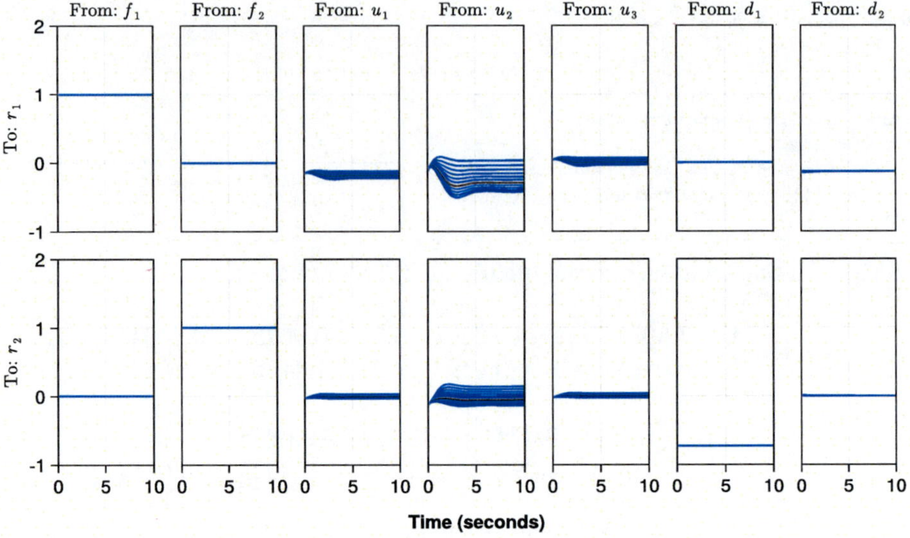

Fig. 8.7 Parametric step responses for the optimal constant filter synthesis

(e.g., when inspecting the step responses of the internal form of the FDI filter), the overall performance in terms of the $\mathcal{H}_\infty$-norms is not better than for a zeroth-order synthesis.

2. The insight obtained by employing the **Procedure EFDI** for the synthesis of zeroth (least)-order FDI filters can be fully exploited to simplify the computation of constant filter feedthrough gains $D_{Q^{(i)}}$ (see Remark 8.1). These gains are uniquely determined and therefore their smooth variations reflect the continuous dependence of model data on the mass variation. As it will be shown in the next section, the smooth dependence of gains on mass is instrumental in obtaining LPV gain-scheduling-based filters by using standard interpolation techniques.

3. The amplitudes of step responses in Fig. 8.7 can be used to roughly estimate the maximum amplitudes of inputs, which still allow the robust detection and isolation of faults of given amplitudes. For example, to detect an AoA sensor fault of amplitude $f_{1,\min}$, the maximum amplitudes of step signals of control inputs must be less than $|u_1| < f_{1,\min}/0.14$, $|u_2| < f_{1,\min}/0.52$, $|u_3| < f_{1,\min}/0.22$, while the disturbance input $|d_2| < f_{1,\min}/0.13$. For $f_{1,\min} = 4$ deg, the maximum amplitudes for the system inputs u_1, u_2, u_3 are, respectively, 33.3, 7.9 and 18.2°, while the maximum amplitude of the disturbance input d_2 is 30 kts. This will allow the robust isolation of AoA faults practically for all physically meaningful wind conditions. Since the two elevators are simultaneously actuated during normal operation, their allowed maximum amplitudes must be halved. To cope with the limited range of allowed elevator and stabilizer deflections, fault isolation can be performed, without false alarms, only during low control activities of these surfaces. Similarly, to detect a VCAS sensor fault of amplitude $f_{2,\min}$, the main limitation arises from the maximum amplitude of the disturbance input $|d_1| < f_{2,\min}/0.73$.

For $f_{2,\min} = 20$ kts, the maximum amplitude of the disturbance input d_1 is about 28 kts. For practical situations, this value may impose a severe limitation on the minimum size of robustly detectable VCAS sensor faults. The use of gain-scheduling techniques (see next section) may substantially improve the robustness of isolation of VCAS sensor faults.

4. For the synthesis of a robust LTI FDI filter it is possible to alternatively employ the **Procedure AFDI**, to solve a suitably formulated AFDIP with fictitious noise inputs generated to account for the effects of mass variations. Also, **Procedure AMMS** can be employed to solve an AMMP with $M_r(s) = I_2$ as reference model.

8.2.2   Robust LPV FDI Filter Synthesis

We show in this section that the robustness of the FDD system can be significantly improved by employing gain-scheduling techniques with respect to mass. This approach is realistic, since there exist several weight and balance calculation procedures, which can produce sufficiently accurate mass estimations during the whole flight of an aircraft. Therefore, instead of using a constant LTI FDI filter $Q(s)$, it is possible to use a parameter dependent filter $Q(s, \Delta m)$, whose matrices in its state-space realization depend smoothly on the actual mass variation Δm. We discuss two methods to determine a parameter dependent feedthrough matrix $D_Q(\Delta m)$ to serve for interpolating at arbitrary values of mass.

The first method relies on building polynomial approximations of the entries of $D_Q(\Delta m)$ by using the N values $D_{Q^{(i)}}$ corresponding to mass variation values Δm_i, for $i = 1, \ldots, N$. The method is a simple parameter fitting approach, which assumes a k-th degree polynomial form

$$D_{Q_k}(\Delta m, \theta) := \sum_{j=0}^{k} D_{Q_{k,j}}(\Delta m)^j , \tag{8.23}$$

with θ a $m_f \times (k+1)(p+m_u)$ parameter matrix defined as

$$\theta = \begin{bmatrix} D_{Q_{k,0}} & \cdots & D_{Q_{k,k}} \end{bmatrix} . \tag{8.24}$$

The best fitting determines the coefficient matrices $D_{Q_{k,0}}, \ldots, D_{Q_{k,k}}$ by solving for the optimal value θ_{opt} of θ, the linear least-squares problem

$$\min_{\theta} \left\| \begin{bmatrix} D_{Q_k}(\Delta m_1, \theta) - D_{Q^{(1)}} \\ \vdots \\ D_{Q_k}(\Delta m_N, \theta) - D_{Q^{(N)}} \end{bmatrix} \right\|_F ,$$

Table 8.5 Robustness analysis results of interpolation-based approximations

Model #	$\left\|E_{Q_0}^{(i)}(s)\right\|_\infty$	$\left\|E_{Q_1}^{(i)}(s)\right\|_\infty$	$\left\|E_{Q_2}^{(i)}(s)\right\|_\infty$	$\left\|E_{Q_3}^{(i)}(s)\right\|_\infty$	$\left\|E_{Q_4}^{(i)}(s)\right\|_\infty$
1	504.82	138.619	16.3228	0.8877	0.0233
2	381.13	45.078	3.5607	0.6994	0.0722
3	262.33	18.410	10.4113	0.6290	0.0059
4	146.88	57.084	9.6294	0.1467	0.0701
5	34.78	75.102	5.1293	0.2185	0.0571
6	75.75	75.757	0.3033	0.3033	0.0175
7	180.99	61.952	4.6064	0.1937	0.0828
8	281.46	36.329	6.6122	0.0755	0.0698
9	376.54	1.473	5.6031	0.1436	0.0331
10	464.75	40.811	1.3518	0.1271	0.1225
11	545.68	88.019	6.0316	0.2083	0.0670

where $\|\cdot\|_F$ denotes the Frobenius matrix norm.

We computed for $k = 0, \ldots, 4$ the optimal k-th degree approximations of the form (8.23) and with $Q_k(s, \Delta m) := D_{Q_k}(\Delta m, \theta_{opt})$ we evaluated the corresponding $\mathcal{H}_\infty$-norms of $E_{Q_k}^{(i)}(s) := Q_k(s, \Delta m_i)G_e^{(i)}(s) - \begin{bmatrix} 0 & 0 & I_2 \end{bmatrix}$. As it can be seen in Table 8.5, the fourth-order polynomial gain approximation provides acceptable performance robustness in terms of the norms $\|E_{Q_4}^{(i)}(s)\|_\infty$, which is near to the performance of the reference solution in all mass grid points.

The step responses of $R^{(i)}(s)$ defined in (8.17) for $Q^{(i)}(s) = Q_4^{(i)}(s)$ are shown in Fig. 8.8, and clearly illustrate that the fourth-order polynomial approximation provides a satisfactory robustness with respect to mass variations, without any noticeable difference to the reference solution.

Listing 8.8 presents the Julia code used for determining a fourth-order polynomial fitting of the individual elements of the filter feedthrough matrix and for the assessment of the robustness of approximations by computing the corresponding error norms $\|E_{Q_k}^{(i)}(s)\|_\infty$. Note that the simplified approach described in Remark 8.1 to determine the filter gains of individual syntheses is also included. This code fully exploits the structural features of the formulated FDIP, by using the insight obtained by applying **Procedure EFDI**.

Listing 8.15 Part 3 of script `CS2_2.jl`: LPV FDI filter synthesis using polynomial data fitting

```
# Part 3 - LPV synthesis using polynomial gain interpolation

using Polynomials
# form extended systems [ Gu[i] Gd[i] Gf[i]; I 0 0]
# and determine constant solutions based on Remark 8.1
syse = similar(Vector{DescriptorStateSpace},N)
D = zeros(T,mf,p+mu,N)
Dref = [zeros(T,mf,n+mu+md) eye(T,mf)]
```

```julia
for i = 1:N
    syse[i] = [syssenf[i].sys;eye(mu,mu+md+mf)]
    cde = [syse[i].C syse[i].D]
    D[:,:,i] = Dref/cde
end


maxdeg = 4      # maximum interpolation order
massi = load("cs2data.jld2", "massi")
Nintp = similar(Matrix{T},N,maxdeg+1)
Qpol = Polynomial.(zeros(T,mf,p+mu))
for deg = 0:maxdeg
    for i = 1:mf
        for j = 1:p+mu
            Qpol[i,j] = fit(view(massi,:), view(D,i,j,:), deg)
        end
    end
    for i = 1:N
        rezi = (massi[i].|>Qpol)*syse[i]-[zeros(mf,mu+md) eye(2)]
        Nintp[i,deg+1] = glinfnorm(rezi)[1]
    end
end
```

The excellent results obtained previously rely on the important assumption of a smooth variation of the underlying data $D_{Q^{(i)}}$, $i = 1, \ldots, N$, which are used for interpolation purposes. This condition is not always fulfilled, especially, in the cases when the numerically computed solutions are not unique. The presence of outliers in the computed results may make the interpolation task difficult, if not impossible. Special techniques, as for example, variable scaling, data normalization, transformations to special coordinate forms, etc., are necessary to enforce the continuity of the

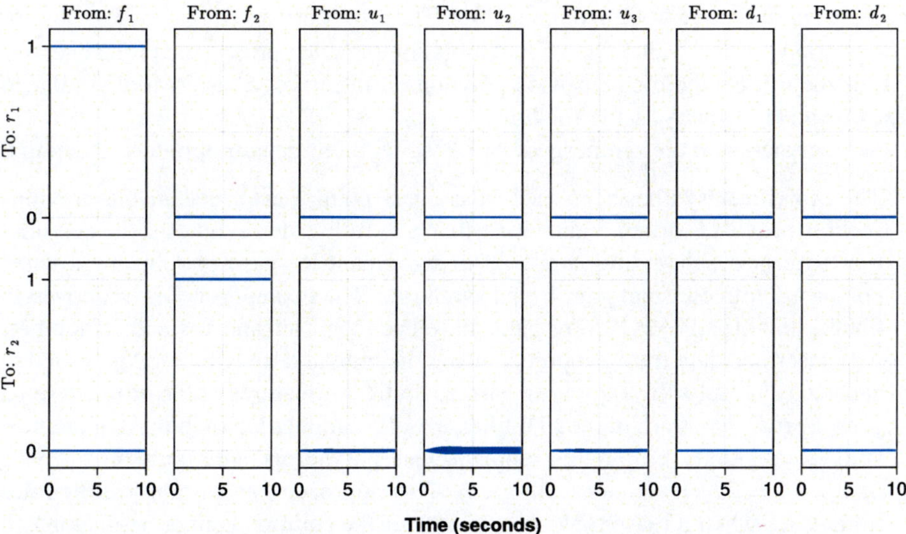

Fig. 8.8 Parametric step responses for LPV-filter synthesis using polynomial interpolation

mapping, provided by the employed computational method, from the input data to the computed results.

A more suitable solution to the synthesis of LPV FDI filters is to use tools, which find the best fitting of the LPV filter parameters working on the original multiple model. For this purpose, consider the normalized grid parameters $x_i = 2(\Delta m_i - 0.5)$ and define the parameterized gain as

$$\widetilde{D}_{Q_k}(x, \theta) := \sum_{j=0}^{k} \widetilde{D}_{Q_{k,j}} x^j , \qquad (8.25)$$

where the LPV filter parameters are collected in

$$\theta = \begin{bmatrix} \widetilde{D}_{Q_{k,0}} & \cdots & \widetilde{D}_{Q_{k,k}} \end{bmatrix} . \qquad (8.26)$$

To determine the optimal parameters, the min-max optimization problem

$$\min_{\theta} \max_{i=1:N} \left\| E_{\widetilde{Q}_k}^{(i)}(s, x_i, \theta) \right\|_{\infty}, \qquad (8.27)$$

can be solved, where $E_{\widetilde{Q}_k}^{(i)}(s, x, \theta) := \widetilde{Q}_k(s, x, \theta) G_e^{(i)}(s) - \begin{bmatrix} 0 & 0 & I_2 \end{bmatrix}$ and $\widetilde{Q}_k(s, x, \theta) = \widetilde{D}_{Q_k}(x, \theta)$, with $\widetilde{D}_{Q_k}(x, \theta)$ defined in (8.25).

In [161] we used the MATLAB tuning tool **systune**, available in the MATLAB *Control System Toolbox*, for determining a third-order polynomial LPV filter by solving the min-max optimization problem (8.27). This function allows to optimally fit the free parameters (e.g., θ in (8.24)) of gain-scheduling system objects using a multiple-model-based optimization setting. The error norms $\left\| E_{\widetilde{Q}_k}^{(i)}(s, x_i, \theta_{opt}) \right\|_{\infty}$ for the resulting optimal value θ_{opt} of θ exhibit comparable performance robustness as that of the 4th degree element-wise polynomial fitting of gains. A tool similar to **systune** is not yet available for Julia.

Some remarks on the synthesis of an LPV FDI filter are appropriate at this point.

1. The individual synthesis, in each mass grid point, performed in the previous section provided strong indications that a zeroth-order synthesis, in conjunction with gain-scheduling, can lead to significant increase of robustness when compared with constant gain-based syntheses. The two approaches employed to obtain suitable LPV FDI filter gains have each their advantages and difficulties. The described data fitting approach using the individually computed gains led to satisfactory results, by exploiting various particular features of the problem (e.g., zeroth-order dynamics of the FDI filter, simple computation of gains to guarantee smooth variations, employing element-wise polynomial interpolation for each of $2 \times 11 = 22$ elements of the gain). This approach has certainly difficulties in the case when a reasonable smoothness of the solution cannot be guaranteed (e.g., due to non-uniqueness of the solution) or when the solution has higher order dynamics. In this latter case, coordinate transformations may be necessary to enforce the continuous mapping of problem data to the results computed by

the synthesis algorithms. The second approach is based on direct tuning methods of fixed structure filter using non-smooth optimization-based tuning tools of the $(2 \times 11 - 4) \times 4 = 72$ free elements of θ in (8.26). This optimization-based approach has a considerable flexibility in choosing the nature of interpolation formulas employed for gain-scheduling, in handling model variabilities (e.g., via multiple models), or in imposing particular structure of the LPV coefficients (e.g., by fixing the values of certain coefficients and excepting them from the optimization). However, this approach needs a careful definition of the optimization problem by translating the synthesis conditions into meaningful optimization criteria or constraints. Also, due to the intrinsic local nature of the employed search techniques, finding initial starting points leading to the best (or even global) minima can be challenging.

2. An issue, which has not been addressed in the synthesis, is the robustness of the gain-scheduling-based filter with respect to estimation errors in the scheduling variable. Estimation errors, as large as 10% of the aircraft mass, are common and can significantly degrade the robustness performance. One way to address this issue is to include this uncertainty in the problem formulation, by requiring to determine the FDI filter, which best fits the uncertainties induced by inaccurate mass estimations. This can be seen as a non-standard formulation of the gain-scheduling-based synthesis and can be addressed, for example, by extending the number of models N, by including for each grid point several models corresponding to different nearby mass values.

3. The fault estimation aspect of the AoA and VCAS sensor faults has been solved in a robust way and allows the reconstruction of correct measurements in the case of failures. This permits the replacement of the failed sensors by virtual, model-based sensors, which can further be employed without impeding the control performance, even after a total failure of all air data sensors (a triple sensor redundancy is usually provided on civil aircraft for each sensor). This is an important aspect when solving air data sensor monitoring problems in the light of several incidents caused by such failures.

8.2.3 Monitoring Air Data Sensor Faults—Industrial Aspects

The current industrial practice for monitoring air data sensor faults is to employ redundant measurements, usually via a triplex redundancy for each AoA and VCAS sensor. The use of triplex sensor redundancy for isolation of single sensor faults has been already discussed in Example 5.10, which illustrates the perfect robustness of this (2-of-3) voting-based scheme. A *consolidated measurement* can be determined as the mean value of three or two healthy sensor measurements. The main limitation of this scheme is due to the potentially erroneous interpretation of the healthy and non-healthy measurements, in the case when two sensors fail simultaneously and produce almost the same wrong measurements. This case may occur, for example, if two AoA sensors freeze similarly due to icing. In such a case, the two wrong measurements are (erroneously) "diagnosed" by the voting-based FDD system as healthy. Their use to form a consolidated measurement may lead to unpredictable effects. Also, inconsistencies in the VCAS measurements (e.g., also caused by icing or dirt) may negatively impact the performance of the aircraft control system.

To enhance the fault isolation capabilities of presently employed voting-based schemes, an alternative monitoring scheme can be employed, where each measurement of AoA and VCAS is individually monitored by a low-order FDI filter. The consolidated AoA and VCAS measurements can be computed as the mean values of all healthy AoA and VCAS sensors, respectively.

In the case of monitoring the AoA sensors, the filter corresponding to the first row of the optimal $\overline{Q}_{opt}(s)$ determined in Sect. 8.2.1 can be employed to build three (identical) zeroth-order FDI filters to monitor the AoA measurements. These filters thus individually detect and isolate the three AoA sensor faults. All healthy sensor measurements can be used for signal consolidation, and even the measurement of a single healthy sensor can be reliably employed. Moreover, in the case of a total failure of all AoA sensors, a virtual sensor can be built as

$$\widehat{\alpha}(t) = \alpha(t) - \hat{f}_1(t),$$

where $\alpha(t)$ is the actual (faulty) AoA measurement and $\hat{f}_1(t)$ is the estimated AoA fault produced by the AoA FDI filter.

Similarly, in the case of monitoring the VCAS sensors, the filter corresponding to the second row of the optimal $\overline{Q}_{opt}(s)$ determined in Sect. 8.2.1 can be employed to build three (identical) zeroth-order FDI filters to monitor the VCAS measurements. However, if sufficiently accurate mass estimations are available, then better FDD performance robustness can be achieved by using—for example, the LPV FDI filter $Q_4(s, \Delta m)$ determined in Sect. 8.2.2, to build gain-scheduling-based FDI filters for the VCAS sensors. In the case of a total failure of all VCAS sensors, a virtual sensor can be built to provide a corrected VCAS measurement $\widehat{V}_c(t)$ as

$$\widehat{V}_c(t) = V_c(t) - \hat{f}_2(t),$$

where $V_c(t)$ is the actual (faulty) VCAS measurement and $\hat{f}_2(t)$ is the estimated VCAS fault produced by the VCAS FDI filter.

The robustness of the FDD system for air data sensor faults during cruise can be enhanced, by switching between several filters, which are robust on smaller ranges of mass variations. This can alleviate the potential difficulties related to inaccuracies in mass estimation. Also, robustness with respect to other typical variabilities for a cruise condition such as limited variations in altitude and speed or variations of the centre of gravity position can be enforced by using the described synthesis techniques. Covering all flight regimes and manoeuvres (e.g., take-off, landing, etc.) across the flight envelope of a civil aircraft can be done by combining several possible approaches, as for example, switching between several robust FDI filters, which cover limited ranges of flight altitudes, or mass variations, or both.

8.2.4 Linearized State-Space Models with Additive Sensor Faults

Each of the linearized models (8.15) with additive sensor faults has been obtained by extracting a minimal realization of the longitudinal dynamics of a linearized aircraft model corresponding to a standard cruise situation (i.e., in straight and level flight). The original linearized aircraft models are the same as that used in Sect. 8.1.4, however with a different set of output measurements and a reduced set of inputs. No actuator models have been included and it was assumed that the control inputs in the model are the corresponding deflections of the attached control surfaces. The linear multiple models of the aircraft have been obtained by the linearization of a nonlinear aircraft model in a specific flight condition (i.e., cruise with constant speed V_0 at constant altitude h_0) and for specific values of certain aircraft parameters (i.e., N values m_i, $i = 1, \ldots, N$ of the aircraft mass m and a constant value of the relative position of its centre of gravity X_{cg}). The numerical values of the entries of the system matrices of any linearized aircraft model depend on such parameters. The employed nonlinear model describes a generic longitudinal model of a civil aircraft with 3 primary control surfaces (2 elevators, 1 stabilizer). The linearized aircraft models with additive sensor faults have the multiple-model form (see Sect. 2.2.3)

$$
\begin{aligned}
\dot{x}^{(i)}(t) &= A^{(i)}x^{(i)}(t) + B_u^{(i)}u(t) + B_d^{(i)}d(t), \\
y^{(i)}(t) &= C^{(i)}x^{(i)}(t) + D_u^{(i)}u(t) + D_d^{(i)}d(t) + D_f f(t),
\end{aligned}
\tag{8.28}
$$

where the dimensions of vectors $x^{(i)}(t)$, $y^{(i)}(t)$, $u(t)$, $d(t)$ and $f(t)$ are respectively, $n = 4$, $p = 8$, $m_u = 3$, $m_d = 2$ and $m_f = 2$. These variables approximate small deviations of the system variables of the aircraft nonlinear model from their equilibrium (or trim) values. The models (8.28) correspond to the $N = 11$ values of the mass parameter, covering the whole range of variation of the mass $[m_{\min}, m_{\max}]$, where $m_{\min}$ and $m_{\max}$ are the smallest and largest values of mass, respectively. The employed grid of mass values is defined as $m_i = m_{\min} + \Delta m_i (m_{\max} - m_{\min})$ for $i = 1, \ldots, N$, where Δm_i is the percentage of mass variation from its whole range. The chosen nominal model corresponds to $\Delta m_7 = 60\%$ of mass variation. For all linearized models, the rest of parameters are constant and equal to their nominal values.

The system variables in the state-space model (8.28) are defined as follows:

$$
y^{(i)} = \begin{pmatrix} pitch\ angle \\ angle\ of\ attack \\ pitch\ rate \\ calibrated\ airspeed \\ ground\ speed\ X\ axis \\ ground\ speed\ Z\ axis \\ acceleration\ X\ axis \\ acceleration\ Z\ axis \end{pmatrix}, \quad x^{(i)} = \begin{pmatrix} residual\ quaternion\ component \\ ground\ speed\ X\ body\ axis \\ ground\ speed\ Z\ body\ axis \\ pitch\ rate \end{pmatrix},
$$

$$
u = \begin{pmatrix} right\ elevator\ deflection \\ stabilizer\ trim\ angle \\ left\ elevator\ deflection \end{pmatrix}, \quad d = \begin{pmatrix} wind\ speed\ X\ axis \\ wind\ speed\ Z\ axis \end{pmatrix},
$$

$$
f = \begin{pmatrix} angle\ of\ attack\ sensor\ fault \\ calibrated\ airspeed\ sensor\ fault \end{pmatrix}.
$$

The basis of the longitudinal aircraft multiple model (8.28) was the same aircraft multiple model as that used for obtaining the multiple model in (8.13). However, this time, a different set of output measurements have been used, which, besides the monitored measurements of AoA and VCAS sensors, includes all relevant longitudinal measurements, which are independent of the measurements of air data sensors. A reduced set of 5 state variables resulted by eliminating all unobservable states of the full model (8.28) with the selected set of measurements. The final models with 4 state equations have been obtained by eliminating an uncontrollable eigenvalue in the origin. This was done by forming a *residual quaternion component* as a linear combination of the 2nd and 4th quaternion components.

The matrices of the nominal state-space model (8.28), for $i = 7$, have the following numerical values:

$$
A^{(7)} = \begin{bmatrix} 0 & 0 & 0 & 0.5 \\ -19.61 & -0.003 & 0.061 & -9.0618 \\ -0.8893 & -0.0777 & -0.8004 & 197.8868 \\ 0 & 0.0001 & -0.0073 & -0.5290 \end{bmatrix},
$$

$$
B_u^{(7)} = \begin{bmatrix} 0 & 0 & 0 \\ 0.0021 & 0.0094 & 0.0021 \\ -0.1360 & -0.6063 & -0.1360 \\ -0.0280 & -0.1250 & -0.0280 \end{bmatrix}, \quad B_d^{(7)} = \begin{bmatrix} 0 & 0 \\ 0.0001 & -0.0314 \\ 0.0586 & 0.4097 \\ 0.0001 & 0.0038 \end{bmatrix},
$$

$$
C^{(7)} = \begin{bmatrix} 114.6505 & 0 & 0 & 0 \\ 0 & -0.0129 & 0.2847 & 0 \\ 0 & 0 & 0 & 57.2958 \\ 0 & 1.4210 & 0.0644 & 0 \\ 0 & 0.9995 & 0.0453 & 0 \\ -402.2337 & -0.0453 & 0.9995 & 0 \\ -1.9990 & -0.0003 & 0.0062 & 0.0046 \\ -0.0907 & -0.0079 & -0.0816 & -0.2977 \end{bmatrix}, \quad D_f = \begin{bmatrix} 0 & 0 \\ 1 & 0 \\ 0 & 0 \\ 0 & 1 \\ 0 & 0 \\ 0 & 0 \\ 0 & 0 \\ 0 & 0 \end{bmatrix},
$$

$$
D_u^{(7)} = \begin{bmatrix} 0 & 0 & 0 \\ 0 & 0 & 0 \\ 0 & 0 & 0 \\ 0 & 0 & 0 \\ 0 & 0 & 0 \\ 0 & 0 & 0 \\ 0.0002 & 0.0010 & 0.0002 \\ -0.0139 & -0.0618 & -0.0139 \end{bmatrix}, \quad D_d^{(7)} = \begin{bmatrix} 0 & 0 \\ 0 & -0.1467 \\ 0 & 0 \\ -0.7321 & 0 \\ 0 & 0 \\ 0 & 0 \\ 0 & -0.0032 \\ 0.0060 & 0.0418 \end{bmatrix}.
$$

Each component model of the linearized aircraft multiple model (8.28) is stable. For example, the eigenvalues of the state matrix $A^{(7)}$ are

$$
\Lambda\big(A^{(7)}\big) = \{-0.6646 \pm 1.1951\mathrm{i}, \, -0.0016 \pm 0.0600\mathrm{i}\}.
$$

8.3   Notes and References

Section 7.1. For the considered two case studies of flight actuator and air data sensor faults, the underlying model is similar to several civil transport aircraft models. The global and local monitoring of primary flight actuator faults has been considered in [145] within the GARTEUR FM-AG16,[1] where the focus was on the synthesis of least-order FDI filters for a linearized nominal aircraft model. A similar work, performed within the IMMUNE Project,[2] has been reported in [149]. The monitoring of several types of actuator faults (jamming, oscillatory failure, disconnection) was one of the benchmark scenarios formulated in the European Union funded project ADDSAFE.[3] Results on the local (actuator level) monitoring of flight actuator faults

---

[1] The European Flight Mechanics Action Group FM-AG(16) on Fault Tolerant Control was established in 2004 and concluded in 2008. It represented a collaboration involving thirteen European partners from industry, universities and research establishments under the auspices of the Group for Aeronautical Research and Technology in Europe (GARTEUR) program.

[2] IMMUNE (Intelligent Monitoring and Management of Unexpected Events) was a ONERA-DLR Joint Research Project started in 2006 and concluded in 2009.

[3] ADDSAFE (Advanced Fault Diagnosis for Safer Flight Guidance and Control) was a EU-funded project in the framework programme FP7, started in 2009 and finished in 2012; see https://cordis.europa.eu/project/id/233815.

have been reported, for example, in [105, 158, 159, 178]. The monitoring of a specific type of flight actuator faults (oscillatory failure case) for an AIRBUS A380 aircraft is discussed in [59], while the AIRBUS practice to address actuator and sensor fault monitoring problems in conjunction with control reconfiguration techniques is described in [60].

Section 8.2. The monitoring of simultaneous air data sensor faults for the AoA and VCAS was one of the benchmark scenarios formulated in the European Union funded project RECONFIGURE.[4] Preliminary results for monitoring the AoA sensor faults have been presented in [160]. The individual monitoring of triplex AoA sensors has been earlier proposed in [104], by using measurements of the VCAS sensor. The reconstruction of air data sensor faults has been also addressed in [1] by using sliding mode techniques. Here, the sensor faults are reformulated as (fictitious) actuator faults for which unknown-input estimation techniques are employed. The resulting filters have orders, which exceed the order of the underlying aircraft model.

---

[4] RECONFIGURE (Reconfiguration of Control in Flight for Integral Global Upset Recovery) was a EU-funded project in the framework programme FP7, started in 2013 and finished in 2016; see https://cordis.europa.eu/project/id/314544.

Part III
Background Material

Part III of the book includes a substantial amount of background material on rational matrices, descriptor systems and computational algorithms for descriptor systems. It starts with the presentation of the theoretical backgrounds of rational matrices and descriptor systems. The presentation exhibits a certain parallelism between the two main sections, intended to ease introducing or recalling to readers the main theoretical concepts. The final chapter starts with the presentation of an overview of numerical linear algebra techniques, which form the basic computational layer for the developed synthesis procedures. These algorithms cover the computation of several important matrix decompositions and condensed forms, as well as the solution of several classes of linear and quadratic matrix equations. The next computational layer consists of several algorithms for descriptor systems, which are instrumental for the developed synthesis procedures. Most of these algorithms have been presented for the first time in the first edition of this book. Their presentations are also included in this edition and partly serve to document the accompanying numerical software, consisting of the **MatrixEquations**, **MatrixPencils** and **DescriptorSystems** packages, which were implemented by the author as part of the second edition book project.

Chapter 9
System Theoretical Concepts

Rational transfer function matrices and linear time-invariant descriptor systems are the two main system representations employed in this book. The purpose of this chapter is to provide a concise statement of the main results on rational matrices and descriptor systems to facilitate the understanding of the mathematical terminology used throughout the book. The treatment in depth of most of the concepts was not possible in the restricted size of this book. Therefore, at the end of this chapter, we indicate the main references covering the presented results and make suggestions for supplementary readings.

9.1 Rational Transfer Function Matrices

In this section, we succinctly present the main concepts and properties of *transfer function matrices* (TFMs) of linear time-invariant systems, which are relevant for this book, especially for the formulation of fault detection problems in Chap. 3 and for the development of synthesis procedures in Chap. 5. We consider only TFMs, which belong to the class of rational matrices for which each entry is expressed as a ratio of two polynomials in a complex indeterminate. The polynomials are assumed to have real coefficients. Polynomial matrices can be seen as a particular type of rational matrices. Many aspects for scalar rational functions, for example, poles and zeros, partial-fraction decompositions or coprime factorizations have nontrivial generalizations for polynomial and rational matrices. Many operations on standard matrices have nice generalizations for rational matrices. Straightforward generalizations are the rank, determinant and inverse. The conjugate transposition of a complex matrix generalizes to the conjugation of a rational matrix, while the inner–outer and spectral factorizations can be seen as generalizations of the familiar QR and Cholesky factorizations, respectively.

9.1.1 Transfer Functions

Transfer functions are used to describe the input–output behaviour of *single-input single-output* (SISO) *linear time-invariant* (LTI) systems by relating the input and output variables via a gain depending on a frequency variable. For a SISO system with input $u(t)$ and output $y(t)$ depending on the continuous time variable t, let $\mathbf{u}(s) := \mathcal{L}(u(t))$ and $\mathbf{y}(s) := \mathcal{L}(y(t))$ denote the Laplace transformed input and output, respectively. Then, the transfer function of the continuous-time LTI system is defined as

$$g(s) := \frac{\mathbf{y}(s)}{\mathbf{u}(s)}$$

and relates the input and output in the form

$$\mathbf{y}(s) = g(s)\mathbf{u}(s) \, . \tag{9.1}$$

The complex variable $s = \sigma + j\omega$ has for $\sigma = 0$ the interpretation of a complex frequency. If the time variable has a discrete variation with equally spaced values with increments given by a sampling-period T, then the transfer function of the discrete-time system is defined using the $\mathcal{Z}$-transforms of the input and output variables $\mathbf{u}(z) := \mathcal{Z}(u(t))$ and $\mathbf{y}(z) := \mathcal{Z}(y(t))$, respectively, as

$$g(z) := \frac{\mathbf{y}(z)}{\mathbf{u}(z)}$$

and relates the input and output in the form

$$\mathbf{y}(z) = g(z)\mathbf{u}(z) \, . \tag{9.2}$$

The complex variable z is related to the complex variable s as $z = e^{sT}$. We will use the variable λ to denote either the s or z variables, depending on the context, continuous- or discrete-time, respectively. Furthermore, we will restrict our discussion to rational transfer functions $g(\lambda)$, which can be expressed as a ratio of two polynomials with real coefficients

$$g(\lambda) = \frac{\alpha(\lambda)}{\beta(\lambda)} = \frac{a_m\lambda^m + a_{m-1}\lambda^{m-1} + \cdots + a_1\lambda + a_0}{b_n\lambda^n + b_{n-1}\lambda^{n-1} + \cdots + b_1\lambda + b_0} \, , \tag{9.3}$$

with $a_m \neq 0$ and $b_n \neq 0$. There are notable exceptions (e.g., continuous-time systems with time delays), which lead to nonrational transfer functions. In such cases, suitable approximations of the nonrational part can be used (e.g., rational Padé approximation) to arrive at a rational expression as above.

Let $\mathbb{R}(\lambda)$ be the set of real rational functions with real coefficients in an indeterminate λ, and let $\mathbb{R}[\lambda]$ be the set of polynomials with real coefficients. Since polynomials can be assimilated with special rational functions with 1 as denominator, $\mathbb{R}[\lambda] \subset \mathbb{R}(\lambda)$. It is easy to show that $\mathbb{R}(\lambda)$ is closed under the addition and multiplication operations. Both operations are associative and commutative, the multiplication is distributive over addition and each operation possesses an identity element in $\mathbb{R}(\lambda)$. Finally, there exist inverses for all elements under addition and for all nonzero elements under multiplication. Therefore, the set $\mathbb{R}(\lambda)$ forms a *field*. The subset of polynomials $\mathbb{R}[\lambda]$ forms only a *ring* (more exactly, a Euclidean domain with identity), because the only invertible elements in $\mathbb{R}[\lambda]$ are the nonzero real constants, which are thus the *units* of the ring.

A transfer function $g(\lambda)$ as in (9.3) is called *proper* if $m \leq n$, *strictly proper* if $m < n$, *biproper* if $m = n$ and *improper* if $m > n$. A polynomial $g(\lambda)$ corresponds to the case when $n = 0$. The subset of proper rational functions forms a ring, whose units are the biproper rational functions. The degree of $g(\lambda)$ in (9.3) is defined as $\deg g(\lambda) = \max(m, n)$, while the difference $n - m$ is called the *relative degree* of $g(\lambda)$. Using the Euclidean polynomial division algorithm, it follows that any improper $g(\lambda)$ can be written as the sum of a proper part and a polynomial part.

9.1.2 Transfer Function Matrices

Transfer function matrices are used to describe the input–output behaviour of *multi-input multi-output* (MIMO) LTI systems by relating the input and output variables via a matrix of gains depending on a frequency variable. Consider a MIMO system with m inputs $u_1(t), \ldots, u_m(t)$, which form the m-dimensional input vector $u(t) = [\, u_1(t), \ldots, u_m(t) \,]^T$, and p outputs $y_1(t), \ldots, y_p(t)$, which form the p-dimensional output vector $y(t) = [\, y_1(t), \ldots, y_p(t) \,]^T$. For a continuous dependence of $u(t)$ and $y(t)$ on the time variable t, let $\mathbf{u}(s)$ and $\mathbf{y}(s)$ be the Laplace-transformed input and output vectors, respectively, while in the case of a discrete dependence on t, we denote $\mathbf{u}(z)$ and $\mathbf{y}(z)$ as the $\mathcal{Z}$-transformed input and output vectors, respectively. We denote with λ the frequency variable, which is either s or z, depending on the nature of the time variation, continuous or discrete, respectively. Let $G(\lambda)$ be the $p \times m$ *transfer function matrix* (TFM) defined as

$$G(\lambda) = \begin{bmatrix} g_{11}(\lambda) & \cdots & g_{1m}(\lambda) \\ \vdots & \ddots & \vdots \\ g_{p1}(\lambda) & \cdots & g_{pm}(\lambda) \end{bmatrix},$$

which relates the m-dimensional input vector u to the p-dimensional output vector y in the form

$$\mathbf{y}(\lambda) = G(\lambda)\mathbf{u}(\lambda). \tag{9.4}$$

The element $g_{ij}(\lambda)$ describes the contribution of the j-th input $u_j(t)$ to the i-th output $y_i(t)$. We assume that each matrix entry $g_{ij}(\lambda) \in \mathbb{R}(\lambda)$, and thus it can be expressed as a ratio of two polynomials $\alpha_{ij}(\lambda)$ and $\beta_{ij}(\lambda)$ with real coefficients as $g_{ij}(\lambda) = \alpha_{ij}(\lambda)/\beta_{ij}(\lambda)$ of the form (9.3).

Each TFM $G(\lambda)$ belongs to the set of rational matrices with real coefficients, thus having elements in the field of real rational functions $\mathbb{R}(\lambda)$. Polynomial matrices, having elements in the ring of polynomials with real coefficients $\mathbb{R}[\lambda]$, can be assimilated in a natural way with special rational matrices with all elements having 1 as denominator. Let $\mathbb{R}(\lambda)^{p \times m}$ and $\mathbb{R}[\lambda]^{p \times m}$ denote the sets of $p \times m$ rational and polynomial matrices with real coefficients, respectively. To simplify the notation, we will also use $G(\lambda) \in \mathbb{R}(\lambda)$ or $G(\lambda) \in \mathbb{R}[\lambda]$ if the dimensions of $G(\lambda)$ are not relevant or are clear from the context.

A rational matrix $G(\lambda) \in \mathbb{R}(\lambda)$ is called *proper* if $\lim_{\lambda \to \infty} G(\lambda) = D$, with D having a finite norm. Otherwise, $G(\lambda)$ is called *improper*. If $D = 0$, then $G(\lambda)$ is *strictly proper*. An invertible $G(\lambda)$ is *biproper* if both $G(\lambda)$ and $G^{-1}(\lambda)$ are proper. A polynomial matrix $U(\lambda) \in \mathbb{R}[\lambda]$ is called *unimodular* if it is invertible and its inverse $U^{-1}(\lambda) \in \mathbb{R}[\lambda]$ (i.e., a polynomial matrix). The determinant of a unimodular matrix is therefore a constant.

The degree of a rational matrix $G(\lambda)$, also known as the McMillan degree, is defined in Sect. 9.1.4. We only give here the definition of the degree of a rational vector $v(\lambda)$. For this, we express first $v(\lambda)$ in the form $v(\lambda) = \widetilde{v}(\lambda)/d(\lambda)$, where $d(\lambda)$ is the monic least common multiple of all denominator polynomials of the elements of $v(\lambda)$, and $\widetilde{v}(\lambda)$ is the corresponding polynomial vector $\widetilde{v}(\lambda) := d(\lambda)v(\lambda)$. Then, $\deg v(\lambda) = \max(\deg \widetilde{v}(\lambda), \deg d(\lambda))$.

9.1.3 *Linear Dependence, Normal Rank and Minimal Basis*

A p-dimensional rational vector $v(\lambda) \in \mathbb{R}(\lambda)^p$ can be seen as either a $1 \times p$ or a $p \times 1$ rational matrix. A set of rational vectors $V(\lambda) := \{v_1(\lambda), \ldots, v_k(\lambda)\}$ is said to be *linearly dependent* over the field $\mathbb{R}(\lambda)$ if there exist k rational functions $\gamma_i(\lambda) \in \mathbb{R}(\lambda)$, $i = 1, \ldots, k$, with $\gamma_i(\lambda) \neq 0$ for at least one i, such that the linear combination

$$\sum_{i=1}^{k} \gamma_i(\lambda)v_i(\lambda) = 0. \tag{9.5}$$

The set of vectors $V(\lambda)$ is *linearly independent* over $\mathbb{R}(\lambda)$ if (9.5) implies that $\gamma_i(\lambda) = 0$ for each $i = 1, \ldots, k$. It is important to note that a linearly dependent set $V(\lambda)$ over $\mathbb{R}(\lambda)$ can be still linearly independent over another field (e.g., the field of reals if $\gamma_i \in \mathbb{R}$).

The *normal rank* of a rational matrix $G(\lambda) \in \mathbb{R}(\lambda)^{p \times m}$, which we also denote by rank $G(\lambda)$, is the maximal number of linearly independent rows over the field of rational functions $\mathbb{R}(\lambda)$. It can be shown that the normal rank of $G(\lambda)$ is the maximally possible rank of the complex matrix $G(\lambda)$ for all values of $\lambda \in \mathbb{C}$ such that $G(\lambda)$ has a finite norm. This interpretation provides a simple way to determine the normal rank by evaluating the maximal rank of $G(\lambda)$ for a few random values of the frequency variable λ.

It is easy to check that the set of p-dimensional rational vectors $\mathbb{R}(\lambda)^p$ forms a vector space with scalars defined over $\mathbb{R}(\lambda)$. If $\mathcal{V}(\lambda) \subset \mathbb{R}(\lambda)^p$ is a vector space, then there exists a set of linearly independent rational vectors $V(\lambda) := \{v_1(\lambda), \ldots, v_{n_b}(\lambda)\} \subset \mathcal{V}(\lambda)$ such that any vector in $\mathcal{V}(\lambda)$ is a linear combination of the vectors in $V(\lambda)$ (equivalently, any set of $n_b + 1$ vectors, including an arbitrary vector from $\mathcal{V}(\lambda)$ and the n_b vectors in $V(\lambda)$, is linearly dependent). The set $V(\lambda)$ is called a *basis* of the vector space $\mathcal{V}(\lambda)$ and n_b is the dimension of $\mathcal{V}(\lambda)$. With a slight abuse of notation, we denote $V(\lambda)$ the matrix formed of the n_b stacked row vectors

$$
V(\lambda) = \begin{bmatrix} v_1(\lambda) \\ \vdots \\ v_{n_b}(\lambda) \end{bmatrix}
$$

or the n_b concatenated column vectors

$$
V(\lambda) = \begin{bmatrix} v_1(\lambda) \cdots v_{n_b}(\lambda) \end{bmatrix} .
$$

Interestingly, as basis vectors we can always use polynomial vectors since we can replace each vector $v_i(\lambda)$ of a rational basis, by $v_i(\lambda)$ multiplied with the least common multiple of the denominators of the components of $v_i(\lambda)$.

The use of polynomial bases allows to define the main concepts related to so-called minimal bases. Let n_i be the degree of the i-th polynomial vector $v_i(\lambda)$ of a polynomial basis $V(\lambda)$ (i.e., n_i is the largest degree of the components of $v_i(\lambda)$). Then, $n := \sum_{i=1}^{n_b} n_i$ is, by definition, the *degree* of the polynomial basis $V(\lambda)$. A *minimal polynomial basis* of $\mathcal{V}(\lambda)$ is one for which n has the least achievable value. For a minimal polynomial basis, n_i, for $i = 1, \ldots, n_b$, are called the *row* or *column minimal indices* (also known as *left* or *right Kronecker indices*, respectively). Two important examples are the left and right nullspace bases of a rational matrix, which are shortly discussed below.

Let $G(\lambda)$ be a $p \times m$ rational matrix $G(\lambda)$ whose normal rank is $r < \min(p, m)$. It is easy to show that the set

$$
\mathcal{N}_L(G(\lambda)) := \{v(\lambda) \in \mathbb{R}(\lambda)^{1 \times p} \mid v(\lambda)G(\lambda) = 0\}
$$

is a linear space called the *left nullspace* of $G(\lambda)$. Analogously,

$$
\mathcal{N}_R(G(\lambda)) := \{v(\lambda) \in \mathbb{R}(\lambda)^{m \times 1} \mid G(\lambda)v(\lambda) = 0\}
$$

is a linear space called the *right nullspace* of $G(\lambda)$. The dimension of $\mathcal{N}_L(G(\lambda))$ $[\mathcal{N}_R(G(\lambda))]$ is $p - r$ $[m - r]$, and therefore, there exist $p - r$ $[m - r]$ linearly independent polynomial vectors, which form a minimal polynomial basis for $\mathcal{N}_L(G(\lambda))$ $[\mathcal{N}_R(G(\lambda))]$. Let $n_{l,i}$ $[n_{r,i}]$ be the *left* [*right*] *minimal indices* and let $n_l := \sum_{i=1}^{p-r} n_{l,i}$

$\left[\, n_r := \sum_{i=1}^{m-r} n_{r,i} \,\right]$ be the least possible degree of the left [right] nullspace basis. The least left and right degrees n_l and n_r, respectively, play an important role in relating the main structural elements of rational matrices (see the discussion of poles and zeros in Sect. 9.1.4).

Some properties of minimal polynomial bases, which are relevant for the synthesis methods presented in this book, are summarized below for the left nullspace bases. Similar results can be given for the right nullspace bases.

Lemma 9.1 *Let $G(\lambda)$ be a $p \times m$ rational matrix of normal rank r and let $N_l(\lambda)$ be a $(p - r) \times p$ minimal polynomial basis of the left nullspace $\mathcal{N}_L(G(\lambda))$ with left minimal (or left Kronecker) indices $n_{l,i}$, $i = 1, \ldots, p - r$. Then the following holds:*

1. *The left minimal indices are unique up to permutations (i.e., if $\widetilde{N}_l(\lambda)$ is another minimal polynomial basis, then $N_l(\lambda)$ and $\widetilde{N}_l(\lambda)$ have the same left minimal indices).*
2. *$N_l(\lambda)$ is irreducible, having full row rank for all $\lambda \in \mathbb{C}$ (i.e., $N_l(\lambda)$ has no finite zeros; see Sect. 9.1.4).*
3. *$N_l(\lambda)$ is row- reduced (i.e., the leading row coefficient matrix formed from the coefficients of the highest row degrees has full row rank.)*

An irreducible and row-reduced polynomial basis is actually a minimal polynomial basis. Irreducibility implies that any polynomial vector $v(\lambda)$ in the space spanned by the rows of $N_l(\lambda)$ can be expressed as a linear combination of basis vectors $v(\lambda) = \phi(\lambda)N_l(\lambda)$, with $\phi(\lambda)$ being a polynomial vector. In particular, assuming the rows of $N_l(\lambda)$ are ordered such that $n_{l,i} \leq n_{l,i+1}$ for $i = 1, \ldots, p - r - 1$, then for any $v(\lambda)$ of degree $n_{l,i}$, the corresponding $\phi(\lambda)$ has as its j-th element a polynomial of degree at most $n_{l,i} - n_{l,j}$ for $j = 1, \ldots, i$, and the rest of the components are zero. This property allows to easily generate left polynomial annihilators of given degrees and can be exploited in the synthesis of least-order residual generators (see Sect. 5.2).

Minimal polynomial bases allow to easily build *simple minimal proper rational bases*, which are the natural counterparts of the minimal polynomial bases. These are proper rational bases having the property that the sum of degrees of the rows [columns] is equal to the least left [right] degree of a minimal polynomial basis n_l [n_r]. A simple minimal proper rational left nullspace basis with arbitrary poles can be constructed by forming $\widetilde{N}_l(\lambda) := M(\lambda)N_l(\lambda)$ with

$$M(\lambda) = \operatorname{diag}\left(1/d_1(\lambda), \ldots, 1/d_{p-r}(\lambda)\right), \tag{9.6}$$

where $d_i(\lambda)$ is a polynomial of degree $n_{l,i}$ with arbitrary roots. Since $N_l(\lambda)$ is row-reduced, it follows that $D_l := \lim_{\lambda \to \infty} \widetilde{N}_l(\lambda)$ has full row rank (i.e., $\widetilde{N}_l(\lambda)$ has no infinite zeros; see Sect. 9.1.4). A simple minimal proper rational left nullspace basis allows to generate, in a straightforward manner, left rational annihilators of given McMillan degrees by forming linear combinations of the basis vectors in $\widetilde{N}_l(\lambda)$ using specially chosen rational vectors $\phi(\lambda)$ (see Sect. 9.1.4 for the definition of the McMillan degree of a rational matrix). This property can be equally exploited in the synthesis of least-order residual generators (see Sect. 5.2).

9.1.4 Poles and Zeros

For a scalar rational function $g(\lambda) \in \mathbb{R}(\lambda)$, the values of λ for which $g(\lambda)$ is infinite are called the *poles* of $g(\lambda)$. If $g(\lambda) = \alpha(\lambda)/\beta(\lambda)$ has the form in (9.3), then the n roots λ_i^p, $i = 1, \ldots, n$, of $\beta(\lambda)$ are the *finite poles* of $g(\lambda)$, and if $m > n$, there are also, by convention, $m - n$ *infinite poles*. The values of λ for which $g(\lambda) = 0$ are called the *zeros* of $g(\lambda)$. The m roots λ_i^z, $i = 1, \ldots, m$, of $\alpha(\lambda)$ are the *finite zeros* of $g(\lambda)$, and if $n > m$, there are also, by convention, $n - m$ *infinite zeros*. It follows that the number of poles is equal to the number of zeros and is equal to $\max(m, n)$, the degree of $g(\lambda)$. The rational function $g(\lambda)$ in (9.3) can be equivalently expressed in terms of its finite poles and zeros in the factorized form

$$g(\lambda) = k_g \frac{\prod_{i=1}^{m}(\lambda - \lambda_i^z)}{\prod_{i=1}^{n}(\lambda - \lambda_i^p)}, \tag{9.7}$$

where $k_g = a_m/b_n$. If $g(\lambda)$ is the transfer function of a SISO LTI system, then we will always assume that $g(\lambda)$ is in a minimal cancelled (irreducible) form, that is, the polynomials $\alpha(\lambda)$ and $\beta(\lambda)$ in (9.3) have 1 as the greatest common divisor. Equivalently, the two polynomials have no common roots, and therefore no pole–zero cancellation may occur in (9.7). Two such polynomials are called *coprime*.

In studying the stability of systems, the poles play a primordial role. Their real parts, in the case of a continuous-time system, or moduli, in the case of a discrete-time system, determine the asymptotic (exponential) decay or divergence speed of the system output. A SISO LTI system with the transfer function $g(\lambda)$ is *exponentially stable* (or equivalently $g(\lambda)$ is stable) if $g(\lambda)$ is proper and has all poles in the appropriate *stability domain* $\mathbb{C}_s$. The system is *unstable* if it has at least one pole outside of the stability domain and *anti-stable* if all poles lie outside of the stability domain. Poles inside the stability domain are called *stable poles*, while those outside of the stability domain are called *unstable poles*. For continuous-time systems, the stability domain is the open left-half complex plane $\mathbb{C}_s = \{s \in \mathbb{C} : \Re(s) < 0\}$, while for discrete-time systems the stability domain is the open unit disc $\mathbb{C}_s = \{z \in \mathbb{C} : |z| < 1\}$. We denote by $\partial\mathbb{C}_s$ the boundary of the stability domain. For continuous-time systems, the boundary of the stability domain is the extended imaginary axis (i.e., including the point at infinity) $\partial\mathbb{C}_s = \{\infty\} \cup \{s \in \mathbb{C} : \Re(s) = 0\}$, while for discrete-time systems the boundary of the stability domain is the unit circle $\partial\mathbb{C}_s = \{z \in \mathbb{C} : |z| = 1\}$. We denote $\overline{\mathbb{C}}_s = \mathbb{C}_s \cup \partial\mathbb{C}_s$ as the closure of the stability domain. We denote the *instability domain* of poles by $\overline{\mathbb{C}}_u$, and it is the complement of $\mathbb{C}_s$ in $\mathbb{C}$, $\overline{\mathbb{C}}_u = \mathbb{C} \setminus \mathbb{C}_s$. It is also the closure of the set denoted by $\mathbb{C}_u$, which, for a continuous-time system, is the open right-half-plane $\mathbb{C}_u = \{s \in \mathbb{C} : \Re(s) > 0\}$, while for a discrete-time system it is the exterior of the unit circle $\mathbb{C}_u = \{z \in \mathbb{C} : |z| > 1\}$. The *stability degree* of poles is defined as the largest real part of the poles in the continuous-time case, or the largest absolute value of the poles in the discrete-time case.

Let $\mathbb{R}_s(\lambda)$ be the set of proper stable transfer functions having poles only in $\mathbb{C}_s$. A transfer function $g(\lambda) \in \mathbb{R}_s(\lambda)$ having only zeros in $\mathbb{C}_s$ is called *minimum-*

phase. Otherwise it is called *non-minimum-phase*. The zeros of $g(\lambda)$ in $\mathbb{C}_s$ are called *minimum-phase zeros*, while those outside $\mathbb{C}_s$ are called *non-minimum-phase zeros*.

There are no straightforward generalizations of poles and zeros of scalar rational functions to the rational matrix case. Instrumental for a rigorous definition are two canonical forms: the *Smith form* for polynomial matrices and the *Smith–McMillan form* for rational matrices. For polynomial matrices, we have the following important result.

Lemma 9.2 *Let $P(\lambda) \in \mathbb{R}[\lambda]^{p \times m}$ be any polynomial matrix. Then, there exist unimodular matrices $U(\lambda) \in \mathbb{R}[\lambda]^{p \times p}$ and $V(\lambda) \in \mathbb{R}[\lambda]^{m \times m}$ such that*

$$U(\lambda)P(\lambda)V(\lambda) = S(\lambda) := \begin{bmatrix} \alpha_1(\lambda) & 0 & \cdots & 0 & \cdots 0 \\ 0 & \alpha_2(\lambda) & \cdots & 0 & \cdots 0 \\ \vdots & \vdots & \ddots & \vdots & \vdots \\ 0 & 0 & \cdots & \alpha_r(\lambda) & \cdots 0 \\ \vdots & \vdots & & \vdots & \vdots \\ 0 & 0 & \cdots & 0 & \cdots 0 \end{bmatrix} \tag{9.8}$$

and $\alpha_i(\lambda)$ divides $\alpha_{i+1}(\lambda)$ for $i = 1, \ldots, r-1$.

The polynomial matrix $S(\lambda)$ is called the *Smith form* of $P(\lambda)$ and r is the normal rank of $P(\lambda)$. The diagonal elements $\alpha_1(\lambda), \ldots, \alpha_r(\lambda)$ are called the *invariant polynomials* of $P(\lambda)$. The roots of the polynomials $\alpha_i(\lambda)$, for $i = 1, \ldots, r$, are called the *finite zeros* of the polynomial matrix $P(\lambda)$. To each distinct finite zero λ_z of $P(\lambda)$, we can associate the multiplicities $\sigma_i(\lambda_z) \geq 0$ of root λ_z in each of the polynomials $\alpha_i(\lambda)$, for $i = 1, \ldots, r$. By convention, $\sigma_i(\lambda_z) = 0$ if λ_z is not a root of $\alpha_i(\lambda)$. The divisibility properties of $\alpha_i(\lambda)$ imply that

$$0 \leq \sigma_1(\lambda_z) \leq \sigma_2(\lambda_z) \leq \cdots \leq \sigma_r(\lambda_z).$$

Any rational matrix $G(\lambda)$ can be expressed as

$$G(\lambda) = \frac{P(\lambda)}{d(\lambda)},$$

where $d(\lambda)$ is the monic least common multiple of the denominator polynomials of the entries of $G(\lambda)$, and $P(\lambda) := d(\lambda)G(\lambda)$ is a polynomial matrix. Then, we have the following straightforward extension of Lemma 9.2 to rational matrices.

Lemma 9.3 *Let $G(\lambda) \in \mathbb{R}(\lambda)^{p \times m}$ be any rational matrix. Then, there exist unimodular matrices $U(\lambda) \in \mathbb{R}[\lambda]^{p \times p}$ and $V(\lambda) \in \mathbb{R}[\lambda]^{m \times m}$ such that*

$$
U(\lambda)G(\lambda)V(\lambda) = H(\lambda) :=
\begin{bmatrix}
\frac{\alpha_1(\lambda)}{\beta_1(\lambda)} & 0 & \cdots & 0 & \cdots & 0 \\
0 & \frac{\alpha_2(\lambda)}{\beta_2(\lambda)} & \cdots & 0 & \cdots & 0 \\
\vdots & \vdots & \ddots & \vdots & & \vdots \\
0 & 0 & \cdots & \frac{\alpha_r(\lambda)}{\beta_r(\lambda)} & \cdots & 0 \\
\vdots & \vdots & & \vdots & & \vdots \\
0 & 0 & \cdots & 0 & \cdots & 0
\end{bmatrix},
\tag{9.9}
$$

with $\alpha_i(\lambda)$ and $\beta_i(\lambda)$ coprime for $i = 1, \ldots, r$ and $\alpha_i(\lambda)$ divides $\alpha_{i+1}(\lambda)$ and $\beta_{i+1}(\lambda)$ divides $\beta_i(\lambda)$ for $i = 1, \ldots, r - 1$.

The rational matrix $H(\lambda)$ is called the *Smith–McMillan form* of $G(\lambda)$ and r is the normal rank of $G(\lambda)$. The Smith–McMillan form is a powerful conceptual tool, which allows to define rigorously the notions of poles and zeros of MIMO LTI systems and to establish several basic factorization results of rational matrices.

The roots of the numerator polynomials $\alpha_i(\lambda)$, for $i = 1, \ldots, r$, are called the *finite zeros* of the rational matrix $G(\lambda)$, and the roots of the denominator polynomials $\beta_i(\lambda)$, for $i = 1, \ldots, r$, are called the *finite poles* of the rational matrix $G(\lambda)$. To each finite λ_z, which is a zero or a pole of $G(\lambda)$ (or both), we can associate its multiplicities $\{\sigma_1(\lambda_z), \ldots, \sigma_r(\lambda_z)\}$, where $\sigma_i(\lambda_z)$ is the multiplicity of λ_z either as a pole or a zero of the ratio $\alpha_i(\lambda)/\beta_i(\lambda)$, for $i = 1, \ldots, r$. By convention, we use negative values for poles and positive values for zeros. The divisibility properties of $\alpha_i(\lambda)$ and $\beta_i(\lambda)$ imply that

$$
\sigma_1(\lambda_z) \le \sigma_2(\lambda_z) \le \cdots \le \sigma_r(\lambda_z) .
$$

The r-tuple of multiplicities $\{\sigma_1(\lambda_z), \ldots, \sigma_r(\lambda_z)\}$ completely characterizes the *pole–zero structure* of $G(\lambda)$ in λ_z.

The relative degrees of $\alpha_i(\lambda)/\beta_i(\lambda)$ do not provide the correct information on the multiplicity of infinite zeros and poles. This is because the used unimodular transformations may have poles and zeros at infinity. To overcome this, the multiplicity of zeros and poles at infinity are defined in terms of multiplicities of poles and zeros of $G(1/\lambda)$ in zero. The *McMillan degree* of a rational matrix $G(\lambda)$, usually denoted by $\delta(G(\lambda))$, is the number n_p of its poles, both finite and infinite, counting all multiplicities. If n_z is the number of zeros (finite and infinite, counting all multiplicities), then we have the following important structural relation for any rational matrix

$$
n_p = n_z + n_l + n_r,
$$

where n_l and n_r are the least degrees of the minimal polynomial bases for the left and right nullspaces of $G(\lambda)$, respectively.

For a given rational matrix $G(\lambda)$, any (finite or infinite) pole of its elements $g_{ij}(\lambda)$ is also a pole of $G(\lambda)$. Therefore, many notions related to poles of SISO LTI systems introduced previously can be extended in a straightforward manner to MIMO LTI systems. For example, the notion of properness of $G(\lambda)$ can be equivalently defined as the nonexistence of infinite poles in the elements of $G(\lambda)$ (i.e., $G(\lambda)$ has only finite poles). The notion of *exponential stability* of a LTI system with a proper TFM $G(\lambda)$ can be defined as the lack of unstable poles in all elements of $G(\lambda)$ (i.e., all poles of $G(\lambda)$ are in the stable domain $\mathbb{C}_s$). A TFM $G(\lambda)$ with only stable poles is called *stable*. Otherwise, $G(\lambda)$ is called *unstable*. A similar definition applies for the *stability degree* of poles.

The zeros of $G(z)$ are also called the *transmission zeros* of the corresponding LTI system. A proper and stable $G(z)$ is *minimum-phase* if all its finite zeros are stable. Otherwise, it is called *non-minimum-phase*.

9.1.5 Additive Decompositions

Consider a disjunct partition of the complex plane $\mathbb{C}$ as

$$\mathbb{C} = \mathbb{C}_g \cup \mathbb{C}_b, \quad \mathbb{C}_g \cap \mathbb{C}_b = \emptyset, \tag{9.10}$$

where both $\mathbb{C}_g$ and $\mathbb{C}_b$ are symmetrically located with respect to the real axis, and $\mathbb{C}_g$ has at least one point on the real axis. Since $\mathbb{C}_g$ and $\mathbb{C}_b$ are disjoint, each pole of any transfer function lies either in $\mathbb{C}_g$ or in $\mathbb{C}_b$. Therefore, the subscripts are sometimes associated with the "good" and "bad" poles of the transfer functions. In applications, $\mathbb{C}_g$ corresponds to a region or a discrete set of points in which we wish to place the poles of a synthesized controller or fault detection filter. In what follows, we will always assume that $\infty \notin \mathbb{C}_g$.

Since $\mathbb{C}_g$ and $\mathbb{C}_b$ are disjoint, each pole of any element $g_{ij}(\lambda)$ of a TFM $G(\lambda)$ lies either in $\mathbb{C}_g$ or in $\mathbb{C}_b$. Therefore, it is easy to see that any $G(\lambda)$ can be additively decomposed as

$$G(\lambda) = G_g(\lambda) + G_b(\lambda), \tag{9.11}$$

where $G_g(\lambda)$ has only poles in $\mathbb{C}_g$, while $G_b(\lambda)$ has only poles in $\mathbb{C}_b$. For such a decomposition of $G(\lambda)$, we always have that

$$\delta\left(G(\lambda)\right) = \delta\left(G_g(\lambda)\right) + \delta\left(G_b(\lambda)\right).$$

For example, if $\mathbb{C}_g = \mathbb{C} \setminus \{\infty\}$ and $\mathbb{C}_b = \{\infty\}$, then (9.11) represents the additive decomposition of a possibly improper rational matrix as the sum of its proper and polynomial parts. This decomposition, in general, is not unique, because an arbitrary constant term can be always added to one term and subtracted from the other one. Another frequently used decomposition is the stable–unstable decomposition of proper rational matrices, when $\mathbb{C}_g = \mathbb{C}_s$ (stability region) and $\mathbb{C}_b = \overline{\mathbb{C}}_u$ (instability region).

9.1.6   *Fractional Factorizations*

For a given disjunct partition (9.10) of the complex plane as $\mathbb{C} = \mathbb{C}_g \cup \mathbb{C}_b$, any rational matrix $G(\lambda)$ can be expressed in a left fractional form

$$G(\lambda) = M^{-1}(\lambda)N(\lambda), \qquad (9.12)$$

or in a right fractional form

$$G(\lambda) = N(\lambda)M^{-1}(\lambda), \qquad (9.13)$$

where both the denominator factor $M(\lambda)$ and the numerator factor $N(\lambda)$ have only poles in $\mathbb{C}_g$. These fractional factorizations over a "good" domain of poles $\mathbb{C}_g$ are important in various observer, fault detection filter, or controller synthesis methods, because they allow to achieve the placement of all poles of a TFM $G(\lambda)$ in the domain $\mathbb{C}_g$ simply, by a premultiplication or postmultiplication of $G(\lambda)$ with a suitable $M(\lambda)$. Fractional factorizations of the form (9.12) or (9.13) are not unique. For example, if $U(\lambda)$ is an invertible rational matrix with poles only in $\mathbb{C}_g$, then both (9.12) and

$$G(\lambda) = (U(\lambda)M(\lambda))^{-1} (U(\lambda)N(\lambda)) \qquad (9.14)$$

are left fractional factorizations of $G(\lambda)$.

Of special interest are the so-called coprime factorizations, where the factors satisfy additional conditions. A fractional representation of the form (9.12) is a *left coprime factorization* (LCF) over $\mathbb{C}_g$ of $G(\lambda)$, if there exist $U(\lambda)$ and $V(\lambda)$ with poles only in $\mathbb{C}_g$, which satisfy the *Bezout identity*

$$M(\lambda)U(\lambda) + N(\lambda)V(\lambda) = I \, .$$

A fractional representation of the form (9.13) is a *right coprime factorization* (RCF) over $\mathbb{C}_g$ of $G(\lambda)$, if there exist $U(\lambda)$ and $V(\lambda)$ with poles only in $\mathbb{C}_g$, which satisfy

$$U(\lambda)M(\lambda) + V(\lambda)N(\lambda) = I \, .$$

Coprime factorizations are not unique as well. For example, if $U(\lambda)$ is an invertible rational matrix with poles and zeros only in $\mathbb{C}_g$ and $G(\lambda) = M^{-1}(\lambda)N(\lambda)$ is a LCF, then (9.14) is also a LCF of $G(\lambda)$.

An important class of coprime factorizations is the class of coprime factorizations with minimum-degree denominators. Recall that $\delta(G(\lambda))$, the McMillan degree of $G(\lambda)$, is defined as the number of poles of $G(\lambda)$, both finite and infinite, counting all multiplicities. It follows that for any $G(\lambda)$, we have $\delta(G(\lambda)) = n_g + n_b$, where n_g and n_b are the number of poles of $G(\lambda)$ in $\mathbb{C}_g$ and $\mathbb{C}_b$, respectively. The denominator factor $M(\lambda)$ has the minimum-degree property if $\delta(M(\lambda)) = n_b$. When determining minimum-degree coprime factorizations, the n_b poles of $M(\lambda)$ can be arbitrarily chosen from $\mathbb{C}_g$. However, the poles of $M^{-1}(\lambda)$ (which are the zeros of $M(\lambda)$) are fixed, and are precisely the n_b poles of $G(\lambda)$ in $\mathbb{C}_b$. An important consequence of this fact is that if $G(\lambda)$ has no poles in a certain region $\Omega \subset \mathbb{C}_b$, then $M(\lambda)$ has no zeros in Ω as well.

Of special interest in FDI-related synthesis procedures is the left factorization with respect to a given partition (9.10) of the complex plane of a row partitioned matrix with n_b block rows

$$G(\lambda) = \begin{bmatrix} G_1(\lambda) \\ G_2(\lambda) \\ \vdots \\ G_{n_b}(\lambda) \end{bmatrix},$$

such that $M(\lambda)$ in (9.12) has a block-diagonal form

$$M(\lambda) = \mathrm{diag}\left(M_1(\lambda), M_2(\lambda), \ldots, M_{n_b}(\lambda)\right), \tag{9.15}$$

where the size of the square diagonal block $M_i(\lambda)$ is equal to the number of rows of $G_i(\lambda)$. Such a factorization can be obtained by determining each diagonal block $M_i(\lambda)$ separately from a LCF of the i-th block row $G_i(\lambda) = M_i^{-1}(\lambda)N_i(\lambda)$. The resulting $N(\lambda)$ is

$$N(\lambda) = \begin{bmatrix} N_1(\lambda) \\ N_2(\lambda) \\ \vdots \\ N_{n_b}(\lambda) \end{bmatrix}. \tag{9.16}$$

If all n_b row blocks are row vectors, then $M(\lambda)$ results diagonal. In this case, the overall left factorization with $M(\lambda)$ in (9.15) and $N(\lambda)$ in (9.16) preserves in the resulting $N(\lambda)$ the zero–nonzero structure of $G(\lambda)$. Although for each block row $G_i(\lambda)$ we used a LCF, in general, the overall factorization (9.12) is, in general, not coprime.

The *conjugate* of the TFM $G(\lambda)$ is denoted by $G^\sim(\lambda)$ and is defined in a continuous-time setting as $G^\sim(s) = G^T(-s)$, while in a discrete-time setting as $G^\sim(z) = G^T(1/z)$. A square TFM $G(\lambda)$ is *all-pass* if $G^\sim(\lambda)G(\lambda) = I$. If $G(\lambda)$ is a stable TFM and satisfies $G^\sim(\lambda)G(\lambda) = I$ then it is called an *inner* TFM, while if it satisfies $G(\lambda)G^\sim(\lambda) = I$ it is called a *co-inner* TFM. Note that an inner or co-inner TFM must not be square, but must have full column rank (injective) or full row rank (surjective), respectively. It is remarkable that each proper TFM $G(\lambda)$ without poles on the boundary of stability domain $\partial\mathbb{C}_s$ has a stable LCF of the form (9.12) or a stable RCF of the form (9.13) with the denominator factor $M(\lambda)$ inner. The McMillan degree of $M(\lambda)$ is equal to the number of the unstable poles of $G(\lambda)$.

A possible approach to compute coprime factorizations with denominator factors of least McMillan degrees employs the additive decomposition (9.11) with respect to a disjunct partition of the complex plane (9.10). Consider the additive decomposition

$$G(\lambda) = G_g(\lambda) + G_b(\lambda), \tag{9.17}$$

where $G_g(\lambda)$ has only poles in $\mathbb{C}_g$, while $G_b(\lambda)$ has only poles in $\mathbb{C}_b$. For such a decomposition of $G(\lambda)$, we always have that $\delta\left(G(\lambda)\right) = \delta\left(G_g(\lambda)\right) + \delta\left(G_b(\lambda)\right)$. Compute now a LCF $G_b(\lambda) = M^{-1}(\lambda)N_g(\lambda)$, such that $M(\lambda)$ and $N_g(\lambda)$ have only poles in $\mathbb{C}_g$ and $\delta(M(\lambda)) = \delta(G_b(\lambda))$ (thus the least possible one). The numerator factor of a LCF with least-order denominator factor $M(\lambda)$ is given by

$$N(\lambda) = M(\lambda)G_g(\lambda) + N_g(\lambda)$$

and has only poles in $\mathbb{C}_g$. Similarly, if $G_b(\lambda) = N_g(\lambda)M^{-1}(\lambda)$ is a RCF with $\delta(M(\lambda)) = \delta(G_b(\lambda))$, then the numerator factor of a RCF with least-order denominator factor $M(\lambda)$ is given by

$$N(\lambda) = G_g(\lambda)M(\lambda) + N_g(\lambda)\,.$$

This factorization approach can be applied also to compute left or right fractional factorizations with least-order diagonal denominators, by applying the above procedure to each row or column of $G(\lambda)$.

For the computation of coprime factorizations with minimum-degree denominators, recursive pole–zero cancellation techniques can be employed. For example, for the computation of the LCF of $G_b(\lambda)$, it is possible to find a sequence of $n_b := \delta(G_b(\lambda))$ nonsingular rational matrices $M_i(\lambda)$, $i = 1, \ldots, n_b$, each of McMillan degree 1, with one (arbitrary) pole in $\mathbb{C}_g$ and one (fixed) zero in $\mathbb{C}_b$, such that the sequence $N_i(\lambda) := M_i(\lambda)N_{i-1}(\lambda)$ for $i = 1, \ldots, n_b$, initialized with $N_0(\lambda) = G_b(\lambda)$, generates the factors $N_g(\lambda) := N_{n_b}(\lambda)$ and $M(\lambda) := M_{n_b}(\lambda) \cdots M_1(\lambda)$ of the LCF $G_b(\lambda) = M^{-1}(\lambda)N_g(\lambda)$. The zero of $M_i(\lambda)$ is chosen to cancel with a pole of $N_{i-1}(\lambda)$ lying in $\mathbb{C}_b$, such that after n_b steps, all poles of $G_b(\lambda)$ are cancelled and dislocated to values in $\mathbb{C}_g$. This approach can be also employed when additionally imposing that all elementary factors $M_i(\lambda)$ are inner. In the case of complex poles, the above technique leads, in general, to complex factorizations. Therefore, to obtain real factorizations, for each complex conjugate pair of poles a second-degree real factor can be used to cancel simultaneously both poles. Second-degree factors may also be necessary when dislocating a pair of real poles into a pair of complex conjugate poles. General formulas for constructing first- and second-degree factors are given in [127]. Similar first-degree factors can be used for cancelling a single infinite pole (to be used in the case of an improper $G(\lambda)$).

The use of the above approach involves several delicate computations involving transfer function manipulations. First, one has to compute the poles of $G(\lambda)$ in order to know which ones have to be cancelled. Then, one has to compute a partial-fraction expansion of $G(\lambda)$ to arrive at the additive decomposition (9.17) and, further, of each $N_{i-1}(\lambda)$ (for $i = 1, \ldots, n_b$). From the coefficient matrices of this expansion, one derives certain vectors needed for the construction of the factors $M_i(\lambda)$. This computation can be very involved when $G_b(\lambda)$ has multiple poles or when real factors have to be enforced by using second-degree real factors to cancel a pair of complex conjugated poles (see [127] for details). Moreover, after each pole cancellation with a factor $M_i(\lambda)$, the expansion of $N_{i-1}(\lambda)$ has to be updated. The state-space-computation-based methods, presented in Sect. 10.3.5, provide numerically appealing alternative approaches for the computation of coprime factorizations.

9.1.7 Norms

The $\mathcal{H}_\infty$-*norm* of a proper and stable TFM $G(\lambda) \in \mathbb{R}_s^{p \times m}(\lambda)$ can be defined as

$$\|G\|_\infty := \sup_{\lambda \in \partial \mathbb{C}_s} \overline{\sigma}\,(G(\lambda))\,, \tag{9.18}$$

where $\overline{\sigma}(\cdot)$ denotes the largest singular value. The set $\mathbb{R}_s^{p \times m}(\lambda)$ together with the norm defined in (9.18) is denoted by $\mathcal{H}_\infty^{p \times m}$ (or simply $\mathcal{H}_\infty$ if the dimensions are not relevant or are clear from the context) and is the Hardy space of TFMs with bounded ∞-norm (i.e., the set of stable and proper TFMs). Let $\mathcal{L}_\infty^{p \times m}$ (or $\mathcal{L}_\infty$) denote the set of proper TFMs without poles on the boundary of the stability domain $\partial \mathbb{C}_s$. The $\mathcal{L}_\infty$-norm of a $G(\lambda) \in \mathcal{L}_\infty$ is defined as in (9.18). Its computation can be reduced to a $\mathcal{H}_\infty$-norm computation by using the stable RCF with inner denominator of $G(\lambda) = \widetilde{G}(\lambda)M_i^{-1}(\lambda)$ and the fact that $M_i^{-1}(\lambda)$ is an all-pass TFM. It follows

$$\|G\|_\infty = \|\widetilde{G}(\lambda)M_i^{-1}(\lambda)\|_\infty = \|\widetilde{G}(\lambda)\|_\infty\,.$$

The $\mathcal{H}_2$-*norm* of TFM $G(\lambda) \in \mathbb{R}_s^{p \times m}(\lambda)$ is defined for a continuous-time system as

$$\|G\|_2 := \left(\frac{1}{2\pi} \int_{-\infty}^{\infty} \text{trace}\left(G^T(-i\omega)G(i\omega)\right) d\omega \right)^{1/2} \tag{9.19}$$

and for a discrete-time system as

$$\|G\|_2 := \left(\frac{1}{2\pi} \int_{0}^{2\pi} \text{trace}\left(G^T(e^{-i\theta})G(e^{i\theta})\right) d\theta \right)^{1/2}. \tag{9.20}$$

$\mathcal{H}_2^{p \times m}$ (or $\mathcal{H}_2$) denotes the Hardy space of TFMs with bounded 2-norm. In the continuous-time case, this is the set of stable and strictly proper TFMs. We define similarly $\mathcal{L}_2^{p \times m}$ (or $\mathcal{L}_2$) the space of TFMs without poles on $\partial \mathbb{C}_s$ such that the integrals in (9.19) or (9.20), depending on the type of the system (continuous- or discrete-time), are bounded. The $\mathcal{L}_2$-norm has formally the same definition as the $\mathcal{H}_2$-norm in (9.19) or (9.20) and can be computed using the stable RCF with inner denominator of $G(\lambda) = \widetilde{G}(\lambda)M_i^{-1}(\lambda)$ and the norm-preserving property of all-pass factors

$$\|G\|_2 = \|\widetilde{G}(\lambda)M_i^{-1}(\lambda)\|_2 = \|\widetilde{G}(\lambda)\|_2\,.$$

The *Hankel norm* of a stable TFM $G(\lambda)$ is a measure of the influence of past inputs on future outputs and will be denoted by $\|G(\lambda)\|_H$. The precise mathematical definition of the Hankel norm involves some advanced concepts from functional analysis (i.e., it is the norm of the Hankel operator Γ_G associated with $G(\lambda)$). For any stable $G(\lambda)$, the relation to $\mathcal{H}_\infty$-norm is $\|G(\lambda)\|_H \leq \|G(\lambda)\|_\infty$. For further details, see [177].

9.1.8 Inner–Outer and Spectral Factorizations

A proper and stable TFM $G(\lambda)$ is *outer* if it is minimum-phase and full row rank (surjective), and is *co-outer* if it is minimum-phase and full column rank (injective). A full row rank (full column rank) proper and stable TFM $G(\lambda)$ is *quasi-outer* (*quasi-co-outer*) if it has only zeros in $\overline{\mathbb{C}}_s$ (i.e., in the stability domain and its boundary). Any stable TFM $G(\lambda)$ without zeros in $\partial\mathbb{C}_s$ has an *inner–outer factorization*

$$G(\lambda) = G_i(\lambda)G_o(\lambda), \tag{9.21}$$

with $G_i(\lambda)$ inner and $G_o(\lambda)$ outer. Similarly, $G(\lambda)$ has a *co-outer–co-inner factorization*

$$G(\lambda) = G_{co}(\lambda)G_{ci}(\lambda), \tag{9.22}$$

with $G_{co}(\lambda)$ co-outer and $G_{ci}(\lambda)$ co-inner. Any stable TFM $G(\lambda)$ has an *inner–quasi-outer factorization* of the form (9.21), where $G_i(\lambda)$ is inner and $G_o(\lambda)$ is quasi-outer, and also has a *quasi-co-outer–co-inner factorization* of the form (9.22), where $G_{ci}(\lambda)$ is co-inner and $G_{co}(\lambda)$ is quasi-co-outer.

Remark 9.1 The inner–outer factorization of a TFM $G(\lambda)$ can be interpreted as a generalization of the orthogonal QR factorization of a real matrix. The inner factor can be seen as the generalization of a matrix with orthonormal columns. Its role in an inner–outer factorization is twofold: to compress the given $G(\lambda)$ to a full row rank TFM and to dislocate all zeros of $G(\lambda)$ lying in $\mathbb{C}_u$ into positions within $\mathbb{C}_s$, which are symmetric (in a certain sense) with respect to $\partial\mathbb{C}_s$. $\qquad\square$

In some applications, instead of the (compact) inner–outer factorization (9.21), an alternative (extended) factorization with square inner factor is desirable. The *extended inner–outer factorization* and *extended inner–quasi-outer factorization* have the form

$$G(\lambda) = \begin{bmatrix} G_i(\lambda) & G_i^\perp(\lambda) \end{bmatrix} \begin{bmatrix} G_o(\lambda) \\ 0 \end{bmatrix}, \tag{9.23}$$

where $G_i^\perp(\lambda)$ is the inner orthogonal complement of $G_i(\lambda)$ such that $\begin{bmatrix} G_i(\lambda) & G_i^\perp(\lambda) \end{bmatrix}$ is square and inner. Similarly, the *extended co-outer–co-inner factorization* and *extended quasi-co-outer–co-inner factorization* have the form

$$G(\lambda) = \begin{bmatrix} G_{co}(\lambda) & 0 \end{bmatrix} \begin{bmatrix} G_{ci}(\lambda) \\ G_{ci}^{\perp}(\lambda) \end{bmatrix}, \qquad (9.24)$$

where $G_{ci}^{\perp}(\lambda)$ is the co-inner orthogonal complement of $G_i(\lambda)$ such that $\begin{bmatrix} G_{ci}(\lambda) \\ G_{ci}^{\perp}(\lambda) \end{bmatrix}$ is square and co-inner (thus also inner).

The outer factor $G_o(\lambda)$ of an $G(\lambda)$ without zeros in $\partial\mathbb{C}_s$ satisfies

$$G^{\sim}(\lambda)G(\lambda) = G_o^{\sim}(\lambda)G_o(\lambda)$$

and, therefore, it is a solution of the *minimum-phase right spectral factorization* problem. Similarly, the co-outer factor $G_{co}(\lambda)$ of an $G(\lambda)$ without zeros on $\partial\mathbb{C}_s$ satisfies

$$G(\lambda)G^{\sim}(\lambda) = G_{co}(\lambda)G_{co}^{\sim}(\lambda)$$

and, therefore, it is a solution of the *minimum-phase left spectral factorization* problem.

Combining the LCF with inner denominator and the inner–outer factorization, we have for an arbitrary $G(\lambda)$, without poles and zeros on the boundary of the stability domain $\partial\mathbb{C}_s$, that

$$G(\lambda) = M_i^{-1}(\lambda)N(\lambda) = M_i^{-1}(\lambda)N_i(\lambda)N_o(\lambda),$$

where $M_i(\lambda)$ and $N_i(\lambda)$ are inner and $N_o(\lambda)$ is outer. It follows that the outer factor $N_o(\lambda)$ is the solution of the *stable minimum-phase right spectral factorization* problem

$$G^{\sim}(\lambda)G(\lambda) = N_o^{\sim}(\lambda)N_o(\lambda).$$

Similarly, by combining the RCF with inner denominator and the co-outer–co-inner factorization, we obtain

$$G(\lambda) = N(\lambda)M_i^{-1}(\lambda) = N_{co}(\lambda)N_{ci}(\lambda)M_i^{-1}(\lambda),$$

with $M_i(\lambda)$ inner, $N_{ci}(\lambda)$ co-inner and $N_{co}(\lambda)$ co-outer. Then, $N_{co}(\lambda)$ is the solution of the *stable minimum-phase left spectral factorization* problem

$$G(\lambda)G^{\sim}(\lambda) = N_{co}(\lambda)N_{co}^{\sim}(\lambda).$$

If $G(\lambda)$ has poles or zeros on the boundary of the stability domain $\partial\mathbb{C}_s$, then we can still achieve the above factorizations by including all poles and zeros of $G(\lambda)$ on $\partial\mathbb{C}_s$ in the resulting spectral factors $N_o(\lambda)$ or $N_{co}(\lambda)$.

9.1.9 *Linear Rational Matrix Equations*

For $G(\lambda) \in \mathbb{R}(\lambda)^{p \times m}$ and $F(\lambda) \in \mathbb{R}(\lambda)^{q \times m}$, consider the solution of the linear rational matrix equation

$$X(\lambda)G(\lambda) = F(\lambda) \tag{9.25}$$

where $X(\lambda) \in \mathbb{R}(\lambda)^{q \times p}$ is the solution we seek. The existence of a solution is guaranteed if the compatibility condition for the linear system (9.25) is fulfilled.

Lemma 9.4 *The rational equation (9.25) has a solution if and only if*

$$\operatorname{rank} G(\lambda) = \operatorname{rank} \begin{bmatrix} G(\lambda) \\ F(\lambda) \end{bmatrix}. \tag{9.26}$$

Let r be the rank of $G(\lambda)$. In the most general case, the solution of (9.25) (if it exists) is not unique and can be expressed as

$$X(\lambda) = X_0(\lambda) + Y(\lambda)N_l(\lambda), \tag{9.27}$$

where $X_0(\lambda)$ is a particular solution of (9.25), $N_l(\lambda) \in \mathbb{R}(\lambda)^{(p-r) \times p}$ is a rational matrix representing a basis of the left nullspace $\mathcal{N}_L(G(\lambda))$ (is empty if $r = p$), while $Y(\lambda) \in \mathbb{R}(\lambda)^{q \times (p-r)}$ is an arbitrary rational matrix. The particular solution $X_0(\lambda)$ can be expressed as $X_0(\lambda) = F(\lambda)G^+(\lambda)$, where $G^+(\lambda)$ is a particular generalized inverse (so-called {1}-inverse) of $G(\lambda)$.

An important aspect is to establish conditions, which ensure the existence of a solution $X(\lambda)$ having only poles in a "good" domain $\mathbb{C}_g$, or equivalently, $X(\lambda)$ having no poles in the "bad" domain $\mathbb{C}_b := \mathbb{C} \setminus \mathbb{C}_g$. Such a condition can be obtained in terms of the pole–zero structures of the rational matrices $G(\lambda)$ and $[\, G^T(\lambda) \; F^T(\lambda) \,]^T$ at a particular value λ_z of the frequency parameter λ.

Lemma 9.5 *The rational equation (9.25) has a solution without poles in $\mathbb{C}_b$ if and only if (9.26) is fulfilled and the rational matrices $G(\lambda)$ and $\begin{bmatrix} G(\lambda) \\ F(\lambda) \end{bmatrix}$ have the same pole–zero structure for all $\lambda_z \in \mathbb{C}_b$.*

All solutions of (9.25), without poles in $\mathbb{C}_b$, can be also expressed in the form (9.27), where all intervening matrices $X_o(\lambda)$, $N_l(\lambda)$ and $Y(\lambda)$ are rational matrices without poles in $\mathbb{C}_b$.

The characterization provided by Lemma 9.5 is relevant when solving synthesis problems of fault detection filters and controllers using an *exact model-matching* approach, where the physical realizability requires the properness and stability of the solutions (i.e., $\mathbb{C}_g = \mathbb{C}_s$). For example, if $G(\lambda)$ has unstable zeros in λ_z, then $F(\lambda)$ must be chosen to have the same or richer zero structure in order to ensure the cancellation of these zeros (appearing now as unstable poles of any generalized inverse $G^+(\lambda)$). The fixed poles in $\mathbb{C}_b$ correspond to those zeros of $G(\lambda)$ for which the above condition is not fulfilled, and thus no complete cancellations take place.

If the solvability condition (9.26) is satisfied, but no solution with poles only in $\mathbb{C}_g$ exists, then from any computed solution $X(\lambda)$ we can compute a LCF on $\mathbb{C}_g$ $X(\lambda) = M^{-1}(\lambda)\widetilde{X}(\lambda)$ with least McMillan degree $M(\lambda)$ such that $\widetilde{X}(\lambda)$ has only poles in $\mathbb{C}_g$ and satisfies $\widetilde{X}(\lambda)G(\lambda) = M(\lambda)F(\lambda)$. In this way, we can determine the least McMillan degree updating factor $M(\lambda)$ of the right-hand side $F(\lambda)$ for which the problem becomes solvable over $\mathbb{C}_g$.

An important aspect is the exploitation of the non-uniqueness of the solution in (9.25) when rank $G(\lambda) < p$, by determining a solution with the least possible McMillan degree. This problem is known in the literature as the *minimum design problem* (MDP) and primarily targets the reduction of the complexity of real-time burden when implementing filters or controllers. Of particular importance are proper and stable solutions, which are suitable for a physical (causal) realization. If the minimal degree solution is not proper and stable, then it is of interest to find a proper and stable solution with the least McMillan degree. Surprisingly, this problem does not have a satisfactory procedural solution; most of proposed approaches involve parametric searches using suitably parameterized solutions of given candidate degrees. The above-mentioned updating of $F(\lambda)$ by replacing it with $M(\lambda)F(\lambda)$ for a suitable factor $M(\lambda)$ allows to compute least McMillan degree stable and proper solutions for the updated problem. This order cannot be achieved if structural constraints on $M(\lambda)$ (e.g., diagonal form) are present.

9.1.10 *Approximate Model-Matching*

The formulation of the approximate model-matching problem can be done as an error minimization problem, where the approximate solution of the rational equation $X(\lambda)G(\lambda) = F(\lambda)$ involves the minimization of some norm (e.g., $\mathcal{L}_2$- or $\mathcal{L}_\infty$-norm) of the error $\mathcal{E}(\lambda) := F(\lambda) - X(\lambda)G(\lambda)$. For example, the standard formulation of the $\mathcal{H}_\infty$ *model-matching problem* ($\mathcal{H}_\infty$-MMP) is as follows: given $G(\lambda), F(\lambda) \in \mathcal{H}_\infty$, find $X(\lambda) \in \mathcal{H}_\infty$, which minimizes $\|\mathcal{E}(\lambda)\|_\infty$. The $\mathcal{H}_2$ *model-matching problem* ($\mathcal{H}_2$-MMP) has a similar formulation. We will use $\|\cdot\|_{\infty/2}$ to denote either the $\mathcal{L}_\infty$- or $\mathcal{L}_2$-norm.

The following results provide sufficient conditions for the solvability of the $\mathcal{H}_\infty$-MMP and $\mathcal{H}_2$-MMP in the standard case.

Lemma 9.6 *An optimal solution $X(\lambda)$ of the $\mathcal{H}_\infty$-MMP exists if $G(\lambda)$ has no zeros on $\partial\mathbb{C}_s$.*

Lemma 9.7 *An optimal solution $X(\lambda)$ of the $\mathcal{H}_2$-MMP exists if $G(\lambda)$ has no zeros on $\partial\mathbb{C}_s$ and, additionally, in the continuous time*

$$\text{rank } G(\infty) = \text{rank} \begin{bmatrix} G(\infty) \\ F(\infty) \end{bmatrix}. \tag{9.28}$$

The rank condition (9.28) merely ensures that the optimal solution $X(\lambda)$ can be chosen such that the resulting $\mathcal{E}(\lambda)$ is strictly proper and therefore $\|\mathcal{E}(\lambda)\|_2$ is finite.

In what follows, we will sketch a general approach for solving approximate model-matching problems based on transforming the error minimization problems into appropriate *least distance problems* (LDPs), by using factorization techniques. When solving $\mathcal{H}_\infty$- or $\mathcal{H}_2$-MMPs, we will additionally assume that $G(\lambda)$ has full row rank.

For the solution of both $\mathcal{H}_\infty$- and $\mathcal{H}_2$-MMPs, we employ the (extended) outer–inner factorization of $G(\lambda)$ to reduce these problems to $\mathcal{H}_\infty$- or $\mathcal{H}_2$-LDPs, respectively. Consider the extended factorization

$$G(\lambda) = \begin{bmatrix} G_o(\lambda) & 0 \end{bmatrix} G_i(\lambda) = \begin{bmatrix} G_o(\lambda) & 0 \end{bmatrix} \begin{bmatrix} G_{i,1}(\lambda) \\ G_{i,2}(\lambda) \end{bmatrix} = G_o(\lambda) G_{i,1}(\lambda),$$

where $G_i(\lambda) := \begin{bmatrix} G_{i,1}(\lambda) \\ G_{i,2}(\lambda) \end{bmatrix}$ is square and inner and $G_o(\lambda)$ is square and outer (therefore invertible in $\mathcal{H}_\infty$). This allows to write successively

$$\begin{aligned}
\|\mathcal{E}(\lambda)\|_{\infty/2} &= \|F(\lambda) - X(\lambda)G(\lambda)\|_{\infty/2} \\
&= \left\| \left(F(\lambda)G_i^\sim(\lambda) - X(\lambda)\begin{bmatrix} G_o(\lambda) & 0 \end{bmatrix} \right) G_i(\lambda) \right\|_{\infty/2} \\
&= \left\| \begin{bmatrix} \widetilde{F}_1(\lambda) - Y(\lambda) & \widetilde{F}_2(\lambda) \end{bmatrix} \right\|_{\infty/2},
\end{aligned}$$

where $Y(\lambda) := X(\lambda)G_o(\lambda) \in \mathcal{H}_\infty$ and

$$F(\lambda)G_i^\sim(\lambda) = \begin{bmatrix} F(\lambda)G_{i,1}^\sim(\lambda) \,\big|\, F(\lambda)G_{i,2}^\sim(\lambda) \end{bmatrix} := \begin{bmatrix} \widetilde{F}_1(\lambda) \,\big|\, \widetilde{F}_2(\lambda) \end{bmatrix}.$$

Thus, the problem of computing a stable $X(\lambda)$, which minimizes the error norm $\|\mathcal{E}(\lambda)\|_{\infty/2}$, has been reduced to a LDP to compute the stable solution $Y(\lambda)$, which minimizes the norm $\left\| \begin{bmatrix} \widetilde{F}_1(\lambda) - Y(\lambda) & \widetilde{F}_2(\lambda) \end{bmatrix} \right\|_{\infty/2}$. The solution of the original MMP is given by

$$X(\lambda) = Y(\lambda)G_o^{-1}(\lambda).$$

In general, we have that $\begin{bmatrix} \widetilde{F}_1(\lambda) & \widetilde{F}_2(\lambda) \end{bmatrix} \notin \mathcal{H}_\infty$. If $\widetilde{F}_2(\lambda)$ is present (i.e., $G(\lambda)$ is not square), we have a *2-block* LDP, while if $G(\lambda)$ is square, then $\widetilde{F}_2(\lambda)$ is not present and we have an *1-block* LDP. In what follows, we discuss shortly the solution approaches for the 1- and 2-block problems. Note that these approaches underlie the approximate synthesis methods of optimal fault detection filters presented in Sect. 5.7.

Solution of the 1-block $\mathcal{H}_\infty$-LDP. In the case of the $\mathcal{H}_\infty$-norm, the stable optimal solution $Y(\lambda)$ of the 1-block problem can be computed by solving an optimal Nehari problem. Let $L_s(\lambda)$ be the stable part and let $L_u(\lambda)$ be the unstable part in the additive decomposition

$$\widetilde{F}_1(\lambda) = L_s(\lambda) + L_u(\lambda). \tag{9.29}$$

Then, for the optimal solution, we have successively

$$\|\mathcal{E}(\lambda)\|_\infty = \|\widetilde{F}_1(\lambda) - Y(\lambda)\|_\infty = \|L_u(\lambda) - Y_s(\lambda)\|_\infty = \|L_u^\sim(\lambda)\|_H \,,$$

where $Y_s(\lambda)$ is the stable optimal Nehari solution and

$$Y(\lambda) = Y_s(\lambda) + L_s(\lambda)\,.$$

Solution of the 2-block $\mathcal{H}_\infty$-LDP. A stable optimal solution $Y(\lambda)$ of the 2-block LDP can be approximately determined as the solution of the suboptimal 2-block LDP

$$\|[\,\widetilde{F}_1(\lambda) - Y(\lambda)\ \ \widetilde{F}_2(\lambda)\,]\|_\infty < \gamma, \tag{9.30}$$

where $\gamma_{opt} < \gamma \le \gamma_{opt} + \varepsilon$, with ε an arbitrary user-specified (accuracy) tolerance for the least achievable value γ_{opt} of γ. The standard solution approach is a bisection-based γ-iteration method, where the solution of the 2-block problem is approximated by successively computed γ-suboptimal solutions of appropriately defined 1-block problems.

Let γ_l and γ_u be lower and upper bounds for γ_{opt}, respectively. Such bounds can be computed, for example, as

$$\gamma_l = \|\widetilde{F}_2(\lambda)\|_\infty, \quad \gamma_u = \big\|[\,\widetilde{F}_1(\lambda)\ \ \widetilde{F}_2(\lambda)\,]\big\|_\infty . \tag{9.31}$$

For a given $\gamma > \gamma_l$, we compute first a stable minimum-phase left spectral factorization

$$\gamma^2 I - \widetilde{F}_2(\lambda)\widetilde{F}_2^\sim(\lambda) = V(\lambda)V^\sim(\lambda), \tag{9.32}$$

where $V(\lambda)$ is biproper, stable and minimum-phase. Further, we compute the additive decomposition

$$V^{-1}(\lambda)\widetilde{F}_1(\lambda) = L_s(\lambda) + L_u(\lambda), \tag{9.33}$$

where $L_s(\lambda)$ is the stable part and $L_u(\lambda)$ is the unstable part. If $\gamma > \gamma_{opt}$, the suboptimal 2-block problem (9.30) is equivalent to the suboptimal 1-block problem

$$\big\|V^{-1}(\lambda)\big(\widetilde{F}_1(\lambda) - Y(\lambda)\big)\big\|_\infty \le 1 \tag{9.34}$$

and $\gamma_H := \|L_u^\sim(\lambda)\|_H < 1$. In this case, we readjust the upper bound to $\gamma_u = \gamma$. If $\gamma \le \gamma_{opt}$, then $\gamma_H \ge 1$ and we read just the lower bound to $\gamma_l = \gamma$. For the bisection value $\gamma = (\gamma_l + \gamma_u)/2$, we redo the factorization (9.32) and decomposition (9.33). This process is repeated until $\gamma_u - \gamma_l \le \varepsilon$.

At the end of iterations, we have either $\gamma_{opt} < \gamma \le \gamma_u$ if $\gamma_H < 1$ or $\gamma_l < \gamma \le \gamma_{opt}$ if $\gamma_H \ge 1$, in which case we set $\gamma = \gamma_u$. We compute the stable solution of (9.34) as

$$Y(\lambda) = V(\lambda)(L_s(\lambda) + Y_s(\lambda)), \tag{9.35}$$

where, for any γ_1 satisfying $1 \geq \gamma_1 > \gamma_H$, $Y_s(\lambda)$ is the stable solution of the optimal Nehari problem

$$\|L_u(\lambda) - Y_s(\lambda)\|_\infty = \|L_u^\sim(\lambda)\|_H. \qquad (9.36)$$

Solution of the $\mathcal{H}_2$-LDP. In the case of $\mathcal{H}_2$-norm, the solution of the LDP is

$$Y(\lambda) = L_s(\lambda),$$

where $L_s(\lambda)$ is the stable projection in (9.29). In the continuous-time case, we take the unstable projection $L_u(s)$ strictly proper. With the above choice, the achieved minimum $\mathcal{H}_2$-norm of $\mathcal{E}(\lambda)$ is

$$\|\mathcal{E}(\lambda)\|_2 = \|[\, L_u(\lambda) \ \widetilde{F}_2(\lambda)\,]\|_2.$$

Since the underlying TFMs are unstable, the $\mathcal{L}_2$-norm is used in the last equation. In the continuous-time case, according to Lemma 9.7, the error norm $\|\mathcal{E}(s)\|_2$ is finite only if $\widetilde{F}_2(s)$ is strictly proper.

9.2 Descriptor Systems

In this section, we present the main concepts and properties of systems in a generalized state-space form

$$\begin{aligned} E\lambda x(t) &= Ax(t) + Bu(t), \\ y(t) &= Cx(t) + Du(t), \end{aligned} \qquad (9.37)$$

where $x(t) \in \mathbb{R}^n$ is the state vector, $u(t) \in \mathbb{R}^m$ is the input vector and $y(t) \in \mathbb{R}^p$ is the output vector, and where λ is the differential operator $\lambda x(t) = \frac{d}{dt}x(t)$ for a continuous-time system and the advance operator $\lambda x(t) = x(t+1)$ for a discrete-time system. In all what follows, we assume E is square and possibly singular, and the pencil $A - \lambda E$ is regular (i.e., $\det(A - \lambda E) \not\equiv 0$). If $E = I_n$, we call the representation (9.37) a *standard state-space system*, while for $E \neq I_n$ we call (9.37) a *descriptor system*. The corresponding input–output representation of the descriptor system (9.37) is

$$\mathbf{y}(\lambda) = G(\lambda)\mathbf{u}(\lambda),$$

where, depending on the system type, $\lambda = s$, the complex variable in the Laplace transform for a continuous-time system, or $\lambda = z$, the complex variable in the $\mathcal{Z}$-transform for a discrete-time system, $\mathbf{y}(\lambda)$ and $\mathbf{u}(\lambda)$ are the Laplace- or $\mathcal{Z}$-transformed output and input vectors, respectively, and $G(\lambda)$ is the rational *transfer function matrix* (TFM) of the system defined as

$$G(\lambda) = C(\lambda E - A)^{-1}B + D. \qquad (9.38)$$

We alternatively denote descriptor systems of the form (9.37) with the quadruple $(A - \lambda E, B, C, D)$ or a standard state-space system with (A, B, C, D) (if $E = I_n$), and use the notation

$$G(\lambda) := \left[\begin{array}{c|c} A - \lambda E & B \\ \hline C & D \end{array} \right], \tag{9.39}$$

to relate the TFM $G(\lambda)$ to a particular descriptor system realization as in (9.37).

It is well known that a descriptor system of the form (9.37) is the most general description for a linear time-invariant system. Continuous-time descriptor systems arise frequently from modelling interconnected systems containing algebraic loops or constrained mechanical systems, which describe contact phenomena. Discrete-time descriptor representations are frequently used to model economic processes.

9.2.1 Descriptor Realizations of Rational Matrices

The main use of the descriptor systems in this book is to allow the manipulation of rational matrices via their descriptor representations. The main result, which allows this, is the following

Theorem 9.1 *For any rational matrix $G(\lambda) \in \mathbb{R}(\lambda)^{p \times m}$, there exist $n \geq 0$ and the real matrices $E, A \in \mathbb{R}^{n \times n}$, $B \in \mathbb{R}^{n \times m}$, $C \in \mathbb{R}^{p \times n}$ and $D \in \mathbb{R}^{p \times m}$, with $A - \lambda E$ regular, such that (9.38) holds.*

If $G(\lambda)$ is proper, this theorem is a well-known result of the realization theory of standard state-space systems. Using this, a simple constructive proof allows to obtain a descriptor realization of an improper $G(\lambda)$ by using the additive decomposition (see Sect. 9.1.5)

$$G(\lambda) = G_p(\lambda) + G_{pol}(\lambda),$$

where $G_p(\lambda)$ is the proper part of $G(\lambda)$ and $G_{pol}(\lambda)$ is its strict polynomial part. The proper part $G_p(\lambda)$ has a standard state-space realization (A_p, B_p, C_p, D_p) and for the strictly proper $\lambda^{-1} G_{pol}(\lambda^{-1})$, we can build another standard state-space realization $(A_{pol}, B_{pol}, C_{pol}, 0)$. Then, we obtain

$$G(\lambda) = \left[\begin{array}{c|c} A - \lambda E & B \\ \hline C & D \end{array} \right] := \left[\begin{array}{cc|c} A_p - \lambda I & 0 & B_p \\ 0 & I - \lambda A_{pol} & B_{pol} \\ \hline C_p & C_{pol} & D_p \end{array} \right].$$

The descriptor realization $(A - \lambda E, B, C, D)$ of a given rational matrix $G(\lambda)$ is not unique. For example, if U and V are invertible matrices of the size n of the square matrix E, then two descriptor realizations $(A - \lambda E, B, C, D)$ and $(\widetilde{A} - \lambda \widetilde{E}, \widetilde{B}, \widetilde{C}, \widetilde{D})$ related by a *system similarity transformation* of the form

$$(\widetilde{A} - \lambda \widetilde{E}, \widetilde{B}, \widetilde{C}, \widetilde{D}) = (UAV - \lambda UEV, UB, CV, D) \tag{9.40}$$

have the same TFM $G(\lambda)$. Moreover, among all possible realizations of a given $G(\lambda)$, with different sizes n, there exist realizations, which have the least dimension. A descriptor realization $(A - \lambda E, B, C, D)$ of the rational matrix $G(\lambda)$ is called *minimal* if the dimension n of the square matrices E and A is the least possible one. The minimal realization of a given $G(\lambda)$ is also not unique, since two minimal realizations related by a system similarity transformation as in (9.40) correspond to the same $G(\lambda)$.

A minimal descriptor system realization $(A - \lambda E, B, C, D)$ is characterized by the following five conditions.

Theorem 9.2 *A descriptor system realization $(A - \lambda E, B, C, D)$ of order n is minimal if the following conditions are fulfilled:*

$$(i) \ \operatorname{rank} \begin{bmatrix} A - \lambda E & B \end{bmatrix} = n, \quad \forall \lambda \in \mathbb{C},$$

$$(ii) \ \operatorname{rank} \begin{bmatrix} E & B \end{bmatrix} = n,$$

$$(iii) \ \operatorname{rank} \begin{bmatrix} A - \lambda E \\ C \end{bmatrix} = n, \quad \forall \lambda \in \mathbb{C},$$

$$(iv) \ \operatorname{rank} \begin{bmatrix} E \\ C \end{bmatrix} = n,$$

$$(v) \ A\mathcal{N}(E) \subseteq \mathcal{R}(E).$$

The conditions (i) and (ii) are known as *finite* and *infinite controllability*, respectively. A system, which fulfils both (i) and (ii), is called *controllable*. Similarly, the conditions (iii) and (iv) are known as *finite* and *infinite observability*, respectively. A system, which fulfils both (iii) and (iv), is called *observable*. A descriptor realization, which satisfies $(i) - (iv)$, is called *irreducible* (also weakly minimal). Condition (v) jointly with conditions $(i) - (iv)$ expresses the absence of non-dynamic modes. The numerical computation of minimal realizations is addressed in Sect. 10.3.1.

9.2.2 Poles, Zeros and Minimal Indices

Consider the irreducible descriptor system $(A - \lambda E, B, C, D)$ with the corresponding TFM $G(\lambda) \in \mathbb{R}(\lambda)^{p \times m}$. Two pencils play a fundamental role in defining the main structural elements of the rational matrix $G(\lambda)$ (see Sects. 9.1.3 and 9.1.4). The regular *pole pencil*

$$P(\lambda) := A - \lambda E \tag{9.41}$$

characterizes the pole structure of $G(\lambda)$, exhibited by the Weierstrass canonical form of the pole pencil $P(\lambda)$. The (singular) system pencil

$$S(\lambda) := \begin{bmatrix} A - \lambda E & B \\ C & D \end{bmatrix} \tag{9.42}$$

characterizes the zero structure of $G(\lambda)$, as well as the right and left singular structures of $G(\lambda)$, which are exhibited by the Kronecker canonical form of the system pencil $S(\lambda)$. Both canonical forms can be defined in terms of strict equivalence of linear pencils. Recall that two pencils $M - \lambda N$ and $\widetilde{M} - \lambda \widetilde{N}$ with $M, N, \widetilde{M}, \widetilde{N} \in \mathbb{C}^{m \times n}$ are *strictly equivalent* if there exist two invertible matrices $U \in \mathbb{C}^{m \times m}$ and $V \in \mathbb{C}^{n \times n}$ such that

$$U(M - \lambda N)V = \widetilde{M} - \lambda \widetilde{N}. \tag{9.43}$$

For a regular pencil, the strict equivalence leads to the (complex) *Weierstrass canonical form*, which is instrumental to characterize the poles of a descriptor system.

Lemma 9.8 *Let $M - \lambda N$ be an arbitrary regular pencil with $M, N \in \mathbb{C}^{n \times n}$. Then, there exist invertible matrices $U \in \mathbb{C}^{n \times n}$ and $V \in \mathbb{C}^{n \times n}$ such that*

$$U(M - \lambda N)V = \begin{bmatrix} J_f - \lambda I & 0 \\ 0 & I - \lambda J_\infty \end{bmatrix}, \tag{9.44}$$

where J_f is in a (complex) Jordan canonical form

$$J_f = \mathrm{diag}\left(J_{s_1}(\lambda_1), J_{s_2}(\lambda_2), \ldots, J_{s_k}(\lambda_k) \right), \tag{9.45}$$

with $J_{s_i}(\lambda_i)$ an elementary $s_i \times s_i$ Jordan block of the form

$$J_{s_i}(\lambda_i) = \begin{bmatrix} \lambda_i & 1 & & \\ & \lambda_i & \ddots & \\ & & \ddots & 1 \\ & & & \lambda_i \end{bmatrix}$$

and J_∞ is nilpotent and has the (nilpotent) Jordan form

$$J_\infty = \mathrm{diag}\left(J_{s_1^\infty}(0), J_{s_2^\infty}(0), \ldots, J_{s_h^\infty}(0) \right). \tag{9.46}$$

The Weierstrass canonical form (9.44) exhibits the finite and infinite eigenvalues of the pencil $M - \lambda N$. Overall, by including all multiplicities, there are $n_f = \sum_{i=1}^k s_i$ *finite eigenvalues* and $n_\infty = \sum_{i=1}^h s_i^\infty$ *infinite eigenvalues*. Infinite eigenvalues with $s_i^\infty = 1$ are called *simple infinite eigenvalues*. We can also express the rank of N as

$$\mathrm{rank}\, N = n_f + \mathrm{rank}\, J_\infty = n_f + \sum_{i=1}^h (s_i^\infty - 1) = n_f + n_\infty - h = n - h.$$

If M and N are real matrices, then there exist real matrices U and V such that the pencil $U(M - \lambda N)V$ is in a *real Weierstrass canonical form*, where the only difference is that J_f is in a real Jordan form. In this form, the elementary real Jordan blocks correspond to pairs of complex conjugate eigenvalues. If $M - \lambda N = A - \lambda I$ (e.g., the pole pencil for a standard state-space system), then all eigenvalues are finite and J_f in the Weierstrass form is simply the (real) Jordan form of A. The transformation matrices can be chosen such that $U = V^{-1}$.

For a general (singular) pencil, the strict equivalence leads to the (complex) *Kronecker canonical form*, which is instrumental to characterize the zeros and singularities of a descriptor system.

Lemma 9.9 *Let $M - \lambda N$ be an arbitrary pencil with $M, N \in \mathbb{C}^{m \times n}$. Then, there exist invertible matrices $U \in \mathbb{C}^{m \times m}$ and $V \in \mathbb{C}^{n \times n}$ such that*

$$
U(M - \lambda N)V = \begin{bmatrix} K_r(\lambda) & & \\ & K_{reg}(\lambda) & \\ & & K_l(\lambda) \end{bmatrix}, \tag{9.47}
$$

where

(1) The full row rank pencil $K_r(\lambda)$ has the form

$$
K_r(\lambda) = \mathrm{diag}\left(L_{\epsilon_1}(\lambda), L_{\epsilon_2}(\lambda), \cdots, L_{\epsilon_{v_r}}(\lambda)\right),
$$

with $L_i(\lambda)$ ($i \geq 0$) an $i \times (i+1)$ bidiagonal pencil of form

$$
L_i(\lambda) = \begin{bmatrix} -\lambda & 1 & & \\ & \ddots & \ddots & \\ & & -\lambda & 1 \end{bmatrix}; \tag{9.48}
$$

(2) The regular pencil $K_{reg}(\lambda)$ is in a Weierstrass canonical form

$$
K_{reg}(\lambda) = \begin{bmatrix} \widetilde{J}_f - \lambda I & \\ & I - \lambda \widetilde{J}_\infty \end{bmatrix},
$$

with $\widetilde{J}_f$ in a (complex) Jordan canonical form as in (9.45) and with $\widetilde{J}_\infty$ in a nilpotent Jordan form as in (9.46);
(3) The full column rank $K_l(\lambda)$ has the form

$$
K_l(\lambda) = \mathrm{diag}\left(L_{\eta_1}^T(\lambda), L_{\eta_2}^T(\lambda), \cdots, L_{\eta_{v_l}}^T(\lambda)\right).
$$

As it is apparent from (9.47), the Kronecker canonical form exhibits the right and left singular structures of the pencil $M - \lambda N$ via the full row rank block $K_r(\lambda)$ and full column rank block $K_l(\lambda)$, respectively, and the eigenvalue structure via the regular pencil $K_{reg}(\lambda)$. The full row rank pencil $K_r(\lambda)$ is $n_r \times (n_r + v_r)$, where $n_r =$

$\sum_{i=1}^{\nu_r} \epsilon_i$, the full column rank pencil $K_l(\lambda)$ is $(n_l + \nu_l) \times n_l$, where $n_l = \sum_{j=1}^{\nu_l} \eta_j$, while the regular pencil $K_{reg}(\lambda)$ is $n_{reg} \times n_{reg}$, with $n_{reg} = \tilde{n}_f + \tilde{n}_\infty$, where $\tilde{n}_f$ is the number of finite eigenvalues in $\Lambda(\tilde{J}_f)$ and $\tilde{n}_\infty$ is the number of infinite eigenvalues in $\Lambda(I - \lambda \tilde{J}_\infty)$ (or equivalently the number of null eigenvalues in $\Lambda(\tilde{J}_\infty)$). The $\epsilon_i \times (\epsilon_i + 1)$ blocks $L_{\epsilon_i}(\lambda)$ with $\epsilon_i \geq 0$ are the right elementary Kronecker blocks, and ϵ_i, for $i = 1, \ldots, \nu_r$, are called the *right Kronecker indices*. The $(\eta_i + 1) \times \eta_i$ blocks $L_{\eta_i}^T(\lambda)$ with $\eta_i \geq 0$ are the left elementary Kronecker blocks, and η_i, for $i = 1, \ldots, \nu_l$, are called the *left Kronecker indices*. The normal rank r of the pencil $M - \lambda N$ results as

$$r := \operatorname{rank}(M - \lambda N) = n_r + \tilde{n}_f + \tilde{n}_\infty + n_l.$$

If $M - \lambda N$ is regular, then there are no left and right Kronecker structures, and the Kronecker canonical form is simply the Weierstrass canonical form.

Remark 9.2 By additional column permutations of the block $K_r(\lambda)$ and row permutations of the block $K_l(\lambda)$ (which can be included in the left and right transformation matrices U and V), we can bring these blocks to the alternative forms

$$K_r(\lambda) = \begin{bmatrix} B_r \; A_r - \lambda I_{n_r} \end{bmatrix}, \qquad K_l(\lambda) = \begin{bmatrix} A_l - \lambda I_{n_l} \\ C_l \end{bmatrix}, \tag{9.49}$$

where the pair (A_r, B_r) is in a Brunovsky controllable form

$$A_r = \begin{bmatrix} A_{r,1} & & & \\ & A_{r,2} & & \\ & & \ddots & \\ & & & A_{r,\nu_r} \end{bmatrix}, \qquad B_r = \begin{bmatrix} b_{r,1} & & & \\ & b_{r,2} & & \\ & & \ddots & \\ & & & b_{r,\nu_r} \end{bmatrix},$$

with $A_{r,i}$ an $\varepsilon_i \times \varepsilon_i$ matrix and $b_{r,i}$ an $\varepsilon_i \times 1$ column vector of the forms

$$A_{r,i} = \begin{bmatrix} 0 & I_{\varepsilon_i - 1} \\ 0 & 0 \end{bmatrix} = J_{\varepsilon_i}(0), \qquad b_{r,i} = \begin{bmatrix} 0 \\ \vdots \\ 0 \\ 1 \end{bmatrix},$$

and the pair (A_l, C_l) is in a Brunovsky observable form

$$A_l = \begin{bmatrix} A_{l,1} & & & \\ & A_{l,2} & & \\ & & \ddots & \\ & & & A_{l,\nu_l} \end{bmatrix}, \qquad C_l = \begin{bmatrix} c_{l,1} & & & \\ & c_{l,2} & & \\ & & \ddots & \\ & & & c_{l,\nu_l} \end{bmatrix}, \tag{9.50}$$

with $A_{l,i}$ an $\eta_i \times \eta_i$ matrix and $c_{l,i}$ a $1 \times \eta_i$ row vector of the forms

$$A_{l,i} = \begin{bmatrix} 0 & 0 \\ I_{\eta_i - 1} & 0 \end{bmatrix} = J_{\eta_i}^T(0), \qquad c_{l,i} = \begin{bmatrix} 0 & \cdots & 0 & 1 \end{bmatrix}.$$

$\square$

The main structural results for the rational TFM $G(\lambda)$ can be stated in terms of its irreducible descriptor system realization $(A - \lambda E, B, C, D)$ of order n. The following facts rely on the Weierstrass canonical form (9.44) of the pole pencil $P(\lambda)$ in (9.41) (see Lemma 9.8) and the Kronecker canonical form (9.47) of the system pencil $S(\lambda)$ in (9.42) (see Lemma 9.9):

(1) The *finite poles* of $G(\lambda)$ are the finite eigenvalues of the pole pencil $P(\lambda)$ and are the eigenvalues (counting multiplicities) of the matrix J_f in the Weierstrass canonical form (9.44) of the pencil $P(\lambda)$.

(2) The *infinite poles* of $G(\lambda)$ have multiplicities defined by the multiplicities of the infinite eigenvalues of the pole pencil $P(\lambda)$ minus 1 and are the dimensions minus 1 of the nilpotent Jordan blocks in the matrix J_∞ in the Weierstrass canonical form (9.44) of the pencil $P(\lambda)$.

(3) The *finite zeros* of $G(\lambda)$ are the n_f finite eigenvalues (counting multiplicities) of the system pencil $S(\lambda)$ and are the eigenvalues (counting multiplicities) of the matrix $\widetilde{J}_f$ in the Kronecker canonical form (9.47) of the pencil $S(\lambda)$.

(4) The *infinite zeros* of $G(\lambda)$ have multiplicities defined by the multiplicities of the infinite eigenvalues of the system pencil $S(\lambda)$ minus 1 and are the dimensions minus 1 of the nilpotent Jordan blocks in the matrix $\widetilde{J}_\infty$ in the Kronecker canonical form (9.47) of the pencil $S(\lambda)$.

(5) The *left minimal indices* of $G(\lambda)$ are pairwise equal to the left Kronecker indices of $S(\lambda)$ and are the row dimensions ε_i of the blocks $L_{\varepsilon_i}(\lambda)$ for $i = 1, \ldots, \nu_r$ in the Kronecker canonical form (9.47) of the pencil $S(\lambda)$.

(6) The *right minimal indices* of $G(\lambda)$ are pairwise equal to the right Kronecker indices of $S(\lambda)$ and are the column dimensions η_i of the blocks $L_{\eta_i}^T(\lambda)$ for $i = 1, \ldots, \nu_l$ in the Kronecker canonical form (9.47) of the pencil $S(\lambda)$.

(7) The *normal rank* of $G(\lambda)$ is $r = \operatorname{rank} S(\lambda) - n = n_r + \tilde{n}_f + \tilde{n}_\infty + n_l - n$.

These facts allow to formulate simple conditions to characterize some pole–zero-related properties of an irreducible descriptor system $(A - \lambda E, B, C, D)$ such as properness, stability or minimum-phase, in terms of the eigenvalues of the pole and system pencils. The descriptor system $(A - \lambda E, B, C, D)$ is *proper* if all infinite eigenvalues of the regular pencil $A - \lambda E$ are simple (i.e., the system has no infinite poles). It is straightforward to show using the Weierstrass canonical form of the pencil $A - \lambda E$ that any irreducible proper descriptor system can be always reduced to a minimal order descriptor system, with the descriptor matrix E invertible, or even to a standard state-space representation with $E = I$ (see Sect. 7.2.2 for numerical procedures). The irreducible descriptor system $(A - \lambda E, B, C, D)$ is *improper* if the regular pencil $A - \lambda E$ has at least one infinite eigenvalue, which is not simple (i.e., has at least one infinite pole). A *polynomial* descriptor system is one for which $A - \lambda E$ has only infinite eigenvalues of which at least one is not simple (i.e., has only infinite poles). The concept of stability involves naturally the properness of the system. The irreducible descriptor system $(A - \lambda E, B, C, D)$ is *exponentially stable* if it has only finite poles and all poles belong to the stable region $\mathbb{C}_s$ (the pencil $A - \lambda E$ still can have simple infinite eigenvalues). The irreducible descriptor system $(A - \lambda E, B, C, D)$ is *unstable* if it has at least one finite pole outside of the stability domain or at least one infinite pole. The finite poles (or finite eigenvalues) inside the stability domain are called *stable poles* (*stable eigenvalues*), while the poles lying outside of the stability domain are called *unstable poles*. The irreducible descriptor system $(A - \lambda E, B, C, D)$ is *minimum-phase* if it has only finite zeros and all finite zeros belong to the stable region $\mathbb{C}_s$.

To check the finite controllability condition (*i*) and finite observability condition (*iii*) of Theorem 9.2, it is sufficient to check that

$$\text{rank} \begin{bmatrix} A - \lambda_i E & B \end{bmatrix} = n \tag{9.51}$$

and, respectively,

$$\text{rank} \begin{bmatrix} A - \lambda_i E \\ C \end{bmatrix} = n \tag{9.52}$$

for all distinct finite eigenvalues λ_i in the Weierstrass canonical form (9.44) of the pencil $A - \lambda E$. A finite eigenvalue λ_i is *controllable* if (9.51) is fulfilled, and *uncontrollable* otherwise. Similarly, a finite eigenvalue λ_i is *observable* if (9.52) is fulfilled, and *unobservable* otherwise. If the rank conditions (9.51) are fulfilled for all $\lambda_i \in \overline{\mathbb{C}}_u$, we call the descriptor system $(A - \lambda E, B, C, D)$ (or equivalently the pair $(A - \lambda E, B)$) *finite stabilizable*. Finite stabilizability guarantees the existence of a state-feedback matrix $F \in \mathbb{R}^{m \times n}$ such that all finite eigenvalues of $A + BF - \lambda E$ lie in $\mathbb{C}_s$. If the rank conditions (9.52) are fulfilled for all $\lambda_i \in \overline{\mathbb{C}}_u$, we call the descriptor system $(A - \lambda E, B, C, D)$ (or equivalently the pair $(A - \lambda E, C)$) *finite detectable*. Finite detectability guarantees the existence of an output injection matrix $K \in \mathbb{R}^{n \times p}$ such that all finite eigenvalues of $A + KC - \lambda E$ lie in $\mathbb{C}_s$.

The notion of strong stabilizability is related to the existence of a state-feedback matrix F such that all finite eigenvalues of $A + BF - \lambda E$ lie in $\mathbb{C}_s$ and all infinite eigenvalues of $A + BF - \lambda E$ are simple. The necessary and sufficient conditions for the existence of such an F is the *strong stabilizability* of the pair $(A - \lambda E, B)$, that is, (1) the finite stabilizability of the pair $(A - \lambda E, B)$; and (2) $\text{rank}[\, E \ AN_\infty \ B \,] = n$, where the columns of N_∞ form a basis of $\mathcal{N}(E)$. Similarly, strong detectability is related to the existence of an output injection matrix K such that all finite eigenvalues of $A + KC - \lambda E$ lie in $\mathbb{C}_s$ and all infinite eigenvalues of $A + KC - \lambda E$ are simple. The necessary and sufficient conditions for the existence of such a K is the *strong detectability* of the pair $(A - \lambda E, C)$, that is, (1) the finite detectability of the pair $(A - \lambda E, C)$; and (2) $\text{rank}[\, E^T \ A^T L_\infty \ C^T \,] = n$, where the columns of L_∞ form a basis of $\mathcal{N}(E^T)$.

Remark 9.3 The reduction of matrix pencils to the Weierstrass or Kronecker canonical forms generally involves the use of non-orthogonal, possibly ill-conditioned, transformation matrices. Therefore, the computation of these forms must be avoided when devising numerically reliable algorithms for descriptor systems. Alternative condensed forms, as the generalized real Schur form or various Kronecker-like forms, can be determined by using exclusively orthogonal transformations and can be often used instead of the Weierstrass or Kronecker canonical forms, respectively. Suitable algorithms for the computations of these alternative forms are discussed in Sects. 10.1.4 and 10.1.6.   □

9.2.3 *Operations with Rational Matrices*

In this section, we present some of the most frequently used operations on the TFMs of descriptor systems. For all operations, we will assume that the conditions to perform these operations (e.g., lack of particular poles, dimensional compatibility) are fulfilled. First, we address operations on a single TFM $G(\lambda)$ with the descriptor realization $(A - \lambda E, B, C, D)$. The *transposed* TFM $G^T(\lambda)$ corresponds to the *dual descriptor system* with the realization

$$G^T(\lambda) = \left[\begin{array}{c|c} A^T - \lambda E^T & C^T \\ \hline B^T & D^T \end{array} \right].$$

An alternative realization of the transpose, which, for example, preserves the upper triangular, upper quasi-triangular or upper block-triangular structure of the pencil $A - \lambda E$, is

$$G^T(\lambda) = \left[\begin{array}{c|c} PA^T P - \lambda PE^T P & PC^T \\ \hline B^T P & D^T \end{array} \right],$$

where P is the particular permutation matrix

$$P = \left[\begin{array}{ccc} 0 & & 1 \\ & \cdot^{\cdot^{\cdot}} & \\ 1 & & 0 \end{array} \right].$$

If $G(\lambda)$ is invertible, then an inversion-free realization of the *inverse* TFM $G^{-1}(\lambda)$ is given by

$$G^{-1}(\lambda) = \left[\begin{array}{cc|c} A - \lambda E & B & 0 \\ C & D & I \\ \hline 0 & -I & 0 \end{array} \right]. \tag{9.53}$$

This realization is not minimal, even if the original realization is minimal. If D is invertible, then an alternative realization of the inverse is

$$G^{-1}(\lambda) = \left[\begin{array}{c|c} A - BD^{-1}C - \lambda E & -BD^{-1} \\ \hline D^{-1}C & D^{-1} \end{array}\right],$$

which is minimal if the original realization is minimal.

The *conjugate* (or *adjoint*) TFM $G^{\sim}(\lambda)$ is defined in the continuous-time case as $G^{\sim}(s) = G^{T}(-s)$ and has the realization

$$G^{\sim}(s) = \left[\begin{array}{c|c} -A^{T} - sE^{T} & C^{T} \\ \hline -B^{T} & D^{T} \end{array}\right], \tag{9.54}$$

while in the discrete-time case $G^{\sim}(z) = G^{T}(1/z)$ and has the realization

$$G^{\sim}(z) = \left[\begin{array}{cc|c} E^{T} - zA^{T} & 0 & -C^{T} \\ zB^{T} & I & D^{T} \\ \hline 0 & I & 0 \end{array}\right]. \tag{9.55}$$

If $G(z)$ has a standard state-space realization (A, B, C, D) with A invertible, then an alternative realization of $G^{\sim}(z)$ is

$$G^{\sim}(z) = \left[\begin{array}{c|c} A^{-T} - zI & -A^{-T}C^{T} \\ \hline B^{T}A^{-T} & D^{T} - B^{T}A^{-T}C^{T} \end{array}\right].$$

The transformed discrete-time $G(z)$ via the bilinear transformation $z = \frac{1+s}{1-s}$ has the continuous-time realization

$$G\left(\tfrac{1+s}{1-s}\right) = \left[\begin{array}{c|c} A - E - s(A + E) & \sqrt{2}B \\ \hline \sqrt{2}C(A + E)^{-1}E & D - C(A + E)^{-1}B \end{array}\right], \tag{9.56}$$

or alternatively

$$G\left(\tfrac{1+s}{1-s}\right) = \left[\begin{array}{c|c} A - E - s(A + E) & \sqrt{2}E(A + E)^{-1}B \\ \hline \sqrt{2}C & D - C(A + E)^{-1}B \end{array}\right]. \tag{9.57}$$

By this transformation, the infinite eigenvalues of the pencil $A - zE$ become the eigenvalues at $s = 1$ of the pencil $A - E - s(A + E)$. It is straightforward to observe that if $A - zE$ has h nilpotent Jordan blocks in its Weierstrass canonical form, then there will be h unobservable or h uncontrollable eigenvalues at $s = 1$ in the realization of $G\left(\tfrac{1+s}{1-s}\right)$ according to which realization is used, (9.56) or (9.57), respectively. Therefore, in general, the realization (9.56) may not be detectable, while the realization (9.57) may not be stabilizable.

Conversely, the transformed continuous-time $G(s)$ obtained via the bilinear transformation $s = \frac{z-1}{z+1}$ has the discrete-time realization

$$G\left(\tfrac{z-1}{z+1}\right) = \left[\begin{array}{c|c} A + E - z(E - A) & \sqrt{2}B \\ \hline \sqrt{2}C(E - A)^{-1}E & D + C(E - A)^{-1}B \end{array}\right], \qquad (9.58)$$

or alternatively

$$G\left(\tfrac{z-1}{z+1}\right) = \left[\begin{array}{c|c} A + E - z(E - A) & \sqrt{2}E(E - A)^{-1}B \\ \hline \sqrt{2}C & D + C(E - A)^{-1}B \end{array}\right]. \qquad (9.59)$$

By this transformation, the infinite eigenvalues of the pencil $A - sE$ become the eigenvalues at $z = -1$ of the pencil $A + E - z(E - A)$. It is straightforward to observe that if $A - sE$ has h nilpotent Jordan blocks in its Weierstrass canonical form, then there will be h unobservable or h uncontrollable eigenvalues at $z = -1$ in the realization of $G\left(\tfrac{z-1}{z+1}\right)$ according to which realization is used, (9.58) or (9.59), respectively. Therefore, in general, the realization (9.58) may not be detectable, while the realization (9.59) may not be stabilizable.

Consider now two systems with TFMs $G_1(\lambda)$ having the descriptor realization $(A_1 - \lambda E_1, B_1, C_1, D_1)$ and $G_2(\lambda)$ having the descriptor realization $(A_2 - \lambda E_2, B_2, C_2, D_2)$. The product $G_1(\lambda)G_2(\lambda)$ represents the *series coupling* of the two systems and has the descriptor realization

$$G_1(\lambda)G_2(\lambda) := \left[\begin{array}{cc|c} A_1 - \lambda E_1 & B_1 C_2 & B_1 D_2 \\ 0 & A_2 - \lambda E_2 & B_2 \\ \hline C_1 & D_1 C_2 & D_1 D_2 \end{array}\right].$$

The *parallel coupling* corresponds to the sum $G_1(\lambda) + G_2(\lambda)$ and has the realization

$$G_1(\lambda) + G_2(\lambda) := \left[\begin{array}{cc|c} A_1 - \lambda E_1 & 0 & B_1 \\ 0 & A_2 - \lambda E_2 & B_2 \\ \hline C_1 & C_2 & D_1 + D_2 \end{array}\right].$$

The *column concatenation* of the two systems corresponds to building $\begin{bmatrix} G_1(\lambda) \\ G_2(\lambda) \end{bmatrix}$ and has the realization

$$\begin{bmatrix} G_1(\lambda) \\ G_2(\lambda) \end{bmatrix} = \left[\begin{array}{cc|c} A_1 - \lambda E_1 & 0 & B_1 \\ 0 & A_2 - \lambda E_2 & B_2 \\ \hline C_1 & 0 & D_1 \\ 0 & C_2 & D_2 \end{array}\right].$$

The *row concatenation* of the two systems corresponds to building $\begin{bmatrix} G_1(\lambda) & G_2(\lambda) \end{bmatrix}$ and has the realization

$$\begin{bmatrix} G_1(\lambda) & G_2(\lambda) \end{bmatrix} = \left[\begin{array}{cc|cc} A_1 - \lambda E_1 & 0 & B_1 & 0 \\ 0 & A_2 - \lambda E_2 & 0 & B_2 \\ \hline C_1 & C_2 & D_1 & D_2 \end{array}\right].$$

The *diagonal stacking* of the two systems corresponds to building $\begin{bmatrix} G_1(\lambda) & 0 \\ 0 & G_2(\lambda) \end{bmatrix}$ and has the realization

$$\begin{bmatrix} G_1(\lambda) & 0 \\ 0 & G_2(\lambda) \end{bmatrix} = \left[\begin{array}{cc|cc} A_1 - \lambda E_1 & 0 & B_1 & 0 \\ 0 & A_2 - \lambda E_2 & 0 & B_2 \\ \hline C_1 & 0 & D_1 & 0 \\ 0 & C_2 & 0 & D_2 \end{array}\right].$$

9.2.4 *Minimal Rational Nullspace Bases*

Let $G(\lambda)$ be a $p \times m$ rational matrix of normal rank r, and let $(A - \lambda E, B, C, D)$ be a descriptor realization of $G(\lambda)$. To determine a basis $N_l(\lambda)$ of the left nullspace of $G(\lambda)$, we can exploit the simple fact that $N_l(\lambda)$ is a left nullspace basis of $G(\lambda)$ if and only if, for a suitable $M_l(\lambda)$,

$$Y_l(\lambda) := \begin{bmatrix} M_l(\lambda) & N_l(\lambda) \end{bmatrix} \tag{9.60}$$

is a left nullspace basis of the associated system pencil $S(\lambda)$ (9.42). Thus, to compute $N_l(\lambda)$ we can determine first a left nullspace basis $Y_l(\lambda)$ for $S(\lambda)$ and then $N_l(\lambda)$ results as

$$N_l(\lambda) = Y_l(\lambda) \begin{bmatrix} 0 \\ I_p \end{bmatrix}.$$

By duality, if $Y_r(\lambda)$ is a right nullspace basis for $S(\lambda)$, then a right nullspace basis of $G(\lambda)$ is given by

$$N_r(\lambda) = \begin{bmatrix} 0 & I_m \end{bmatrix} Y_r(\lambda).$$

The Kronecker canonical form (9.47) of the system pencil $S(\lambda)$ in (9.42) allows to easily determine left and right nullspace bases of $G(\lambda)$. Let $\overline{S}(\lambda) = US(\lambda)V$ be the Kronecker canonical form (9.47) of $S(\lambda)$, where U and V are the respective left and right transformation matrices. If $\overline{Y}_l(\lambda)$ is a left nullspace basis of $\overline{S}(\lambda)$ then

$$N_l(\lambda) = \overline{Y}_l(\lambda)U \begin{bmatrix} 0 \\ I_p \end{bmatrix}. \tag{9.61}$$

Similarly, if $\overline{Y}_r(\lambda)$ is a right nullspace basis of $\overline{S}(\lambda)$ then

$$N_r(\lambda) = \begin{bmatrix} 0 & I_m \end{bmatrix} V \overline{Y}_r(\lambda). \tag{9.62}$$

We choose $\overline{Y}_l(\lambda)$ of the form

$$\overline{Y}_l(\lambda) = \begin{bmatrix} 0 & \overline{Y}_{l,3}(\lambda) \end{bmatrix}, \tag{9.63}$$

where $\overline{Y}_{l,3}(\lambda)$ satisfies $\overline{Y}_{l,3}(\lambda)K_l(\lambda) = 0$. Similarly, we choose $\overline{Y}_r(\lambda)$ of the form

$$\overline{Y}_r(\lambda) = \begin{bmatrix} \overline{Y}_{r,1}(\lambda) \\ 0 \end{bmatrix}, \tag{9.64}$$

where $\overline{Y}_{r,1}(\lambda)$ satisfies $K_r(\lambda)\overline{Y}_{r,1}(\lambda) = 0$. Both $\overline{Y}_{l,3}(\lambda)$ and $\overline{Y}_{r,1}(\lambda)$ can be determined as polynomial or rational matrices and the resulting bases are polynomial or rational as well.

Example 9.1 We show exemplarily how to determine a minimal left nullspace basis. We can choose $\overline{Y}_{l,3}(\lambda)$ in (9.63) a $\nu_l \times (\nu_l + n_l)$ polynomial matrix with a block-diagonal form

$$\overline{Y}_{l,3}(\lambda) = \begin{bmatrix} w_1(\lambda) & & & \\ & w_2(\lambda) & & \\ & & \ddots & \\ & & & w_{\nu_l}(\lambda) \end{bmatrix},$$

with $w_i(\lambda)$ a $1 \times (\eta_i + 1)$ least degree polynomial vector satisfying $w_i(\lambda)L_{\eta_i}^T(\lambda) = 0$. Each vector $w_i(\lambda)$ can be chosen of the form

$$w_i(\lambda) = \begin{bmatrix} 1 & -\lambda & \lambda^2 & \cdots & (-\lambda)^{\eta_i} \end{bmatrix}.$$

If we denote $v_i(\lambda)$ as the i-th row vector of the nullspace basis $N_l(\lambda)$, it follows, by taking into account (9.61) and (9.63), that $v_i(\lambda)$ is a polynomial vector of degree η_i. The resulting $N_l(\lambda)$ has degree n_l (i.e., the sum of row degrees), and thus is a minimal polynomial basis for the left nullspace of $G(\lambda)$. From linear algebra arguments, we have that the number ν_l of basis vectors in $N_l(\lambda)$ is $\nu_l = p - r$. $\Diamond$

We can construct simple minimal proper rational nullspace bases by assuming that U and V have been updated to include row and column permutations such that $K_r(\lambda)$ and $K_l(\lambda)$ have the forms in (9.49). We compute first

$$U \begin{bmatrix} 0 \\ I_p \end{bmatrix} = \begin{bmatrix} * \\ B_l \\ D_l \end{bmatrix}, \tag{9.65}$$

where $B_l \in \mathbb{R}^{n_l \times p}$ and $D_l \in \mathbb{R}^{\nu_l \times p}$. Then, by choosing $\overline{Y}_{l,3}(\lambda)$ in (9.63) as

$$\overline{Y}_{l,3}(\lambda) = \begin{bmatrix} C_l(\lambda I - A_l)^{-1} & I \end{bmatrix},$$

we obtain from (9.61) with (9.65)

$$N_l(\lambda) = C_l(\lambda I - A_l)^{-1}B_l + D_l. \tag{9.66}$$

By direct computation, we can verify that the i-th row of $N_l(\lambda)$ has McMillan degree η_i; thus the total order of the realization (A_l, B_l, C_l, D_l) is n_l. It follows that $N_l(\lambda)$ is a simple minimal proper rational left nullspace basis.

Consider the transformation matrix

$$\overline{U} = \begin{bmatrix} I_{n_r+n_{reg}} & 0 & 0 \\ 0 & I_{n_l} & K \\ 0 & 0 & I_{v_l} \end{bmatrix} =: \begin{bmatrix} I_{n_r+n_{reg}} & 0 \\ 0 & U_l \end{bmatrix} \tag{9.67}$$

and compute $\widetilde{S}(\lambda) := \overline{U} U S_G(\lambda) V$, which has the updated full column rank block

$$U_l K_l(\lambda) = \begin{bmatrix} A_l + KC_l - \lambda I_{n_l} \\ C_l \end{bmatrix}.$$

Since the pair (A_l, C_l) is observable, we can assign the eigenvalues of $A_l + KC_l$ to arbitrary values (e.g., in a "good" domain $\mathbb{C}_g \subset \mathbb{C}$). From

$$\overline{U} U \begin{bmatrix} 0 \\ I_p \end{bmatrix} = \begin{bmatrix} * \\ B_l + KD_l \\ D_l \end{bmatrix}, \tag{9.68}$$

we obtain an alternative expression of the nullspace basis

$$\widetilde{N}_l(\lambda) = C_l(\lambda I - A_l - KC_l)^{-1}(B_l + KD_l) + D_l. \tag{9.69}$$

Similarly, we can compute

$$\begin{bmatrix} D_r & C_r & * \end{bmatrix} = \begin{bmatrix} 0 & I_m \end{bmatrix} V, \tag{9.70}$$

where $D_r \in \mathbb{R}^{m \times v_r}$ and $C_r \in \mathbb{R}^{m \times n_r}$. Then, by choosing $\overline{Y}_{r,1}(\lambda)$ in (9.64) as

$$\overline{Y}_{r,1}(\lambda) = \begin{bmatrix} I \\ (\lambda I - A_r)^{-1} B_r \end{bmatrix},$$

we obtain from (9.62) with (9.70)

$$N_r(\lambda) = C_r(\lambda I - A_r)^{-1} B_r + D_r, \tag{9.71}$$

which is a simple minimal proper rational right nullspace basis as well. We can also obtain a dual alternative form (which corresponds to (9.69))

$$\widetilde{N}_r(\lambda) = (C_r + D_r F)(\lambda I - A_r - B_r F)^{-1} B_r + D_r. \tag{9.72}$$

A numerically reliable computational approach to compute proper minimal nullspace bases of rational matrices is described in Sect. 10.3.2 and relies on using

Kronecker-like forms (instead of the Kronecker form), which can be determined by using exclusively orthogonal similarity transformations.

9.2.5 Additive Decompositions

Consider a disjunct partition of the complex plane $\mathbb{C}$ as

$$\mathbb{C} = \mathbb{C}_g \cup \mathbb{C}_b, \quad \mathbb{C}_g \cap \mathbb{C}_b = \emptyset, \tag{9.73}$$

where both $\mathbb{C}_g$ and $\mathbb{C}_b$ are symmetrically located with respect to the real axis, and $\mathbb{C}_g$ has at least one point on the real axis. Since $\mathbb{C}_g$ and $\mathbb{C}_b$ are disjoint, each pole of any transfer function lies either in $\mathbb{C}_g$ or in $\mathbb{C}_b$. Let $G(\lambda)$ be a rational TFM (possibly improper) with a descriptor system representation $G(\lambda) = (A - \lambda E, B, C, D)$. Using a general similarity transformation using two invertible matrices U and V, we can determine an equivalent representation of $G(\lambda)$ with partitioned system matrices of the form

$$G(\lambda) = \left[\begin{array}{c|c} UAV - \lambda UEV & UB \\ \hline CV & D \end{array} \right] = \left[\begin{array}{cc|c} A_g - \lambda E_g & 0 & B_g \\ 0 & A_b - \lambda E_b & B_b \\ \hline C_g & C_b & D \end{array} \right], \tag{9.74}$$

where $\Lambda(A_g - \lambda E_g) \subset \mathbb{C}_g$ and $\Lambda(A_b - \lambda E_b) \subset \mathbb{C}_b$. It follows that $G(\lambda)$ can be additively decomposed as

$$G(\lambda) = G_g(\lambda) + G_b(\lambda), \tag{9.75}$$

where

$$G_g(\lambda) = \left[\begin{array}{c|c} A_g - \lambda E_g & B_g \\ \hline C_g & D \end{array} \right], \quad G_b(\lambda) = \left[\begin{array}{c|c} A_b - \lambda E_b & B_b \\ \hline C_b & 0 \end{array} \right], \tag{9.76}$$

and $G_g(\lambda)$ has only poles in $\mathbb{C}_g$, while $G_b(\lambda)$ has only poles in $\mathbb{C}_b$. The spectral separation in (9.74) is automatically provided by the Weierstrass canonical form of the pencil $A - \lambda E$, where the diagonal Jordan blocks are suitably permuted to correspond to the desired eigenvalue splitting. This approach automatically leads to partial-fraction expansions of $G_g(\lambda)$ and $G_b(\lambda)$.

For the computation of additive spectral decompositions, a numerically reliable procedure is presented in Sect. 10.3.4.

9.2.6 Coprime Factorizations

Consider a disjunct partition of the complex plane as $\mathbb{C} = \mathbb{C}_b \cup \mathbb{C}_g$, $\mathbb{C}_b \cap \mathbb{C}_g = \emptyset$, where $\mathbb{C}_b$ and $\mathbb{C}_g$ denote the "bad" and "good" regions of $\mathbb{C}$, respectively. Let $G(\lambda)$ be a $p \times m$ rational matrix with an n-th order descriptor system realization

$$G(\lambda) = \left[\begin{array}{c|c} A - \lambda E & B \\ \hline C & D \end{array}\right]. \tag{9.77}$$

We say the descriptor system (9.77) (or equivalently the pair $(A - \lambda E, B)$) is $\mathbb{C}_b$-*stabilizable* if rank $\left[\,A - \lambda E\ \ B\,\right] = n$ for all finite $\lambda \in \mathbb{C}_b$ and rank$[\,E\ AN_\infty\ B\,] = n$, where the columns of N_∞ form a basis of $\mathcal{N}(E)$. The descriptor system (9.77) (or equivalently the pair $(A - \lambda E, C)$) is $\mathbb{C}_b$-*detectable* if rank $\left[\begin{array}{c} A - \lambda E \\ C \end{array}\right] = n$ for all finite $\lambda \in \mathbb{C}_b$ and rank$[\,E^T\ A^T L_\infty\ C^T\,] = n$, where the columns of L_∞ form a basis of $\mathcal{N}(E^T)$. The strong stabilizability and strong detectability properties of the descriptor system (9.77), defined in Sect. 9.2.2, correspond to the choice $\mathbb{C}_b = \overline{\mathbb{C}}_u$.

For any $\mathbb{C}_b$-*stabilizable* descriptor realization $(A - \lambda E, B, C, D)$ of $G(\lambda)$, we can construct a *right coprime factorization* (RCF) $G(\lambda) = N(\lambda)M^{-1}(\lambda)$, where $N(\lambda)$ and $M(\lambda)$ are proper rational matrices with only poles in $\mathbb{C}_g$ and are mutually coprime (see Sect. 9.1.6 for definitions). For this, it is sufficient to determine a state-feedback matrix F such that all finite eigenvalues in $\Lambda(A + BF - \lambda E)$ belong to $\mathbb{C}_g$ and all infinite eigenvalues in $\Lambda(A + BF - \lambda E)$ are simple. The descriptor realizations of the factors are given by

$$\left[\begin{array}{c} N(\lambda) \\ M(\lambda) \end{array}\right] = \left[\begin{array}{c|c} A + BF - \lambda E & B \\ \hline C + DF & D \\ F & I_m \end{array}\right]. \tag{9.78}$$

For the computation of a suitable state-feedback matrix F, the S-stabilization algorithm proposed in [133] can be employed. Similarly, for a $\mathbb{C}_b$-*detectable* descriptor realization $(A - \lambda E, B, C, D)$ of $G(\lambda)$, we can construct a *left coprime factorization* (LCF) $G(\lambda) = M^{-1}(\lambda)N(\lambda)$, where $N(\lambda)$ and $M(\lambda)$ are proper rational matrices with only poles in $\mathbb{C}_g$ and are mutually coprime. For this, it is sufficient to determine an output injection matrix K such that all finite eigenvalues in $\Lambda(A + KC - \lambda E)$ belong to $\mathbb{C}_g$ and all infinite eigenvalues of $\Lambda(A + KC - \lambda E)$ are simple. The descriptor realizations of the factors are given by

$$\left[\,N(\lambda)\ M(\lambda)\,\right] = \left[\begin{array}{c|cc} A + KC - \lambda E & B + KD & K \\ \hline C & D & I_p \end{array}\right]. \tag{9.79}$$

For the computation of a suitable output injection matrix K, the S-stabilization algorithm of [133] can be applied to the dual pair $(A^T - \lambda E^T, C^T)$ to obtain K^T.

The RCF and LCF can be seen as techniques to dislocate the poles of a given TFM $G(\lambda)$ by postmultiplication or premultiplication with a suitable TFM $M(\lambda)$, respectively. For both RCF and LCF, it is possible to determine the denominator factor $M(\lambda)$ of least McMillan degree, which dislocates the minimum number of poles. The least McMillan degree of $M(\lambda)$ is equal to the number of generalized eigenvalues of the pair (E, A), which lie in $\mathbb{C}_b$. In what follows, we give the descriptor system representation-based version of the conceptual approach of Sect. 9.1.6 to determine coprime factorizations with minimum-degree denominator factors. The additive decomposition of $G(\lambda)$ as $G(\lambda) = G_g(\lambda) + G_b(\lambda)$ can be obtained (see Sect. 9.2.5) with

$$G_g(\lambda) = \left[\begin{array}{c|c} A_g - \lambda E_g & B_g \\ \hline C_g & D \end{array}\right], \quad G_b(\lambda) = \left[\begin{array}{c|c} A_b - \lambda E_b & B_b \\ \hline C_b & 0 \end{array}\right], \tag{9.80}$$

and $G_g(\lambda)$ has only poles in $\mathbb{C}_g$ (i.e., $\Lambda(A_g - \lambda E_g) \subset \mathbb{C}_g$), while $G_b(\lambda)$ has only poles in $\mathbb{C}_b$ (i.e., $\Lambda(A_b - \lambda E_b) \subset \mathbb{C}_b$).

The RCF $G_b(\lambda) = N_g(\lambda)M^{-1}(\lambda)$ can be obtained with the factors

$$\left[\begin{array}{c} N_g(\lambda) \\ M(\lambda) \end{array}\right] = \left[\begin{array}{c|c} A_b + B_bF - \lambda E_b & B_b \\ \hline C_b & 0 \\ F & I_m \end{array}\right], \tag{9.81}$$

where F has been determined such that $M(\lambda)$ and also $N_g(\lambda)$ have poles only in $\mathbb{C}_g$ (i.e., all finite eigenvalues of the pencil $A_b + B_bF - \lambda E_b$ are in $\mathbb{C}_g$ and all its infinite eigenvalues are simple). The numerator factor is given by

$$N(\lambda) = (G_b(\lambda) + G_g(\lambda))M(\lambda) = \left[I_p \ \ G_g(\lambda) \right] \left[\begin{array}{c} N_g(\lambda) \\ M(\lambda) \end{array}\right]$$

and has the descriptor realization

$$N(\lambda) = \left[\begin{array}{cc|c} A_g - \lambda E_g & B_gF & B_g \\ 0 & A_b + B_bF - \lambda E_b & B_b \\ \hline C_g & C_b + DF & D \end{array}\right]. \tag{9.82}$$

Similarly, the LCF $G_b(\lambda) = M^{-1}(\lambda)N_g(\lambda)$ can be obtained with the factors

$$\left[N_g(\lambda) \ \ M(\lambda) \right] = \left[\begin{array}{c|cc} A_b + KC_b - \lambda E_b & B_b \ \ K \\ \hline C_b & 0 \ \ I_p \end{array}\right],$$

where K has been determined such that $M(\lambda)$ and also $N_g(\lambda)$ have poles only in $\mathbb{C}_g$ (i.e., all finite eigenvalues of the pencil $A_b + KC_b - \lambda E_b$ are in $\mathbb{C}_g$ and all its infinite eigenvalues are simple). The numerator factor is given by

$$N(\lambda) = M(\lambda)(G_b(\lambda) + G_g(\lambda)) = \begin{bmatrix} N_g(\lambda) & M(\lambda) \end{bmatrix} \begin{bmatrix} I_m \\ G_g(\lambda) \end{bmatrix}$$

and has the descriptor realization

$$N(\lambda) = \left[\begin{array}{cc|c} A_b + KC_b - \lambda E_b & KC_g & B_b + KD \\ 0 & A_g - \lambda E_g & B_g \\ \hline C_b & C_g & D \end{array} \right] . \qquad (9.83)$$

The above approach can be employed to compute a RCF $G(\lambda) = N(\lambda)M^{-1}(\lambda)$ with least-order *inner* denominator $M(\lambda)$. For this, define $\mathbb{C}_g = \mathbb{C}_s$ and $\mathbb{C}_b = \mathbb{C} \setminus \mathbb{C}_s$. Assume also that $G(\lambda)$ has no poles in $\partial\mathbb{C}_s$, including infinity in the continuous-time case. In the continuous-time case, we use the additive decomposition $G(s) = G_g(s) + G_b(s)$, with the two terms given in (9.80) for $\lambda = s$, and we compute first a stable and proper RCF $G_b(\lambda) = N_g(\lambda)M^{-1}(\lambda)$ such that $M(\lambda)$ is inner. The factors $N_g(s)$ and $M(s)$ have for $\lambda = s$ the realizations in (9.81), with F computed as

$$F = -B_b^T E_b^{-T} Y^{-1},$$

where Y is the positive definite solution of the generalized Lyapunov equation

$$A_b Y E_b^T + E_b Y A_b^T - B_b B_b^T = 0.$$

The resulting realization of $N(s)$ is given by (9.82) for $\lambda = s$.

In the discrete-time case, we use the additive decomposition $G(z) = G_g(z) + G_b(z)$, where the realization $(A_b - zE_b, B_b, C_b, 0)$ of $G_b(z)$ includes, in the most general case, all infinite poles among the unstable poles. Therefore, the natural setting is to allow for E_b to be singular. Unfortunately, explicit formulas for the factors similar to (9.78) exist only in the case when E_b is invertible. Therefore, to address the general case, we can use the bilinear transformation technique to be able to employ the explicit formulas stated above for the continuous-time case. For this, we perform first the bilinear transformation $z = \frac{1+s}{1-s}$ to obtain $\widetilde{G}_b(s) := G_b(\frac{1+s}{1-s})$ with the continuous-time descriptor system realization $(\widetilde{A}_b - s\widetilde{E}_b, \widetilde{B}_b, \widetilde{C}_b, \widetilde{D}_b)$, which we assume to be minimal. (*Note: This is always the case if E_b is nonsingular. However, for a singular E_b, the pair $(\widetilde{A}_b - s\widetilde{E}_b, \widetilde{B}_b)$ has uncontrollable eigenvalues at $s = 1$ (thus it is not stabilizable) if the realization is computed using (9.57) or the pair $(\widetilde{A}_b - s\widetilde{E}_b, \widetilde{C}_b)$ has unobservable eigenvalues at $s = 1$ (thus it is not detectable) if the realization is computed using (9.56). The non-minimal part can be eliminated by using a suitable algorithm to remove the uncontrollable or unobservable eigenvalues, as presented in Sect. 10.3.1. Therefore, we assume that this operation has been already performed and the realization $(\widetilde{A}_b - s\widetilde{E}_b, \widetilde{B}_b, \widetilde{C}_b, \widetilde{D}_b)$ is minimal.*) Then, we apply the above procedure, for continuous-time systems, to obtain the RCF $\widetilde{G}(s) = \widetilde{N}_g(s)\widetilde{M}^{-1}(s)$ with the inner denominator $\widetilde{M}(s)$. Assume that $N_g(z) := \widetilde{N}_g(\frac{z-1}{z+1})$ and $M(z) := \widetilde{M}(\frac{z-1}{z+1})$ have the realizations

$$\begin{bmatrix} N_g(z) \\ M(z) \end{bmatrix} = \left[\begin{array}{c|c} \overline{A}_g - z\overline{E}_g & \overline{B}_g \\ \hline \overline{C}_g & \overline{D}_g \\ \overline{F} & \overline{W} \end{array} \right],$$

where $\Lambda(\overline{A}_g - z\overline{E}_g) \subset \mathbb{C}_s$. Then, the numerator factor results as

$$N(z) = \begin{bmatrix} I_p & G_g(z) \end{bmatrix} \begin{bmatrix} N_g(z) \\ M(z) \end{bmatrix} = \left[\begin{array}{cc|c} A_g - zE_g & B_g\overline{F} & B_g\overline{W} \\ 0 & \overline{A}_g - z\overline{E}_g & \overline{B}_g \\ \hline C_g & \overline{C}_g + D_g\overline{F} & \overline{D}_g + D_g\overline{W} \end{array} \right].$$

A similar approach can be devised to determine a LCF with inner denominator. Numerically reliable procedures, which avoid the computation of the potentially sensitive Weierstrass canonical form (by working instead with the generalized real Schur form), are described in Sect. 10.3.5.

9.2.7 Norms

Consider in this section, a TFM $G(\lambda)$ without poles in $\partial\mathbb{C}_s$ (the boundary of stability domain including infinity in the continuous-time case) having a descriptor state-space realization $(A - \lambda E, B, C, D)$. The definition (9.18) of the $\mathcal{H}_\infty$-norm of $G(\lambda)$ (and also of the $\mathcal{L}_\infty$-norm) in terms of the largest singular value of the frequency-response gain (see Sect. 9.1.7) is already an indication for the lack of closed-form formulas to express this norm in terms of matrices of a state-space realization. Efficient numerical algorithms for the computation of the $\mathcal{H}_\infty$- and $\mathcal{L}_\infty$-norms are iterative (e.g., employ bisection-based approximation techniques) and rely on state-space representations of $G(\lambda)$. This latter aspect is instrumental for both efficiency and numerical reliability of the computational algorithms. For efficient numerical algorithms for the computation of the $\mathcal{H}_\infty$- and $\mathcal{L}_\infty$-norms, see the literature cited in Sect. 9.3.

For the computation of the $\mathcal{H}_2$-norm of a stable and proper TFM $G(\lambda)$, closed-form formulas are available for both continuous- and discrete-time systems (see the definitions (9.19) or (9.20) in Sect. 9.1.7).

Lemma 9.10 *Let $G(s)$ be a strictly proper and stable TFM of a continuous-time system, and let $(A - sE, B, C, 0)$ be an irreducible descriptor realization with E invertible. Then, the $\mathcal{H}_2$-norm of $G(s)$ can be evaluated as*

$$\|G(s)\|_2 = \sqrt{\text{trace}\,(B^T Q B)},$$

where Q is the observability Gramian satisfying the following generalized Lyapunov equation:

$$A^T Q E + E^T Q A + C^T C = 0.$$

Alternatively, the $\mathcal{H}_2$-norm of $G(s)$ can be evaluated as

$$\|G(s)\|_2 = \sqrt{\text{trace}\,(CPC^T)},$$

where P is the controllability Gramian satisfying the following generalized Lyapunov equation

$$APE^T + EPA^T + BB^T = 0.$$

Lemma 9.11 *Let $G(z)$ be a strictly proper and stable TFM of a discrete-time system, and let $(A - zE, B, C, D)$ be an irreducible descriptor realization with E invertible. Then, the $\mathcal{H}_2$-norm of $G(z)$ can be evaluated as*

$$\|G(z)\|_2 = \sqrt{\text{trace}\,(B^T QB + D^T D)},$$

where Q is the observability Gramian satisfying the following generalized Stein equation:

$$A^T QA - E^T QE + C^T C = 0.$$

Alternatively, the $\mathcal{H}_2$-norm of $G(z)$ can be evaluated as

$$\|G(z)\|_2 = \sqrt{\text{trace}\,(CPC^T + D^T D)},$$

where P is the controllability Gramian satisfying the following generalized Stein equation:

$$APA^T - EPE^T + BB^T = 0.$$

The $\mathcal{L}_2$-norm of a proper TFM $G(\lambda)$ without poles on the boundary of stability domain $\partial\mathbb{C}_s$ has formally the same definition as the $\mathcal{H}_2$-norm in (9.19) or (9.20). This norm can be computed using, for example, the stable RCF with inner denominator $G(\lambda) = \widetilde{G}(\lambda)M_i^{-1}(\lambda)$ with $M_i(\lambda)$ inner, and exploiting the norm-preserving property of all-pass factors

$$\|G\|_2 = \|\widetilde{G}(\lambda)M_i^{-1}(\lambda)\|_2 = \|\widetilde{G}(\lambda)\|_2.$$

Alternatively, we can use the stable–unstable decomposition $G(\lambda) = G_s(\lambda) + G_u(\lambda)$ and compute

$$\|G\|_2 = \sqrt{\|G_s(\lambda)\|_2^2 + \|G_u^\sim(\lambda)\|_2^2}.$$

For the evaluation of the Hankel norm of a stable TFM $G(\lambda)$, the state-space representation allows to use explicit formulas.

Lemma 9.12 *Let $G(s)$ be a proper and stable TFM of a continuous-time system, and let $(A - sE, B, C, D)$ be an irreducible descriptor realization with E invertible. Then, the Hankel norm of $G(s)$ can be evaluated as*

$$\|G(s)\|_H = \overline{\sigma}(RES),$$

where $P = SS^T$ is the controllability Gramian and $Q = R^T R$ is the observability Gramian, which satisfy the following generalized Lyapunov equations

$$APE^T + EPA^T + BB^T = 0,$$
$$A^T QE + E^T QA + C^T C = 0.$$

Lemma 9.13 *Let $G(z)$ be a proper and stable TFM of a continuous-time system, and let $(A - zE, B, C, D)$ be an irreducible descriptor realization with E invertible. Then, the Hankel norm of $G(z)$ can be evaluated as*

$$\|G(z)\|_H = \overline{\sigma}(RES),$$

where $P = SS^T$ is the controllability Gramian and $Q = R^T R$ is the observability Gramian, which satisfy the following generalized Stein equations:

$$APA^T - EPE^T + BB^T = 0,$$
$$A^T QA - E^T QE + C^T C = 0.$$

For the solution of the generalized Lyapunov and Stein equations in this section, computational procedures, discussed in Sect. 10.2.1, can be employed.

9.2.8 Inner–Outer and Spectral Factorizations

Recall from Sect. 9.1.8 that any stable proper TFM $G(\lambda)$ without zeros in $\partial\mathbb{C}_s$ (the boundary of the appropriate stability domain) has an *inner–outer factorization*

$$G(\lambda) = G_i(\lambda)G_o(\lambda), \tag{9.84}$$

where $G_i(\lambda)$ is the inner factor and $G_o(\lambda)$ is the outer factor. Similarly, $G(\lambda)$ has a *co-outer–co-inner factorization*

$$G(\lambda) = G_{co}(\lambda)G_{ci}(\lambda), \tag{9.85}$$

where $G_{co}(\lambda)$ is the co-outer factor and $G_{ci}(\lambda)$ is the co-inner factor. In view of the applications of this factorization in solving the synthesis problems of fault detection and isolation filters, using the procedures presented in Chap. 5, we only consider the particular case, when $G(\lambda)$ has full column rank or full row rank, in which case the outer factor or co-outer factor result invertible, respectively.

Assume that the stable proper TFM $G(\lambda)$ has an irreducible descriptor realization

$$G(\lambda) = \left[\begin{array}{c|c} A - \lambda E & B \\ \hline C & D \end{array}\right],\tag{9.86}$$

with E invertible. We have the following standard result for a continuous-time system.

Theorem 9.3 *If $G(s)$ is a proper and stable, full column rank TFM without zeros in $\partial\mathbb{C}_s$, then $G(s)$ has an inner–outer factorization $G(s) = G_i(s)G_o(s)$, with the particular realizations of the factors*

$$G_i(s) = \left[\begin{array}{c|c} A + BF - sE & BH^{-1} \\ \hline C + DF & DH^{-1} \end{array}\right], \qquad G_o(s) = \left[\begin{array}{c|c} A - sE & B \\ \hline -HF & H \end{array}\right],$$

where H is an invertible matrix satisfying $D^T D = H^T H$, and F is given by

$$F = -(D^T D)^{-1}(B^T X_s E + D^T C),$$

with $X_s \geq 0$ being the stabilizing solution of the generalized continuous-time algebraic Riccati equation (GCARE)

$$A^T X E + E^T X A - (E^T X B + C^T D)(D^T D)^{-1}(B^T X E + D^T C) + C^T C = 0.$$

A similar result for a discrete-time system is as follows.

Theorem 9.4 *If $G(z)$ is a proper and stable, full column rank TFM without zeros in $\partial\mathbb{C}_s$, then $G(z)$ has an inner–outer factorization $G(z) = G_i(z)G_o(z)$, with the particular realizations of the factors*

$$G_i(z) = \left[\begin{array}{c|c} A + BF - zE & BH^{-1} \\ \hline C + DF & DH^{-1} \end{array}\right], \qquad G_o(z) = \left[\begin{array}{c|c} A - zE & B \\ \hline -HF & H \end{array}\right],$$

where H is an invertible matrix satisfying $D^T D + B^T X_s B = H^T H$, and F is given by

$$F = -(H^T H)^{-1}(B^T X_s A + D^T C),$$

with $X_s \geq 0$ being the stabilizing solution of the generalized discrete-time algebraic Riccati equation (GDARE)

$$A^T X A - E^T X E - (A^T X B + C^T D)(D^T D + B^T X B)^{-1}(B^T X A + D^T C) + C^T C = 0.$$

Instead of the (compact) inner–outer factorization (9.84), the *extended inner–outer factorization*

$$G(\lambda) = \left[\, G_i(\lambda)\ G_i^\perp(\lambda) \,\right]\left[\begin{array}{c} G_o(\lambda) \\ 0 \end{array}\right]\tag{9.87}$$

is sometimes desirable, where $G_i^\perp(\lambda)$ is the inner orthogonal complement of $G_i(\lambda)$ such that $\begin{bmatrix} G_i(\lambda) & G_i^\perp(\lambda) \end{bmatrix}$ is square and inner. In the continuous-time case, a descriptor realization of $G_i^\perp(\lambda)$ is given by

$$G_i^\perp(s) = \left[\begin{array}{c|c} A + BF - sE & -X_s^\dagger E^{-T} C^T D^\perp \\ \hline C + DF & D^\perp \end{array} \right],$$

where $D^\perp$ is an orthogonal complement chosen such that $\begin{bmatrix} DH^{-1} & D^\perp \end{bmatrix}$ is square and orthogonal. In the discrete-time case, we have

$$G_i^\perp(z) = \left[\begin{array}{c|c} A + BF - zE & Y \\ \hline C + DF & W \end{array} \right],$$

where Y and W satisfy

$$\begin{aligned} A^T X_s Y + C^T W &= 0, \\ B^T X_s Y + D^T W &= 0, \\ W^T W + Y^T X_s Y &= I. \end{aligned}$$

Similar results for the co-outer–co-inner factorization (or the extended co-outer–co-inner factorization) can be easily obtained by considering the inner–outer factorization (or its extended version) for the dual system with the TFM $G^T(\lambda)$ having the descriptor realization $(A^T - \lambda E^T, C^T, B^T, D^T)$.

Remark 9.4 The more general case with $G(\lambda)$ having zeros in $\partial\mathbb{C}_s$ is relevant for solving synthesis problems of fault detection filters by using approximate model-matching techniques (see Sect. 5.7). A computational approach for this case is presented in Sect. 10.3.6. The resulting quasi-outer factor $G_o(\lambda)$ has all zeros in $\overline{\mathbb{C}}_s$, which include all zeros of $G(\lambda)$ in $\partial\mathbb{C}_s$. $\qquad\square$

Recall from Sect. 9.1.8 that the outer factor $G_o(\lambda)$ corresponding to a $G(\lambda)$ without zeros in $\partial\mathbb{C}_s$ is a solution of the *minimum-phase right spectral factorization* problem, while the co-outer factor $G_{co}(\lambda)$ is a solution of the *minimum-phase left spectral factorization* problem. By combining the LCF (RCF) with inner denominator and the inner–outer factorization, we can obtain a solution of the *stable minimum-phase right (left) spectral factorization* problem. For a discussion of these aspects, see Sect. 9.1.8.

A special factorization encountered when solving the AMMP (see Eq. (9.32) in Sect. 9.1.10) is the following: for a given TFM $G(\lambda)$ without poles in $\partial\mathbb{C}_s$ and a given bound $\gamma > \|G(\lambda)\|_\infty$, compute a stable and minimum-phase TFM $G_o(\lambda)$ such that

$$\gamma^2 I - G(\lambda)G^\sim(\lambda) = G_o(\lambda)G_o^\sim(\lambda).$$

This computation can be addressed in two steps. In the first step, we compute a RCF $G(\lambda) = N(\lambda)M^{-1}(\lambda)$, with the denominator factor $M(\lambda)$ inner. It follows that

$$\gamma^2 I - G(\lambda)G^\sim(\lambda) = \gamma^2 I - N(\lambda)N^\sim(\lambda),$$

where $N(\lambda)$ is proper and has only poles in $\mathbb{C}_s$. In the second step, we determine the stable and minimum-phase $G_o(\lambda)$, which satisfies

$$\gamma^2 I - N(\lambda)N^\sim(\lambda) = G_o(\lambda)G_o^\sim(\lambda). \tag{9.88}$$

The first step has been already discussed in Sect. 9.2.6, and therefore we assume that for an irreducible descriptor realization $(A - \lambda E, B, C, D)$ of $G(\lambda)$, we determined a stable $N(\lambda)$ with a descriptor realization $(\widetilde{A} - \lambda\widetilde{E}, \widetilde{B}, \widetilde{C}, \widetilde{D})$.

In the continuous-time case, we can compute the spectral factor $G_o(s)$ by using the following result.

Lemma 9.14 *Let $N(s)$ be a stable TFM and let $(\widetilde{A} - s\widetilde{E}, \widetilde{B}, \widetilde{C}, \widetilde{D})$ be its descriptor system realization. For $\gamma > \|N(s)\|_\infty$, a realization of a stable and minimum-phase spectral factor $G_o(s)$, satisfying (9.88) for $\lambda = s$, is given by*

$$G_o(s) = \left[\begin{array}{c|c} \widetilde{A} - s\widetilde{E} & -K_s R^{1/2} \\ \hline \widetilde{C} & R^{1/2} \end{array}\right],$$

where

$$R = \gamma^2 I - \widetilde{D}\widetilde{D}^T,$$

$$K_s = (\widetilde{E}Y_s\widetilde{C}^T + \widetilde{B}\widetilde{D}^T)R^{-1},$$

and Y_s is the stabilizing solution of the GCARE

$$\widetilde{A}Y\widetilde{E}^T + \widetilde{E}Y\widetilde{A}^T + (\widetilde{E}Y\widetilde{C}^T + \widetilde{B}\widetilde{D}^T)R^{-1}(\widetilde{C}Y\widetilde{E}^T + \widetilde{D}\widetilde{B}^T) + \widetilde{B}\widetilde{B}^T = 0.$$

We have the following analogous result in the discrete-time case.

Lemma 9.15 *Let $N(z)$ be a stable TFM and let $(\widetilde{A} - z\widetilde{E}, \widetilde{B}, \widetilde{C}, \widetilde{D})$ be its descriptor realization. For $\gamma > \|N(z)\|_\infty$, a realization of a stable and minimum-phase spectral factor $G_o(z)$, satisfying (9.88) for $\lambda = z$, is given by*

$$G_o(z) = \left[\begin{array}{c|c} \widetilde{A} - \lambda\widetilde{E} & -K_s R^{1/2} \\ \hline \widetilde{C} & R^{1/2} \end{array}\right],$$

where

$$R_D = \gamma^2 I - \widetilde{D}\widetilde{D}^T,$$

$$R = R_D - \widetilde{C}Y_s\widetilde{C}^T,$$

$$K_s = (\widetilde{A}Y_s\widetilde{C}^T + \widetilde{B}\widetilde{D}^T)R^{-1},$$

and Y_s is the stabilizing solution of the GDARE

$$\widetilde{A}Y\widetilde{A}^T - \widetilde{E}Y\widetilde{E}^T - (\widetilde{A}Y\widetilde{C}^T + \widetilde{B}\widetilde{D}^T)(-R_D + \widetilde{C}Y\widetilde{C}^T)^{-1}(\widetilde{C}Y\widetilde{A}^T + \widetilde{D}\widetilde{B}^T) + \widetilde{B}\widetilde{B}^T = 0.$$

9.2.9   Linear Rational Equations

For $G(\lambda) \in \mathbb{R}(\lambda)^{p \times m}$ and $F(\lambda) \in \mathbb{R}(\lambda)^{q \times m}$, consider the solution of the linear rational matrix equation

$$X(\lambda)G(\lambda) = F(\lambda), \tag{9.89}$$

where $X(\lambda) \in \mathbb{R}(\lambda)^{q \times p}$ is the solution we seek. The existence of a solution is guaranteed if the compatibility condition for linear systems is fulfilled. Recall from Lemma 9.4 that the rational equation (9.89) has a solution if and only if

$$\operatorname{rank} G(\lambda) = \operatorname{rank} \begin{bmatrix} G(\lambda) \\ F(\lambda) \end{bmatrix}. \tag{9.90}$$

An equivalent condition can be derived in terms of descriptor system representations of $G(\lambda)$ and $F(\lambda)$, which we assume to be of the form

$$G(\lambda) = \left[\begin{array}{c|c} A - \lambda E & B \\ \hline C_G & D_G \end{array} \right], \qquad F(\lambda) = \left[\begin{array}{c|c} A - \lambda E & B \\ \hline C_F & D_F \end{array} \right]. \tag{9.91}$$

Such representations, which share the pair $(A - \lambda E, B)$, can be easily obtained by determining a descriptor realization of the compound rational matrix $\begin{bmatrix} G(\lambda) \\ F(\lambda) \end{bmatrix}$. It is easy to observe that any solution of (9.89) is also part of the solution of the linear polynomial equation

$$Y(\lambda) \begin{bmatrix} A - \lambda E & B \\ C_G & D_G \end{bmatrix} = \begin{bmatrix} C_F & D_F \end{bmatrix}, \tag{9.92}$$

where $Y(\lambda) = \begin{bmatrix} W(\lambda) & X(\lambda) \end{bmatrix}$. Therefore, alternatively to solving (9.89), we can solve instead (9.92) for $Y(\lambda)$ and compute $X(\lambda)$ as

$$X(\lambda) = Y(\lambda) \begin{bmatrix} 0 \\ I_p \end{bmatrix}. \tag{9.93}$$

Define the system pencils corresponding to $G(\lambda)$ and the compound $\begin{bmatrix} G(\lambda) \\ F(\lambda) \end{bmatrix}$ as

$$S_G(\lambda) := \begin{bmatrix} A - \lambda E & B \\ C_G & D_G \end{bmatrix}, \qquad S_{G,F}(\lambda) := \begin{bmatrix} A - \lambda E & B \\ C_G & D_G \\ C_F & D_F \end{bmatrix}. \tag{9.94}$$

We have the following result similar to Lemma 9.4.

Lemma 9.16 *The rational equation (9.89) with $G(\lambda)$ and $F(\lambda)$ having the descriptor realizations in (9.91) has a solution if and only if*

$$\operatorname{rank} S_G(\lambda) = \operatorname{rank} S_{G,F}(\lambda). \tag{9.95}$$

Let $\mathbb{C}_b$ be the "bad" domain of the complex plane, where the solution $X(\lambda)$ must not have poles. We have the following result similar to Lemma 9.5.

Lemma 9.17 *The rational equation* (9.89) *with* $G(\lambda)$ *and* $F(\lambda)$ *having the descriptor realizations in* (9.91) *has a solution without poles in* $\mathbb{C}_b$ *if and only if the matrix pencils* $S_G(\lambda)$ *and* $S_{G,F}(\lambda)$ *defined in* (9.94) *fulfil* (9.95) *and have the same zero structure for all zeros of* $G(\lambda)$ *in* $\mathbb{C}_b$.

In what follows, we give a constructive procedure for the computation of a solution $X(\lambda)$ of the rational equation (9.89), which can be seen also as a proof of the above lemmas. To establish the solvability condition (9.95) and to solve (9.92), we reduce the system pencil $S_G(\lambda)$ to a suitable Kronecker form. Let U and V be invertible matrices to reduce $S_G(\lambda)$ to the alternative Kronecker form (see Remark 9.2)

$$\overline{S}_G(\lambda) = U S_G(\lambda) V = \begin{bmatrix} A_r - \lambda E_r & 0 & 0 \\ 0 & A_{reg} - \lambda E_{reg} & 0 \\ 0 & 0 & A_l - \lambda I_{n_l} \\ 0 & 0 & C_l \end{bmatrix}, \tag{9.96}$$

where $A_r - \lambda E_r$ $(=: K_r(\lambda))$ has full row rank n_r, $A_{reg} - \lambda E_{reg}$ $(=: K_{reg}(\lambda))$ is an $n_{reg} \times n_{reg}$ regular subpencil in a Weierstrass canonical form, and $\begin{bmatrix} A_l - \lambda I_{n_l} \\ C_l \end{bmatrix}$ $(=: K_l(\lambda))$ has full column rank n_l with the pair $(A_l - \lambda I_{n_l}, C_l)$ observable. We have immediately that if $\overline{Y}(\lambda)$ is a solution of the reduced equation

$$\overline{Y}(\lambda)\overline{S}_G(\lambda) = \begin{bmatrix} C_F & D_F \end{bmatrix} V, \tag{9.97}$$

then $Y(\lambda) = \overline{Y}(\lambda)U$ is a solution of (9.92) and thus

$$X(\lambda) = \overline{Y}(\lambda)U \begin{bmatrix} 0 \\ I_p \end{bmatrix} \tag{9.98}$$

is a solution of Eq. (9.89).

Partition now

$$\begin{bmatrix} C_F & D_F \end{bmatrix} V := \begin{bmatrix} -\overline{C}_1 & -\overline{C}_2 & -\overline{C}_3 \end{bmatrix}$$

in accordance with the column structure of $\overline{S}_G(\lambda)$ in (9.96). We choose $\overline{Y}(\lambda)$ of the form

$$\overline{Y}(\lambda) = \begin{bmatrix} \overline{Y}_1(\lambda) & \overline{Y}_2(\lambda) & \overline{Y}_3(\lambda) & \overline{Y}_4(\lambda) \end{bmatrix},$$

where the column partitioning of $\overline{Y}(\lambda)$, in four blocks, corresponds to the row partitioning of $\overline{S}_G(\lambda)$ in (9.96). From (9.97), we obtain the equations fulfilled by the blocks of $\overline{Y}(\lambda)$:

$$\overline{Y}_1(\lambda)(A_r - \lambda E_r) = -\overline{C}_1,$$
$$\overline{Y}_2(\lambda)(A_{reg} - \lambda E_{reg}) = -\overline{C}_2,$$
$$\overline{Y}_3(\lambda)(A_l - \lambda I_{n_l}) + \overline{Y}_4(\lambda)C_l = -\overline{C}_3.$$

The equation satisfied by $\overline{Y}_1(\lambda)$ has a solution if and only if

$$\text{rank}(A_r - \lambda E_r) = \text{rank}\begin{bmatrix} A_r - \lambda E_r \\ -\overline{C}_1 \end{bmatrix}.$$

Since $A_r - \lambda E_r$ has full row rank n_r, this is possible if and only if $\overline{C}_1 = 0$ and the corresponding solution is $\overline{Y}_1(\lambda) = 0$. For the rest of the blocks, we have

$$\overline{Y}_2(\lambda) = \overline{C}_2(\lambda E_{reg} - A_{reg})^{-1},$$

while $\overline{Y}_3(\lambda)$ and $\overline{Y}_4(\lambda)$ jointly satisfy

$$\overline{Y}_3(\lambda) = \overline{C}_3(\lambda I_{n_l} - A_l)^{-1} + \overline{Y}_4(\lambda)C_l(\lambda I_{n_l} - A_l)^{-1}.$$

Thus, the component $\overline{Y}_4(\lambda)$ can be freely chosen. The condition $\overline{C}_1 = 0$ makes that

$$\text{rank}\, S_{G,F}(\lambda) = \text{rank}\begin{bmatrix} \overline{S}_G(\lambda) \\ \begin{bmatrix} C_F\ D_F \end{bmatrix} V \end{bmatrix} = n_r + n_{reg} + n_l = \text{rank}\, S_G(\lambda).$$

This proves the existence condition of Lemma 9.16 (which is also part of the existence conditions in Lemma 9.17).

We can compute now the solution $X(\lambda)$ according to (9.98). Let partition

$$U\begin{bmatrix} 0 \\ I_p \end{bmatrix} = \begin{bmatrix} B_r \\ B_{reg} \\ B_l \\ D_l \end{bmatrix} \tag{9.99}$$

in accordance with the row structure of $\overline{S}_G(\lambda)$. Then, we obtain the solution in the general form

$$X(\lambda) = X_0(\lambda) + \overline{Y}_4(\lambda)X_N(\lambda), \tag{9.100}$$

where

$$X_0(\lambda) := \overline{C}_2(\lambda E_{reg} - A_{reg})^{-1}B_{reg} + \overline{C}_3(\lambda I_{n_l} - A_l)^{-1}B_l$$

can be interpreted as a particular solution, the term

$$X_N(\lambda) := C_l(\lambda I_{n_l} - A_l)^{-1}B_l + D_l$$

is a $\nu_l \times p$ proper rational left nullspace basis of $G(\lambda)$ (see (9.66) in Sect. 9.2.4), while $\overline{Y}_4(\lambda)$ is an arbitrary $q \times \nu_l$ rational matrix. Recall that $\nu_l = p - \text{rank } G(\lambda)$.

We now discuss shortly the conditions for the existence of a solution $X(\lambda)$ with all poles in the desired "good" domain $\mathbb{C}_g \subset \mathbb{C}$. We assume $\mathbb{C}_g$ includes only finite regions of $\mathbb{C}$. Observe first that any particular solution $X_0(\lambda)$ has generally two components $X_0(\lambda) = X_{0,z}(\lambda) + X_{0,l}(\lambda)$, where

$$X_{0,z}(\lambda) = \overline{C}_2(\lambda E_{reg} - A_{reg})^{-1} B_{reg} \tag{9.101}$$

and

$$X_{0,l}(\lambda) = \overline{C}_3(\lambda I_{n_l} - A_l)^{-1} B_l \, .$$

The term $X_{0,z}(\lambda)$ has poles, which originate from the zeros of $G(\lambda)$, while $X_{0,l}(\lambda)$ has poles, which originate from the left structure of $G(\lambda)$. These latter poles are called the "spurious" poles of $X(\lambda)$ and we will show that they can be freely chosen to lie in $\mathbb{C}_g$. If we use the additional left transformation $\overline{U}$ in (9.67), then using (9.68) and similar developments as in Sect. 9.2.4, we obtain instead $X_{0,l}(\lambda)$, the updated term

$$\widetilde{X}_{0,l}(\lambda) = \overline{C}_3(\lambda I_{n_l} - A_l - KC_l)^{-1}(B_l + KD_l),$$

where K is a constant gain, which, due to the observability of the pair (A_l, C_l), can be chosen such that $\Lambda(A_l + KC_l) \subset \mathbb{C}_g$. Therefore, $\widetilde{X}_{0,l}(\lambda)$ has only poles in $\mathbb{C}_g$.

The poles of the term $X_{0,z}(\lambda)$ are fixed, and represent a subset of $\Lambda(A_{reg} - \lambda E_{reg})$. More precisely, these poles are the controllable and observable eigenvalues in $\Lambda(A_{reg} - \lambda E_{reg})$ of the descriptor system realization $(A_{reg} - \lambda E_{reg}, B_{reg}, \overline{C}_2, 0)$. If we assume that the descriptor realizations in (9.91) share the *controllable* pair $(A - \lambda E, B)$, then the pair $(A_{reg} - \lambda E_{reg}, B_{reg})$ is controllable as well. Therefore, to guarantee that all poles of $X_{0,z}(\lambda)$ lie in $\mathbb{C}_g$, all eigenvalues of $A_{reg} - \lambda E_{reg}$ outside $\mathbb{C}_g$ must be unobservable. Assume that the regular pencil $A_{reg} - \lambda E_{reg}$ exhibits the following spectral separation:

$$A_{reg} - \lambda E_{reg} = \begin{bmatrix} A_b - \lambda E_b & 0 \\ 0 & A_g - \lambda E_g \end{bmatrix},$$

where $A_b - \lambda E_b$ contains the diagonal blocks in the Weierstrass canonical form, which have finite eigenvalues in $\mathbb{C}_b$ or infinite eigenvalues with multiplicities at least 2, while $A_g - \lambda E_g$ contains the diagonal blocks in the Weierstrass canonical form, which have finite eigenvalues in $\mathbb{C}_g$ or infinite eigenvalues with multiplicities equal to 1 (this can be easily arranged by reordering the diagonal blocks in the Weierstrass canonical form). Partition correspondingly $\overline{C}_2$ and B_{reg} as

$$\overline{C}_2 = \begin{bmatrix} \overline{C}_{2,b} & \overline{C}_{2,g} \end{bmatrix}, \qquad B_{reg} = \begin{bmatrix} B_{reg,b} \\ B_{reg,g} \end{bmatrix} \, .$$

All eigenvalues in $\Lambda(A_b - \lambda E_b)$ are unobservable if and only if $\overline{C}_{2,b} = 0$. It follows

$$X_{0,z}(\lambda) = \overline{C}_{2,g}(\lambda E_g - A_g)^{-1} B_{reg,g} \, ,$$

which has poles only in $\mathbb{C}_g$ (because all infinite eigenvalues are simple).

The condition $\overline{C}_{2,b} = 0$ guarantees that $S_G(\lambda)$ and $S_{G,F}(\lambda)$ have the same eigenvalue (zero) structure for all $\lambda_z \in \Lambda(A_b - \lambda E_b) \subset \mathbb{C}_b$, which is the statement of Lemma 9.17.

The existing freedom to choose $\overline{Y}_4(\lambda)$ in (9.100) can be exploited to determine solutions $X(\lambda)$ with least McMillan degree. A numerically reliable computational approach to determine such a least-order solution of the dual equation $G(\lambda)X(\lambda) = F(\lambda)$ is described in Sect. 10.3.7.

9.3 Notes and References

Section 9.1. The material in this section is covered in several textbooks of which we mention only the two widely cited books of Kailath [72] and Vidyasagar [168]. Many useful information are also presented in [56]. Among them, we mention an elementary proof of Lemma 9.2 on the existence and uniqueness of the Smith form of a polynomial matrix. The results of Lemma 9.1 on the properties of polynomial bases have been established by Forney [45], and are taken from [72]. The concept of *simple minimal proper rational basis* has been introduced by Vardulakis and Karcanias in [128] as the natural counterpart of a minimal polynomial basis. The recursive pole dislocation-based approach to compute coprime factorizations has been inspired by the ideas of Belevitch [10], and refined in the work of Vandewalle and Dewilde [127]. See also the work of Van Dooren [124] for further discussions of this and similar recursive factorization techniques. A basic treatment of the $\mathcal{H}_\infty$-model-matching problem, including the bisection-based γ-iteration approach, is given by Francis in [46]. The main aspects related to spectral factorizations, solving minimum distance problems (Nehari problem), as well as $\mathcal{H}_2$-optimal control, are covered in [177].

Section 9.2. Linear descriptor systems (also known in the literature as linear differential-algebraic-equations-based systems or generalized state-space systems or singular systems) are discussed, to different depths and with different focus, in several books [22, 26, 39, 76]. The simple realization procedure of general rational matrices (representing the proof of Theorem 9.1) has been presented in [167]. The equivalence theory of linear matrix pencils is covered in [49]. For the real Jordan form of a matrix, see, for example, [64, Sect. 3.4]. The concept of simple minimal proper rational bases has been introduced in [128]. Necessary and sufficient conditions for the existence of a feedback matrix, which places $r = \mathrm{rank}\, E$ finite eigenvalues in $\Lambda(A + BF - \lambda E)$ into a "good" region $\mathbb{C}_g$, are given in [90] and suitable stabilization algorithms are discussed in [133]. Computational methods for the evaluation of $\mathcal{H}_\infty$- or $\mathcal{L}_\infty$-norm have been proposed for standard continuous-time systems in [18, 21], and extended to both continuous- and discrete-time descriptor systems in [16]. Recent developments for proper continuous-time descriptor systems, proposed in [13], target the reduction of computational effort and improvement of the numerical accuracy, by employing sophisticated structure preserving matrix pencil reduction techniques. Lemma 9.14 extends to proper descriptor system the formulas developed in [177, Corollary 13.22]. Lemma 9.15 extends to proper descriptor system the formulas developed in [177, Theorem 21.26] for the solution of the dual factorization $\gamma^2 I - N^{\sim}(z)N(z) = G_o^{\sim}(z)G_o(z)$.

Chapter 10
Computational Algorithms and Software

This chapter presents, in detail, the main algorithms for descriptor systems, which underlie the computational methods used in the synthesis procedures considered in this book. The core computations in these algorithms involve several matrix decompositions and condensed forms, which are obtainable using orthogonal transformations and, therefore, are provably numerical stable. Important applications of the condensed form are in developing numerically stable computational algorithms for the solution of several generalized matrix equations (Lyapunov, Stein, Sylvester, Riccati), which are frequently encountered in addressing the solution of synthesis problems in the fields of control and fault detection. The use of condensed forms, obtainable using orthogonal transformations (instead of using the potentially highly sensitive Weierstrass, Kronecker or Brunovsky canonical forms), is also instrumental in developing numerically reliable procedures for the solution of several basic computational problems for descriptor systems as well as in some, rather specialized, algorithms for proper descriptor systems. Although this chapter is primarily intended for numerical experts having interests in control-related numerical techniques, it also serves to highlight the complexity of the underlying computations, which are necessary to address the synthesis problems of fault detection and isolation filters in a numerically sound way. A collection of software tools implements the algorithms presented in this chapter and can be employed to reproduce all computational results presented in this book.

10.1 Matrix Decompositions and Condensed Forms

The condensed forms of matrices play an important role in solving many control-related computational problems. A widely used computational paradigm in solving many computational problems consists of three main steps: (1) transform the original problem into a simpler one by reducing the problem data to condensed forms; (2) solve the transformed problem by using specially devised methods for the respective condensed forms and (3) recover the solution of the original problem using back transformation to the original form. In this section we present several basic matrix decompositions, obtainable using orthogonal transformations, which involve several condensed forms of matrices, pairs of matrices or even triples of matrices.

The use of orthogonal transformations is a widely accepted approach to promote numerical reliability of computations with finite precision. These transformations are perfectly conditioned with respect to inversion and, therefore, have the very desirable property that they do not amplify the existing uncertainties in the data. This feature is very important, since uncertainties in problem data are ubiquitous, representing inherent inaccuracies in data (e.g., truncation or discretization errors), or roundoff errors that occurred in previous computational steps, or both. When using orthogonal transformations to transform problem data, it is often possible to bound the roundoff errors resulting as an effect of performed transformations on the data and even to show that the computed results are the exact solution of a problem with slightly perturbed data. Numerical algorithms exhibiting such a property are called (backward) numerically stable and underlie many algorithms for basic linear algebra computations. The use of numerically stable algorithms guarantees that the computed solution is accurate, if the computational problem is well conditioned.

In what follows, we present several matrix decompositions involving particular condensed forms, which can be obtained using exclusively orthogonal transformations. These decompositions are the basis for many numerically stable algorithms employed by the synthesis procedures presented in this book. We will not address detailed algorithms for the computation of these forms, because they are described in detail in several numerical linear algebra textbooks. However, we will indicate the associated computational complexity, by giving an estimation of the number of performed *floating-point computations* (*flops*) by a typical algorithm. For each decomposition we mention several straightforward applications, which often represent the building blocks of more complex numerical algorithms.

10.1.1 Singular Value Decomposition

The *singular value decomposition* (SVD) is a fundamental matrix factorization, which plays an important conceptual and computational role in linear algebra. The computation of the SVD can be interpreted as the reduction of a given rectangular matrix to a "diagonal" form using pre- and post-multiplications with orthogonal matrices. The main theoretical result regarding the SVD is the following theorem.

Theorem 10.1 *For any matrix $A \in \mathbb{R}^{m \times n}$, there exist orthogonal matrices $U \in \mathbb{R}^{m \times m}$ and $V \in \mathbb{R}^{n \times n}$ such that*

$$A = U \Sigma V^T,$$

where $\Sigma = \mathrm{diag}(\Sigma_r, 0)$ with $\Sigma_r = \mathrm{diag}(\sigma_1, \sigma_2, \ldots, \sigma_r)$ and $\sigma_1 \geq \sigma_2 \geq \cdots \geq \sigma_r > 0$.

The value of r defines obviously the *rank* of A. If we partition $U = \begin{bmatrix} U_1 & U_2 \end{bmatrix}$ and $V = \begin{bmatrix} V_1 & V_2 \end{bmatrix}$ column-wise compatible with the row and column partitions of Σ, respectively, then

$$A = \begin{bmatrix} U_1 & U_2 \end{bmatrix} \begin{bmatrix} \Sigma_r & 0 \\ 0 & 0 \end{bmatrix} \begin{bmatrix} V_1^T \\ V_2^T \end{bmatrix} = U_1 \Sigma_r V_1^T, \tag{10.1}$$

which can be interpreted as a full rank factorization of A. We denote with $\sigma_i(A)$, $i = 1, \ldots, p$, the $p := \min(m, n)$ *singular values* of A, which are formed of the r nonzero singular values $\sigma_1, \ldots, \sigma_r$ together with $p - r$ zero singular values. The largest singular value $\overline{\sigma}(A) := \sigma_1$ is equal to $\|A\|_2$, the 2-norm of matrix A. For a square invertible matrix of order n, the 2-norm condition number with respect to inversion can be computed as $\kappa_2(A) := \|A\|_2 \|A^{-1}\|_2 = \sigma_1/\sigma_n$. The *Moore–Penrose pseudo-inverse* of A can be computed as $A^{\dagger} = V_1 \Sigma_r^{-1} U_1^T$. The minimum norm solution of the linear least-squares problem $\min_{x \in \mathbb{R}^n} \|Ax - b\|_2$ is simply $x = A^{\dagger}b = V_1 \Sigma_r^{-1} U_1^T b$.

Remark 10.1 The SVD is considered the primary tool to reliably determine the rank of a matrix. However, by applying any of the available numerically stable algorithms to compute the SVD, there will be almost always p nonzero singular values because of the incurred roundoff errors. If the original matrix A has the "mathematical rank" equal to r, then we can expect that $p - r$ of the numerically computed singular values to be "small". Thus, to determine the rank of A correctly, we need to choose a tolerance $\varepsilon > 0$ and define the "numerical rank" of A as r if the r-th and $r + 1$-th computed singular values satisfy

$$\sigma_r > \varepsilon \geq \sigma_{r+1}. \tag{10.2}$$

Such a rank decision can be seen "reliable" if the gap $\sigma_r - \sigma_{r-1}$ is "large". It is important to note that the significance of the terms "small" and "large" is always in direct relation with the actual magnitudes of the matrix elements. The choice of the tolerance ε should be consistent with both the machine precision (i.e., $\varepsilon \geq \mathbf{u}\overline{\sigma}(A)$, where $\mathbf{u} = 2^{-52} \approx 2.22 \cdot 10^{-16}$ is the unit roundoff for the IEEE double precision floating-point representation), but also with the relative errors in the data (i.e., $\varepsilon \geq 10^{-k}\overline{\sigma}(A)$, where k is the number of correct decimal digits in the entries of A). We call the rank r determined such as (10.2) holds the ε-*rank* of A. $\qquad\square$

A typical numerical algorithm for the computation of the full SVD (i.e., Σ, U and V) requires, for $m \geq n$, about $4m^2n + 8mn^2 + 9m^3$ *flops*, but only $4mn^2 - 4n^3/4$ *flops* for rank determination (i.e., computation of only Σ). For a properly implemented SVD algorithm, it can be shown that the computed diagonal matrix $\overline{\Sigma}$ is exact for a slightly perturbed A, in the following sense

$$U^T(A + E)V = \overline{\Sigma},$$

where $U^T U = I$, $V^T V = I$, $\|E\|_2 = \mathcal{O}(\mathbf{u}\|A\|_2)$, and the computed $\overline{U}$ and $\overline{V}$ are almost orthogonal satisfying $\left\|U - \overline{U}\right\|_2 = \mathcal{O}(\mathbf{u})$ and $\left\|V - \overline{V}\right\|_2 = \mathcal{O}(\mathbf{u})$.

In the rest of this section, we discuss some straightforward applications of the SVD. We assume the SVD of A has the partitioned form in (10.1), where r represents the ε-rank for a given tolerance ε satisfying (10.2) (i.e., all singular values of A satisfying $\sigma_i(A) \leq \varepsilon$ are considered equal to zero). The partitioned SVD (10.1) can be used to define orthogonal bases for the range and kernel of the matrix A as

$$\mathcal{R}(A) = \mathcal{R}(U_1), \qquad \mathcal{N}(A) = \mathcal{R}(V_2),$$

as well as for its transpose A^T as

$$\mathcal{R}(A^T) = \mathcal{R}(V_1), \qquad \mathcal{N}(A^T) = \mathcal{R}(U_2).$$

The orthogonal projections on the respective subspaces can be computed as

$$P_{\mathcal{R}(A)} = U_1 U_1^T, \quad P_{\mathcal{N}(A)} = V_2 V_2^T,$$

$$P_{\mathcal{R}(A^T)} = V_1 V_1^T, \quad P_{\mathcal{N}(A^T)} = U_2 U_2^T,$$

where $P_{\mathcal{X}}$ denotes the orthogonal projection on a subspace $\mathcal{X}$.

Several row and column compressions can be easily obtained in terms of the elements of the SVD (10.1). Let Π_c and Π_r be permutation matrices defined as

$$\Pi_c = \begin{bmatrix} 0 & I_r \\ I_{n-r} & 0 \end{bmatrix}, \qquad \Pi_r = \begin{bmatrix} 0 & I_{m-r} \\ I_r & 0 \end{bmatrix}. \tag{10.3}$$

Then

$$U^T A = \begin{bmatrix} \Sigma_r V_1^T \\ 0 \end{bmatrix}, \qquad \Pi_r U^T A = \begin{bmatrix} 0 \\ \Sigma_r V_1^T \end{bmatrix},$$

represent two widely used row compressions of A to full row rank matrices via orthogonal transformations. Similarly,

$$AV = \begin{bmatrix} U_1 \Sigma_r & 0 \end{bmatrix}, \qquad AV\Pi_c = \begin{bmatrix} 0 & U_1 \Sigma_r \end{bmatrix}$$

are column compressions of A to full column rank matrices via orthogonal transformations.

10.1.2 QR Factorization

The QR factorization of a rectangular matrix in a product of an orthogonal matrix and an upper triangular matrix has many applications, which are similar to those of the SVD. Since the associated computational burden for the determination of the QR factorization is significantly smaller than for the computation of the SVD, it is almost always advantageous to employ the QR factorization instead the SVD, whenever this is possible. We cautiously remark that this gain of efficiency may sometime involve a certain loss of reliability in problems involving rank determinations. Fortunately, this may only occur for some rather "exotic" matrices and, therefore, QR factorization-based techniques are generally preferred to SVD-based methods in many control-oriented algorithms.

The main result of the QR factorization is the following one.

Theorem 10.2 *For any matrix $A \in \mathbb{R}^{m \times n}$, there exists an orthogonal matrix $Q \in \mathbb{R}^{m \times m}$ and an upper triangular matrix $R \in \mathbb{R}^{m \times n}$ such that*

$$A = QR.$$

Specifically, if $m > n$, then R has the form $R = \begin{bmatrix} R_{11} \\ 0 \end{bmatrix}$ with $m - n$ trailing zero rows, while if $n \geq m$ then $R = \begin{bmatrix} R_{11} & R_{12} \end{bmatrix}$. In both cases, R_{11} is a $p \times p$ upper triangular matrix, with $p = \min(m, n)$.

The QR factorization (some authors prefer the term QR decomposition) is the basic tool to solve the linear least-squares problem $\min_{x \in \mathbb{R}^n} \| Ax - b \|_2$, in the case when A is a full column rank matrix. The least-squares solution is simply $x = \begin{bmatrix} R_{11}^{-1} & 0 \end{bmatrix} Q^T b$. Furthermore, if R_{11} is chosen with positive diagonal elements, then R_{11} is the upper triangular factor of the *Cholesky factorization* of $A^T A$ as $A^T A = R_{11}^T R_{11}$. Another application in the case $m > n$ is the computation of the SVD using bidiagonalization-based methods. These techniques can exploit the upper triangular shape of R_{11} to improve the overall computational efficiency.

We have a similar result for the so-called *RQ factorization*, which is mainly relevant for the case $m \leq n$.

Theorem 10.3 *For any matrix $A \in \mathbb{R}^{m \times n}$, with $m \leq n$, there exists an orthogonal matrix $Q \in \mathbb{R}^{n \times n}$ and an upper triangular matrix $R \in \mathbb{R}^{m \times m}$ such that*

$$A = \begin{bmatrix} 0 & R \end{bmatrix} Q.$$

If $r = \operatorname{rank} A < \min(m, p)$, the rank information cannot be usually read out from the resulting upper triangular factor R of the QR factorization. An alternative rank-revealing factorization can be used which allows the determination of rank. The *QR factorization with column pivoting* has the form

$$A = Q \begin{bmatrix} R_{11} & R_{12} \\ 0 & 0 \end{bmatrix} \varPi =: Q \begin{bmatrix} R_1 \\ 0 \end{bmatrix} \varPi, \tag{10.4}$$

where Q is orthogonal, $R_{11} \in \mathbb{R}^{r \times r}$ is upper triangular and invertible, and $\varPi$ is a permutation matrix. Obviously $r = \operatorname{rank} A$. The role of the column permutations is to enforce the invertibility of the leading block R_{11}. The term *column pivoting* indicates a specific column permutation strategy which tries to additionally enforce that R_{11} is well conditioned (with respect to inversion).

Remark 10.2 The rank determination using the QR factorization with column pivoting can be performed during the computation of this factorization. The factorization procedure iteratively constructs the upper triangular matrix R_{11} in the leading position. After $r = \operatorname{rank} A$ iterations, we have the partial decomposition

$$A = \widehat{Q} \begin{bmatrix} \widehat{R}_{11} & \widehat{R}_{12} \\ 0 & \widehat{R}_{22} \end{bmatrix} \widehat{\Pi},$$

where we expect that $\widehat{R}_{22}$ has a suitably small norm. A typical termination criterion might be

$$\|\widehat{R}_{22}\|_2 \leq \varepsilon, \tag{10.5}$$

where $\varepsilon = \varepsilon_1 \|A\|_2$ for some small parameter ε_1 depending on the machine round-off unit $\mathbf{u}$ and the relative errors in the elements of A. If the above condition is fulfilled, then the matrix has "numerical rank" r (also called ε-rank). Surprisingly, there exist some artificially constructed examples (e.g., the Kahan matrices), for which the nearly rank deficiency cannot be detected in this way. Nevertheless, in practice, the QR factorization with column pivoting is almost as reliable as the SVD in determining matrix ranks. Therefore, it is widely used in many algorithms which involve repeated rank determinations (see, for example, the staircase algorithms in Sect. 10.3.1). Here, the repeated use of the full SVD would increase tremendously the computational complexity, due to the need to explicitly compute the involved orthogonal transformation matrices at each reduction step. $\qquad\square$

A typical numerical algorithm for the computation of the QR factorization with column pivoting is based on the Householder QR factorization technique combined with column permutations, and requires about $4mnr - 2r^2(m + n) + 4r^3/3$ *flops*. Therefore, this algorithm is much more efficient than the algorithms for the computation of the SVD. Using the Householder reduction, the orthogonal transformation matrix Q is determined in a factored form $Q = H_1 H_2 \cdots H_r$, where H_i for $i = 1, \ldots, r$, are elementary orthogonal Householder transformation matrices (also known as Householder reflectors). Therefore, it is possible to avoid the explicit building of Q when computing products as $Q^T B$ or CQ, where B and C are arbitrary matrices of compatible dimensions. For the Householder QR algorithm without pivoting, it can be shown that the computed $\overline{R}$ is exact for a nearby A in the sense

$$Q^T (A + E) = \overline{R},$$

where $Q^T Q = I$ and $\|E\|_2 = \mathcal{O}(\mathbf{u}\|A\|_2)$. The computed $\overline{Q}$ is almost orthogonal in the sense that $\|Q - \overline{Q}\|_2 = \mathcal{O}(\mathbf{u})$. A similar statement is obviously valid for the QR factorization with column pivoting.

In the rest of this section we discuss some straightforward applications of the QR factorization, which parallel those of the SVD. We assume the QR factorization with column pivoting of A has the partitioned form in (10.4), where r represents the ε-rank for a given tolerance ε satisfying (10.5) (i.e., the trailing $m - r$ rows of $Q^T A$ are considered equal to zero). Assume the orthogonal matrix Q in (10.4) is partitioned as $Q = \begin{bmatrix} Q_1 & Q_2 \end{bmatrix}$, where $Q_1 \in \mathbb{R}^{m \times r}$ and $Q_2 \in \mathbb{R}^{m \times (m-r)}$. We can determine orthogonal bases for the range of matrix A and the kernel of the matrix A^T (which is also its orthogonal complement) as

$$\mathcal{R}(A) = \mathcal{R}(Q_1), \qquad \mathcal{N}(A^T) = \mathcal{R}(A)^\perp = \mathcal{R}(Q_2).$$

The orthogonal projections on these subspaces can be computed as $P_{\mathcal{R}(A)} = Q_1 Q_1^T$ and $P_{\mathcal{N}(A^T)} = Q_2 Q_2^T$, respectively. Obviously, orthogonal bases for $\mathcal{R}(A^T)$ and $\mathcal{N}(A)$ can be determined in terms of the QR factorization with column pivoting of the transposed matrix A^T.

The row and column compressions can be obtained similarly as for the SVD. Let Π_r be the permutation matrix defined in (10.3). The row compressions of A to full row rank matrices, via orthogonal transformations, can be obtained in one of the following forms:

$$Q^T A = \begin{bmatrix} R_1 \Pi \\ 0 \end{bmatrix}, \qquad \Pi_r Q^T A = \begin{bmatrix} 0 \\ R_1 \Pi \end{bmatrix}.$$

Column compressions can be computed from the row compressions of the transposed matrix A^T, or, in the case of full row rank matrices, by using directly the RQ factorization (see Theorem 10.3).

10.1.3 Real Schur Decomposition

The *real Schur decomposition* (RSD) of a square real matrix A is a basic matrix decomposition which reveals the eigenvalues of A, by determining its *real Schur form* (RSF) (an upper quasi-triangular form) using orthogonal similarity transformations. The following theorem is the main theoretical result regarding the RSD.

Theorem 10.4 *For any $A \in \mathbb{R}^{n \times n}$ there exists an orthogonal $Q \in \mathbb{R}^{n \times n}$ such that $S = Q^T A Q$ is upper quasi-triangular of the form*

$$S = Q^T A Q = \begin{bmatrix} S_{11} & S_{12} & \cdots & S_{1k} \\ 0 & S_{22} & \cdots & S_{2k} \\ \vdots & \vdots & \ddots & \vdots \\ 0 & 0 & \cdots & S_{kk} \end{bmatrix}, \tag{10.6}$$

where each S_{ii} for $i = 1, \ldots, k$ is either a 1×1 or a 2×2 matrix having complex conjugate eigenvalues.

From the RSF (10.6), the eigenvalues of A result simply as

$$\Lambda(A) = \bigcup_{i=1}^{k} \Lambda(S_{ii}) .$$

The RSF also plays an important role in solving various linear matrix equations (Lyapunov, Stein, Sylvester), while the associated transformation matrix Q can be used to compute orthogonal bases of invariant subspaces (see below), which are useful in solving quadratic matrix Riccati equations.

An important property of the RSF is that the order of eigenvalues (and thus of the associated diagonal blocks) is arbitrary. The reordering of diagonal blocks (thus also of corresponding eigenvalues) can be simply done by interchanging two adjacent diagonal blocks of the RSF. For the swapping of such two blocks orthogonal similarity transformations can be used. Thus, any arbitrary reordering of blocks (and thus of the corresponding eigenvalues) can be achieved in this way. An important application of this fact is the computation of orthogonal bases for the invariant subspaces of A corresponding to a particular eigenvalue or a particular set of eigenvalues.

Consider a disjunct partition of the complex plane as $\mathbb{C} = \mathbb{C}_g \cup \mathbb{C}_b$, $\mathbb{C}_g \cap \mathbb{C}_b = \emptyset$, where $\mathbb{C}_g$ and $\mathbb{C}_b$ denote the "good" and "bad" regions of $\mathbb{C}$ for the location of eigenvalues of A, respectively. The ordered RSF is frequently employed in computational algorithms to exhibit a separation of eigenvalues into two sets, namely, all eigenvalues located in $\mathbb{C}_g$ gathered in the leading diagonal block of the RSF and all eigenvalues located in $\mathbb{C}_b$ gathered in the trailing diagonal block of the RSF. Overall we can achieve the orthogonal reduction of A to an ordered RSF matrix S in the form

$$S = Q^T A Q = \begin{bmatrix} A_g & A_{gb} \\ 0 & A_b \end{bmatrix},$$

where $\Lambda(A_g) \subset \mathbb{C}_g$ and $\Lambda(A_b) \subset \mathbb{C}_b$. If we partition Q as $Q = \begin{bmatrix} Q_1 & Q_2 \end{bmatrix}$ compatibly with the structure of the above S, then we can write

$$A Q_1 = Q_1 A_g .$$

It follows that

$$A \mathcal{R}(Q_1) \subset \mathcal{R}(Q_1)$$

and thus $\mathcal{R}(Q_1)$ is an *invariant subspace* corresponding to the eigenvalues of A lying in $\mathbb{C}_g$.

For the computation of the RSD the so-called Francis QR algorithm (or one of its modern variants) is usually used. This algorithm requires about $25n^3$ *flops* if both Q and S are computed. If the eigenvalue reordering is necessary, for example, to move p eigenvalues in the leading diagonal block of the RSF, then additionally at most $12n(n-p)p$ *flops* are necessary (e.g., $3n^3$ flops if $p=n/2$). If only the eigenvalues are desired, then $10n^3$ *flops* are necessary. The roundoff properties of the QR algorithm are what one would expect of any orthogonal matrix technique. The computed RSF $\overline{S}$ is orthogonally similar to a matrix near to A, that is,

$$Q^T(A+E)Q = \overline{S},$$

where $Q^T Q = I$ and $\|E\|_2 = \mathcal{O}(\mathbf{u}\|A\|_2)$. The computed $\overline{Q}$ is almost orthogonal, in the sense that, $\left\| I - \overline{Q}^T\overline{Q} \right\|_2 = \mathcal{O}(\mathbf{u})$. These relations are valid also in the case of employing eigenvalue reordering.

If A is a symmetric real matrix, then all eigenvalues of A are real and the symmetric real Schur form S is the diagonal form formed from the (real) eigenvalues. If additionally A is positive semi-definite, then all eigenvalues of A are non-negative and we have the following simple formula for the square-root of A

$$A^{\frac{1}{2}} = QS^{\frac{1}{2}}Q^T,$$

where $S^{\frac{1}{2}}$ is the diagonal matrix formed from the square-roots of the eigenvalues. We can even compute the factor R of a Cholesky-like decomposition $A = R^T R$ as

$$R = S^{\frac{1}{2}}Q^T.$$

Such a factor is sometimes (improperly) called the square-root of A.

10.1.4 *Generalized Real Schur Decomposition*

The eigenvalue structure of a regular pencil $A - \lambda E$ is completely described by the Weierstrass canonical form (see Lemma 9.8). However, the computation of this canonical form involves the use of (potentially ill-conditioned) general invertible transformations, and therefore numerical reliability cannot be guaranteed. Fortunately, the computation of Weierstrass canonical form can be avoided in almost all computations, and alternative "less" condensed forms can be employed instead, which can be computed by employing exclusively orthogonal similarity transformations. The *generalized real Schur decomposition* (GRSD) of a matrix pair (A, E) reveals the eigenvalues of the regular pencil $A - \lambda E$, by determining the *generalized real Schur form* (GRSF) of the pair (A, E) (a quasi-triangular–triangular form) using orthogonal similarity transformations on the pencil $A - \lambda E$. The main theoretical result regarding the GRSD is the following theorem.

Theorem 10.5 *Let $A - \lambda E$ be an $n \times n$ regular pencil, with A and E real matrices. Then, there exist orthogonal transformation matrices Q and Z such that*

$$S - \lambda T := Q^T(A - \lambda E)Z = \begin{bmatrix} S_{11} & \cdots & S_{1k} \\ & \ddots & \vdots \\ 0 & & S_{kk} \end{bmatrix} - \lambda \begin{bmatrix} T_{11} & \cdots & T_{1k} \\ & \ddots & \vdots \\ 0 & & T_{kk} \end{bmatrix}, \qquad (10.7)$$

where each diagonal subpencil $S_{ii} - \lambda T_{ii}$, for $i = 1, \ldots, k$, is either of dimension 1×1 in the case of a finite real or infinite eigenvalue of the pencil $A - \lambda E$ or of dimension 2×2, with T_{ii} upper triangular, in the case of a pair of finite complex conjugate eigenvalues of $A - \lambda E$.

The pair (S, T) in (10.7) is in a GRSF and the eigenvalues of $A - \lambda E$ (or the generalized eigenvalues of the pair (A, E)) are given by

$$\Lambda(A - \lambda E) = \bigcup_{i=1}^{k} \Lambda(S_{ii} - \lambda T_{ii}).$$

If $E = I$, then we can always choose $Q = Z$, $T = I$ and S is the RSF of A.

Similar to the RSF, the order of eigenvalues (and thus of the associated pairs of diagonal blocks) of the reduced pencil $S - \lambda T$ is arbitrary. The reordering of the pairs of diagonal blocks (thus also of corresponding eigenvalues) can be done by interchanging two adjacent pairs of diagonal blocks of the GRSF. For the swapping of such two pairs of blocks orthogonal similarity transformations can be used. Thus, any arbitrary reordering of pairs of blocks (and thus of the corresponding eigenvalues) can be achieved in this way. An important application of this fact is the computation of orthogonal bases for the deflating subspaces of the pencil $A - \lambda E$ corresponding to a particular eigenvalue or a particular set of eigenvalues.

For the computation of the GRSD the so-called QZ algorithm is usually used. This algorithm requires about $66n^3$ *flops* if all matrices S, T, Q and Z are computed. If the eigenvalue reordering is necessary, for example, to move p eigenvalues in the leading diagonal blocks of the GRSF, then additionally at most $24n(n - p)p$ *flops* are necessary (e.g., $6n^3$ flops if $p = n/2$). If only the eigenvalues are desired, then $30n^3$ *flops* are necessary. The roundoff properties of the QZ algorithm are what one would expect of any orthogonal matrix technique. The computed pair $(\overline{S}, \overline{T})$, in GRSF, is orthogonally similar to a matrix pair near to (A, E) and satisfies

$$Q^T(A + F)Z = \overline{S}, \qquad Q^T(E + G)Z = \overline{T},$$

where $Q^T Q = I$, $Z^T Z = I$, $\|F\|_2 = \mathcal{O}(\mathbf{u}\|A\|_2)$ and $\|G\|_2 = \mathcal{O}(\mathbf{u}\|E\|_2)$. The computed $\overline{Q}$ and $\overline{Z}$ are almost orthogonal, in the sense that, $\left\| I - \overline{Q}^T \overline{Q} \right\|_2 = \mathcal{O}(\mathbf{u})$ and $\left\| I - \overline{Z}^T \overline{Z} \right\|_2 = \mathcal{O}(\mathbf{u})$. These relations are valid also in the case of employing eigenvalue reordering.

Consider a disjunct partition of the complex plane as $\mathbb{C} = \mathbb{C}_g \cup \mathbb{C}_b$, $\mathbb{C}_g \cap \mathbb{C}_b = \emptyset$, where $\mathbb{C}_g$ and $\mathbb{C}_b$ denote the "good" and "bad" regions of $\mathbb{C}$, respectively. We assume that $\mathbb{C}_g$, and therefore also $\mathbb{C}_b$, are symmetric with respect to the real axis. Then, it is possible to determine the orthogonal transformation matrices Q and Z such that

$$Q^T (A - \lambda E) Z = \begin{bmatrix} A_g - \lambda E_g & A_{gb} - \lambda E_{gb} \\ 0 & A_b - \lambda E_b \end{bmatrix} \tag{10.8}$$

is in a GRSF, where $\Lambda(A_g - \lambda E_g) \subset \mathbb{C}_g$ and $\Lambda(A_b - \lambda E_b) \subset \mathbb{C}_b$. Frequently used eigenvalue splittings are the stable-unstable splitting (i.e., $\mathbb{C}_g = \mathbb{C}_s$ and $\mathbb{C}_b = \mathbb{C} \backslash \mathbb{C}_s$) or the finite-infinite splitting (i.e., $\mathbb{C}_g = \mathbb{C} \backslash \{\infty\}$ and $\mathbb{C}_b = \{\infty\}$). More complicated splittings are possible by combining two or more partitions (see below).

The eigenvalue splitting achieved in the ordered GRSF (10.8) is the main tool for determining deflating subspaces corresponding to the eigenvalues of the pencil $A - \lambda E$. The subspaces $\mathcal{X}$ and $\mathcal{Y}$ form a *deflating pair* for the eigenvalues of $A - \lambda E$ if

$$\dim \mathcal{X} = \dim \mathcal{Y}$$

and

$$A \mathcal{X} \subset \mathcal{Y}, \quad E \mathcal{X} \subset \mathcal{Y},$$

where $\dim \mathcal{S}$ denotes the dimension of the subspace $\mathcal{S}$. If we partition Q and Z compatibly with the structure of the GRSF (10.8) as $Q = \begin{bmatrix} Q_1 & Q_2 \end{bmatrix}$ and, respectively, $Z = \begin{bmatrix} Z_1 & Z_2 \end{bmatrix}$, then we can write

$$A Z_1 = Q_1 A_g, \quad E Z_1 = Q_1 E_g.$$

It follows that $\dim \mathcal{R}(Q_1) = \dim \mathcal{R}(Z_1)$ and

$$A \mathcal{R}(Z_1) \subset \mathcal{R}(Q_1), \quad E \mathcal{R}(Z_1) \subset \mathcal{R}(Q_1).$$

Thus, $\mathcal{R}(Q_1)$ and $\mathcal{R}(Z_1)$ form a pair of (left and right) deflating subspaces associated with the eigenvalues of $A_g - \lambda E_g$. Deflating subspaces generalize the notion of invariant subspaces. If E is invertible, then the (right) deflating subspace $\mathcal{R}(Z_1)$ is an invariant subspace of $E^{-1}A$ corresponding to the eigenvalues of $E^{-1}A$ lying in $\mathbb{C}_g$. An important application of deflating subspaces is the solution of generalized Riccati equations, which can be equivalently formulated as the problem of determining orthogonal bases of the right deflating subspace corresponding to the stable eigenvalues of suitably defined regular pencils (see Sect. 10.2.2).

We describe now a special splitting of eigenvalues, which is instrumental for the computation of the proper and stable coprime factorizations using the methods described in Sect. 10.3.5. Assume $\mathbb{C}_g$ is a finite region of $\mathbb{C}$, symmetric with respect to the real axis, and $\mathbb{C}_b$ is its complement including also the point at infinity. The eigenvalue splitting in question involves the reduction of $A - \lambda E$ to the form

$$\widetilde{A} - \lambda \widetilde{E} = Q^T (A - \lambda E) Z = \begin{bmatrix} A_\infty & * & * \\ 0 & A_g - \lambda E_g & * \\ 0 & 0 & A_b - \lambda E_b \end{bmatrix}, \tag{10.9}$$

where A_∞ is an $(n - r) \times (n - r)$ invertible (upper triangular) matrix, with $r = \text{rank} E$, $\Lambda(A_g - \lambda E_g) \subset \mathbb{C}_g$ and $\Lambda(A_b - \lambda E_b) \subset \mathbb{C}_b$. The leading pair $(A_\infty, 0)$ contains all infinite eigenvalues of $A - \lambda E$ corresponding to first-order eigenvectors, while the rest of infinite eigenvalues are included in $A_b - \lambda E_b$.

The **Procedure GSORSF**, presented in what follows, computes the specially ordered GRSF in (10.9). The same procedure can be also used to obtain a reverse ordering of the diagonal blocks of $Q^T(A - \lambda E)Z$ in (10.9). For this, we apply the procedure to the transposed pencil $A^T - \lambda E^T$ to obtain Q_1 and Z_1 such that $Q_1^T(A^T - \lambda E^T)Z_1$ is in a form, as in the right side of (10.9). Let P be the permutation matrix

$$P = \begin{bmatrix} & & 1 \\ & \cdot^{\cdot^{\cdot}} & \\ 1 & & 0 \end{bmatrix}. \tag{10.10}$$

Then, with $Q = Z_1 P$ and $Z = Q_1 P$ we obtain $Q^T(A - \lambda E)Z$ in the form

$$Q^T(A - \lambda E)Z = \begin{bmatrix} A_b - \lambda E_b & * & * \\ 0 & A_g - \lambda E_g & * \\ 0 & 0 & A_\infty \end{bmatrix}. \tag{10.11}$$

Procedure GSORSF: Specially ordered generalized real Schur form

Inputs : $A - \lambda E$ regular, $\mathbb{C}_g$ and $\mathbb{C}_b$ such that $\mathbb{C} = \mathbb{C}_g \cup \mathbb{C}_b$, $\mathbb{C}_g \cap \mathbb{C}_b = \emptyset$
Outputs: $Q, Z, \widetilde{A} - \lambda\widetilde{E} = Q^T(A - \lambda E)Z$ in (10.9)

(1) Compute an orthogonal Z_1 such that $EZ_1 = \begin{bmatrix} 0 & E_2 \end{bmatrix}$, with E_2 full column rank $r = \text{rank} E$; compute the conformably partitioned $AZ_1 = \begin{bmatrix} A_1 & A_2 \end{bmatrix}$, with A_1 having full column rank $n - r$.

(2) Compute an orthogonal Q_1 such that $Q_1^T A_1 = \begin{bmatrix} A_\infty \\ 0 \end{bmatrix}$, with A_∞ an $(n - r) \times (n - r)$ invertible upper triangular matrix; compute the conformably partitioned matrices

$$Q_1^T A_2 = \begin{bmatrix} A_{12} \\ A_{22} \end{bmatrix}, \quad Q_1^T E_2 = \begin{bmatrix} E_{12} \\ E_{22} \end{bmatrix}.$$

(3) Compute orthogonal Q_2 and Z_2 such that

$$Q_2^T(A_{22} - \lambda E_{22})Z_2 = \begin{bmatrix} A_g - \lambda E_g & A_{gb} - \lambda E_{gb} \\ 0 & A_b - \lambda E_b \end{bmatrix}$$

is in a GRSF, where $\Lambda(A_g - \lambda E_g) \subset \mathbb{C}_g$ and $\Lambda(A_b - \lambda E_b) \subset \mathbb{C}_b$. Compute $A_{12}Z_2 = \begin{bmatrix} A_{\infty,g} & A_{\infty,b} \end{bmatrix}$ and $E_{12}Z_2 = \begin{bmatrix} E_{\infty,g} & E_{\infty,b} \end{bmatrix}$ conformably partitioned with $Q_2^T(A_{22} - \lambda E_{22})Z_2$.

(4) Set $Q = Q_1 \text{diag}(I_{n-r}, Q_2)$, $Z = Z_1 \text{diag}(I_{n-r}, Z_2)$ and define $\widetilde{A}$ and $\widetilde{E}$ from the pencil

$$\widetilde{A} - \lambda\widetilde{E} = \begin{bmatrix} A_\infty & A_{\infty,g} - \lambda E_{\infty,g} & A_{\infty,b} - \lambda E_{\infty,b} \\ 0 & A_g - \lambda E_g & A_{gb} - \lambda E_{gb} \\ 0 & 0 & A_b - \lambda E_b \end{bmatrix}.$$

10.1.5 Controllability and Observability Staircase Forms

Staircase forms represent a large family of block upper triangular condensed forms, which arise from various algorithms which "compress" the numerical data available in single matrices or matrix pairs. All forms already studied such as the diagonal form (originated from the SVD), upper triangular form (originated from the QR factorization), the RSF (originated from the Francis QR algorithm) or the GRSF of a matrix pair (originated from the QZ algorithm), can be interpreted as particular staircase forms. For a general rectangular linear pencil, several Kronecker-like staircase forms (see next section) are obtainable using strict pencil similarity transformations using orthogonal transformations. In this section, we discuss two particular staircase forms, the controllability and observability staircase forms, which appear as parts of this form. However, due to their special importance for the computation of irreducible representation of descriptor systems, we dedicate a separate section for the discussion of their properties and also give a numerically stable computational procedure for their determination.

We have the following main result regarding the controllability staircase form.

Theorem 10.6 *Consider the pair* $(A - \lambda E, B)$, *with* $A, E \in \mathbb{R}^{n \times n}$ *and* $B \in \mathbb{R}^{n \times m}$, *and assume the pencil* $A - \lambda E$ *is regular. Then, there exist orthogonal transformation matrices* Q *and* Z *such that*

$$\left[\, \widehat{B} \,\middle|\, \widehat{A} - \lambda \widehat{E} \,\right] := \left[\, Q^T B \,\middle|\, Q^T A Z - \lambda Q^T E Z \,\right] = \left[\begin{array}{c|cc} B_c & A_c - \lambda E_c & * \\ 0 & 0 & A_{\bar{c}} - \lambda E_{\bar{c}} \end{array}\right],$$

$$(10.12)$$

is in a generalized controllability staircase form with

$$\left[\, B_c \,\middle|\, A_c \,\right] = \left[\begin{array}{c|ccccc} A_{1,0} & A_{1,1} & A_{12} & \cdots & A_{1,k-1} & A_{1,k} \\ 0 & A_{2,1} & A_{22} & \cdots & A_{2,k-1} & A_{2,k} \\ 0 & 0 & A_{32} & \cdots & A_{3,k-1} & A_{3,k} \\ \vdots & \vdots & \vdots & \ddots & \vdots & \vdots \\ 0 & 0 & 0 & \cdots & A_{k,k-1} & A_{k,k} \end{array}\right], \qquad (10.13)$$

where $A_{j,j-1} \in \mathbb{R}^{\nu_j \times \nu_{j-1}}$, *with* $\nu_0 = m$, *are full row rank matrices for* $j = 1, \ldots, k$, *and the resulting upper triangular matrix* E_c *has a similar block partitioned form*

$$E_c = \left[\begin{array}{ccccc} E_{1,1} & E_{1,2} & \cdots & E_{1,k-1} & E_{1,k} \\ 0 & E_{2,2} & \cdots & E_{2,k-1} & E_{2,k} \\ \vdots & \vdots & \ddots & \vdots & \vdots \\ 0 & 0 & \cdots & E_{k-1,k-1} & E_{k-1,k} \\ 0 & 0 & \cdots & 0 & E_{k,k} \end{array}\right], \qquad (10.14)$$

where $E_{j,j} \in \mathbb{R}^{\nu_j \times \nu_j}$. The resulting block dimensions ν_j, $j = 0, 1, \ldots, k$, satisfy

$$m = \nu_0 \geq \nu_1 \geq \cdots \geq \nu_k > 0.$$

The $n_c \times (m + n_c)$ pencil $\left[B_c \, | \, A_c - \lambda E_c \right]$, with $n_c := \sum_{j=1}^{k} \nu_j$, has full row rank for any finite $\lambda \in \mathbb{C}$, and therefore the pair $(A_c - \lambda E_c, B_c)$ is finite controllable. If $n_c < n$, then the $(n - n_c) \times (n - n_c)$ regular pencil $A_{\bar{c}} - \lambda E_{\bar{c}}$ contains the finite uncontrollable eigenvalues of $A - \lambda E$ (and also possibly some infinite ones).

If $m = 1$, then all subdiagonal blocks $A_{j,j-1}$ of A_c are 1×1 and A_c is in a *Hessenberg form*. The pair (A_c, E_c) with A_c in Hessenberg form and E_c upper triangular is in a so-called *generalized Hessenberg form* (GHF). If $m > 1$, then A_c is in a so-called *block Hessenberg form*. If $E = I$, then we can choose $Q = Z$ such that $\widehat{E} = I$.

Remark 10.3 If we partition Q and Z compatibly with the structure of the staircase form (10.12) as $Q = \left[Q_1 \; Q_2 \right]$ and, respectively, $Z = \left[Z_1 \; Z_2 \right]$, then we can write $A Z_1 = Q_1 A_c$ and $E Z_1 = Q_1 E_c$. It follows that $\dim \mathcal{R}(Q_1) = \dim \mathcal{R}(Z_1)$ and $A \mathcal{R}(Z_1) \subset \mathcal{R}(Q_1)$, $E \mathcal{R}(Z_1) \subset \mathcal{R}(Q_1)$. Thus, $\mathcal{R}(Q_1)$ and $\mathcal{R}(Z_1)$ form a pair of (left and right) deflating subspaces associated with the eigenvalues of $A_c - \lambda E_c$. Additionally we have

$$\mathcal{R}(B) \subset A \mathcal{R}(Z_1) + E \mathcal{R}(Z_1) \tag{10.15}$$

and $\mathcal{C}_f := \mathcal{R}(Z_1)$ is a deflating subspace with least possible dimension satisfying (10.15). We call $\mathcal{C}_f$ the *finite controllability subspace* of the pair $(A - \lambda E, B)$. The pair $(A - \lambda E, B)$ is *finite controllable* if the dimension of $\mathcal{C}_f$ is n. $\qquad\square$

We also have the dual result to Theorem 10.6 for the observability staircase form.

Theorem 10.7 *Consider the pair $(A - \lambda E, C)$, with $A, E \in \mathbb{R}^{n \times n}$ and $C \in \mathbb{R}^{p \times n}$, and assume the pencil $A - \lambda E$ is regular. Then, there exist orthogonal transformation matrices Q and Z such that*

$$\left[\frac{\widehat{A} - \lambda \widehat{E}}{\widehat{C}} \right] := \left[\frac{Q^T A Z - \lambda Q^T E Z}{C Z} \right] = \left[\begin{array}{cc} A_{\bar{o}} - \lambda E_{\bar{o}} & * \\ 0 & A_o - \lambda E_o \\ \hline 0 & C_o \end{array} \right], \tag{10.16}$$

is in a generalized observability staircase form with

$$\left[\frac{A_o}{C_o} \right] = \left[\begin{array}{ccccc} A_{\ell,\ell} & A_{\ell,\ell-1} & \cdots & A_{\ell,2} & A_{\ell,1} \\ A_{\ell-1,\ell} & A_{\ell-1,\ell-1} & \cdots & A_{\ell-1,2} & A_{\ell-1,1} \\ 0 & A_{\ell-2,\ell-1} & \cdots & A_{\ell-2,2} & A_{\ell-2,1} \\ \vdots & \vdots & \ddots & \vdots & \vdots \\ 0 & 0 & \cdots & A_{1,2} & A_{1,1} \\ \hline 0 & 0 & \cdots & 0 & A_{0,1} \end{array} \right], \tag{10.17}$$

where $A_{j-1,j} \in \mathbb{R}^{\mu_{j-1} \times \mu_j}$, with $\mu_0 = p$, are full column rank matrices for $j = 1, \ldots, \ell$, and the resulting upper triangular matrix E_o has a similar block partitioned form

$$E_o = \begin{bmatrix} E_{\ell,\ell} & E_{\ell,\ell-1} & \cdots & E_{\ell,2} & E_{\ell,1} \\ 0 & E_{\ell-1,\ell-1} & \cdots & E_{\ell-1,2} & E_{\ell-1,1} \\ \vdots & \vdots & \ddots & \vdots & \vdots \\ 0 & 0 & \cdots & E_{2,2} & E_{2,1} \\ 0 & 0 & \cdots & 0 & E_{1,1} \end{bmatrix}, \tag{10.18}$$

with $E_{j,j} \in \mathbb{R}^{\mu_j \times \mu_j}$. The resulting block dimensions μ_j, $j = 0, 1, \ldots, \ell$, satisfy

$$p = \mu_0 \geq \mu_1 \cdots \geq \mu_\ell > 0 \,.$$

The $(n_o + p) \times n_o$ pencil $\left[\dfrac{A_o - \lambda E_o}{C_o} \right]$, with $n_o := \sum_{j=1}^{\ell} \mu_j$, has full column rank for any finite $\lambda \in \mathbb{C}$, and therefore the pair $(A_o - \lambda E_o, C_o)$ is finite observable. If $n_o < n$, then the $(n - n_o) \times (n - n_o)$ regular pencil $A_{\bar{o}} - \lambda E_{\bar{o}}$ contains the finite unobservable eigenvalues of $A - \lambda E$ (and also possibly some infinite ones).

Remark 10.4 If we partition Q and Z compatibly with the structure of the staircase form (10.16) as $Q = \begin{bmatrix} Q_1 & Q_2 \end{bmatrix}$ and, respectively, $Z = \begin{bmatrix} Z_1 & Z_2 \end{bmatrix}$, then we can write $AZ_1 = Q_1 A_{\bar{o}}$ and $EZ_1 = Q_1 E_{\bar{o}}$. It follows that $\dim \mathcal{R}(Q_1) = \dim \mathcal{R}(Z_1)$ and $A\mathcal{R}(Z_1) \subset \mathcal{R}(Q_1)$, $E\mathcal{R}(Z_1) \subset \mathcal{R}(Q_1)$. Thus, $\mathcal{R}(Q_1)$ and $\mathcal{R}(Z_1)$ form a pair of (left and right) deflating subspaces associated with the eigenvalues of $A_{\bar{o}} - \lambda E_{\bar{o}}$. Additionally, $\overline{\mathcal{O}}_f := \mathcal{R}(Z_1)$ is a deflating subspace with the largest dimension satisfying $\mathcal{R}(Z_1) \subset \mathcal{N}(C)$. We call $\overline{\mathcal{O}}_f$ the *finite unobservable subspace* of the pair $(A - \lambda E, C)$. The pair $(A - \lambda E, C)$ is *finite observable* if $\overline{\mathcal{O}}_f$ is empty. $\qquad\square$

The following procedure to compute the staircase form (10.12) can be seen as a constructive proof of Theorem 10.6. In view of the main application of this procedure (see Sect. 10.3.1), we included a matrix $C \in \mathbb{R}^{p \times n}$ on which all transformations to the right are also applied.

Procedure GCSF: Generalized controllability staircase form

Inputs : $(A - \lambda E, B, C)$
Outputs: $Q, Z, (A - \lambda E, B, C) := (Q^T A Z - \lambda Q^T E Z, Q^T B, C Z); v_j, j = 1, \ldots, \ell$

(1) Compute an orthogonal matrix Z such that EZ is upper triangular;
    compute $A \leftarrow AZ, E \leftarrow EZ, C \leftarrow CZ$.
(2) Set $j = 1, n_c = 0, v_0 = m, A^{(0)} = A, E^{(0)} = E, B^{(0)} = B, Q = I_n$.
(3) Compute orthogonal matrices W and U such that

$$W^T B^{(j-1)} := \begin{bmatrix} A_{j,j-1} \\ 0 \end{bmatrix} \begin{matrix} v_j \\ \rho \end{matrix} ,$$
$$\begin{matrix} v_{j-1} \end{matrix}$$

with $A_{j,j-1}$ full row rank and $W^T E^{(j-1)} U$ upper triangular.
(4) Compute and partition

$$W^T A^{(j-1)} U := \begin{bmatrix} A_{j,j} & A_{j,j+1} \\ B^{(j)} & A^{(j)} \end{bmatrix} \begin{matrix} v_j \\ \rho \end{matrix} , \quad W^T E^{(j-1)} U := \begin{bmatrix} E_{j,j} & E_{j,j+1} \\ O & E^{(j)} \end{bmatrix} \begin{matrix} v_j \\ \rho \end{matrix}$$
$$\qquad\quad \begin{matrix} v_j & \rho \end{matrix} \qquad\qquad\qquad\qquad \begin{matrix} v_j & \rho \end{matrix}$$

(5) For $i = 1, \ldots, j - 1$ compute and partition

$$A_{i,j} U := \begin{bmatrix} A_{i,j} & A_{i,j+1} \end{bmatrix}, \quad E_{i,j} U := \begin{bmatrix} E_{i,j} & E_{i,j+1} \end{bmatrix}$$
$$\qquad\quad \begin{matrix} v_j & \rho \end{matrix} \qquad\qquad\qquad \begin{matrix} v_j & \rho \end{matrix}$$

(6) $Q \leftarrow Q \operatorname{diag}(I_{n_c}, W), Z \leftarrow Z \operatorname{diag}(I_{n_c}, U), C \leftarrow C \operatorname{diag}(I_{n_c}, U)$.
(7) $n_c \leftarrow n_c + v_j$; if $\rho = 0$ then $\ell = j$ and **Exit**.
(8) If $v_j > 0$, then $j \leftarrow j + 1$ and go to Step (3); else, $\ell = j - 1$, and **Exit**.

If the **Procedure GCSF** exits at Step (7), then the original pair $(A - \lambda E, B)$ is finite controllable. However, if the **Procedure GCSF** exits at Step (8), then the original pair $(A - \lambda E, B)$ is not finite controllable. In this case, the trailing $\rho \times \rho$ pencil $A^{(\ell+1)} - \lambda E^{(\ell+1)} =: A_{\bar{c}} - \lambda E_{\bar{c}}$, with $\rho = n - n_c$, contains all uncontrollable finite eigenvalues of $A - \lambda E$.

The **Procedure GCSF** can be implemented such that at Step (1) it exploits any particular shape in the lower triangle of E (e.g., E lower banded). In particular, if E is upper triangular, then the resulting Z is simply $Z = I$ and no further computations are performed at this step. The row compressions at Step (3) are usually performed using rank-revealing QR factorizations with column pivoting (see Sect. 10.1.2). The reductions can be performed using sequences of Givens rotations (instead Householder reflectors), which allow to simultaneously perform the column transformations accumulated in U to maintain the upper triangular form of $E^{(j-1)}$. This reduction technique is described in detail in [131] and is similar to the reduction of a matrix pair to a generalized Hessenberg form. By using this technique, the numerical complexity of **Procedure GCSF** is $\mathcal{O}(n^3)$ (for $m, p \ll n$), provided all transformations are immediately applied without accumulating explicitly W and U. Note that the usage of the more robust rank determinations based on singular values decompositions would increase the overall complexity to $\mathcal{O}(n^4)$ due to the need to accumulate explicitly each W and U. Regarding the numerical properties of **Procedure GCSF**, it is possible to show that the resulting system matrices $\widehat{A}, \widehat{E}, \widehat{B}$,

$\widehat{C}$ are exact for slightly perturbed original data A, E, B, C, while Q an Z are nearly orthogonal matrices. It follows that the **Procedure GCSF** is numerically stable.

To compute the observability staircase form of a pair $(A - \lambda E, C)$, the **Procedure GCSF** can be applied to the dual pair $(A^T - \lambda E^T, C^T)$ to obtain the transformed pair $(\widehat{A}^T - \lambda \widehat{E}^T, \widehat{C}^T)$ in a controllability staircase form. Then, the pair $(P\widehat{A}P - P\widehat{E}P, \widehat{C}P)$, where P is the permutation matrix (10.10), is in an observability staircase form.

10.1.6 Kronecker-Like Forms

Consider the reduction of a general rectangular (or singular) pencil $M - \lambda N$, with $M, N \in \mathbb{R}^{m \times n}$ using strict similarity transformations of the form

$$\widehat{M} - \lambda \widehat{N} = U(M - \lambda N)V,$$

where U and V are invertible matrices. From Lemma 9.9, recall that, by using general invertible transformations, we can determine the Kronecker-canonical form (9.47) of the pencil $M - \lambda N$, which basically characterizes the right and left singular structure and the eigenvalue structure of the pencil. The computation of the Kronecker-canonical form may involve the use of ill-conditioned transformations and, therefore, is potentially numerically unstable. Fortunately, alternative staircase forms, called *Kronecker-like forms*, allow to obtain basically the same (or only a part of) structural information on the pencil $M - \lambda N$ by employing exclusively orthogonal transformations (i.e., $U^T U = I$ and $V^T V = I$).

The following result concerns with one of the main Kronecker-like forms.

Theorem 10.8 *Let $M \in \mathbb{R}^{m \times n}$ and $N \in \mathbb{R}^{m \times n}$ be arbitrary real matrices. Then, there exist orthogonal $U \in \mathbb{R}^{m \times m}$ and $V \in \mathbb{R}^{n \times n}$, such that*

$$U(M - \lambda N)V = \begin{bmatrix} M_r - \lambda N_r & * & * \\ 0 & M_{reg} - \lambda N_{reg} & * \\ 0 & 0 & M_l - \lambda N_l \end{bmatrix}, \tag{10.19}$$

where

(1) The $n_r \times (m_r + n_r)$ pencil $M_r - \lambda N_r$ has full row rank, n_r, for all $\lambda \in \mathbb{C}$ and is in a controllability staircase form

$$M_r - \lambda N_r = \begin{bmatrix} B_r & A_r - \lambda E_r \end{bmatrix}, \tag{10.20}$$

with $B_r \in \mathbb{R}^{n_r \times m_r}$, $A_r, E_r \in \mathbb{R}^{n_r \times n_r}$, and E_r invertible.

(2) The $n_{reg} \times n_{reg}$ pencil $M_{reg} - \lambda N_{reg}$ is regular and its eigenvalues are the eigenvalues of pencil $M - \lambda N$. The pencil $M_{reg} - \lambda N_{reg}$ may be chosen in a GRSF, with arbitrary ordered diagonal blocks.

(3) *The* $(p_l + n_l) \times n_l$ *pencil* $M_l - \lambda N_l$ *has full column rank,* n_l*, for all* $\lambda \in \mathbb{C}$ *and is in a observability staircase form*

$$M_l - \lambda N_l = \begin{bmatrix} A_l - \lambda E_l \\ C_l \end{bmatrix}, \tag{10.21}$$

with $C_l \in \mathbb{R}^{p_l \times n_l}$, A_l, $E_l \in \mathbb{R}^{n_l \times n_l}$, *and* E_l *invertible.*

Let ν_i, $i = 1, \ldots, k$ be the dimensions of the diagonal blocks of $A_r - \lambda E_r$ in the controllability staircase form $\begin{bmatrix} B_r & A_r - \lambda E_r \end{bmatrix}$ and define $\nu_0 = m_r$. These dimensions completely determine the right Kronecker structure of $M - \lambda N$ as follows: there are $\nu_{i-1} - \nu_i$ blocks $L_{i-1}(\lambda)$ of size $(i-1) \times i$, $i = 1, \ldots, k$. Analogously, let μ_i, $i = 1, \ldots, \ell$ be the dimensions of the diagonal blocks of $A_l - \lambda E_l$ in the observability staircase form $\begin{bmatrix} A_l - \lambda E_l \\ C_l \end{bmatrix}$ and define $\mu_0 = p_l$. These dimensions completely determine the left Kronecker structure of $M - \lambda N$ as follows: there are $\mu_{i-1} - \mu_i$ blocks $L_{i-1}^T(\lambda)$ of size $i \times (i-1)$, $i = 1, \ldots, \ell$. We have $n_r = \sum_{i=1}^{k} \nu_i$ and $n_l = \sum_{i=1}^{\ell} \mu_i$, and the normal rank of $M - \lambda N$ is $n_r + n_{reg} + n_l$. The finite Smith zeros of $M - \lambda N$ are the finite eigenvalues of the regular pencil $M_{reg} - \lambda N_{reg}$ and represent the finite values of λ for which $M - \lambda N$ drops its rank below its normal rank.

In Sect. 10.3, several applications of the Kronecker-like forms are presented such as the computation of minimal nullspace basis, system zeros, inner-outer factorizations and the solution of linear rational equations.

For the computation of the Kronecker-like form (10.19), the standard approach is to achieve successive separations of the structural elements and eigenvalues of the pencil $M - \lambda N$. A typical pencil reduction procedure, as **Procedure PREDUCE** presented in this section, uses two orthogonal transformation matrices Q and Z to achieve the following separation:

$$\widetilde{M} - \lambda \widetilde{N} := Q(M - \lambda N)Z = \left[\begin{array}{c|c} M_{r,\infty} - \lambda N_{r,\infty} & * \\ \hline 0 & M_{f,l} - \lambda N_{f,l} \end{array} \right], \tag{10.22}$$

where the $m_{r,\infty} \times n_{r,\infty}$ pencil $M_{r,\infty} - \lambda N_{r,\infty}$ has full row rank for all $\lambda \in \mathbb{C}$ excepting possibly a finite set of infinite values of λ, and the $m_{f,l} \times n_{f,l}$ pencil $M_{f,l} - \lambda N_{f,l}$ has full column rank for all $\lambda \in \mathbb{C}$ excepting possibly a finite set of finite values of λ. Moreover, the pencil $\widetilde{M} - \lambda \widetilde{N}$ is in the following staircase form:

$$\tilde{M} - \lambda\tilde{N} = \begin{bmatrix} M_{1,1} & M_{1,2}-\lambda N_{1,2} & \cdots & M_{1,k}-\lambda N_{1,k} & * \\ 0 & M_{2,2} & \cdots & M_{2,k}-\lambda N_{2,k} & * \\ \vdots & \vdots & \ddots & \vdots & \vdots \\ 0 & 0 & \cdots & M_{k,k} & * \\ 0 & 0 & \cdots & 0 & M_{f,l}-\lambda N_{f,l} \end{bmatrix} \begin{matrix} \} \nu_1 \\ \} \nu_2 \\ \\ \} \nu_k \\ \} m_{f,l} \end{matrix} , \quad (10.23)$$

$$\underbrace{\qquad}_{\mu_1} \underbrace{\qquad}_{\mu_2} \underbrace{\qquad}_{\mu_k} \underbrace{\qquad}_{n_{f,l}}$$

$$(10.23)$$

where $M_{i,i} \in \mathbb{R}^{\nu_i \times \mu_i}$ are full row rank matrices for $i = 1, \ldots, k$, and $N_{i-1,i} \in \mathbb{R}^{\nu_{i-1} \times \mu_i}$ are full column rank matrices for $i = 2, \ldots, k$. The dimensions ν_i and μ_i of the blocks in (10.23) satisfy

$$\mu_1 \geq \nu_1 \geq \mu_2 \geq \nu_2 \geq \cdots \mu_k \geq \nu_k \geq 0$$

and completely determine the right Kronecker structure and the eigenstructure at infinity of the pencil $M - \lambda N$ as follows: there are $\mu_i - \nu_i$ $(i = 1, \ldots, k)$ blocks $L_{i-1}(\lambda)$ of size $(i-1) \times i$ and, with $\mu_{k+1} := 0$, there are $\nu_i - \mu_{i+1}$ $(i = 1, \ldots, k)$ nilpotent Jordan blocks $J_i(0)$ of dimension i (see (9.46)) which correspond to the infinite eigenvalues. The row and column dimensions of pencil $M_{r,\infty} - \lambda N_{r,\infty}$ are given by $m_{r,\infty} = \sum_{i=1}^{k} \nu_i$ and $n_{r,\infty} = \sum_{i=1}^{k} \mu_i$, respectively.

Procedure PREDUCE: Pencil reduction to staircase form

Inputs : $M - \lambda N$ with $M, N \in \mathbb{R}^{m \times n}$
Outputs: $Q, Z, \tilde{M} - \lambda\tilde{N} = Q(M - \lambda N)Z$ in the staircase form (10.23).

(0) Compute orthogonal matrices Q and Z such that

$$\tilde{M} = QMZ = \begin{bmatrix} B^{(0)} & A^{(0)} \\ D^{(0)} & C^{(0)} \end{bmatrix}, \qquad \tilde{N} = QNZ = \begin{bmatrix} 0 & E^{(0)} \\ 0 & 0 \end{bmatrix},$$

where $E^{(0)} \in \mathbb{R}^{\tilde{n} \times \tilde{n}}$ is upper triangular and invertible with $\tilde{n} = \operatorname{rank} N$, $A^{(0)} \in \mathbb{R}^{\tilde{n} \times \tilde{n}}$, $B^{(0)} \in \mathbb{R}^{\tilde{n} \times \tilde{m}}$ with $\tilde{m} = n - \tilde{n}$, $C^{(0)} \in \mathbb{R}^{\tilde{p} \times \tilde{n}}$ with $\tilde{p} = m - \tilde{n}$, $D \in \mathbb{R}^{\tilde{p} \times \tilde{m}}$; set $m_{r,\infty} = 0, n_{r,\infty} = 0, i = 0$.

Step–i: while $\tilde{m} > 0$, do:

(1) Compute orthogonal matrices W and Y such that $WD^{(i)}Y = \begin{bmatrix} D_1^{(i)} & D_2^{(i)} \\ 0 & 0 \end{bmatrix}$

where $D_1^{(i)} \in \mathbb{R}^{\tau_i \times \tau_i}$ is invertible and upper triangular. Obtain

$$\begin{bmatrix} B_1^{(i)} & B_2^{(i)} & A^{(i)}-\lambda E^{(i)} \\ D_1^{(i)} & D_2^{(i)} & C_1^{(i)} \\ 0 & 0 & C_2^{(i)} \end{bmatrix} := \operatorname{diag}(I_{\tilde{n}}, W) \begin{bmatrix} B^{(i)} & A^{(i)}-\lambda E^{(i)} \\ D^{(i)} & C^{(i)} \end{bmatrix} \operatorname{diag}(Y, I_{\tilde{n}}) \cdot$$

Procedure PREDUCE: Pencil reduction to staircase form (continued)

---

(2) Compress the rows of $\begin{bmatrix} B_1^{(i)} & E^{(i)} \\ D_1^{(i)} & 0 \end{bmatrix}$ with orthogonal X such that

$$X \begin{bmatrix} B_1^{(i)} & E^{(i)} \\ D_1^{(i)} & 0 \end{bmatrix} = \begin{bmatrix} B_{11}^{(i)} & E_1^{(i)} \\ 0 & E_2^{(i)} \end{bmatrix},$$

with $B_{11}^{(i)} \in \mathbb{R}^{\tau_i \times \tau_i}$ and $E_2^{(i)} \in \mathbb{R}^{\tilde{n} \times \tilde{n}}$ invertible and upper triangular. Obtain

$$\left[\begin{array}{cc|c} B_{11}^{(i)} & B_{12}^{(i)} & A_1^{(i)} - \lambda E_1^{(i)} \\ 0 & B_{22}^{(i)} & A_2^{(i)} - \lambda E_2^{(i)} \\ \hline 0 & 0 & C_2^{(i)} \end{array} \right] := \mathrm{diag}(X, I_{\tilde{p}-\tau_i}) \left[\begin{array}{cc|c} B_1^{(i)} & B_2^{(i)} & A^{(i)} - \lambda E^{(i)} \\ D_1^{(i)} & D_2^{(i)} & C_1^{(i)} \\ \hline 0 & 0 & C_2^{(i)} \end{array} \right].$$

(3) Compress the rows of $B_{22}^{(i)}$ with orthogonal U such that $U B_{22}^{(i)} = \begin{bmatrix} \widetilde{B}_{22}^{(i)} \\ 0 \end{bmatrix}$, with $\widetilde{B}_{22}^{(i)} \in \mathbb{R}^{\rho_i \times (\tilde{m}-\tau_i)}$ full row rank, and compute orthogonal V such that $U E_2^{(i)} V$ is upper triangular. Obtain

$$\left[\begin{array}{cc|cc} B_{11}^{(i)} & B_{12}^{(i)} & A_{11}^{(i)} - \lambda E_{11}^{(i)} & * \\ \hline 0 & \widetilde{B}_{22}^{(i)} & A_{21}^{(i)} - \lambda E_{21}^{(i)} & * \\ \hline 0 & 0 & A_{31}^{(i)} & A_{32}^{(i)} - \lambda E_{32}^{(i)} \\ \hline 0 & 0 & C_{21}^{(i)} & C_{22}^{(i)} \end{array} \right] :=$$

$$\mathrm{diag}\left(I_{\tau_i}, U, I_{\tilde{p}-\tau_i}\right) \left[\begin{array}{cc|c} B_{11}^{(i)} & B_{12}^{(i)} & A_1^{(i)} - \lambda E_1^{(i)} \\ 0 & B_{22}^{(i)} & A_2^{(i)} - \lambda E_2^{(i)} \\ \hline 0 & 0 & C_2^{(i)} \end{array} \right] \mathrm{diag}(I_{\tilde{m}}, V),$$

with $E_{21}^{(i)} \in \mathbb{R}^{\rho_i \times \rho_i}$ and $E_{32}^{(i)} \in \mathbb{R}^{(\tilde{n}-\rho_i) \times (\tilde{n}-\rho_i)}$ invertible and upper triangular.

(4) Form $Q^{(i)} = \mathrm{diag}\left(I_{m_{r,\infty}}, \widetilde{Q}\right)$ and $Z^{(i)} = \mathrm{diag}\left(I_{n_{r,\infty}}, \widetilde{Z}\right)$ with

$$\widetilde{Q} = \mathrm{diag}\left(I_{\tau_i}, U, I_{\tilde{p}-\tau_i}\right) \mathrm{diag}\left(X, I_{\tilde{p}-\tau_i}\right) \mathrm{diag}\left(I_{\tilde{n}}, W\right), \qquad \widetilde{Z} = \mathrm{diag}\left(Y, V\right),$$

and update $\widetilde{M} \leftarrow Q^{(i)} \widetilde{M} Z^{(i)}$, $\widetilde{N} \leftarrow Q^{(i)} \widetilde{N} Z^{(i)}$, $Q \leftarrow Q^{(i)} Q$, $Z \leftarrow Z Z^{(i)}$.
Set $\nu_{i+1} = \rho_i + \tau_i$, $\mu_{i+1} = \tilde{m}$ and define

$$M_{i+1,i+1} := \left[\begin{array}{c|c} B_{11}^{(i)} & B_{12}^{(i)} \\ \hline 0 & \widetilde{B}_{22}^{(i)} \end{array} \right], \qquad M_{i+1,i+2} - \lambda N_{i+1,i+2} := \left[\begin{array}{c} A_{11}^{(i)} - \lambda E_{11}^{(i)} \\ \hline A_{21}^{(i)} - \lambda E_{21}^{(i)} \end{array} \right],$$

with $M_{i+1,i+1} \in \mathbb{R}^{\nu_{i+1} \times \mu_{i+1}}$ full row rank and $N_{i+1,i+2} \in \mathbb{R}^{\nu_{i+1} \times \rho_i}$ full column rank, and

$$\left[\begin{array}{c|c} B^{(i+1)} & A^{(i+1)} - \lambda E^{(i+1)} \\ \hline D^{(i+1)} & C^{(i+1)} \end{array} \right] := \left[\begin{array}{c|c} A_{31}^{(i)} & A_{32}^{(i)} - \lambda E_{32}^{(i)} \\ \hline C_{21}^{(i)} & C_{22}^{(i)} \end{array} \right].$$

(5) Update $m_{r,\infty} \leftarrow m_{r,\infty} + \nu_{i+1}$, $n_{r,\infty} \leftarrow n_{r,\infty} + \mu_{i+1}$, $\tilde{n} \leftarrow \tilde{n} - \rho_i$, $\tilde{m} \leftarrow \rho_i$, $\tilde{p} \leftarrow \tilde{p} - \tau_i$.
(6) $i \leftarrow i + 1$ and *go to* **Step–i**.

At the end of **Procedure PREDUCE** we obtain the $m_{f,l} \times n_{f,l}$ pencil

$$M_{f,l} - \lambda N_{f,l} := \left[\begin{array}{c} A^{(i)} - \lambda E^{(i)} \\ \hline C^{(i)} \end{array} \right],$$

$$(10.24)$$

with $m_{f,l} = \tilde{n} + \tilde{p}$ and $n_{f,l} = \tilde{n}$, and with $E^{(i)}$ upper triangular and invertible. It follows that the pencil $M_{f,l} - \lambda N_{f,l}$ has only finite and left structure. The number of diagonal blocks $M_{j,j}$ of $M_{r,\infty} - \lambda N_{r,\infty}$ in the staircase form (10.23) is $k = i - 1$, where i is the resulting final value of i at the exit of **Procedure PREDUCE**.

The **Procedure PREDUCE** performs exclusively orthogonal transformations on the matrix pair (M, N). It is possible to show that the resulting pair $(\widetilde{M}, \widetilde{N})$ is exact for a slightly perturbed original pair, while Q and Z are nearly orthogonal matrices. It follows that the **Procedure PREDUCE** is numerically stable.

The computational complexity of **Procedure PREDUCE** mainly depends on the details of the computations performed at Step (2) to obtain $E_2^{(i)}$ in an upper triangular form and, at Step (3), to preserve the upper triangular form of $U E_2^{(i)} V$ and to obtain $E_{21}^{(i)}$ and $E_{32}^{(i)}$ invertible and upper triangular. If the transformation matrices U and V are accumulated (e.g., by performing SVD-based row compressions), the worst-case computational complexity of **Procedure PREDUCE** is $\mathcal{O}(n^4)$ (assuming $n \geq m$), which, for large values of n, is unacceptable. However, using the techniques described in [101], these operations can be performed such that a worst-case computational complexity of $\mathcal{O}(n^3)$ can be guaranteed. The main computational ingredients are specially tailored QR factorizations with column pivoting, which provide almost the same reliability as the rank determinations based on the use of SVD. Using specialized QR factorizations, it is possible to implement the row compressions at Steps (2) and (3) such that the preservation of the upper triangular shape of $E^{(i)}$ is simultaneously possible, without the need to explicitly accumulate the intervening transformations. For the rest of necessary row and column compressions at Step (0) and Step (1), the safer SVD-based computations can be still employed, without increasing excessively the computational complexity.

A straightforward application of the **Procedure PREDUCE** is to perform the infinite-finite separation of the eigenvalues of a regular pencil $M - \lambda N$ (i.e., without right and left structures). Since $M - \lambda N$ has no right structure, $M_{r,\infty} - \lambda N_{r,\infty}$ has only infinite eigenvalues. Similarly, since $M - \lambda N$ has no left structure, $M_{f,l} - \lambda N_{f,l}$ contains all finite eigenvalues of the pencil.

A complementary separation of the pencil $M - \lambda N$ can be achieved by applying **Procedure PREDUCE** to the transposed pencil $M^T - \lambda N^T$ and pertranspose the resulted pencil. Recall that the pertranspose M^P of a matrix $M \in \mathbb{R}^{m \times n}$ is defined as $M^P := P_n M^T P_m$, where P_k denotes the $k \times k$ permutation matrix of the form (10.10). The net effect of applying P_n from left is to reverse the order of rows of a matrix, while the application of P_m from right reverses the order of columns of the matrix. If Q and Z are the orthogonal matrices used to reduce $M^T - \lambda N^T$, then overall we obtain

$$P_m Z^T (M - \lambda N) Q^T P_n = \begin{bmatrix} M_{r,f} - \lambda N_{r,f} & * \\ 0 & M_{\infty,l} - \lambda N_{\infty,l} \end{bmatrix},$$

where $M_{r,f} - \lambda N_{r,f}$ contains the right and finite structure and $M_{\infty,l} - \lambda N_{\infty,l}$ contains the infinite and left structure. Moreover, $M_{\infty,l} - \lambda N_{\infty,l}$ is in a dual staircase form, which is obtained by reversing the orders of the blocks in the staircase form (10.23). Sometimes, it is more advantageous to apply **Procedure PREDUCE** to $M^P - \lambda N^P$ instead of $M^T - \lambda N^T$ (e.g., already existing upper block structures are preserved by pertransposition and thus can be further exploited).

For the computation of the complete Kronecker-like form (10.19) of the pencil $M - \lambda N$ we can employ **Procedure PREDUCE** to perform the first separation in (10.22). Then, by applying **Procedure PREDUCE** to the pertransposed pencil $M_{r,\infty}^P - \lambda N_{r,\infty}^P$, we obtain the separation of the right and infinite structures in the form

$$Q_1(M_{r,\infty} - \lambda N_{r,\infty})Z_1 = \begin{bmatrix} M_r - \lambda N_r & * \\ 0 & M_\infty - \lambda N_\infty \end{bmatrix}, \tag{10.25}$$

where Q_1 and Z_1 are orthogonal matrices, the full row rank pencil $M_r - \lambda N_r$ is in the form (10.20) and the regular pencil $M_\infty - \lambda N_\infty$, with M_∞ invertible and N_∞ nilpotent, contains the infinite eigenvalues. Similarly, by applying **Procedure PREDUCE** to the pertransposed pencil $M_{f,l}^P - \lambda N_{f,l}^P$, we obtain the separation of the finite and left structures in the form

$$Q_2(M_{f,l} - \lambda N_{f,l})Z_2 = \begin{bmatrix} M_f - \lambda N_f & * \\ 0 & M_l - \lambda N_l \end{bmatrix}, \tag{10.26}$$

where Q_2 and Z_2 are orthogonal matrices, the regular pencil $M_f - \lambda N_f$ with N_f invertible contains the finite eigenvalues and the full column rank pencil $M_l - \lambda N_l$ is in an observability staircase form (10.21). Overall we achieved

$$\text{diag}(Q_1, Q_2)Q(M - \lambda N)Z\text{diag}(Z_1, Z_2) = \begin{bmatrix} M_r - \lambda N_r & * & * & * \\ 0 & M_\infty - \lambda N_\infty & * & * \\ 0 & 0 & M_f - \lambda N_f & * \\ 0 & 0 & 0 & M_l - \lambda N_l \end{bmatrix}$$

from which the regular part $M_{reg} - \lambda N_{reg}$ in (10.19) can be immediately read out. For this separation, it is possible to exploit the structure of the pencil $M_{f,l} - \lambda N_{f,l}$ in (10.24) which results when applying **Procedure PREDUCE**. Since in the pertransposed pencil $M_{f,l}^P - \lambda N_{f,l}^P = \left[(C^{(i)})^P \ (A^{(i)})^P - \lambda (E^{(i)})^P \right]$, the invertible matrix $(E^{(i)})^P$ is already upper triangular, therefore when applying **Procedure PREDUCE** to $M_{f,l}^P - \lambda N_{f,l}^P$ the preliminary reduction at Step (0) is not necessary anymore. Alternatively, the **Procedure GCSF** can be employed to obtain $M_l - \lambda N_l$ in an observability staircase form (10.21). This computation is needed to be additionally performed, to obtain $M_r - \lambda N_r$ in a controllability staircase form (10.20).

The above Kronecker-like form exhibits the main structural elements of an arbitrary pencil $M - \lambda N$. However, in some applications, as the computation of rational left nullspace bases in Sect. 7.4, it is necessary only to know the left Kronecker structure. For this purpose, it is sufficient to apply **Procedure PREDUCE** twice, to obtain the basic separation (10.22) and then the splitting of finite and left structures as in (10.26) to obtain the required form

$$
\operatorname{diag}(I, Q_2) Q (M - \lambda N) Z \operatorname{diag}(I, Z_2) =
\begin{bmatrix}
M_{r,\infty} - \lambda N_{r,\infty} & * & * \\
0 & M_f - \lambda N_f & * \\
0 & 0 & M_l - \lambda N_l
\end{bmatrix}.
$$

On the other hand, when all structural details of the Kronecker-like form are necessary, as for example, when solving linear rational equations in Sect. 9.2.9, the separation of right and infinite structure of the pencil $M_{r,\infty} - \lambda N_{r,\infty}$ is necessary. An alternative way to perform this separation is to employ a computational approach proposed in [9] (see Algorithms 3.3.1 and 3.3.2 in [9]). These algorithms exploit all structural information in the staircase form (10.23) and perform the separation of right and infinite structure by employing exclusively orthogonal transformations, however without making any rank decisions. The resulting subpencils $M_r - \lambda N_r$ and $M_\infty - \lambda N_\infty$ are in staircase forms and the dimensions of the resulting diagonal blocks automatically reveal the right Kronecker indices and infinite eigenvalue structure.

10.2 Solution of Matrix Equations

There are several linear and quadratic matrix equations which play an important role in control theory. In this section, we discuss the computational solutions of some of the main equations and give the conditions for the existence of a solution.

10.2.1 Linear Matrix Equations

We discuss the computational solution of two main classes of linear matrix equations. In the first class, we consider the *generalized Sylvester equation* (GSE) of the form

$$
AXG + EXF + Q = 0, \tag{10.27}
$$

where $A, E \in \mathbb{R}^{n \times n}$, $F, G \in \mathbb{R}^{m \times m}$, $Q \in \mathbb{R}^{n \times m}$, and the desired solution is $X \in \mathbb{R}^{n \times m}$. Equation (10.27) has a unique solution if and only if the matrix pencils $A - \lambda E$ and $F - \lambda G$ are regular and $\Lambda(A - \lambda E) \cap \Lambda(F + \lambda G) = \emptyset$.

Two special cases of Eq. (10.27) are of particular interest in this book: the *generalized continuous-time Lyapunov equation* (GCLE) of the form

$$
AXE^T + EXA^T + Q = 0 \tag{10.28}
$$

and the *generalized discrete-time Lyapunov equation* (GDLE) (also called generalized Stein equation)

$$AXA^T - EXE^T + Q = 0, \tag{10.29}$$

where Q, and hence also X, are symmetric. The solvability condition of Eq. (10.28) requires that E is invertible and $\lambda_i + \lambda_j \neq 0$, for all $\lambda_i, \lambda_j \in \Lambda(A - \lambda E)$. The solvability condition of Eq. (10.29) requires that $\lambda_i \lambda_j \neq 1$, for all $\lambda_i, \lambda_j \in \Lambda(A - \lambda E)$. In both cases, of special interest are (semi-)positive definite solutions in the case when Q has the form $Q = BB^T \geq 0$ and $\Lambda(A - \lambda E) \in \mathbb{C}_s$. In this case, the solution X can be directly obtained in a Cholesky-factored form $X = SS^T$, with S upper triangular.

For the numerical solution of the above matrix equations the transformation method (developed initially by Bartels and Stewart to solve the Sylvester equation $AX + XB + C = 0$) can be used. Let Q_1 and Z_1 be orthogonal matrices such that the pair $(P, S) := (Q_1^T A Z_1, Q_1^T E Z_1)$ is in a GRSF, and let Q_2 and Z_2 be orthogonal matrices such that the pair $(T, R) := (Q_2^T F Z_2, Q_2^T G Z_2)$ is in a GRSF. The matrices Q_1 and Z_1, and, Q_2 and Z_2, can be obtained by applying the QZ algorithm to the matrix pairs (A, E) and (F, G), respectively. If we define $Y = Z_1^T X Q_2$ and $H = Q_1^T Q Z_2$, then Eq. (10.27) can be rewritten as

$$PYR + SYT + H = 0.$$

By exploiting the quasi-upper triangular–upper triangular structures of the pairs (P, S) and (T, R), this equation can be solved by a special (back substitution) technique to obtain the solution Y [50, 57]. Then, the solution of (10.27) is computed as $X = Z_1 Y Q_2^T$. The overall computational effort to solve Eq. (10.27) is $\mathcal{O}(n^3 + m^3) + \mathcal{O}(n^2 m + nm^2)$. With obvious simplifications, this approach can be used to solve the GCLE (10.28) and the GDLE (10.29) as well. The overall computational effort to solve these equations is $\mathcal{O}(n^3)$.

The second class of linear equation is the *generalized Sylvester system of equations* (GSSE)

$$\begin{aligned} AX + YF &= C, \\ EX + YG &= D, \end{aligned} \tag{10.30}$$

where $A, E \in \mathbb{R}^{n \times n}$, $F, G \in \mathbb{R}^{m \times m}$, $C, D \in \mathbb{R}^{n \times m}$, and the desired solution is $X, Y \in \mathbb{R}^{n \times m}$. Equation (10.30) has a unique solution if and only if the matrix pencils $A - \lambda E$ and $F - \lambda G$ are regular and $\Lambda(A - \lambda E) \cap \Lambda(F - \lambda G) = \emptyset$. A transformation method (which is similar to that used for solving (10.27)) can be employed to reduce (10.30) to a simpler form. Let Q_1 and Z_1 be orthogonal matrices such that the pair $(P, S) := (Q_1^T A Z_1, Q_1^T E Z_1)$ is in a GRSF, and let Q_2 and Z_2 be orthogonal matrices such that the pair $(R, T) := (Q_2^T F Z_2, Q_2^T G Z_2)$ is in a GRSF. The matrices Q_1 and Z_1, and Q_2 and Z_2 can be obtained by applying the QZ algorithms to the matrix pairs (A, E) and (F, G), respectively. If we define $X_1 = Z_1^T X Z_2$, $Y_1 = Q_1^T Y Q_2$, $C_1 = Q_1^T C Z_2$ and $D_1 = Q_1^T D Z_2$, then the system (10.30) can be rewritten as

$$PX_1 + Y_1 R = C_1,$$
$$SX_1 + Y_1 T = D_1.$$

By exploiting the quasi-upper triangular–upper triangular structures of the pairs (P, S) and (R, T), this system of equations can be efficiently solved using methods proposed in [71]. After solving the transformed system for X_1 and Y_1, we obtain the solution of (10.30) as $X = Z_1 X_1 Z_2^T$ and $Y = Q_1 Y_1 Q_2^T$. The overall computational effort to solve these equations is $\mathcal{O}(n^3 + m^3) + \mathcal{O}(n^2 m + n m^2)$.

10.2.2 *Generalized Algebraic Riccati Equations*

In this section, we address the numerical solution of a class of generalized Riccati equations which appear in various algorithms as the computation of inner-outer factorization (see Sect. 10.3.6) or in spectral factorization problems discussed in Sect. 7.8. We consider a sextuple of matrices (A, E, B, Q, S, R), with the following properties of component matrices: $A \in \mathbb{R}^{n \times n}$, $E \in \mathbb{R}^{n \times n}$ invertible, $B \in \mathbb{R}^{n \times m}$, $Q \in \mathbb{R}^{n \times n}$ symmetric positive semi-definite, $R \in \mathbb{R}^{m \times m}$ symmetric and invertible, and $S \in \mathbb{R}^{n \times m}$. We seek the symmetric positive semi-definite stabilizing solution $X_s \in \mathbb{R}^{n \times n}$ of the *generalized continuous-time algebraic Riccati equation* (GCARE)

$$A^T X E + E^T X A - (E^T X B + S) R^{-1} (B^T X E + S^T) + Q = 0$$

and the corresponding stabilizing state feedback gain $F_s \in \mathbb{R}^{m \times n}$, given by

$$F_s = -R^{-1}(B^T X_s E + S^T) \, ,$$

such that all generalized eigenvalues of the pair $(A + B F_s, E)$ have negative real parts. Similarly, we seek the symmetric positive semi-definite stabilizing solution $X_s \in \mathbb{R}^{n \times n}$ of the *generalized discrete-time algebraic Riccati equation* (GDARE)

$$A^T X A - E^T X E - (A^T X B + S)(R + B^T X B)^{-1}(B^T X A + S^T) + Q = 0$$

and the corresponding stabilizing state feedback gain $F_s \in \mathbb{R}^{m \times n}$, given by

$$F_s = -(R + B^T X_s B)^{-1}(B^T X_s A + S^T) \, ,$$

such that all generalized eigenvalues of the pair $(A + B F_s, E)$ have moduli less than one. Since E is invertible, it is possible to reduce both the GCARE and GDARE to standard Riccati equations, for which there exist standard solution methods. However, to avoid possible accuracy losses due to the need to explicitly invert E, we will indicate methods which directly tackle the above equations, without inverting E.

A unified approach to determine the solutions of the GCARE and GDARE relies on determining an orthogonal basis of the stable deflating subspace of a suitably defined regular matrix pencil $L - \lambda P$. For the solution of the GCARE, we have

$$L = \begin{bmatrix} A & 0 & B \\ -Q & -A^T & -S \\ S^T & B^T & R \end{bmatrix}, \qquad P = \begin{bmatrix} E & 0 & 0 \\ 0 & E^T & 0 \\ 0 & 0 & 0 \end{bmatrix}, \tag{10.31}$$

while for the solution of the GDARE we have

$$L = \begin{bmatrix} A & 0 & B \\ -Q & E^T & -S \\ S^T & 0 & R \end{bmatrix}, \qquad P = \begin{bmatrix} E & 0 & 0 \\ 0 & A^T & 0 \\ 0 & -B^T & 0 \end{bmatrix}. \tag{10.32}$$

Under fairly standard assumptions (e.g., the stabilizability of the pair $(A - \lambda E, B)$ and detectability of the pair $(A - \lambda E, Q - SR^{-1}S^T)$), the existence of the positive semi-definite stabilizing solution X_s is guaranteed. For computational purposes, the main property of the regular pencil $L - \lambda P$ is the existence of an n dimensional (right) deflating subspace corresponding to the stable eigenvalues of $L - \lambda P$. If this subspace is spanned by a $(2n + m) \times n$ matrix Z_1, then we have that $LZ_1 = PZ_1W$, where W is an $n \times n$ matrix such that $\Lambda(W) \in \mathbb{C}_s$. If we partition Z_1 in accordance with the block column structure of the pencil $L - \lambda P$ as

$$Z_1 = \begin{bmatrix} Z_{11} \\ Z_{21} \\ Z_{31} \end{bmatrix}, \tag{10.33}$$

then the stabilizing positive definite solution X_s of both the GCARE and GDARE, and the corresponding stabilizing feedback F_s can be computed as

$$X_s = Z_{21}(EZ_{11})^{-1}, \qquad F_s = Z_{31}Z_{11}^{-1}.$$

To compute Z_1, we can employ the QZ algorithm to determine an ordered GRSF of the pair (L, P) in the form

$$U^T(L - \lambda P)Z = \begin{bmatrix} L_{11} - \lambda P_{11} & L_{12} - \lambda P_{12} \\ 0 & L_{22} - \lambda P_{22} \end{bmatrix}, \tag{10.34}$$

where U and Z are orthogonal transformation matrices, the $n \times n$ subpencil $L_{11} - \lambda P_{11}$ has only stable eigenvalues, i.e., $\Lambda(L_{11} - \lambda P_{11}) \subset \mathbb{C}_s$, and $\Lambda(L_{22} - \lambda P_{22}) \subset \mathbb{C} \backslash \mathbb{C}_s$. Then, Z_1 is formed from the first n columns of the orthogonal matrix Z. Using this approach, the overall computational effort for solving both the GCARE and GDARE is $\mathcal{O}((n + m)^3)$.

10.3 Algorithms for Descriptor Systems

In this section, we present computational procedures for the solution of several basic computational problems for descriptor systems. The theoretical aspects of these problems have been succinctly addressed in Sect. 9.2, where several canonical forms (e.g., Weierstrass, Kronecker) played an important conceptual role in their solutions. However, these canonical forms are not suited to develop reliable numerical algorithms, due to the need of using potentially ill-conditioned transformations for their computation. We present reliable numerical algorithms, which rely on the alternative condensed forms discussed in Sect. 10.1. These forms can be computed using exclusively orthogonal transformations. Therefore, these algorithms are intrinsically numerically reliable and some of them are even numerically stable.

10.3.1 Minimal Realization

Consider a $p \times m$ rational matrix $G(\lambda)$ and let $(A - \lambda E, B, C, D)$ be an n-th order descriptor system realization satisfying

$$G(\lambda) = C(\lambda E - A)^{-1}B + D,$$

with $A - \lambda E$ an $n \times n$ regular pencil. If $Q, Z \in \mathbb{R}^{n \times n}$ are invertible matrices, then it is easy to check that two realizations $(A - \lambda E, B, C, D)$ and $(\widehat{A} - \lambda\widehat{E}, \widehat{B}, \widehat{C}, D)$, whose matrices are related by a similarity transformation of the form

$$\widehat{A} - \lambda\widehat{E} = Q(A - \lambda E)Z, \quad \widehat{B} = QB, \quad \widehat{C} = CZ,$$

have the same TFM $G(\lambda)$. Similarity transformations with Q and Z orthogonal matrices can be used to obtain various staircase forms of the system matrices, which allow to extract lower dimensional descriptor realizations of $G(\lambda)$, and finally to arrive at a minimal order realization with the least possible order n.

A minimal realization $(A - \lambda E, B, C, D)$ is characterized by the five conditions (i)–(v) of Theorem 9.2. An irreducible realization fulfils only conditions (i)–(iv) and is thus controllable and observable. In what follows, we describe a two-stage approach which first constructs an irreducible realization of lower order by successively removing the uncontrollable and unobservable eigenvalues of $A - \lambda E$, and in a second stage removes the non-dynamics modes (i.e., the simple infinite eigenvalues of $A - \lambda E$).

The first reduction stage is accomplished in four steps, by employing repeatedly **Procedure GCSF** to successively remove the finite uncontrollable, infinite uncontrollable, finite unobservable and infinite unobservable eigenvalues of $A - \lambda E$. At the first step of this reduction stage, we apply **Procedure GCSF** to the triple

$(A - \lambda E, B, C)$ to obtain the orthogonal transformation matrices Q_1 and Z_1, such that the equivalent descriptor realization of $G(\lambda)$ has the form

$$\left[\begin{array}{c|c} Q_1^T(A - \lambda E)Z_1 & Q_1^T B \\ \hline CZ_1 & D \end{array} \right] = \left[\begin{array}{cc|c} A_c^f - \lambda E_c^f & * & B_c^f \\ 0 & A_{\bar{c}}^f - \lambda E_{\bar{c}}^f & 0 \\ \hline C_c^f & C_{\bar{c}}^f & D \end{array} \right]. \tag{10.35}$$

The finite controllable descriptor system $(A_c^f - \lambda E_c^f, B_c^f, C_c^f, D)$ has the same TFM $G(\lambda)$ and its order $n_c^f \leq n$. By this step, we can remove the $n - n_c^f$ uncontrollable eigenvalues of $A_{\bar{c}}^f - \lambda E_{\bar{c}}^f$ from the original descriptor system representation $(A - \lambda E, B, C, D)$. Besides all finite uncontrollable eigenvalues, $\Lambda(A_{\bar{c}}^f - \lambda E_{\bar{c}}^f)$ may also contain some of infinite uncontrollable eigenvalues of $A - \lambda E$.

At the second step of the reduction stage, we apply **Procedure GCSF** to the triple $(E_c^f - \lambda A_c^f, B_c^f, C_c^f)$ (note that A_c^f and E_c^f are interchanged) to obtain the orthogonal transformation matrices Q_2 and Z_2, such that the equivalent descriptor realization of $G(\lambda)$ has the form

$$\left[\begin{array}{c|c} Q_2^T(A_c^f - \lambda E_c^f)Z_2 & Q_2^T B_c^f \\ \hline C_c^f Z_2 & D \end{array} \right] = \left[\begin{array}{cc|c} A_c - \lambda E_c & * & B_c \\ 0 & A_{\bar{c}}^\infty - \lambda E_{\bar{c}}^\infty & 0 \\ \hline C_c & C_{\bar{c}}^\infty & D \end{array} \right]. \tag{10.36}$$

As before, the controllable descriptor system $(A_c - \lambda E_c, B_c, C_c, D)$ has the same TFM $G(\lambda)$ and its order $n_c \leq n_c^f$. By this step, we can remove the $n_c^f - n_c$ uncontrollable infinite eigenvalues of $A_{\bar{c}}^\infty - \lambda E_{\bar{c}}^\infty$ (or equivalently the uncontrollable zero eigenvalues of $E_{\bar{c}}^\infty - \lambda A_{\bar{c}}^\infty$) from the original descriptor system representation $(A - \lambda E, B, C, D)$.

In the third step, we apply **Procedure GCSF** to the dual triple $(E_c^T - \lambda A_c^T, C_c^T, B_c^T)$ to obtain the orthogonal transformation matrices Z_3 and Q_3 (note the changed order), such that the equivalent descriptor realization of $G(\lambda)$ has the form

$$\left[\begin{array}{c|c} P_3 Q_3^T(A_c - \lambda E_c)Z_3 P_3 & P_3 Q_3^T B_c \\ \hline C_c Z_3 P_3 & D \end{array} \right] = \left[\begin{array}{cc|c} A_{c\bar{o}}^f - \lambda E_{c\bar{o}}^f & * & B_{c\bar{o}}^f \\ 0 & A_{co}^f - \lambda E_{co}^f & B_{co}^f \\ \hline 0 & C_{co}^f & D \end{array} \right], \tag{10.37}$$

where P_3 is the permutation matrix (10.10) of appropriate size. The controllable and finite observable descriptor system $(A_{co}^f - \lambda E_{co}^f, B_{co}^f, C_{co}^f, D)$ has the same TFM $G(\lambda)$ and its order $n_{co}^f \leq n_c$. By this step we can remove the $n_c - n_{co}^f$ unobservable eigenvalues of $A_{c\bar{o}}^f - \lambda E_{c\bar{o}}^f$ from the original descriptor system representation $(A - \lambda E, B, C, D)$.

Finally, at the fourth step, we apply **Procedure GCSF** to the dual triple $((E_{co}^f)^T - \lambda(A_{co}^f)^T, (C_{co}^f)^T, (B_{co}^f)^T)$ (note that A_{co}^f and E_{co}^f are interchanged) to obtain the orthogonal transformation matrices Z_4 and Q_4, such that the equivalent descriptor realization of $G(\lambda)$ has the form

$$
\left[\begin{array}{c|c}
P_4 Q_4^T (A_{co}^f - \lambda E_{co}^f) Z_4 P_4 & P_4 Q_4^T B_{co}^f \\
\hline
C_{co}^f Z_4 P_4 & D
\end{array}\right]
=
\left[\begin{array}{cc|c}
A_{c\bar{o}}^\infty - \lambda E_{c\bar{o}}^\infty & * & B_{c\bar{o}}^\infty \\
0 & A_{co} - \lambda E_{co} & B_{co} \\
\hline
0 & C_{co} & D
\end{array}\right],
$$

$$\tag{10.38}$$

where P_4 is a permutation matrix as in (10.10) of appropriate size. The irreducible (i.e., controllable and observable) descriptor system $(A_{co} - \lambda E_{co}, B_{co}, C_{co}, D)$ has the same TFM $G(\lambda)$ and its order $n_{co} \leq n_{co}^f$. By this step we can remove the $n_{co}^f - n_{co}$ unobservable infinite eigenvalues of $A_{c\bar{o}}^\infty - \lambda E_{c\bar{o}}^\infty$ from the original descriptor system representation $(A - \lambda E, B, C, D)$.

With the overall transformation matrices defined as

$$
Q := Q_1 \mathrm{diag}(Q_2, I) \mathrm{diag}(Q_3 P_3, I) \mathrm{diag}(I, Q_4 P_4, I),
$$

$$
Z := Z_1 \mathrm{diag}(Z_2, I) \mathrm{diag}(Z_3 P_3, I) \mathrm{diag}(I, Z_4 P_4, I),
$$

we obtained the orthogonally similar system representation

$$
(\widetilde{A} - \lambda \widetilde{E}, \widetilde{B}, \widetilde{C}, D) := (Q^T A Z - \lambda Q^T E Z, Q^T B, C Z, D),
$$

with

$$
\left[\begin{array}{c|c}
\widetilde{A} - \lambda \widetilde{E} & \widetilde{B} \\
\hline
\widetilde{C} & D
\end{array}\right]
=
\left[\begin{array}{ccccc|c}
A_{c\bar{o}}^f - \lambda E_{c\bar{o}}^f & * & * & * & * & B_{c\bar{o}}^f \\
0 & A_{c\bar{o}}^\infty - \lambda E_{c\bar{o}}^\infty & * & * & * & B_{c\bar{o}}^\infty \\
0 & 0 & A_{co} - \lambda E_{co} & * & * & B_{co} \\
0 & 0 & 0 & A_{\bar{c}}^\infty - \lambda E_{\bar{c}}^\infty & * & 0 \\
0 & 0 & 0 & 0 & A_{\bar{c}}^f - \lambda E_{\bar{c}}^f & 0 \\
\hline
0 & 0 & C_{co} & C_{\bar{c}}^\infty & C_{\bar{c}}^f & D
\end{array}\right].
$$

This form, obtained using exclusively orthogonal similarity transformations, represents a particular instance of a generalized Kalman decomposition of the descriptor system matrices from which an irreducible realization $(A_{co} - \lambda E_{co}, B_{co}, C_{co}, D)$ can be readily extracted. There are various ways to improve the efficiency of computations. For example, if the original realization corresponds to a proper system, then the second and fourth steps (i.e., removing uncontrollable or unobservable infinite eigenvalues) can be skipped. Similar simplifications are possible—for example, if the original system description corresponds to a polynomial matrix, or if the original system representation is known to be controllable or observable, or if $A - \lambda E$ has no zero eigenvalues. In the latter case, only the second and fourth steps need to be performed.

The whole computational approach is summarized in the following procedure, which computes for a given triple $(A - \lambda E, B, C)$ an irreducible (i.e., controllable and observable) triple $(A_{co} - \lambda E_{co}, B_{co}, C_{co})$.

Procedure GIR: Generalized irreducible realization algorithm

Input : $(A - \lambda E, B, C)$
Output: Irreducible $(A_{co} - \lambda E_{co}, B_{co}, C_{co})$

(1) Perform **Procedure GCSF** on the triple $(A - \lambda E, B, C)$ and extract the finite controllable triple $(A_c^f - \lambda E_c^f, B_c^f, C_c^f)$.
(2) Perform **Procedure GCSF** on the triple $(E_c^f - \lambda A_c^f, B_c^f, C_c^f)$ and extract the controllable triple $(A_c - \lambda E_c, B_c, C_c)$.
(3) With P an appropriate permutation matrix as in (10.10), perform **Procedure GCSF** on the triple $(P A_c^T P - \lambda P E_c^T P, P C_c^T, B_c^T P)$ and extract the controllable and finite observable triple $(A_{co}^f - \lambda E_{co}^f, B_{co}^f, C_{co}^f)$.
(4) With P an appropriate permutation matrix as in (10.10), perform **Procedure GCSF** on the triple $(P(E_{co}^f)^T P - \lambda P(A_{co}^f)^T P, P(C_{co}^f)^T, (B_{co}^f)^T P)$ and build the irreducible triple $(A_{co} - \lambda E_{co}, B_{co}, C_{co})$.

At the end of Step (1), A_c^f is in an upper block Hessenberg form and E_c^f is upper triangular. The upper block Hessenberg shape of A_c^f at Step (2) can be exploited by the **Procedure GCSF**, to reduce the computational burden at the initial reduction of A_c^f to an upper triangular form. The resulting A_c at Step (2) is therefore upper triangular, while E_c is upper block Hessenberg. At Step (3), the use of $P E_c^T P$ instead of E_c^T allows to preserve the upper block Hessenberg form of E_c obtained at the previous step. This is also the case Step (4), where the upper block Hessenberg structure of A_{co}^f is preserved when using $P(A_{co}^f)^T P$ instead.

The computational effort for **Procedure GIR** is $\mathcal{O}(n^3)$ for $m, p \ll n$. It is possible to show that the computed irreducible descriptor system $(A_{co} - \lambda E_{co}, B_{co}, C_{co}, D)$ is exact for a slightly perturbed original system. Therefore, the **Procedure GIR** can be considered numerically stable.

In the second stage, we have to remove the simple infinite eigenvalues of $A_{co} - \lambda E_{co}$ from the resulting irreducible descriptor representation. For this purpose, we isolate the simple infinite eigenvalues by employing two SVDs. First, we compute the SVD of E_{co} such that

$$U_1^T E_{co} V_1 = \begin{bmatrix} E_{11} & 0 \\ 0 & 0 \end{bmatrix},$$

with U_1 and V_1 orthogonal matrices and E_{11} a (diagonal) invertible matrix of rank r. Applying the same transformations to A we obtain

$$U_1^T A_{co} V_1 = \begin{bmatrix} A_{11} & \widetilde{A}_{12} \\ \widetilde{A}_{21} & \widetilde{A}_{22} \end{bmatrix}.$$

Now, we compute the SVD of $\widetilde{A}_{22}$ such that

$$U_2^T \widetilde{A}_{22} V_2 = \begin{bmatrix} A_{22} & 0 \\ 0 & 0 \end{bmatrix},$$

with U_2 and V_2 orthogonal matrices and A_{22} a (diagonal) invertible matrix of rank q. With $U = U_1 \text{diag}(I_r, U_2)$ and $V = V_1 \text{diag}(I_r, V_2)$ we have the equivalent descriptor realization

$$\left[\begin{array}{c|c} U^T A_{co} V - \lambda U^T E_{co} V & U^T B_{co} \\ C_{co} V & D \end{array} \right] = \left[\begin{array}{ccc|c} A_{11} - \lambda E_{11} & A_{12} & A_{13} & B_1 \\ A_{21} & A_{22} & 0 & B_2 \\ A_{31} & 0 & 0 & B_3 \\ \hline C_1 & C_2 & C_3 & D \end{array} \right].$$

At this step, we have the transformed state vector $\widetilde{x}(t) := V^T x(t)$ partitioned into three components

$$\widetilde{x}(t) = \begin{bmatrix} x_1(t) \\ x_2(t) \\ x_3(t) \end{bmatrix},$$

which correspond to the column structure of $U^T A V$. We can eliminate the second component $x_2(t)$ as

$$x_2(t) = -A_{22}^{-1} A_{21} x_1(t) - A_{22}^{-1} B_2 u(t)$$

and obtain a descriptor representation with the reduced state vector $\overline{x}(t) = \begin{bmatrix} x_1(t) \\ x_3(t) \end{bmatrix}$ and the corresponding minimal realization $(\overline{A} - \lambda \overline{E}, \overline{B}, \overline{C}, \overline{D})$ of $G(\lambda)$, with the matrices given by

$$\overline{A} - \lambda \overline{E} = \begin{bmatrix} A_{11} - A_{12} A_{22}^{-1} A_{21} - \lambda E_{11} & A_{13} \\ A_{31} & 0 \end{bmatrix}, \qquad \overline{B} = \begin{bmatrix} B_1 - A_{12} A_{22}^{-1} B_2 \\ B_3 \end{bmatrix},$$

$$\overline{C} = \begin{bmatrix} C_1 - C_2 A_{22}^{-1} A_{21} & C_3 \end{bmatrix}, \qquad \overline{D} = D - C_2 A_{22}^{-1} B_2.$$

This final elimination step involves non-orthogonal matrix operations, which can lead to unstable computations if the norm of the intervening matrices is too large or A_{22} is ill-conditioned. Fortunately, in most computational algorithms for descriptor systems presented in this book, the elimination of simple infinite eigenvalues is not necessary and we can work with irreducible realizations instead minimal ones. Therefore, we can almost always delay the computation of minimal realizations for the final results of whole computational cycles.

10.3.2 *Minimal Proper Rational Nullspace Bases*

Let $G(\lambda)$ be a $p \times m$ rational matrix of normal rank r. A proper rational basis of the left nullspace $\mathcal{N}_L(G(\lambda))$ (see Sect. 9.1.3) is any $(p - r) \times p$ proper rational matrix $N_l(\lambda)$ of full row rank such that $N_l(\lambda)G(\lambda) = 0$. Similarly, a proper rational basis of the right nullspace $\mathcal{N}_R(G(\lambda))$ is an $m \times (m - r)$ proper rational matrix $N_r(\lambda)$ of full column rank such that $G(\lambda)N_r(\lambda) = 0$. Of special interest are the *minimal* proper rational bases, which have the least McMillan degree. Assume $G(\lambda)$ has an n-th order descriptor system realization $(A - \lambda E, B, C, D)$, with $A - \lambda E$ regular. In this section, we present a numerically reliable computational approach to determine a descriptor system realization of a proper rational left nullspace basis $N_l(\lambda)$ of $G(\lambda)$ and discuss conditions for its minimality. The same approach can be used to determine $N_r(\lambda)$, a proper rational right nullspace basis of $G(\lambda)$, by determining $N_r^T(\lambda)$ as a proper rational left nullspace basis of $G^T(\lambda)$.

The proposed computational approach relies on the fact that $N_l(\lambda)$ is a left nullspace basis of $G(\lambda)$ if and only if, for a suitable $(p - r) \times n$ rational matrix $M_l(\lambda)$,

$$Y_l(\lambda) := [\, M_l(\lambda)\ N_l(\lambda) \,] \tag{10.39}$$

is a left nullspace basis of the system matrix

$$S(\lambda) = \begin{bmatrix} A - \lambda E & B \\ C & D \end{bmatrix}. \tag{10.40}$$

Thus, to compute $N_l(\lambda)$ we can first determine a left nullspace basis $Y_l(\lambda)$ for $S(\lambda)$ and then $N_l(\lambda)$ simply results as

$$N_l(\lambda) = Y_l(\lambda) \begin{bmatrix} 0 \\ I_p \end{bmatrix}. \tag{10.41}$$

As it will be apparent below, the main appeal of this approach is that for the computation of $Y_l(\lambda)$ we can employ powerful pencil manipulation techniques via orthogonal similarity transformations.

Let U and V be orthogonal matrices such that the transformed pencil $\widetilde{S}(\lambda) := U S(\lambda) V$ is in the Kronecker-like staircase form (see Sect. 10.1.6)

$$\widetilde{S}(\lambda) = \left[\begin{array}{c|c} A_r - \lambda E_r & A_{r,l} - \lambda E_{r,l} \\ \hline 0 & A_l - \lambda E_l \\ \hline 0 & C_l \end{array} \right], \tag{10.42}$$

where the descriptor pair $(A_l - \lambda E_l, C_l)$ is observable, E_l is invertible, and $A_r - \lambda E_r$ has full row rank excepting possibly a finite set of values of λ (i.e., the invariant zeros of $S(\lambda)$). As explained in Sect. 10.1.6, the reduction of $S(\lambda)$ to the form (10.42) can be obtained by using (twice) the **Procedure PREDUCE**.

A left nullspace $\widetilde{Y}_l(\lambda)$ of $\widetilde{S}(\lambda)$ in (10.42) can be chosen in the form

$$\widetilde{Y}_l(\lambda) = \left[\begin{array}{c|c|c} 0 & C_l(\lambda E_l - A_l)^{-1} & I \end{array}\right]. \tag{10.43}$$

Then, the left nullspace of $S(\lambda)$ is $Y_l(\lambda) = \widetilde{Y}_l(\lambda)U$ and can be obtained easily after partitioning suitably U as

$$U = \left[\begin{array}{cc} \widehat{B}_{r,l} & B_{r,l} \\ \widehat{B}_l & B_l \\ \widehat{D}_l & D_l \end{array}\right],$$

where the row partitioning corresponds to the column partitioning of $\widetilde{Y}_l(\lambda)$ in (10.43), while the column partitioning corresponds to the row partitioning of $S(\lambda)$ in (10.40). We obtain

$$Y_l(\lambda) = \left[\begin{array}{c|cc} A_l - \lambda E_l & \widehat{B}_l & B_l \\ \hline C_l & \widehat{D}_l & D_l \end{array}\right] \tag{10.44}$$

and the nullspace of $G(\lambda)$ is

$$N_l(\lambda) = \left[\begin{array}{c|c} A_l - \lambda E_l & B_l \\ \hline C_l & D_l \end{array}\right]. \tag{10.45}$$

To obtain this representation of the nullspace basis, we performed exclusively orthogonal transformations on the system matrices. We can prove that all computed matrices are exact for a slightly perturbed original system matrix (10.40). It follows that this method for the computation of the nullspace basis is numerically backward stable.

When using **Procedure PREDUCE**, as described in Sect. 10.1.6, to determine the Kronecker-like form (10.42), we can assume that the resulting subpencil

$$\left[\begin{array}{c} A_o - \lambda E_o \\ C_o \end{array}\right] := \left[\begin{array}{c} A_l - \lambda E_l \\ C_l \end{array}\right], \tag{10.46}$$

which characterizes the left structure of $S(\lambda)$, has the pair $(A_o - \lambda E_o, C_o)$ in an observability staircase form as in (10.17) and (10.18). Let μ_i, $i = 1, \ldots, \ell$ be the dimensions of the diagonal blocks of A_o in (10.17) (and also of E_o in (10.18)), and define $\mu_0 := p_l$ and $\mu_{\ell+1} := 0$ (which corresponds to a fictive full column rank diagonal block $A_{\ell,\ell+1} \in \mathbb{R}^{\mu_\ell \times \mu_{\ell+1}}$ in the leading position of A_o). These dimensions completely determine the left Kronecker structure of $S(\lambda)$ as follows: there are $\mu_{i-1} - \mu_i$ blocks $L_{i-1}^T(\lambda)$ of size $i \times (i-1)$, $i = 1, \ldots, \ell + 1$ (see (9.48)). The row dimension of $N_l(\lambda)$ (i.e., the number of linearly independent basis vectors) is given by the total number of $L_{\eta_i}^T(\lambda)$ blocks (see Example 9.1), thus $\sum_{i=1}^{\ell+1}(\mu_{i-1} - \mu_i) = \mu_0$ (i.e., the row dimension of C_l). Applying standard linear algebra results, it follows that $\mu_0 := p - r$.

The following result shows that the resulting staircase form (10.46) provides the complete structural information on any minimal polynomial basis (and also on any simple proper basis constructed from it, see Sect. 9.1.3).

Proposition 10.1 *If the realization $(A - \lambda E, B, C, D)$ of $G(\lambda)$ is controllable and if μ_i, $i = 1, \ldots, \ell$ are the dimensions of the diagonal blocks of A_o in (10.17) (and also of E_o in (10.18)), and $\mu_0 := p_l$ and $\mu_{\ell+1} := 0$, then a minimal polynomial basis of the left nullspace of $G(\lambda)$ has degree $n_l = \sum_{i=1}^{\ell} \mu_i$ and is formed of $\mu_{i-1} - \mu_i$ polynomial vectors of degree $i - 1$, for $i = 1, \ldots, \ell + 1$.*

Proof The controllability of the descriptor realization ensures that the left Kronecker structure of $G(\lambda)$ and of $S(\lambda)$ are characterized by the same left Kronecker indices. A minimal polynomial basis for the left nullspace of $\widetilde{S}(\lambda)$ can be determined of the form

$$\widehat{Y}_l(\lambda) = \left[\ 0 \ \middle| \ \widehat{N}_l(\lambda) \ \right], \tag{10.47}$$

where $\widehat{N}_l(\lambda)$ is a minimal polynomial basis for the left nullspace of $\begin{bmatrix} A_l - \lambda E_l \\ C_l \end{bmatrix}$.
To construct $\widehat{N}_l(\lambda)$, the basis vectors can be determined by exploiting the staircase form of this pencil. It was shown in [8, Sect. 4.6.4], in a dual context, that a minimal polynomial basis can be computed by selecting $\mu_{i-1} - \mu_i$ polynomial basis vectors of degree $i - 1$, for $i = 1, \ldots, \ell + 1$. The degree of this polynomial basis is

$$\begin{aligned}
\sum_{i=1}^{\ell+1} (\mu_{i-1} - \mu_i)(i - 1) &= \sum_{i=1}^{\ell+1} \mu_{i-1}(i - 1) - \sum_{i=1}^{\ell+1} \mu_i(i - 1) \\
&= \sum_{i=1}^{\ell} \mu_i i - \sum_{i=1}^{\ell} \mu_i(i - 1) \\
&= \sum_{i=1}^{\ell} \mu_i,
\end{aligned}$$

which is equal to n_l, the dimension of the square matrices A_l and E_l.   ∎

A straightforward consequence of Proposition 10.1 is the following result.

Proposition 10.2 *If the realization $(A - \lambda E, B, C, D)$ of $G(\lambda)$ is controllable, then the rational matrix $N_l(\lambda)$ defined in (10.45) is a minimal proper rational basis of the left nullspace of $G(\lambda)$.*

Proof According to the definition of a minimal proper rational basis (see Sect. 9.1.3), its McMillan degree is given by the degree of a minimal polynomial basis (i.e., the sum of the left minimal indices). By Proposition 10.1, the degree of a minimal polynomial basis is $n_l := \sum_{i=1}^{\ell} \mu_i$, which is thus equal to the dimension of the square matrices A_l and E_l. Therefore, we only need to show that the realization (10.45) is irreducible and $N_l(\lambda)$ defined in (10.45) has no zeros.

The pair $(A_l - \lambda E_l, C_l)$ is observable, by the construction of the Kronecker-like form (10.42). To show that the pair $(A_l - \lambda E_l, B_l)$ is controllable, observe that due to the controllability of the pair $(A - \lambda E, B)$, the subpencil $[\, A - \lambda E \ B \,]$ of $S(\lambda)$ in (10.40) has full row rank for all $\lambda \in \mathbb{C}$, and thus the reduced pencil

$$
U \begin{bmatrix} A - \lambda E & B & 0 \\ C & D & I_p \end{bmatrix} \begin{bmatrix} V & 0 \\ 0 & I_p \end{bmatrix} = \begin{bmatrix} A_r - \lambda E_r & A_{r,l} - \lambda E_{r,l} & B_{r,l} \\ 0 & A_l - \lambda E_l & B_l \\ 0 & C_l & D_l \end{bmatrix}
$$

has full row rank for all $\lambda \in \mathbb{C}$ as well. It follows that for all $\lambda \in \mathbb{C}$

$$
\mathrm{rank} \begin{bmatrix} A_l - \lambda E_l & B_l \end{bmatrix} = n_l
$$

and thus the pair $(A_l - \lambda E_l, B_l)$ is controllable.

Since, we also have that

$$
\mathrm{rank} \begin{bmatrix} A_l - \lambda E_l & B_l \\ C_l & D_l \end{bmatrix} = n_l + p - r
$$

for all $\lambda \in \mathbb{C}$, it follows that $N_l(\lambda)$ has no finite or infinite zeros. Thus, D_l has full row rank $p - r$ and the computed basis is column reduced at $\lambda = \infty$ [128]. ∎

In the case, when the realization of $G(\lambda)$ is not controllable, the realization of $N_l(\lambda)$ is not guaranteed to be controllable. The uncontrollable eigenvalues of $A - \lambda E$ may turn partly up as eigenvalues of $A_r - \lambda E_r$ (i.e., invariant zeros) or of $A_l - \lambda E_l$. In the latter case, the resulting proper nullspace basis has not the least possible McMillan degree. Interestingly, a minimal basis cannot be always obtained by simply eliminating the uncontrollable part of the pair $(A_l - \lambda E_l, B_l)$. The reason for this is the lack of the maximal controllability property (see Proposition 10.3).

We can always determine a proper nullspace basis with arbitrarily assigned poles. To show this, consider the transformation matrix

$$
\widehat{U} = \begin{bmatrix} I & 0 & 0 \\ 0 & I & K \\ 0 & 0 & I \end{bmatrix} \tag{10.48}
$$

and compute $\widehat{S}(\lambda) := \widehat{U}\widetilde{S}(\lambda)$ as

$$
\widehat{S}(\lambda) = \left[\begin{array}{c|cc} A_r - \lambda E_r & A_{r,l} - \lambda E_{r,l} \\ \hline 0 & A_l + KC_l - \lambda E_l \\ \hline 0 & C_l \end{array} \right]. \tag{10.49}
$$

We also compute

$$
\widehat{U}U \begin{bmatrix} 0 \\ I_p \end{bmatrix} = \begin{bmatrix} B_{r,l} \\ B_l + KD_l \\ D_l, \end{bmatrix}
$$

and obtain the proper rational left nullspace basis with the alternative realization

$$
\widetilde{N}_l(\lambda) = \left[\begin{array}{c|c} A_l + KC_l - \lambda E_l & B_l + KD_l \\ \hline C_l & D_l \end{array} \right]. \tag{10.50}
$$

Since the descriptor pair $(A_l - \lambda E_l, C_l)$ is completely observable, there exists an output injection matrix K such that the pair $(A_l + KC_l, E_l)$ has arbitrary assigned generalized eigenvalues. According to Proposition 10.2, the basis (10.50) is minimal provided the realization $(A - \lambda E, B, C, D)$ of $G(\lambda)$ is controllable.

This construction shows that the placement of poles of the left nullspace basis (10.50) can be simply achieved by additionally performing a particular similarity transformation on the reduced pencil $\widetilde{S}(\lambda)$. As a consequence, the output injection may move some of uncontrollable generalized eigenvalues of the pair (A_l, E_l) to other locations and make them controllable. It follows that determining a minimal nullspace basis from a non-minimal one may involve the determination of suitable injection matrix, which makes a maximum number of eigenvalues uncontrollable.

Remark 10.5 The alternative proper left nullspace basis $\widetilde{N}_l(\lambda)$ can be interpreted as the numerator factor of the left coprime factorization

$$N_l(\lambda) = \widetilde{M}_l^{-1}(\lambda)\widetilde{N}_l(\lambda) \, ,$$

where $\widetilde{M}_l(\lambda)$ has the descriptor system realization

$$\widetilde{M}_l(\lambda) = \left[\begin{array}{c|c} A_l + KC_l - \lambda E_l & K \\ \hline C_l & I_{p-r} \end{array} \right]$$

$\square$

The following result shows that a minimal proper basis of the form (10.45) has the nice property of being *maximally controllable*, that is, the alternative basis (10.50) remains controllable for an arbitrary output injection matrix K, or equivalently, the pair $(A_l + KC_l - \lambda E_l, B_l + K D_l)$ is controllable for all K.

Proposition 10.3 *If the realization $(A - \lambda E, B, C, D)$ of $G(\lambda)$ is controllable, then the realization of $N_l(\lambda)$ defined in (10.45) is maximally controllable.*

Proof We have to show that for an arbitrary output injection matrix K, the pair $(A_l + KC_l - \lambda E_l, B_l + K D_l)$ is controllable. Let K be an arbitrary injection matrix and construct the alternative proper left nullspace basis $\widetilde{N}_l(\lambda)$ with the realization given in (10.50). Since according to Proposition 10.2, $N_l(\lambda)$ is a minimal nullspace basis, the alternative nullspace basis $\widetilde{N}_l(\lambda)$, with the same McMillan degree, is a minimal basis as well. Therefore, the pair $(A_l + KC_l - \lambda E_l, B_l + K D_l)$ is controllable. $\blacksquare$

Even if the resulting rational basis (10.45) has the least possible McMillan degree, and thus is minimal, still, in general, this basis is not a simple basis. The properties of simple proper minimal bases resemble, in many aspects, the properties of minimal polynomial bases. For our purposes, the main use of simple proper nullspace bases is in the nullspace-based synthesis methods of least-order fault detection filters. As it will be shown below, it is possible to obtain a simple basis starting from a non-simple one.

Consider the proper minimal left nullspace basis $N_l(\lambda)$ of $G(\lambda)$, with the descriptor realization given in (10.45), and we denote with $c_{l,i}$ and $d_{l,i}$ the i-th rows of matrices

C_l and D_l, respectively. The approach to construct a simple minimal proper rational left nullspace basis is based on the following result.

Proposition 10.4 *For each $i = 1, \ldots, p - r$, let K_i be an output injection matrix such that*

$$v_i(\lambda) := c_{l,i}(\lambda E_l - A_l - K_i C_l)^{-1}(B_l + K_i D_l) + d_{l,i} \tag{10.51}$$

has the least possible McMillan degree. Then, $\widetilde{N}_l(\lambda)$ formed by stacking the $p - r$ rational row vectors $v_i(\lambda)$ is a simple minimal proper rational left nullspace basis.

Proof According to Proposition 10.3, the realization (10.45) of $N_l(\lambda)$ is maximally controllable, i.e., the pair $(A_l + K_i C_l - \lambda E_l, B_l + K_i D_l)$ is controllable for arbitrary K_i. Therefore, the maximal order reduction of the McMillan degree of $v_i(\lambda)$ can be achieved by making the pair $(A_l + K_i C_l - \lambda E_l, c_{l,i})$ maximally unobservable via an appropriate choice of K_i. For each $i = 1, \ldots, p - r$, the achievable least McMillan degree of $v_i(\lambda)$ is the corresponding minimal index n_i, representing, in a dual setting, the dimension of the least-order controllability subspace of the standard pair $(E_l^{-T} A_l^T, E_l^{-T} C_l^T)$ containing span $(E_l^{-T} c_{l,i}^T)$. This result is the statement of Lemma 6 in [172]. It is easy to check that $v_i(\lambda)G(\lambda) = 0$, thus $\widetilde{N}_l(\lambda)$ is a left annihilator of $G(\lambda)$. Furthermore, the set of vectors $\{v_1(\lambda), \ldots, v_{p-r}(\lambda)\}$ is linearly independent since the realization of $\widetilde{N}_l(\lambda)$ has the same full row rank matrix D_l as that of $N_l(\lambda)$. It follows that $\widetilde{N}_l(\lambda)$ is a proper left nullspace basis of least dimension $\sum_{i=1}^{p-r} n_i$, with each row $v_i(\lambda)$ of McMillan degree n_i. It follows that $\widetilde{N}_l(\lambda)$ is simple. ∎

Let us assume that each rational vector $v_i(\lambda$ has a descriptor realization of the form

$$v_i(\lambda) = \left[\begin{array}{c|c} \widetilde{A}_{l,i} - \lambda \widetilde{E}_{l,i} & \widetilde{B}_{l,i} \\ \hline \widetilde{c}_{l,i} & d_{l,i} \end{array} \right]. \tag{10.52}$$

Then, the simple minimal proper rational basis $\widetilde{N}_l(\lambda)$, constructed by stacking all $v_i(\lambda)$, for $i = 1, \ldots, r$, has the realization

$$\widetilde{N}_l(\lambda) = \left[\begin{array}{c|c} \widetilde{A}_l - \lambda \widetilde{E}_l & \widetilde{B}_l \\ \hline \widetilde{C}_l & D_l \end{array} \right], \tag{10.53}$$

with

$$\widetilde{A}_l - \lambda \widetilde{E}_l = \begin{bmatrix} \widetilde{A}_{l,1} - \lambda \widetilde{E}_{l,1} & & \\ & \ddots & \\ & & \widetilde{A}_{l,p-r} - \lambda \widetilde{E}_{l,p-r} \end{bmatrix}, \quad \widetilde{B}_l = \begin{bmatrix} \widetilde{B}_{l,1} \\ \vdots \\ \widetilde{B}_{l,p-r} \end{bmatrix},$$

$$\widetilde{C}_l = \begin{bmatrix} \widetilde{c}_{l,1} & & \\ & \ddots & \\ & & \widetilde{c}_{l,p-r} \end{bmatrix}.$$

Remark 10.6 The poles of the simple minimal proper rational left nullspace basis $\widetilde{N}_l(\lambda)$ can be arbitrarily placed by performing left coprime rational factorizations using the realizations in (10.52) (see Remark 10.5)

$$v_i(\lambda) = m_i(\lambda)^{-1}\hat{v}_i(\lambda), \tag{10.54}$$

where $m_i(\lambda)$ are polynomials with arbitrary roots in $\mathbb{C}_s$. Therefore, the resulting alternative simple basis $\widehat{N}_l(\lambda) := [\,\hat{v}_1^T(\lambda), \ \ldots, \ \hat{v}_{p-r}^T(\lambda)\,]^T$ can have arbitrarily assigned poles. In particular, a *special simple basis* can be constructed such that each $m_i(\lambda)$ divides $m_j(\lambda)$, if $j < i$. $\qquad\qquad\qquad\qquad\qquad\qquad\qquad\qquad\qquad\qquad\qquad\quad\square$

Simple rational bases are direct correspondents of polynomial bases and, hence, all operations on polynomial bases have analogous operations on simple rational bases. An important operation (with applications in the synthesis of least-order fault detection filters) is building linear combinations of basis vectors up to a certain McMillan degree. For example, by using the special simple basis in Remark 10.6, any linear combination $\sum_{i=1}^{k} h_i\hat{v}_i(\lambda)$ with constant coefficients h_i of the basis vectors of McMillan degree up to a certain value k, has McMillan degree at most k.

Consider the proper left nullspace basis $N_l(\lambda)$ constructed in (10.45). From the details of the resulting staircase form (10.46) of the pair $(A_l - \lambda E_l, C_l)$, recall that it is possible to obtain the full column rank matrices $A_{i-1,i} \in \mathbb{R}^{\mu_{i-1} \times \mu_i}$ in the form

$$A_{i-1,i} = \begin{bmatrix} R_{i-1,i} \\ 0 \end{bmatrix},$$

where $R_{i-1,i}$ is an upper triangular invertible matrix of order μ_i. The row dimension $\mu_{i-1} - \mu_i$ of the zero block of $A_{i-1,i}$ gives the number of polynomial vectors of degree $i - 1$ in a minimal polynomial basis [8, Sect. 4.6] and thus, also the number of vectors of McMillan degree $i - 1$ in a simple basis. It is straightforward to show the following result.

Corollary 10.1 *For a given minimal proper rational left nullspace basis $N_l(\lambda)$ in the form (10.45), let i be a given index such that $1 \leq i < p - r$, and let h be a $(p - r)$-dimensional row vector having only the trailing i components nonzero. Then, a linear combination of the simple proper rational basis vectors, with McMillan degree at most n_i, can be generated as*

$$v(\lambda) := hC_l(\lambda E_l - A_l - KC_l)^{-1}(B_l + KD_l) + hD_l\,, \tag{10.55}$$

where K is an output injection matrix such that $v(\lambda)$ has the least possible McMillan degree.

This result shows that the determination of a linear combination of vectors of a simple proper rational basis up to a given order n_i is possible directly from a proper rational basis determined in the form (10.45). The matrix K together with a minimal realization of $v(\lambda)$ can be computed efficiently using minimal dynamic cover techniques presented in Sect. 10.4.2. The same approach can be applied repeatedly to determine the basis vectors $v_i(\lambda)$, $i = 1, \ldots, p - r$, of a simple basis by using the particular choices $h = e_i^T$, where e_i is the i-th column of I_{p-r}.

10.3.3 Poles and Zeros Computation

The computation of poles of a rational matrix $G(\lambda)$, with an irreducible descriptor system realization $(A - \lambda E, B, C, D)$, comes down to compute the eigenvalues of the regular pole pencil $A - \lambda E$. This can be achieved by computing the eigenvalues of $A - \lambda E$ from the GRSF of the pair (A, E). The finite poles are the n_p^f finite eigenvalues of $A - \lambda E$, while there are $n_p^\infty = \mathrm{rank}\, E - n_p^f$ infinite poles (recall that the multiplicities of infinite eigenvalues are in excess with one with respect to the multiplicities of infinite poles). The McMillan degree of $G(\lambda)$ results as

$$\delta\big(G(\lambda)\big) = n_p^f + n_p^\infty = \mathrm{rank}\, E\,.$$

A straightforward application of the Kronecker-like form is the computation of the system zeros. Let $G(\lambda)$ be a rational matrix, with an irreducible descriptor system representation $(A - \lambda E, B, C, D)$. The system zeros are those values of λ, where the system pencil

$$S(\lambda) = \begin{bmatrix} A - \lambda E & B \\ C & D \end{bmatrix} := M - \lambda N$$

drops its rank below its normal rank. Thus, the system zeros can be determined from the eigenvalues of the regular pencil $M_{reg} - \lambda N_{reg}$ in the Kronecker-like form (10.19) of the pencil $M - \lambda N$. This can be achieved by computing the eigenvalues of $M_{reg} - \lambda N_{reg}$ from the GRSF of the pair (M_{reg}, N_{reg}). If $M_{reg} - \lambda N_{reg}$ has n_z^f finite eigenvalues, these are the n_z^f finite transmission zeros of the system. Additionally, there are $n_z^\infty = \mathrm{rank}\, N_{reg} - n_z^f$ infinite zeros (recall that the multiplicities of infinite eigenvalues are in excess with one with respect to the multiplicities of infinite zeros).

10.3.4 Additive Decompositions

Consider a disjunct partition of the complex plane $\mathbb{C}$ as $\mathbb{C} = \mathbb{C}_g \cup \mathbb{C}_b$, where both $\mathbb{C}_g$ and $\mathbb{C}_b$ are symmetrically located with respect to the real axis, $\mathbb{C}_g$ has at least one point on the real axis, and $\mathbb{C}_g \cap \mathbb{C}_b = \emptyset$. Let $G(\lambda)$ be a rational TFM with a descriptor system realization $(A - \lambda E, B, C, D)$. We describe a state-space approach to compute the additive decomposition

$$G(\lambda) = G_g(\lambda) + G_b(\lambda), \tag{10.56}$$

where $G_g(\lambda)$ has only poles in $\mathbb{C}_g$, while $G_b(\lambda)$ has only poles in $\mathbb{C}_b$.

The additive spectral decomposition (10.56) can be computed using a block diagonalization technique of the pole pencil $A - \lambda E$ (9.74). The basic computation is to determine the two invertible matrices U and Z to bring the matrices of the transformed pair (UEZ, UAZ) in suitable block diagonal forms. The following procedure computes the additive decomposition (10.56) using the descriptor realization $(A - \lambda E, B, C, D)$ of $G(\lambda)$, by determining the descriptor realizations $(A_g - \lambda E_g, B_g, C_g, D_g)$ of $G_g(\lambda)$ and $(A_b - \lambda E_b, B_b, C_b, D_b)$ of $G_b(\lambda)$.

Procedure GSDEC: Generalized additive spectral decomposition

Inputs : $G(\lambda) = (A - \lambda E, B, C, D)$, $\mathbb{C}_g$
Outputs: $G_g(\lambda) = (A_g - \lambda E_g, B_g, C_g, D_g)$, $G_b(\lambda) = (A_b - \lambda E_b, B_b, C_b, D_b)$

(1) By using the QZ-algorithm, compute orthogonal U_1 and V_1, such that the matrix pair $(U_1 A V_1, U_1 E V_1)$ is in an ordered GRSF

$$
U_1 A V_1 = \begin{bmatrix} A_g & A_{gb} \\ 0 & A_b \end{bmatrix}, \quad U_1 E V_1 = \begin{bmatrix} E_g & E_{gb} \\ 0 & E_b \end{bmatrix},
$$

such that $\Lambda(A_g - \lambda E_g) \subset \mathbb{C}_g$ and $\Lambda(A_b - \lambda E_b) \subset \mathbb{C} \setminus \mathbb{C}_g$.

(2) Compute the left and right transformation matrices, U_2 and V_2, respectively, of the form

$$
U_2 = \begin{bmatrix} I & Y \\ 0 & I \end{bmatrix}, \quad V_2 = \begin{bmatrix} I & X \\ 0 & I \end{bmatrix},
$$

where X and Y satisfy the Sylvester system of equations

$$
\begin{aligned}
A_g X + Y A_b &= -A_{gb}, \\
E_g X + Y E_b &= -E_{gb}.
\end{aligned}
$$

(3) Compute

$$
\begin{bmatrix} B_g \\ B_b \end{bmatrix} = U_2 U_1 B, \quad \begin{bmatrix} C_g & C_b \end{bmatrix} = C V_1 V_2, \quad D_g = D, \quad D_b = 0,
$$

where the row partitioning of $U_2 U_1 B$ and column partitioning of $C V_1 V_2$ are analogous to the row and column partitioning of $U_1 A V_1$.

The resulting pencil $U_2 U_1 (A - \lambda E) V_1 V_2$ at Step (2) is block diagonal. The existence of a unique solution (X, Y) of the Sylvester system to be solved at Step (2) is guaranteed by $\Lambda(A_g - \lambda E_g) \cap \Lambda(A_b - \lambda E_b) = \emptyset$. An efficient solution method, which exploits the GRSFs of the pairs (A_g, E_g) and (A_b, E_b), has been proposed in [71].

10.3.5 Coprime Factorizations

Consider a $p \times m$ rational matrix $G(\lambda)$ having a descriptor system realization $(A - \lambda E, B, C, D)$, for which we will not assume further properties (e.g., minimality or irreducibility). Consider also a disjunct partition of the complex plane as $\mathbb{C} = \mathbb{C}_b \cup \mathbb{C}_g$, $\mathbb{C}_b \cap \mathbb{C}_g = \emptyset$, where $\mathbb{C}_b$ and $\mathbb{C}_g$ denote the "bad" and "good" regions of $\mathbb{C}$, respectively. In this section we present algorithms for the computation of a *right coprime factorization* (RCF) of $G(\lambda)$ in the form $G(\lambda) = N(\lambda)M^{-1}(\lambda)$, where $N(\lambda)$ and $M(\lambda)$ are proper rational matrices with all poles in $\mathbb{C}_g$ and are mutually coprime (see Sect. 9.1.6 for definitions). A special case relevant for many applications is when $\mathbb{C}_g = \mathbb{C}_s$ and $\mathbb{C}_b = \mathbb{C}\backslash\mathbb{C}_g$ and we additionally impose that the denominator factor $M(\lambda)$ is inner. The algorithms to compute RCFs can be equally employed to determine a *left coprime factorization* (LCF) of $G(\lambda)$ in the form $G(\lambda) = M^{-1}(\lambda)N(\lambda)$, where $N(\lambda)$ and $M(\lambda)$ are coprime proper rational matrices with all poles in $\mathbb{C}_g$. We can determine the factors of a LCF factorization from those of a RCF of $G^T(\lambda) = N^T(\lambda)\left(M^T(\lambda)\right)^{-1}$. Therefore, we only discuss algorithms for the computation of RCFs.

The presented algorithms compute RCFs with minimum-degree denominators, by employing a recursive pole dislocation technique (see Sect. 9.1.6), by which all poles of $G(\lambda)$ situated in $\mathbb{C}_b$ are successively moved into $\mathbb{C}_g$, via recursive pole–zero cancellations with elementary denominator factors. To cancel a real pole $\beta \in \mathbb{C}_b$ of $G(\lambda)$, we multiply $G(\lambda)$ from right with an elementary invertible proper factor $\widetilde{M}(\lambda)$ of McMillan degree one, which has β as a zero and $\gamma \in \mathbb{C}_g$ as a pole. For a complex pole, the corresponding $\widetilde{M}(\lambda)$ would contain complex coefficients. Fortunately, we can simultaneously cancel a pair of complex conjugate poles $\beta, \bar{\beta} \in \mathbb{C}_b$ of $G(\lambda)$, by post-multiplying $G(\lambda)$ with an elementary invertible proper factor $\widetilde{M}(\lambda)$ of McMillan degree two, having only real coefficients. This factor has β and $\bar{\beta}$ as zeros and $\gamma_1, \gamma_2 \in \mathbb{C}_g$ as poles (either two real poles or a pair of complex conjugate poles). This pole–zero cancellation technique can be successively employed to dislocate all n_b poles of $G(\lambda)$. The resulting denominator factor can be represented in a product form as

$$M(\lambda) = \widetilde{M}_1(\lambda)\widetilde{M}_2(\lambda)\cdots\widetilde{M}_k(\lambda), \tag{10.57}$$

where each $\widetilde{M}_i(\lambda)$ $(i = 1, \ldots, k)$ is an invertible elementary proper factor with McMillan degree equal to one or two. The computational procedure can be formalized as k successive applications of the updating formula

$$\begin{bmatrix} N_i(\lambda) \\ M_i(\lambda) \end{bmatrix} = \begin{bmatrix} N_{i-1}(\lambda) \\ M_{i-1}(\lambda) \end{bmatrix} \widetilde{M}_i(\lambda), \quad i = 1, \ldots, k, \tag{10.58}$$

initialized with $N_0(\lambda) = G(\lambda)$ and $M_0(\lambda) = I_m$. Then, $N(\lambda) = N_k(\lambda)$ and $M(\lambda) = M_k(\lambda)$. By this approach, it is automatically achieved that the resulting $M(\lambda)$ has the least achievable McMillan degree n_b.

We can derive state-space formulas for the efficient implementation of the updating operations in (10.58). Assume $N_{i-1}(\lambda)$ and $M_{i-1}(\lambda)$ have the descriptor realizations

$$\begin{bmatrix} N_{i-1}(\lambda) \\ M_{i-1}(\lambda) \end{bmatrix} = \left[\begin{array}{cc|c} A_{11} - \lambda E_{11} & A_{12} - \lambda E_{12} & B_1 \\ 0 & A_{22} - \lambda E_{22} & B_2 \\ \hline C_{N,1} & C_{N,2} & D_N \\ C_{M,1} & C_{M,2} & D_M \end{array}\right] =: \left[\begin{array}{c|c} \widetilde{A} - \lambda \widetilde{E} & \widetilde{B} \\ \hline \widetilde{C}_N & \widetilde{D}_N \\ \widetilde{C}_M & \widetilde{D}_M \end{array}\right], \quad (10.59)$$

where $\Lambda(A_{22} - \lambda E_{22}) \subset \mathbb{C}_b$. We assume that $A_{22} - \lambda E_{22}$ is a 1×1 pencil in the case when $A_{22} - \lambda E_{22}$ has a real or an infinite eigenvalue, or is a 2×2 pencil, in the case when $A_{22} - \lambda E_{22}$ has a pair of complex conjugate eigenvalues. This form automatically results if the pair $(\widetilde{A}, \widetilde{E})$ is in the specially ordered *generalized real Schur form* (GRSF) determined by using **Procedure GSORSF** in Sect. 10.1.4. If $B_2 = 0$, then the eigenvalue(s) of $A_{22} - \lambda E_{22}$ is (are) not controllable, and thus can be removed to obtain realizations of $N_{i-1}(\lambda)$ and $M_{i-1}(\lambda)$ of reduced orders

$$\begin{bmatrix} N_{i-1}(\lambda) \\ M_{i-1}(\lambda) \end{bmatrix} = \left[\begin{array}{c|c} A_{11} - \lambda E_{11} & B_1 \\ \hline C_{N,1} & D_N \\ C_{M,1} & D_M \end{array}\right]. \quad (10.60)$$

After suitable reordering of diagonal blocks of $A_{11} - \lambda E_{11}$ using orthogonal similarity transformations (see Sect. 10.1.4), a new realization $N_{i-1}(\lambda)$ and $M_{i-1}(\lambda)$ can be determined with the matrices again in the form (10.59). If $B_2 \neq 0$, then we have two cases, which are separately discussed in what follows.

If the pencil $A_{22} - \lambda E_{22}$ has finite eigenvalues (i.e., E_{22} is invertible), then the pair $(A_{22} - \lambda E_{22}, B_2)$ is (finite) controllable and there exists F_2 such that the eigenvalues of $A_{22} + B_2 F_2 - \lambda E_{22}$ can be placed in arbitrary locations in $\mathbb{C}_g$. Assume that such an F_2 has been determined and define the elementary factor $\widetilde{M}_i(\lambda) = (A_{22} + B_2 F_2 - \lambda E_{22}, B_2 W, F_2, W)$, where W is chosen to ensure the invertibility of $\widetilde{M}_i(\lambda)$. To compute stable and proper RCFs, the choice $W = I_m$ is always possible. However, alternative choices of W are necessary to ensure, for example, that $\widetilde{M}_i(\lambda)$ is inner. It is easy to check that the updated factors $N_i(\lambda)$ and $M_i(\lambda)$ in (10.58) have the realizations

$$\begin{bmatrix} N_i(\lambda) \\ M_i(\lambda) \end{bmatrix} := \begin{bmatrix} N_{i-1}(\lambda) \\ M_{i-1}(\lambda) \end{bmatrix} \widetilde{M}_i(\lambda) = \left[\begin{array}{cc|c} A_{11} - \lambda E_{11} & A_{12} + B_1 F_2 - \lambda E_{12} & B_1 W \\ 0 & A_{22} + B_2 F_2 - \lambda E_{22} & B_2 W \\ \hline C_{N,1} & C_{N,2} + D_N F_2 & D_N W \\ C_{M,1} & C_{M,2} + D_M F_2 & D_M W \end{array}\right].$$

If we denote $\widetilde{F} = \begin{bmatrix} 0 & F_2 \end{bmatrix}$, then the above relations lead to the following updating formulas:

$$
\begin{aligned}
\widetilde{A} &\leftarrow \widetilde{A} + \widetilde{B}\widetilde{F}, \\
\widetilde{B} &\leftarrow \widetilde{B}W, \\
\widetilde{C}_N &\leftarrow \widetilde{C}_N + \widetilde{D}_N\widetilde{F}, \\
\widetilde{C}_M &\leftarrow \widetilde{C}_M + \widetilde{D}_M\widetilde{F}, \\
\widetilde{D}_N &\leftarrow \widetilde{D}_N W, \\
\widetilde{D}_M &\leftarrow \widetilde{D}_M W.
\end{aligned}
\tag{10.61}
$$

If the 1×1 pencil $A_{22} - \lambda E_{22}$ has an infinite eigenvalue (i.e., $E_{22} = 0$), then we choose the elementary factor $\widetilde{M}_i(\lambda) = (\gamma - \lambda,\ B_2,\ F_2,\ W)$, where γ is an arbitrary real eigenvalue in $\mathbb{C}_g$, W is a projection matrix chosen such $B_2 W = 0$ and $\operatorname{rank}\begin{bmatrix} B_2 \\ W \end{bmatrix} = m$, and F_2 has been chosen such that $B_2 F_2 = -A_{22}$ and $\operatorname{rank}[\,F_2\ W\,] = m$ (the rank conditions guarantee the invertibility of $\widetilde{M}_i(\lambda)$). Straightforward choices of F_2 and W are, for example, $F_2 = -B_2^T(B_2 B_2^T)^{-1} A_{22}$ and $W = I_m - B_2^T(B_2 B_2^T)^{-1} B_2$. By this choice of $\widetilde{M}_i(\lambda)$, we made the infinite eigenvalue in the realization of the updated factors $N_i(\lambda)$ and $M_i(\lambda)$ simple, and after its elimination, we obtain the realizations

$$
\begin{bmatrix} N_i(\lambda) \\ M_i(\lambda) \end{bmatrix} := \begin{bmatrix} N_{i-1}(\lambda) \\ M_{i-1}(\lambda) \end{bmatrix} \widetilde{M}_i(\lambda) = \left[\begin{array}{cc|c} A_{11} - \lambda E_{11} & A_{12} + B_1 F_2 - \lambda E_{12} & B_1 W \\ 0 & \gamma - \lambda & B_2 \\ \hline C_{N,1} & C_{N,2} + D_N F_2 & D_N W \\ C_{M,1} & C_{M,2} + D_M F_2 & D_M W \end{array} \right].
$$

The above relations lead to the following updating formulas:

$$
\begin{aligned}
\widetilde{A} &\leftarrow \begin{bmatrix} A_{11} & A_{12} + B_1 F_2 \\ 0 & \gamma \end{bmatrix}, \\[4pt]
\widetilde{E} &\leftarrow \begin{bmatrix} E_{11} & E_{12} \\ 0 & 1 \end{bmatrix}, \\[4pt]
\widetilde{B} &\leftarrow \begin{bmatrix} B_1 W \\ B_2 \end{bmatrix}, \\[4pt]
\widetilde{C}_N &\leftarrow \begin{bmatrix} C_{N,1} & C_{N,2} + D_N F_2 \end{bmatrix}, \\[4pt]
\widetilde{C}_M &\leftarrow \begin{bmatrix} C_{M,1} & C_{M,2} + D_M F_2 \end{bmatrix}, \\[4pt]
\widetilde{D}_N &\leftarrow D_N W, \\[4pt]
\widetilde{D}_M &\leftarrow D_M W.
\end{aligned}
\tag{10.62}
$$

The updating techniques relying on the formulas (10.61) and (10.62) ensure that, if the original pair $(\widetilde{A}, \widetilde{E})$ was in a GRSF, then the updated pair will have a similar form, possibly with $\widetilde{A} - \lambda\widetilde{E}$ having a 2×2 trailing block which corresponds to two real generalized eigenvalues (to recover the GRSF, such a block can be further split into two 1×1 blocks using an orthogonal similarity transformation). By reordering the diagonal blocks in the GRSF of the updated pair $(\widetilde{A}, \widetilde{E})$, we can bring in the trailing position new blocks whose generalized eigenvalues lie in $\mathbb{C}_b$. The described eigenvalue dislocation process is repeated until all eigenvalues are moved into $\mathbb{C}_g$, using suitably chosen elementary denominators.

The following procedure computes a proper and stable RCF of an arbitrary rational TFM $G(\lambda)$ with respect to a given partition $\mathbb{C} = \mathbb{C}_b \cup \mathbb{C}_g$ as $G(\lambda) = N(\lambda)M^{-1}(\lambda)$, where the resulting factors $N(\lambda)$ and $M(\lambda)$ have the realizations $N(\lambda) = (\widetilde{A} - \lambda\widetilde{E}, \widetilde{B}, \widetilde{C}_N, \widetilde{D}_N)$ and $M(\lambda) = (\widetilde{A} - \lambda\widetilde{E}, \widetilde{B}, \widetilde{C}_M, \widetilde{D}_M)$.

Procedure GRCF: Generalized stable right coprime factorization

---

Inputs   : $G(\lambda) = (A - \lambda E, B, C, D)$ with $A, E \in \mathbb{R}^{n \times n}$, $B \in \mathbb{R}^{n \times m}$, $C \in \mathbb{R}^{p \times n}$,
$\quad\quad\quad\quad D \in \mathbb{R}^{p \times m}$; $\mathbb{C}_g$ and $\mathbb{C}_b$, such that $\mathbb{C} = \mathbb{C}_b \cup \mathbb{C}_g$, $\mathbb{C}_b \cap \mathbb{C}_g = \emptyset$

Outputs: $N(\lambda) = (\widetilde{A} - \lambda\widetilde{E}, \widetilde{B}, \widetilde{C}_N, \widetilde{D}_N)$ and $M(\lambda) = (\widetilde{A} - \lambda\widetilde{E}, \widetilde{B}, \widetilde{C}_M, \widetilde{D}_M)$ such that
$\quad\quad\quad\quad G(\lambda) = N(\lambda)M^{-1}(\lambda)$, all finite eigenvalues of $\widetilde{A} - \lambda\widetilde{E}$ are in $\mathbb{C}_g$ and all infinite
$\quad\quad\quad\quad$ eigenvalues of $\widetilde{A} - \lambda\widetilde{E}$ are simple

(1) Compute, using **Procedure GSORSF**, the orthogonal matrices Q and Z to reduce the pair (A, E) to the special ordered GRSF

$$\widetilde{A} := Q^T A Z = \begin{bmatrix} A_\infty & * & * \\ 0 & A_g & * \\ 0 & 0 & A_b \end{bmatrix}, \quad \widetilde{E} := Q^T E Z = \begin{bmatrix} 0 & * & * \\ 0 & E_g & * \\ 0 & 0 & E_b \end{bmatrix},$$

where $A_\infty \in \mathbb{R}^{(n-r) \times (n-r)}$ is invertible and upper triangular, with
$r = \operatorname{rank} E$, $\Lambda(A_g - \lambda E_g) \subset \mathbb{C}_g$ with $A_g, E_g \in \mathbb{R}^{n_g \times n_g}$ and $\Lambda(A_b - \lambda E_b) \subset \mathbb{C}_b$ with
$A_b, E_b \in \mathbb{R}^{n_b \times n_b}$. Compute $\widetilde{B} := Q^T B$, $\widetilde{C}_N := CZ$, $\widetilde{C}_M = 0$, $\widetilde{D}_N = D$, $\widetilde{D}_M = I_m$. Set
$q = n - n_b$.

(2) If $q = n$, **Exit**.

(3) Let (A_{22}, E_{22}) be the last $k \times k$ diagonal blocks of the GRSF of $(\widetilde{A}, \widetilde{E})$ (with
$k = 1$ or $k = 2$) and let B_2 be the $k \times m$ matrix formed from the last k rows of $\widetilde{B}$.
If $\|B_2\| \le \varepsilon$ (a given tolerance), then remove the parts corresponding to the uncontrollable
eigenvalues $\Lambda(A_{22} - \lambda E_{22})$: $\widetilde{A} \leftarrow \widetilde{A}(1 : n - k, 1 : n - k)$, $\widetilde{E} \leftarrow \widetilde{E}(1 : n - k, 1 : n - k)$,
$\widetilde{B} \leftarrow \widetilde{B}(1 : n - k, 1 : m)$, $\widetilde{C}_N \leftarrow \widetilde{C}_N(1 : p, 1 : n - k)$, $\widetilde{C}_M \leftarrow \widetilde{C}_M(1 : p, 1 : n - k)$;
update $n \leftarrow n - k$ and go to Step (2).

(4) If $E_{22} \neq 0$, determine F_2 such that $\Lambda(A_{22} + B_2 F_2 - \lambda E_{22}) \subset \mathbb{C}_g$.
Set $\widetilde{F} = [\, 0 \;\; F_2 \,]$ and compute $\widetilde{A} \leftarrow \widetilde{A} + \widetilde{B}\widetilde{F}$, $\widetilde{C}_N \leftarrow \widetilde{C}_N + \widetilde{D}_N \widetilde{F}$, $\widetilde{C}_M \leftarrow \widetilde{C}_M + \widetilde{D}_M \widetilde{F}$.

(5) If $E_{22} = 0$, compute $F_2 = -B_2^T (B_2 B_2^T)^{-1} A_{22}$ and $W = I_m - B_2^T (B_2 B_2^T)^{-1} B_2$.
Choose $\gamma \in \mathbb{C}_g$ and update $\widetilde{A}, \widetilde{E}, \widetilde{B}, \widetilde{C}_N, \widetilde{D}_N, \widetilde{C}_M$ and $\widetilde{D}_M$ using (10.62).

(6) Compute the orthogonal matrices $\widetilde{Q}$ and $\widetilde{Z}$ to move the last blocks of $(\widetilde{A}, \widetilde{E})$ to
positions $(q + 1, q + 1)$ by interchanging the diagonal blocks of the GRSF. Compute
$\widetilde{A} \leftarrow \widetilde{Q}^T \widetilde{A}\widetilde{Z}$, $\widetilde{E} \leftarrow \widetilde{Q}^T \widetilde{E}\widetilde{Z}$, $\widetilde{B} \leftarrow \widetilde{Q}^T \widetilde{B}$, $\widetilde{C}_N \leftarrow \widetilde{C}_N \widetilde{Z}$, $\widetilde{C}_M \leftarrow \widetilde{C}_M \widetilde{Z}$.
Put $q \leftarrow q + k$ and go to Step (2).

This algorithm is completely general, being applicable regardless of the original descriptor realization is $\mathbb{C}_b$-stabilizable or not, is infinite controllable or not. The resulting pair $(\widetilde{A}, \widetilde{E})$ is in a special GRSF with $n - r$ simple infinite eigenvalues in the leading $n - r$ positions (no such block exists if E is invertible). A minimal realization of the least McMillan degree denominator $M(\lambda)$ can be easily determined. The resulting $\widetilde{C}_M$ has always the form

$$\widetilde{C}_M = [\,0 \ \ \widetilde{C}_{M,2}\,], \tag{10.63}$$

where the number of columns of $\widetilde{C}_{M,2}$ is equal to the number of controllable generalized eigenvalues of the pair (A, E) lying in $\mathbb{C}_b$. By partitioning accordingly the resulting $\widetilde{E}$, $\widetilde{A}$ and $\widetilde{B}$

$$\widetilde{A} = \begin{bmatrix} A_{11} & A_{12} \\ 0 & A_{22} \end{bmatrix}, \quad \widetilde{E} = \begin{bmatrix} E_{11} & E_{12} \\ 0 & E_{22} \end{bmatrix}, \quad \widetilde{B} = \begin{bmatrix} B_1 \\ B_2 \end{bmatrix}, \tag{10.64}$$

then $(A_{22} - \lambda E_{22}, B_2, \widetilde{C}_{M,2}, \widetilde{D}_M)$ is a minimal descriptor system realization of $M(\lambda)$. Notice however that the order of the minimal realization of $M(\lambda)$ can be higher than the least possible McMillan degree if some eigenvalues of $A - \lambda E$ in $\mathbb{C}_b$ are controllable but not observable.

The **Procedure GRCF** can be interpreted as an extension of the generalized pole assignment algorithm of [133], which generalizes the pole assignment algorithm of [129] for standard systems. The roundoff error analysis of this latter algorithm [130] revealed that if each gain matrix F_2 computed at Step (4) or Step (5) satisfies $\|F_2\| \leq \kappa \|A\|/\|B\|$, with κ having moderate values (say $\kappa < 100$), then the standard pole assignment algorithm is numerically backward stable. This condition is also applicable in our case, because it is independent of the presence of E. We note however that, unfortunately, this condition cannot be always fulfilled if large gains are necessary to stabilize the system. This can arise either if the unstable poles are too "far" from the stable region or if these poles are weekly controllable. Nevertheless, the **Procedure GRCF** can be considered a reliable algorithm, since the above condition can be checked at each computational step and therefore the potential loss of numerical stability can be easily detected.

A similar recursive procedure can be developed to compute RCFs with inner denominators. In this case, we use the partition of the complex plane with $\mathbb{C}_g = \mathbb{C}_s$ and $\mathbb{C}_b = \mathbb{C}\backslash\mathbb{C}_s$. A necessary and sufficient condition for the existence of such a factorization is that $G(\lambda)$ has no poles in $\partial\mathbb{C}_s$ (the boundary of $\mathbb{C}_s$). In the continuous-time case, this means that the pencil $A - sE$ has no finite eigenvalues on the imaginary axis and all infinite eigenvalues of $A - sE$ are simple. In the discrete-time case, $A - zE$ has no eigenvalues on the unit circle centred in the origin. However, for the sake of generality, $G(z)$ can be improper, thus $A - zE$ may have multiple infinite eigenvalues.

For the computation of the RCF with inner denominators we use a similar recursive pole dislocation technique as in the case of a general RCF, by using elementary inner factors. The denominator factor results in the factored form (10.57), where each $\widetilde{M}_i(\lambda)$ $(i = 1, \ldots, k)$ is an elementary inner factor with McMillan degree equal to one or two. These factors are used to reflect the unstable poles of $G(\lambda)$ to stable symmetric positions with respect to the imaginary axis, in the case of a continuous-time system, or with respect to the unit circle in the origin, in the case of a discrete-time system.

In what follows, we give the formulas to determine the elementary inner factors to be used in (10.57) and derive appropriate updating formulas of the factors. We assume $N_{i-1}(\lambda)$ and $M_{i-1}(\lambda)$ have the descriptor realizations in (10.59) and $B_2 \neq 0$ (otherwise the uncontrollable part $A_{22} - \lambda E_{22}$ can be removed from the realization, see (10.60)). In the case when $A_{22} - \lambda E_{22}$ has finite eigenvalues (i.e., E_{22} is invertible) we choose the elementary inner factor as $\widetilde{M}_i(\lambda) = (A_{22} + B_2 F_2 - \lambda E_{22},\ B_2 W,\ F_2,\ W)$. The updating formulas for this case are the same as those employed in **Procedure GRCF** given in (10.61). For the computation of F_2 and W, we have the following results.

Lemma 10.1 *Let $(A_{22} - s E_{22}, B_2)$ be a controllable continuous-time descriptor pair with E_{22} invertible and $\Lambda(A_{22} - s E_{22}) \subset \mathbb{C}_u$. Then the elementary denominator factor $\widetilde{M}_i(s) = (A_{22} + B_2 F_2 - s E_{22},\ B_2 W,\ F_2,\ W)$ is inner by choosing F_2 and W as*

$$A_{22} Y E_{22}^T + E_{22} Y A_{22}^T - B_2 B_2^T = 0,$$
$$F_2 = -B_2^T (Y E_{22}^T)^{-1}, \quad W = I_m.$$

Lemma 10.2 *Let $(A_{22} - z E_{22}, B_2)$ be a controllable discrete-time descriptor pair with E_{22} invertible and $\Lambda(A_{22} - z E_{22}) \subset \mathbb{C}_u$. Then the elementary denominator factor $\widetilde{M}_i(z) = (A_{22} + B_2 F_2 - z E_{22},\ B_2 W,\ F_2,\ W)$ is inner by choosing F_2 and W as*

$$A_{22} Y A_{22}^T - B_2 B_2^T = E_{22} Y E_{22}^T,$$
$$F_2 = -B_2^T (Y A_{22}^T)^{-1},$$
$$W^T (I + B_2^T (E_{22} Y E_{22}^T)^{-1} B_2) W = I.$$

If the 1×1 pencil $A_{22} - z E_{22}$ has an infinite eigenvalue (i.e., $E_{22} = 0$), then we have the following result for the choice of the elementary inner factor.

Lemma 10.3 *Let $(A_{22} - z E_{22}, B_2)$ be an infinite controllable discrete-time descriptor pair with $E_{22} = 0$, and A_{22} nonzero. Then the elementary denominator factor $\widetilde{M}_i(z) = (0 + z A_{22},\ B_2,\ F_2,\ W)$ is inner by choosing F_2 and W as*

$$F_2 = -B_2^T (B_2 B_2^T)^{-1} A_{22},$$
$$W = I - B_2^T (B_2 B_2^T)^{-1} B_2.$$

By this choice of $\widetilde{M}_i(z)$, we made the infinite eigenvalue in the realization of the updated factors $N_i(z)$ and $M_i(z)$ simple, and after its elimination, we obtain the realizations

$$\begin{bmatrix} N_i(z) \\ M_i(z) \end{bmatrix} := \begin{bmatrix} N_{i-1}(z) \\ M_{i-1}(z) \end{bmatrix} \widetilde{M}_i(z) = \left[\begin{array}{cc|c} A_{11} - z E_{11} & A_{12} + B_1 F_2 - z E_{12} & B_1 W \\ 0 & z A_{22} & B_2 \\ \hline C_{N,1} & C_{N,2} + D_N F_2 & D_N W \\ C_{M,1} & C_{M,2} + D_M F_2 & D_M W \end{array} \right].$$

The above relations lead to the following updating formulas:

$$\widetilde{A} \;\leftarrow\; \begin{bmatrix} A_{11} & A_{12} + B_1 F_2 \\ 0 & 0 \end{bmatrix}, \quad \widetilde{E} \leftarrow \begin{bmatrix} E_{11} & E_{12} \\ 0 & -A_{22} \end{bmatrix}, \quad \widetilde{B} \leftarrow \begin{bmatrix} B_1 W \\ B_2 \end{bmatrix},$$

$$\widetilde{C}_N \;\leftarrow\; \begin{bmatrix} C_{N,1} & C_{N,2} + D_N F_2 \end{bmatrix}, \quad \widetilde{C}_M \leftarrow \begin{bmatrix} C_{M,1} & C_{M,2} + D_M F_2 \end{bmatrix}, \tag{10.65}$$

$$\widetilde{D}_N \;\leftarrow\; D_N W, \quad \widetilde{D}_M \leftarrow D_M W.$$

The following procedure computes a stable RCF with inner denominator of a rational TFM $G(\lambda)$, without poles in $\partial\mathbb{C}_s$, as $G(\lambda) = N(\lambda)M^{-1}(\lambda)$, where the resulting factors $N(\lambda)$ and $M(\lambda)$ have the realizations $N(\lambda) = (\widetilde{A} - \lambda\widetilde{E}, \widetilde{B}, \widetilde{C}_N, \widetilde{D}_N)$ and $M(\lambda) = (\widetilde{A} - \lambda\widetilde{E}, \widetilde{B}, \widetilde{C}_M, \widetilde{D}_M)$.

Procedure GRCFID: Generalized RCF with inner denominator

Inputs : $G(\lambda) = (A - \lambda E, B, C, D)$ with $A, E \in \mathbb{R}^{n \times n}$, $B \in \mathbb{R}^{n \times m}$, $C \in \mathbb{R}^{p \times n}$,
$\quad\quad\quad D \in \mathbb{R}^{p \times m}$

Outputs: $N(\lambda) = (\widetilde{A} - \lambda\widetilde{E}, \widetilde{B}, \widetilde{C}_N, \widetilde{D}_N)$ and $M(\lambda) = (\widetilde{A} - \lambda\widetilde{E}, \widetilde{B}, \widetilde{C}_M, \widetilde{D}_M)$ such that
$\quad\quad\quad G(\lambda) = N(\lambda)M^{-1}(\lambda)$, $M(\lambda)$ is inner, all finite eigenvalues of $\widetilde{A} - \lambda\widetilde{E}$ are in $\mathbb{C}_s$
$\quad\quad\quad$ and all infinite eigenvalues of $\widetilde{A} - \lambda\widetilde{E}$ are simple

(1) Compute using **Procedure GSORSF**, the orthogonal matrices Q and Z to
reduce the pair (A, E) to the special ordered GRSF

$$\widetilde{A} := Q^T A Z = \begin{bmatrix} A_\infty & * & * \\ 0 & A_s & * \\ 0 & 0 & A_u \end{bmatrix}, \quad \widetilde{E} := Q^T E Z = \begin{bmatrix} 0 & * & * \\ 0 & E_s & * \\ 0 & 0 & E_u \end{bmatrix},$$

where $A_\infty \in \mathbb{R}^{(n-r) \times (n-r)}$ is invertible and upper triangular, with $r = \operatorname{rank} E$,
$\Lambda(A_s - \lambda E_s) \subset \mathbb{C}_s$ with $A_s, E_s \in \mathbb{R}^{n_s \times n_s}$ and $\Lambda(A_u - \lambda E_u) \subset \mathbb{C}_u$ with
$A_u, E_u \in \mathbb{R}^{n_u \times n_u}$. Compute $\widetilde{B} := Q^T B$, $\widetilde{C}_N := CZ$, $\widetilde{C}_M = 0$, $\widetilde{D}_N = D$, $\widetilde{D}_M = I_m$.
Set $q = n - n_u$.

(2) If $q = n$, **Exit**.

(3) Let (A_{22}, E_{22}) be the last $k \times k$ diagonal blocks of the GRSF of $(\widetilde{A}, \widetilde{E})$ (with
$k = 1$ or $k = 2$) and let B_2 be the matrix formed from the last k rows of $\widetilde{B}$.
If $\|B_2\| \leq \varepsilon$ (a given tolerance), then remove the parts corresponding to the
uncontrollable eigenvalues $\Lambda(A_{22} - \lambda E_{22})$: $\widetilde{A} \leftarrow \widetilde{A}(1 : n - k, 1 : n - k)$,
$\widetilde{E} \leftarrow \widetilde{E}(1 : n - k, 1 : n - k)$, $\widetilde{B} \leftarrow \widetilde{B}(1 : n - k, 1 : m)$, $\widetilde{C}_N \leftarrow \widetilde{C}_N(1 : p, 1 : n - k)$,
$\widetilde{C}_M \leftarrow \widetilde{C}_M(1 : p, 1 : n - k)$; update $n \leftarrow n - k$ and go to Step (2).

(4) If $E_{22} \neq 0$, compute F_2 and W according to Lemma 10.1 in the
continuous-time case or according to Lemma 10.2 in the discrete-time case.
Set $\widetilde{F} = [\,0\;\; F_2\,]$ and update $\widetilde{A}, \widetilde{B}, \widetilde{C}_N, \widetilde{D}_N, \widetilde{C}_M$ and $\widetilde{D}_M$ using (10.61).

(5) If $E_{22} = 0$, compute $F_2 = -B_2^T (B_2 B_2^T)^{-1} A_{22}$ and $W = I_m - B_2^T (B_2 B_2^T)^{-1} B_2$,
and update $\widetilde{A}, \widetilde{E}, \widetilde{B}, \widetilde{C}_N, \widetilde{D}_N, \widetilde{C}_M$ and $\widetilde{D}_M$ using (10.65).

(6) Compute the orthogonal matrices $\widetilde{Q}$ and $\widetilde{Z}$ to move the last blocks of $(\widetilde{A}, \widetilde{E})$ to
positions $(q + 1, q + 1)$ by interchanging the diagonal blocks of the GRSF. Compute
$\widetilde{A} \leftarrow \widetilde{Q}^T \widetilde{A} \widetilde{Z}$, $\widetilde{E} \leftarrow \widetilde{Q}^T \widetilde{E} \widetilde{Z}$, $\widetilde{B} \leftarrow \widetilde{Q}^T \widetilde{B}$, $\widetilde{C}_N \leftarrow \widetilde{C}_N \widetilde{Z}$, $\widetilde{C}_M \leftarrow \widetilde{C}_M \widetilde{Z}$.
Put $q \leftarrow q + k$ and go to Step (2).

The resulting inner factor $M(\lambda)$ has least McMillan degree, only if all unstable generalized eigenvalues of the pair (E, A) are observable. A minimal realization of $M(\lambda)$ can be explicitly determined as $(A_{22} - \lambda E_{22}, B_2, \widetilde{C}_{M,2}, \widetilde{D}_M)$, where the matrices of the realization are defined in (10.64) and (10.63).

The numerical properties of **Procedure GRCFID** are similar to those of **Procedure GRCF**, as long as the matrix gains $\|F_2\|$ at Steps (4) and (5) are reasonably small. However, this condition for numerical reliability may not always be fulfilled due to the lack of freedom in assigning the poles. Recall that the unstable poles are reflected in symmetrical position with respect to $\partial\mathbb{C}_s$, and this may occasionally require large gains.

10.3.6   *Inner-Outer Factorization*

In the light of the needs of the synthesis algorithms presented in Chap. 5, we discuss the computation of the inner-outer factorization of a particular $p \times m$ rational matrix $G(\lambda)$, namely which is proper and has full column rank. Assume that $G(\lambda)$ has an n-th order descriptor system realization

$$G(\lambda) = \left[\begin{array}{c|c} A - \lambda E & B \\ \hline C & D \end{array}\right], \tag{10.66}$$

with E an invertible $n \times n$ matrix. Consider the disjunct partition of the complex plane as $\mathbb{C} = \mathbb{C}_u \cup \overline{\mathbb{C}}_s$. We discuss the computation of the inner-outer factorization of $G(\lambda)$ either in the compact form

$$G(\lambda) = G_{i,1}(\lambda)G_o(\lambda), \tag{10.67}$$

or in the extended form

$$G(\lambda) = \left[\, G_{i,1}(\lambda)\ G_{i,2}(\lambda)\,\right] \begin{bmatrix} G_o(\lambda) \\ 0 \end{bmatrix} = G_i(\lambda) \begin{bmatrix} G_o(\lambda) \\ 0 \end{bmatrix}, \tag{10.68}$$

where $G_i(\lambda) := \left[\, G_{i,1}(\lambda)\ G_{i,2}(\lambda)\,\right]$ is a square inner TFM (i.e., with $G_{i,1}(\lambda)$ inner too), and $G_o(\lambda)$ is an invertible *quasi*-outer TFM, having all zeros in $\overline{\mathbb{C}}_s$. The stability of $G_o(\lambda)$ is ensured, provided $G(\lambda)$ is stable. The component $G_{i,2}(\lambda)$ is a complementary inner factor (also called an "orthogonal" complement of $G_{i,1}(\lambda)$) (see Sect. 9.1.8).

For the computation of inner-outer factorization of $G(\lambda)$, a special reduced form of the system matrix will be instrumental.

Proposition 10.5 *Let $G(\lambda)$ be a $p \times m$ proper rational matrix of full column rank with a stabilizable realization given in (10.66). Then, there exist orthogonal matrices U and Z such that*

$$\begin{bmatrix} U & 0 \\ 0 & I \end{bmatrix} \left[\begin{array}{c|c} A - \lambda E & B \\ \hline C & D \end{array} \right] Z = \left[\begin{array}{c|cc} A_s - \lambda E_s & * & * \\ \hline 0 & A_{u\ell} - \lambda E_{u\ell} & B_{u\ell} \\ \hline 0 & C_{u\ell} & D_{u\ell} \end{array} \right], \qquad (10.69)$$

where

(a) *The regular pencil $A_s - \lambda E_s$ contains the zeros of $G(\lambda)$ in $\overline{\mathbb{C}}_s$;*
(b) *The descriptor system defined by*

$$G_{u\ell}(\lambda) = \left[\begin{array}{c|c} A_{u\ell} - \lambda E_{u\ell} & B_{u\ell} \\ \hline C_{u\ell} & D_{u\ell} \end{array} \right] \qquad (10.70)$$

is proper, with the $n_\ell \times n_\ell$ matrix $E_{u\ell}$ invertible, is stabilizable, has full column rank, and has only zeros in $\mathbb{C}_u$.

Proof This proposition is a simplified version of a slight variation of Theorem 3.1 of [103] combined with Theorem 2.2 in [100], where constructive proofs are also given to determine the orthogonal matrices U and Z, as well as the condensed form (10.69), using numerically stable computational algorithms. For convenience, we describe the main computational steps of this reduction for the considered particular case. Let denote the initial system pencil as

$$S_0(\lambda) := \left[\begin{array}{c|c} A - \lambda E & B \\ \hline C & D \end{array} \right]$$

and observe that $S_0(\lambda)$ has full column rank $n + m$ and furthermore $[\, A - \lambda E \;\; B \,]$ has full row rank for all $\lambda \in \mathbb{C}_u$. The reduction algorithm has three computational steps, which are presented in what follows.

(1) Compute orthogonal Z_1 such that

$$[\, C \;\; D \,] Z_1 = [\, 0 \;\; C_2^{(1)} \,],$$

with $C_2^{(1)}$ of full column rank and define

$$S_1(\lambda) := S_0(\lambda) Z_1 = \left[\begin{array}{c|c} A_{11}^{(1)} - \lambda E_{11}^{(1)} & A_{12}^{(1)} - \lambda E_{12}^{(1)} \\ \hline 0 & C_2^{(1)} \end{array} \right].$$

Since E is invertible, it follows that $[\, E_{11}^{(1)} \;\; E_{12}^{(1)} \,]$ has full row rank n.

(2) Compute orthogonal U and Z_2 to reduce the pencil $A_{11}^{(1)} - \lambda E_{11}^{(1)}$ to a Kronecker-like form (see Sect. 10.1.6)

$$U(A_{11}^{(1)} - \lambda E_{11}^{(1)}) Z_2 = \begin{bmatrix} A_s - \lambda E_s & * & * \\ 0 & A_u - \lambda E_u & * \\ 0 & 0 & \overline{A}_\ell - \lambda \overline{E}_\ell \end{bmatrix} := \begin{bmatrix} A_s - \lambda E_s & * \\ 0 & \overline{A}_{u\ell} - \lambda \overline{E}_{u\ell} \end{bmatrix},$$

where $\Lambda(A_s - \lambda E_s) \subset \overline{\mathbb{C}}_s$, $\Lambda(A_u - \lambda E_u) \subset \mathbb{C}_u$, and $\overline{A}_\ell - \lambda \overline{E}_\ell$ has full column rank for all $\lambda \in \mathbb{C}$. Define

$$S_2(\lambda) := \mathrm{diag}(U, I)S_1(\lambda)\mathrm{diag}(Z_2, I) = \left[\begin{array}{c|cc} A_s - \lambda E_s & * & * \\ \hline 0 & \overline{A}_{u\ell} - \lambda \overline{E}_{u\ell} & \overline{B}_{u\ell} - \lambda \overline{F}_{u\ell} \\ \hline 0 & 0 & C_2^{(1)} \end{array}\right].$$

It is easy to show that $[\,\overline{A}_{u\ell} - \lambda \overline{E}_{u\ell}\ \ \overline{B}_{u\ell} - \lambda \overline{F}_{u\ell}\,]$ has full row rank for all $\lambda \in \mathbb{C}_u$ and also $[\,\overline{E}_{u\ell}\ \ \overline{F}_{u\ell}\,]$ has full row rank.

(3) Compute orthogonal Z_3 such that

$$\left[\begin{array}{cc} \overline{E}_{u\ell} & \overline{F}_{u\ell} \\ \hline 0 & C_2^{(1)} \end{array}\right] Z_3 = \left[\begin{array}{cc} E_{u\ell} & 0 \\ \hline C_{u\ell} & D_{u\ell} \end{array}\right],$$

with $E_{u\ell}$ invertible. Define

$$S_3(\lambda) = S_2(\lambda)\mathrm{diag}(I, Z_3) = \left[\begin{array}{c|cc} A_s - \lambda E_s & * & * \\ \hline 0 & A_{u\ell} - \lambda E_{u\ell} & B_{u\ell} \\ \hline 0 & C_{u\ell} & D_{u\ell} \end{array}\right].$$

The properties (a) and (b) follow immediately from the above properties of the blocks of the reduced final form. The overall transformation matrix Z is defined as

$$Z = Z_1\mathrm{diag}(Z_2, I)\mathrm{diag}(I, Z_3).$$

∎

Remark 10.7 This proposition extracts from the original system (10.66) a proper system (10.70) which has a standard inner-outer factorization. It can be shown that there exists an invertible $G_r(\lambda)$ with zeros only in $\overline{\mathbb{C}}_s$ such that

$$G_{u\ell}(\lambda)G_r(\lambda) = G(\lambda).$$

It follows that $G_{u\ell}(\lambda)$ and $G(\lambda)$ have the same inner factor. Assume $G_i(\lambda)$ is a square inner TFM such that

$$G_{u\ell}(\lambda) = G_i(\lambda)\left[\begin{array}{c} G_{o,1}(\lambda) \\ 0 \end{array}\right]$$

is an extended standard inner-outer factorization, where $G_{o,1}(\lambda)$ has only zeros in $\mathbb{C}_s$. Then with $G_o(\lambda) := G_{o,1}(\lambda)G_r(\lambda)$ we immediately obtain an inner-quasi-outer factorization of $G(\lambda)$ in the form (10.68). $\qquad\square$

We discuss now the computation of the inner-outer factorization separately for the continuous-time and discrete-times cases.

In the continuous-time case, we can further refine the reduced form (10.69) by observing that $D_{u\ell}$ is full column rank (otherwise $G_{u\ell}(s)$ would have infinite zeros).

Therefore, we can compress $D_{u\ell}$ to a full row rank matrix using an orthogonal transformation matrix V, such that

$$V^T D_{u\ell} = \begin{bmatrix} D_\ell \\ 0 \end{bmatrix}, \qquad C_\ell := V^T C_{u\ell} = \begin{bmatrix} C_{\ell,1} \\ C_{\ell,2} \end{bmatrix}, \tag{10.71}$$

where D_ℓ is invertible. With this, we have the following result from [103].

Proposition 10.6 *Let $G(s)$ be a $p \times m$ proper full column rank rational matrix with a stabilizable realization (10.66), let U and Z be orthogonal transformation matrices such that (10.69) holds and let V be an orthogonal transformation matrix which compresses $D_{u\ell}$ as in (10.71). Let X_s be the positive definite stabilizing solution of the generalized continuous-time Riccati equation (GCARE)*

$$A_{u\ell}^T X E_{u\ell} + E_{u\ell}^T X A_{u\ell} - (E_{u\ell}^T X B_{u\ell} + C_{u\ell}^T D_{u\ell})$$

$$\times (D_{u\ell}^T D_{u\ell})^{-1}(B_{u\ell}^T X E_{u\ell} + D_{u\ell}^T C_{u\ell}) + C_{u\ell}^T C_{u\ell} = 0 \tag{10.72}$$

and let F_s be the corresponding stabilizing feedback

$$F_s = -R^{-1}(B_{u\ell}^T X_s E_{u\ell} + D_{u\ell}^T C_{u\ell}),$$

with $R := D_{u\ell}^T D_{u\ell} > 0$. Then, the factors of the inner-quasi-outer factorization (10.68) are given by

$$G_i(s) = \begin{bmatrix} G_{i,1}(s) & G_{i,2}(s) \end{bmatrix} = V \left[\begin{array}{c|cc} A_{u\ell} + B_{u\ell}F_s - sE_{u\ell} & B_{u\ell}D_\ell^{-1} & -X_s^{-1}E_{u\ell}^{-T}C_{\ell,2}^T \\ \hline C_{\ell,1} + D_\ell F_s & I & 0 \\ C_{\ell,2} & 0 & I \end{array} \right]$$

and

$$G_o(s) = \left[\begin{array}{c|c} A - sE & B \\ \hline \widetilde{C} & \widetilde{D} \end{array} \right],$$

where $\begin{bmatrix} \widetilde{C} & \widetilde{D} \end{bmatrix} := R^{1/2} \begin{bmatrix} 0 & F_s & I \end{bmatrix} Z^T$.

In the discrete-time case, we have the following result from [100].

Proposition 10.7 *Let $G(z)$ be a $p \times m$ proper full column rank rational matrix with a stabilizable realization (10.66), let U and Z be orthogonal transformation matrices such that (10.69) holds. Let X_s be the stabilizing solution of the generalized discrete-time Riccati equation (GDARE)*

$$A_{u\ell}^T X A_{u\ell} - E_{u\ell}^T X E_{u\ell} - (A_{u\ell}^T X B_{u\ell} + C_{u\ell}^T D_{u\ell})$$

$$\times (D_{u\ell}^T D_{u\ell} + B_{u\ell}^T X B_{u\ell})^{-1}(B_{u\ell}^T X A_{u\ell} + D_{u\ell}^T C_{u\ell}) + C_{u\ell}^T C_{u\ell} = 0 \tag{10.73}$$

and let F_s be the corresponding stabilizing feedback

$$F_s = -R^{-1}(B_{u\ell}^T X_s A_{u\ell} + D_{u\ell}^T C_{u\ell}),$$

with $R := D_{u\ell}^T D_{u\ell} + B_{u\ell}^T X_s B_{u\ell} > 0$. Then, the factors of the inner-quasi-outer factorization (10.67) are given by

$$G_{i,1}(z) = \left[\begin{array}{c|c} A_{u\ell} + B_{u\ell} F_s - z E_{u\ell} & B_{u\ell} R^{-\frac{1}{2}} \\ \hline C_{u\ell} + D_{u\ell} F_s & D_{u\ell} R^{-\frac{1}{2}} \end{array}\right]$$

and

$$G_o(z) = \left[\begin{array}{c|c} A - z E & B \\ \hline \widetilde{C} & \widetilde{D} \end{array}\right],$$

where $\left[\widetilde{C}\ \widetilde{D}\right] := R^{1/2}\left[0\ F_s\ I\right]Z^T$.

Remark 10.8 The complementary inner factor $G_{i,2}(z)$ can be computed in the form [177]

$$G_{i,2}(z) = \left[\begin{array}{c|c} A_{u\ell} + B_{u\ell} F_s - z E_{u\ell} & Y \\ \hline C_{u\ell} + D_{u\ell} F_s & W \end{array}\right],$$

where Y and W satisfy

$$A_{u\ell}^T X_s Y + C_{u\ell}^T W = 0,$$
$$B_{u\ell}^T X_s Y + D_{u\ell}^T W = 0,$$
$$W^T W + Y^T X_s Y = I.$$

To compute Y and W, we can determine first an orthogonal nullspace basis $\left[\begin{array}{c} \widetilde{Y} \\ \widetilde{W} \end{array}\right]$
satisfying

$$\left[\begin{array}{cc} A_{u\ell}^T X_s & C_{u\ell}^T \\ B_{u\ell}^T X_s & D_{u\ell}^T \end{array}\right]\left[\begin{array}{c} \widetilde{Y} \\ \widetilde{W} \end{array}\right] = 0$$

and then compute $Y = \widetilde{Y} L^{-1}$ and $W = \widetilde{W} L^{-1}$, where L is a Cholesky factor satisfying

$$\widetilde{W}^T \widetilde{W} + \widetilde{Y}^T X_s \widetilde{Y} = L^T L.$$

A numerically reliable way to compute the orthogonal nullspace is via the singular value decomposition

$$\left[\begin{array}{cc} A_{u\ell}^T X_s & C_{u\ell}^T \\ B_{u\ell}^T X_s & D_{u\ell}^T \end{array}\right] = \left[U_1\ U_2\right]\left[\begin{array}{cc} \Sigma & 0 \\ 0 & 0 \end{array}\right]\left[V_1\ V_2\right]^T,$$

where Σ is an invertible $k \times k$ diagonal matrix and $\begin{bmatrix} U_1 & U_2 \end{bmatrix}$ and $\begin{bmatrix} V_1 & V_2 \end{bmatrix}$ are compatibly partitioned orthogonal matrices. Then we can set

$$\begin{bmatrix} \widetilde{Y} \\ \widetilde{W} \end{bmatrix} = V_2 ,$$

where V_2 is a matrix whose orthonormal columns span the right nullspace basis. $\square$

10.3.7 Linear Rational Matrix Equations

Several synthesis algorithms presented in Chap. 5 (see Sect. 7.9) involve the solution of linear rational equations of the form

$$G(\lambda)X(\lambda) = F(\lambda), \tag{10.74}$$

where $G(\lambda)$ and $F(\lambda)$ are given $p \times m$ and $p \times q$ rational matrices, respectively, and $X(\lambda)$ is the $m \times q$ rational matrix sought, which must have the least possible McMillan degree. It is a well-known fact that the system (10.74) has a solution provided the rank condition

$$\operatorname{rank} G(\lambda) = \operatorname{rank}[\, G(\lambda) \ F(\lambda) \,] \tag{10.75}$$

is fulfilled. We assume in what follows that this condition holds.

The general solution of (10.74) can be expressed as

$$X(\lambda) = X_0(\lambda) + X_N(\lambda)Y(\lambda), \tag{10.76}$$

where $X_0(\lambda)$ is any particular solution of (10.74), $X_N(\lambda)$ is a rational matrix whose columns form a basis for the right nullspace of $G(\lambda)$, and $Y(\lambda)$ is an arbitrary rational matrix with compatible dimensions. In the case when both $X_0(\lambda)$ and $X_N(\lambda)$ are proper, a possible approach to compute a solution $X(\lambda)$ of least McMillan degree is to determine a suitable proper $Y(\lambda)$ to achieve this goal. A geometric control theoretic method for this purpose has been developed in [93], based on computing minimum dynamic covers. This method has been turned into an efficient and numerically reliable state-space computational approach in [139], which can be used to determine a least McMillan degree solution of (10.74) for this particular case.

Since $X_N(\lambda)$ can always be chosen proper (see Sect. 7.4), the main difficulty in using the above approach is the computation of an appropriate $Y(\lambda)$ in the case when there is no proper solution of (10.74), and thus $X_0(\lambda)$ cannot be chosen proper. To overcome this difficulty we can determine $X_0(\lambda)$ so that its polynomial part corresponds to a minimal number of infinite poles. These infinite poles originate from the intrinsic improper nature of any solution of (10.74) and are related to the common infinite zeros of $G(\lambda)$ and $F(\lambda)$. In what follows, we show how to determine a special particular solution $X_0(\lambda)$ with minimum number of infinite poles. Then, we determine a rational basis $X_N(\lambda)$ for the right nullspace of $G(\lambda)$ which will serve to determine a solution $X(\lambda)$ of least McMillan degree. This goal is achieved by employing an approach similar to that of [93] to determine a proper $Y(\lambda)$ to reduce the McMillan degree of the proper part of $X_0(\lambda)$. This approach relies on the generalized minimum cover algorithm of [142].

Computation of $X_0(\lambda)$

Let us assume that the rational matrices $G(\lambda)$ and $F(\lambda)$ have descriptor realizations of order n of the forms

$$G(\lambda) := \left[\begin{array}{c|c} A - \lambda E & B_G \\ \hline C & D_G \end{array}\right], \quad F(\lambda) := \left[\begin{array}{c|c} A - \lambda E & B_F \\ \hline C & D_F \end{array}\right], \tag{10.77}$$

where we only assume that the pencil $A - \lambda E$ is regular. Such realizations, which share the pair $(A - \lambda E, C)$, automatically result from a minimal realization of the compound TFM $\left[\, G(\lambda)\ F(\lambda)\,\right]$.

Let $S_G(\lambda)$ and $S_F(\lambda)$ be the system matrix pencils associated with the realizations of $G(\lambda)$ and $F(\lambda)$

$$S_G(\lambda) = \left[\begin{array}{cc} A - \lambda E & B_G \\ C & D_G \end{array}\right], \quad S_F(\lambda) = \left[\begin{array}{cc} A - \lambda E & B_F \\ C & D_F \end{array}\right].$$

Using the straightforward relations

$$\left[\begin{array}{cc} A - \lambda E & B_G \\ 0 & G(\lambda) \end{array}\right] = \left[\begin{array}{cc} I_n & 0 \\ -C(A - \lambda E)^{-1} & I_p \end{array}\right] S_G(\lambda),$$

$$\left[\begin{array}{cc} A - \lambda E & B_F \\ 0 & F(\lambda) \end{array}\right] = \left[\begin{array}{cc} I_n & 0 \\ -C(A - \lambda E)^{-1} & I_p \end{array}\right] S_F(\lambda),$$

it is easy to see that $X(\lambda)$ is a solution of $G(\lambda)X(\lambda) = F(\lambda)$ if and only if

$$Y(\lambda) = \left[\begin{array}{cc} Y_{11}(\lambda) & Y_{12}(\lambda) \\ Y_{21}(\lambda) & X(\lambda) \end{array}\right]$$

is a solution of

$$S_G(\lambda)Y(\lambda) = S_F(\lambda). \tag{10.78}$$

The existence of the solution of (10.78) is guaranteed by (10.75), which is equivalent to

$$\mathrm{rank}\, S_G(\lambda) = \mathrm{rank}[\, S_G(\lambda)\ S_F(\lambda)\,]. \tag{10.79}$$

It follows that, instead of solving the rational equation $G(\lambda)X(\lambda) = F(\lambda)$, we can solve the polynomial equation (10.78) and take

$$X(\lambda) = \begin{bmatrix} 0 & I_m \end{bmatrix} Y(\lambda) \begin{bmatrix} 0 \\ I_q \end{bmatrix}.$$

In fact, since we are only interested in the second block column $Y_2(\lambda)$ of $Y(\lambda)$, we need only to solve

$$\begin{bmatrix} A - \lambda E & B_G \\ C & D_G \end{bmatrix} Y_2(\lambda) = \begin{bmatrix} B_F \\ D_F \end{bmatrix} \tag{10.80}$$

and compute $X(\lambda)$ as

$$X(\lambda) = \begin{bmatrix} 0 & I_m \end{bmatrix} Y_2(\lambda).$$

The condition (10.79) for the existence of a solution becomes

$$\operatorname{rank} \begin{bmatrix} A - \lambda E & B_G \\ C & D_G \end{bmatrix} = \operatorname{rank} \begin{bmatrix} A - \lambda E & B_G & B_F \\ C & D_G & D_F \end{bmatrix}. \tag{10.81}$$

To solve (10.80), we isolate a full rank part of $S_G(\lambda)$ by reducing it to a particular Kronecker-like form. Let Q and Z be orthogonal matrices to reduce $S_G(\lambda)$ to the Kronecker-like form

$$\overline{S}_G(\lambda) := Q S_G(\lambda) Z = \begin{bmatrix} B_r & A_r - \lambda E_r & A_{r,reg} - \lambda E_{r,reg} & * \\ 0 & 0 & A_{reg} - \lambda E_{reg} & * \\ 0 & 0 & 0 & A_l - \lambda E_l \end{bmatrix}, \tag{10.82}$$

where $A_{reg} - \lambda E_{reg}$ is a regular subpencil, the pair $(A_r - \lambda E_r, B_r)$ is controllable with E_r invertible and the subpencil $A_l - \lambda E_l$ has full column rank for all $\lambda \in \mathbb{C}$. The above reduction can be computed by employing numerically stable algorithms, as those described in Sect. 10.1.6.

If $\overline{Y}_2(\lambda)$ is a solution of the reduced equation

$$\overline{S}_G(\lambda) \overline{Y}_2(\lambda) = Q \begin{bmatrix} B_F \\ D_F \end{bmatrix}, \tag{10.83}$$

then $Y_2(\lambda) = Z\overline{Y}_2(\lambda)$, and thus

$$X(\lambda) = \begin{bmatrix} 0 & I_m \end{bmatrix} Z \overline{Y}_2(\lambda)$$

is a solution of the equation $G(\lambda)X(\lambda) = F(\lambda)$. Partition

$$Q\begin{bmatrix} -B_F \\ -D_F \end{bmatrix} = \begin{bmatrix} \overline{B}_1 \\ \overline{B}_2 \\ \overline{B}_3 \end{bmatrix}$$

in accordance with the row structure of $\overline{S}_G(\lambda)$. Since $A_l - \lambda E_l$ has full column rank, it follows from (10.81) that $\overline{B}_3 = 0$ (otherwise no solution exists). Thus, $\overline{Y}_2(\lambda)$ has the form

$$\overline{Y}_2(\lambda) = \begin{bmatrix} \overline{Y}_{12}(\lambda) \\ \overline{Y}_{22}(\lambda) \\ \overline{Y}_{32}(\lambda) \\ 0 \end{bmatrix},$$

where the partitioning of $\overline{Y}_2(\lambda)$ corresponds to the column partitioning of $\overline{S}_G(\lambda)$. To determine a particular solution $X_0(\lambda)$, we can freely choose $\overline{Y}_{12}(\lambda) = 0$ and determine $\overline{Y}_{22}(\lambda)$ and $\overline{Y}_{32}(\lambda)$ by solving

$$\begin{bmatrix} \overline{Y}_{22}(\lambda) \\ \overline{Y}_{32}(\lambda) \end{bmatrix} = \begin{bmatrix} \lambda E_r - A_r & \lambda E_{r,reg} - A_{r,reg} \\ 0 & \lambda E_{reg} - A_{reg} \end{bmatrix}^{-1} \begin{bmatrix} \overline{B}_1 \\ \overline{B}_2 \end{bmatrix}.$$

Let partition $[\, 0 \; I_m \,]Z$ in accordance with the column structure of $S_G(\lambda)$ as

$$[\, 0 \; I_m \,]Z = [\, D_r \; C_r \; C_{reg} \; C_l \,] \tag{10.84}$$

and denote

$$\overline{A} - \lambda \overline{E} = \begin{bmatrix} A_r - \lambda E_r & A_{r,reg} - \lambda E_{r,reg} \\ 0 & A_{reg} - \lambda E_{reg} \end{bmatrix}, \quad \overline{B} = \begin{bmatrix} \overline{B}_1 \\ \overline{B}_2 \end{bmatrix}, \quad \overline{C} = [\, C_r \; C_{reg} \,].$$

$$\tag{10.85}$$

Then, a particular solution $X_0(\lambda)$ of the equation $G(\lambda)X(\lambda) = F(\lambda)$ can be determined with the descriptor system realization

$$X_0(\lambda) := \left[\begin{array}{c|c} \overline{A} - \lambda \overline{E} & \overline{B} \\ \hline \overline{C} & 0 \end{array} \right]. \tag{10.86}$$

Some properties of $X_0(\lambda)$ can be easily deduced from the computed Kronecker-like form. The pair $(\overline{C}, \overline{A} - \lambda \overline{E})$ is always observable, but, in general, the pair $(\overline{A} - \lambda \overline{E}, \overline{B})$ may be uncontrollable. The poles of $X_0(\lambda)$ are among the generalized eigenvalues of the pair $(\overline{A}, \overline{E})$ and are partly freely assignable and partly fixed. The generalized eigenvalues of the pair (A_r, E_r) are called the "spurious" poles, and they originate from the column singularity of $G(\lambda)$. These poles are freely assignable by appropriate choice of a (non-orthogonal) right transformation matrix [137]. The fixed poles are the controllable eigenvalues of the pair $(A_{reg} - \lambda E_{reg}, \overline{B}_2)$. If $G(\lambda)$

and $F(\lambda)$ have no common poles and zeros then the pair $(A_{reg} - \lambda E_{reg}, \overline{B}_2)$ is controllable. In this case $X_0(\lambda)$ has the minimum possible poles at infinity.

According to the dual of Lemma 9.5, there exists a solution $X_0(\lambda)$ without a pole in γ (finite or infinite) if the pole and zero structures of $G(\lambda)$ and $[G(\lambda)\ F(\lambda)]$ at γ coincide. For practical computations, this implies that some or all of common poles and zeros of $G(\lambda)$ and $[G(\lambda)\ F(\lambda)]$ will cancel. This cancellation can be done explicitly by removing the uncontrollable eigenvalues (finite and infinite) of the pair $(A_{reg} - \lambda E_{reg}, \overline{B}_2)$.

Removing the uncontrollable eigenvalues of the pair $(A_{reg} - \lambda E_{reg}, \overline{B}_2)$ can be done using the generalized controllability staircase form algorithm of **Procedure GCSF** described in Sect. 10.1.5 (see also Sect. 10.3.1). By applying this algorithm, two orthogonal matrices Q_{reg} and Z_{reg} are determined such that all uncontrollable finite eigenvalues are separated in the trailing part of the transformed regular pencil $Q_{reg}(A_{reg} - \lambda E_{reg})Z_{reg}$, while the corresponding rows of $Q_{reg}\overline{B}_2$ are zero. The uncontrollable part of the triple $(\overline{A} - \lambda\overline{E}, \overline{B}, \overline{C})$ can be thus eliminated by removing the appropriate trailing rows and columns from the matrices of the transformed triple $(\overline{Q}(\overline{A} - \lambda\overline{E})\overline{Z}, \overline{Q}\overline{B}, \overline{C}\overline{Z})$, where $\overline{Q} = \text{diag}(I, Q_{reg})$ and $\overline{Z} = \text{diag}(I, Z_{reg})$. The same technique can be used to remove the uncontrollable infinite eigenvalues by simply interchanging the roles of matrices $\overline{A}$ and $\overline{E}$, thus working on the triple $(\overline{E} - \lambda\overline{A}, \overline{B}, \overline{C})$. For the sake of simplicity we reuse the same notation (with bar) by assuming that the pair $(\overline{A} - \lambda\overline{E}, \overline{B})$ is already controllable, thus the resulting $X_0(\lambda)$ fulfils the requirement for a minimal number of poles at infinity.

To compute the particular solution $X_0(\lambda)$ we employed exclusively orthogonal similarity transformations to determine the matrices of a descriptor realization in (10.86). Therefore, this computation is numerically stable, because we can easily show that the computed system matrices in the presence of roundoff errors are exact for an original problem with slightly perturbed data.

In view of the order reduction step described later, we need to enforce a block diagonal descriptor matrix $\overline{E}$ in (10.85) (i.e., with $E_{r,reg} = 0$). This can be easily achieved by performing an additional non-orthogonal column transformation using the transformation matrix

$$V = \begin{bmatrix} I & -E_r^{-1}E_{r,reg} \\ 0 & I \end{bmatrix}.$$

The transformed system $(\overline{A}V - \lambda\overline{E}V, \overline{B}, \overline{C}V, 0)$, representing also $X_0(\lambda)$, has thus a block diagonal descriptor matrix $\overline{E}V$. To simplify the presentation we will reuse the notation with bar and assume in what follows that $E_{r,reg} = 0$ in (10.85).

Computation of $X_N(\lambda)$

Using the same reduction of $S_G(\lambda)$ to $\overline{S}_G(\lambda)$ as in (10.82), a right nullspace basis $X_N(\lambda)$ of $G(\lambda)$ can be computed from a right nullspace basis $\overline{Y}_N(\lambda)$ of $\overline{S}_G(\lambda)$ as

$$X_N(\lambda) = [\, 0 \ I_m \,] Z \overline{Y}_N(\lambda) \,.$$

We can determine $\overline{Y}_N(\lambda)$ in the form

$$\overline{Y}_N(\lambda) = \begin{bmatrix} I \\ (\lambda E_r - A_r)^{-1} B_r \\ 0 \\ 0 \end{bmatrix} \,.$$

With C_r and D_r defined in (10.84), we obtain a descriptor realization of $X_N(\lambda)$ as

$$X_N(\lambda) := \left[\begin{array}{c|c} A_r - \lambda E_r & B_r \\ \hline C_r & D_r \end{array} \right] \,.$$

Obviously $X_N(\lambda)$ is proper and controllable. Furthermore, according to Proposition 10.2 applied to the dual realization of $X_N^T(\lambda)$, the realization of $X_N(\lambda)$ is observable, provided the realization of $G(\lambda)$ in (10.77) is observable. Moreover, the poles of $X_N(\lambda)$ are freely assignable by appropriately choosing the transformation matrices Q and Z to reduce the system pencil $S_G(\lambda)$. Note that, to obtain this nullspace basis, we performed exclusively orthogonal transformations on the system matrices. We can prove that all computed matrices are exact for a slightly perturbed original system. It follows that the algorithm to compute the nullspace basis is numerically stable.

Computation of a least-order solution $X(\lambda)$

We can represent $X_N(\lambda)$ to have the same state, descriptor and output matrices as $X_0(\lambda)$. Let these realizations of $X_0(\lambda)$ and $X_N(\lambda)$ be

$$\left[\, X_0(\lambda) \ X_N(\lambda) \,\right] := \left[\begin{array}{c|cc} \overline{A} - \lambda \overline{E} & \overline{B} & \overline{B}_r \\ \hline \overline{C} & \overline{D} & \overline{D}_r \end{array} \right] := \left[\begin{array}{cc|cc} A_r - \lambda E_r & A_{r,reg} & \overline{B}_1 & B_r \\ 0 & A_{reg} - \lambda E_{reg} & \overline{B}_2 & 0 \\ \hline C_r & C_{reg} & 0 & D_r \end{array} \right] ,$$

$$\tag{10.87}$$

where E_r is invertible.

We consider first the case when $X_0(\lambda)$ is proper, that is, all eigenvalues of the pencil $A_{reg} - \lambda E_{reg}$ are finite and thus $\overline{E}$ is invertible. In this case, it was shown in [93] that a solution with least McMillan degree can be determined as $X(\lambda) = X_0(\lambda) + X_N(\lambda)Y(\lambda)$ by choosing an appropriate proper $Y(\lambda)$. This can be done by determining a suitable feedback matrix $\overline{F}_r$ and a feedforward matrix $\overline{L}_r$ to cancel the maximum number of unobservable and uncontrollable poles of

$$X(\lambda) := \left[\begin{array}{c|c} \overline{A} + \overline{B}_r \overline{F}_r - \lambda \overline{E} & \overline{B} + \overline{B}_r \overline{L}_r \\ \hline \overline{C} + \overline{D}_r \overline{F}_r & \overline{D} + \overline{D}_r \overline{L}_r \end{array} \right]. \tag{10.88}$$

It can be shown that if we start with a minimal realization of $[\, G(\lambda)\ F(\lambda)\,]$, then we cannot produce any unobservable poles in $X(\lambda)$ via state feedback. Therefore, we only need to determine the matrices $\overline{F}_r$ and $\overline{L}_r$ to cancel the maximum number of uncontrollable poles.

This problem has been solved in [93] by reformulating it as a minimal order dynamic cover design problem. We denote $\widetilde{A} := \overline{E}^{-1}\overline{A}$, $\widetilde{B} := \overline{E}^{-1}\overline{B}$, and $\widetilde{B}_r := \overline{E}^{-1}\overline{B}_r$, and also $\widetilde{\mathcal{B}} = \mathrm{span}\, \widetilde{B}$ and $\widetilde{\mathcal{B}}_r = \mathrm{span}\, \widetilde{B}_r$. Consider the set

$$\mathcal{J} = \{ \mathcal{V} : \widetilde{\mathcal{B}} + \widetilde{A}\mathcal{V} \subset \widetilde{\mathcal{B}}_r + \mathcal{V} \},$$

and let $\mathcal{J}^*$ denote the set of subspaces in $\mathcal{J}$ of least dimension. If $\mathcal{V} \in \mathcal{J}^*$, then a pair $(\overline{F}_r, \overline{L}_r)$ can be determined such that

$$(\widetilde{A} + \widetilde{B}_r \overline{F}_r)\mathcal{V} + \mathrm{span}\,(\widetilde{B} + \widetilde{B}_r \overline{L}_r) \subset \mathcal{V}.$$

Thus, determining a minimal dimension $\mathcal{V}$ is equivalent to a minimal order cover design problem, and a conceptual geometric approach to solve it has been indicated in [93]. The outcome of his method is, besides $\mathcal{V}$, the pair $(\overline{F}_r, \overline{L}_r)$ which achieves a maximal order reduction by forcing pole–zero cancellations. This approach, in the case of standard systems (i.e., $\overline{E} = I$), has been turned into a numerically reliable procedure in [139] and extended to the descriptor case with invertible $\overline{E}$ in [142]. In this latter procedure, $\overline{F}_r$ and $\overline{L}_r$ are determined from a special controllability staircase form of the pair $(\overline{A} - \lambda \overline{E}, [\,\overline{B}_r\ \overline{B}\,])$ obtained using a numerically reliable method relying on both orthogonal and non-orthogonal similarity transformations. Details of this algorithm are given in Sect. 10.4.3.

It is possible to refine this approach by exploiting the structure of matrices in (10.87). Assuming $\overline{F}_r = [\, F_r\ F_{reg}\,]$ is partitioned according to the structure of $\overline{A}$, we get from (10.88)

$$X(\lambda) := \left[\begin{array}{cc|c} A_r + B_r F_r - \lambda E_r & A_{r,reg} + B_r F_{reg} & \overline{B}_1 + B_r \overline{L}_r \\ 0 & A_{reg} - \lambda E_{reg} & \overline{B}_2 \\ \hline C_r + D_r F_r & C_{reg} + D_r F_{reg} & \overline{D} + D_r \overline{L}_r \end{array} \right].$$

Since the eigenvalues of $A_{reg} - \lambda E_{reg}$ are not controllable via $\overline{B}_r$, the state feedback $\overline{F}_r$ affects only the blocks $A_r - \lambda E_r$ and $A_{r,reg}$. To make a maximum number of eigenvalues of $A_r + B_r F_r - \lambda E_r$ uncontrollable, we can alternatively solve a minimum dynamic cover problem of lower dimension for the system

$$\left[\, X_{0,r}(\lambda)\ X_N(\lambda)\,\right] := \left[\begin{array}{c|c} A_r - \lambda E_r & \left[\, A_{r,reg}\ \overline{B}_1\,\right]\ B_r \\ \hline C_r & \left[\, C_{r,reg}\ \overline{D}\,\right]\ D_r \end{array}\right],$$

by determining an appropriate state feedback matrix F_r and a feedforward matrix $[\, F_{reg}\ \overline{L}_r\,]$. Besides lower size of the computational problem, the main advantage of this approach is that it is applicable regardless $A_{reg} - \lambda E_{reg}$ has infinite eigenvalues or not.

10.4 Special Algorithms

In this section we describe several algorithms, which are instrumental in addressing least-order synthesis problems of fault detection and isolation filters and the solution of the Nehari problem, which is encountered in solving least distance problems.

10.4.1 Special Controllability Staircase Form Algorithm

The computational methods of minimum dynamic covers, presented in Sects. 10.4.2 and 10.4.3, rely on a special controllability staircase form (see Sect. 10.1.5) involving a controllable descriptor pair $(A - \lambda E, [\, B_1\ B_2\,])$, where $A, E \in \mathbb{R}^{n \times n}$ with E invertible, $B_1 \in \mathbb{R}^{n \times m_1}$, $B_2 \in \mathbb{R}^{n \times m_2}$. The main difference to the reduction performed in **Procedure GCSF** is in exploiting, at the j-th reduction step, the partitioned form of the matrix $B^{(j-1)} := [\, B_1^{(j-1)}\ B_2^{(j-1)}\,]$, by compressing its rows in two steps. In the first step, the rows of $B_1^{(j-1)}$ are compressed, while in the second step, those columns of the updated $B_2^{(j-1)}$ are compressed, which are linearly independent of the columns of $B_1^{(j-1)}$. All row compressions can be performed using orthogonal similarity transformations.

The following procedure determines for a descriptor triple $(A - \lambda E, [\, B_1\ B_2\,], C)$, two orthogonal transformation matrices Q and Z such that for the resulting triple $(Q^T A Z - \lambda Q^T E Z, [\, Q^T B_1\ Q^T B_2\,], C Z)$, the pencil $[\, Q^T B_1\ Q^T B_2\ Q^T A Z - \lambda Q^T E Z\,]$ is in a special controllability staircase form with $Q^T E Z$ upper triangular.

Procedure GSCSF: Generalized special controllability staircase form

Input   : $(A - \lambda E, [\, B_1 \ B_2\,], C)$

Outputs: $Q, Z,$

$\quad\quad (A - \lambda E, [\, B_1 \ B_2\,], C) := (Q^T A Z - \lambda Q^T E Z, [\, Q^T B_1 \ \ Q^T B_2\,], C Z),$

$\quad\quad (v_{1,j}, v_{2,j}), \ j = 1, \ldots, \ell$

(1) Compute an orthogonal matrix Q such that $Q^T E$ is upper triangular; compute $A \leftarrow Q^T A,\ E \leftarrow Q^T E,\ B_1 \leftarrow Q^T B_1,\ B_2 \leftarrow Q^T B_2$. Set $Z = I_n$.

(2) Set $j = 1,\ r = 0,\ v_{1,0} = m_1,\ v_{2,0} = m_2,\ A^{(0)} = A,\ E^{(0)} = E,\ B_1^{(0)} = B_1,$ $B_2^{(0)} = B_2$.

(3) Compute orthogonal matrices W and U such that

$$
W^T \left[\, B_1^{(j-1)} \,\middle|\, B_2^{(j-1)} \,\right] :=
\left[
\begin{array}{c|c}
A_{2j-1,2j-3} & A_{2j-1,2j-2} \\
0 & A_{2j,2j-2} \\
0 & 0
\end{array}
\right]
\begin{array}{l}
v_{1,j} \\
v_{2,j} \\
\rho
\end{array}
$$
$$
\qquad\qquad\quad v_{1,j-1} \qquad\ v_{2,j-1}
$$

with $A_{2j-1,2j-3}$ and $A_{2j,2j-2}$ full row rank matrices and $W^T E^{(j-1)} U$ is upper triangular.

(4) Compute and partition

$$
W^T A^{(j-1)} U :=
\left[
\begin{array}{ccc}
A_{2j-1,2j-1} & A_{2j-1,2j} & A_{2j-1,2j+1} \\
A_{2j,2j-1} & A_{2j,2j} & A_{2j,2j+1} \\
B_1^{(j)} & B_2^{(j)} & A^{(j)}
\end{array}
\right]
\begin{array}{l}
v_{1,j} \\
v_{2,j} \\
\rho
\end{array}
$$
$$
\qquad\qquad\quad v_{1,j} \qquad\quad v_{2,j} \qquad\quad \rho
$$

$$
W^T E^{(j-1)} U :=
\left[
\begin{array}{ccc}
E_{2j-1,2j-1} & E_{2j-1,2j} & E_{2j-1,2j+1} \\
0 & E_{2j,2j} & E_{2j,2j+1} \\
0 & 0 & E^{(j)}
\end{array}
\right]
\begin{array}{l}
v_{1,j} \\
v_{2,j} \\
\rho
\end{array}
$$
$$
\qquad\qquad\quad v_{1,j} \qquad\quad v_{2,j} \qquad\quad \rho
$$

(5) For $i = 1, \ldots, 2j - 2$ compute and partition

$$
A_{i,2j-1} U := [\, A_{i,2j-1} \ A_{i,2j} \ A_{i,2j+1}\,]
$$
$$
\qquad\qquad\quad v_{1,j} \qquad v_{2,j} \qquad \rho
$$

$$
E_{i,2j-1} U := [\, E_{i,2j-1} \ E_{i,2j} \ E_{i,2j+1}\,]
$$
$$
\qquad\qquad\quad v_{1,j} \qquad v_{2,j} \qquad \rho
$$

(6) $Q \leftarrow Q\,\mathrm{diag}(I_r, W),\ Z \leftarrow Z\,\mathrm{diag}(I_r, U),\ C \leftarrow C\,\mathrm{diag}(I_r, U)$.

(7) $r \leftarrow r + v_{1,j} + v_{2,j}$; if $\rho = 0$, then $\ell = j$ and **Exit**;

$\quad$ else, $j \leftarrow j + 1$ and go to Step (3).

At the end of this algorithm we have $\widehat{A} - \lambda \widehat{E} := Q^T (A - \lambda E) Z,\ \widehat{B} := [\, Q^T B_1$ $Q^T B_2\,], \widehat{C} := C Z, \widehat{E}$ is upper triangular, and the pair $(\widehat{A}, \widehat{B})$ is in the *special staircase form*

$$
[\widehat{B}|\widehat{A}] =
\begin{bmatrix}
A_{1,-1} & A_{1,0} & A_{11} & A_{12} & \cdots & A_{1,2\ell-3} & A_{1,2\ell-2} & A_{1,2\ell-1} & A_{1,2\ell} \\
0 & A_{2,0} & A_{21} & A_{22} & \cdots & A_{2,2\ell-3} & A_{2,2\ell-2} & A_{2,2\ell-1} & A_{2,2\ell} \\
0 & 0 & A_{31} & A_{32} & \cdots & A_{3,2\ell-3} & A_{3,2\ell-2} & A_{3,2\ell-1} & A_{3,2\ell} \\
0 & 0 & 0 & A_{42} & \cdots & A_{4,2\ell-3} & A_{4,2\ell-2} & A_{4,2\ell-1} & A_{4,2\ell} \\
\vdots & \vdots & \vdots & \vdots & \ddots & \vdots & \vdots & \vdots & \vdots \\
0 & 0 & 0 & 0 & \cdots & A_{2\ell-1,2\ell-3} & A_{2\ell-1,2\ell-2} & A_{2\ell-1,2\ell-1} & A_{2\ell-1,2\ell} \\
0 & 0 & 0 & 0 & \cdots & 0 & A_{2\ell,2\ell-2} & A_{2\ell,2\ell-1} & A_{2\ell,2\ell}
\end{bmatrix},
$$

$$(10.89)$$

where $A_{2j-1,2j-3} \in \mathbb{R}^{\nu_{1,j} \times \nu_{1,j}}$ and $A_{2j,2j-2} \in \mathbb{R}^{\nu_{2,j} \times \nu_{2,j}}$ are full row rank matrices for $j = 1, \ldots, \ell$. The resulting upper triangular matrix $\widehat{E}$ has a similar block partitioned form

$$
\widehat{E} =
\begin{bmatrix}
E_{11} & E_{12} & \cdots & E_{1,2\ell-1} & E_{1,2\ell} \\
0 & E_{22} & \cdots & E_{2,2\ell-1} & E_{2,2\ell} \\
\vdots & \vdots & \ddots & \vdots & \vdots \\
0 & 0 & \cdots & E_{2\ell-1,2\ell-1} & E_{2\ell-1,2\ell} \\
0 & 0 & \cdots & 0 & E_{2\ell,2\ell}
\end{bmatrix}.
$$

$$(10.90)$$

The resulting block dimensions $(\nu_{1,j}, \nu_{2,j})$, $j = 1, \ldots, \ell$, satisfy

$$
m_1 = \nu_{1,0} \geq \nu_{1,1} \geq \cdots \geq \nu_{1,\ell} \geq 0
$$

and

$$
m_2 = \nu_{2,0} \geq \nu_{2,1} \geq \cdots \geq \nu_{2,\ell} \geq 0
$$

and represents the dimensions $n_1 := \sum_{i=1}^{\ell} \nu_{1,j}$ and $n_2 := \sum_{j=1}^{\ell} \nu_{2,j}$ of two subspaces, which underlie the computation of appropriate minimal dynamic covers in the next sections.

When implementing **Procedure GSCSF**, the row compressions at Step (3) are usually performed using rank-revealing QR factorizations with column pivoting. This computation can be done in two steps, first by compressing the r rows of $B_1^{(j-1)}$ to a full row rank matrix $A_{2j-1,2j-3}$ using an orthogonal matrix W_1 (i.e., as $W_1^T B_1^{(j-1)}$), and then by compressing the trailing $r - \nu_{1,j}$ rows of $W_1^T B_2^{(j-1)}$ to a full row rank matrix $A_{2j,2j-2}$ using a second orthogonal matrix W_2. The overall transformation W at Step (3) results as $W = W_1 \mathrm{diag}(I_{\nu_{1,j}}, W_2)$. Both reductions can be performed using sequences of Givens rotations, which allow to simultaneously perform the column transformations accumulated in U to maintain the upper triangular form of $E^{(j-1)}$. This reduction technique is described in detail in [131]. Using this technique, the numerical complexity of **Procedure GSCSF** is $\mathcal{O}(n^3)$, provided all transformations are immediately applied without accumulating explicitly W and U. The usage of the more robust rank determinations based on singular values decompositions would increase the overall complexity to $\mathcal{O}(n^4)$ due to the need to accumulate explicitly W and U. Regarding the numerical properties of **Procedure GSCSF**, it is possible to show that the resulting system matrices $\widehat{A}, \widehat{E}, \widehat{B}, \widehat{C}$ are exact for slightly perturbed original data A, E, B, C, while Q and Z are nearly orthogonal matrices. It follows that the **Procedure GSCSF** is numerically stable. In the standard case we have $E = I$, and therefore $Q = Z$ and $\widehat{E} = I$.

Example 10.1 For $\ell = 3$, $[\,\widehat{B}\ \ \widehat{A}\,]$ and $\widehat{E}$ have similarly block partitioned forms

$$
[\,\widehat{B}\mid\widehat{A}\,] =
\begin{bmatrix}
A_{1,-1} & A_{1,0} & A_{11} & A_{12} & A_{13} & A_{14} & A_{15} & A_{16} \\
0 & A_{2,0} & A_{21} & A_{22} & A_{23} & A_{24} & A_{25} & A_{26} \\
0 & 0 & A_{31} & A_{32} & A_{33} & A_{34} & A_{35} & A_{36} \\
0 & 0 & 0 & A_{42} & A_{43} & A_{44} & A_{45} & A_{46} \\
0 & 0 & 0 & 0 & A_{53} & A_{54} & A_{55} & A_{56} \\
0 & 0 & 0 & 0 & 0 & A_{64} & A_{65} & A_{66}
\end{bmatrix}, \quad
\widehat{E} =
\begin{bmatrix}
E_{11} & E_{12} & \cdots & E_{16} \\
0 & E_{22} & \cdots & E_{26} \\
\vdots & \vdots & \ddots & \vdots \\
O & O & \cdots & E_{66}
\end{bmatrix}.
$$

$\Diamond$

10.4.2 Order Reduction Using Minimum Dynamic Covers of Type I

The computational problem which we address in this section is the following: given a descriptor pair $(A - \lambda E, B)$ with $A, E \in \mathbb{R}^{n \times n}$ and E invertible, $B \in \mathbb{R}^{n \times m}$, and B partitioned as $B = [\,B_1\ B_2\,]$ with $B_1 \in \mathbb{R}^{n \times m_1}$, $B_2 \in \mathbb{R}^{n \times m_2}$, determine the matrix $F \in \mathbb{R}^{m_2 \times n}$ such that the pair $(A + B_2 F - \lambda E, B_1)$ is *maximally uncontrollable* (i.e., $A + B_2 F - \lambda E$ has maximal number of uncontrollable eigenvalues).

This computation is useful to determine least-order solutions of linear rational equations using state feedback techniques. Consider the compatible linear rational system of equations $G(\lambda)X(\lambda) = F(\lambda)$, where $G(\lambda)$ and $F(\lambda)$ are given and $X(\lambda)$ is sought. Assume $X_1(\lambda)$ and $X_2(\lambda)$ are two proper TFMs, which generate all solutions of the rational system of equation $G(\lambda)X(\lambda) = F(\lambda)$ in the form

$$
X(\lambda) = X_1(\lambda) + X_2(\lambda)Y(\lambda), \tag{10.91}
$$

where $X_1(\lambda)$ is any particular solution satisfying $G(\lambda)X_1(\lambda) = F(\lambda)$, $X_2(\lambda)$ is a proper rational basis of the right nullspace of $G(\lambda)$ (i.e., $G(\lambda)X_2(\lambda) = 0$), and $Y(\lambda)$ is arbitrary, having appropriate dimensions. Assume $X_1(\lambda)$ and $X_2(\lambda)$ have the descriptor system realizations

$$
[\,X_1(\lambda)\ \ X_2(\lambda)\,] =
\left[
\begin{array}{c|cc}
A - \lambda E & B_1 & B_2 \\
\hline
C & D_1 & D_2
\end{array}
\right], \tag{10.92}
$$

with the descriptor pair $(A - \lambda E, [\,B_1\ B_2\,])$ controllable and E invertible. Let F be a state feedback gain and define the TFMs

$$
[\,\widetilde{X}_1(\lambda)\ \ \widetilde{X}_2(\lambda)\,] :=
\left[
\begin{array}{c|cc}
A + B_2 F - \lambda E & B_1 & B_2 \\
\hline
C + D_2 F & D_1 & D_2
\end{array}
\right]. \tag{10.93}
$$

It is straightforward to check that

$$\widetilde{X}_1(\lambda) = X_1(\lambda) + X_2(\lambda)Y(\lambda), \quad \widetilde{X}_2(\lambda) = X_2(\lambda)\widetilde{Y}(\lambda), \tag{10.94}$$

where $Y(\lambda)$ and $\widetilde{Y}(\lambda)$ have the descriptor system realizations

$$\left[\, Y(\lambda)\ \widetilde{Y}(\lambda)\,\right] = \left[\begin{array}{c|cc} A + B_2 F - \lambda E & B_1 & B_2 \\ \hline F & 0 & I \end{array}\right]. \tag{10.95}$$

Therefore, $\widetilde{X}_1(\lambda)$ and $\widetilde{X}_2(\lambda)$ also generate all solutions, because $\widetilde{X}_1(\lambda)$ is another particular solution, while $\widetilde{X}_2(\lambda)$ is another right nullspace basis, because $\widetilde{Y}(\lambda)$ is invertible. If F is determined such that the pair $(A + B_2 F - \lambda E, B_1)$ is maximally uncontrollable, then the resulting realization of $\widetilde{X}_1(\lambda)$ contains a maximum number of uncontrollable eigenvalues which can be eliminated using minimal realization techniques. Thus, $\widetilde{X}_1(\lambda)$ represents another particular solution with a reduced McMillan degree.

Remark 10.9 The above approach achieves the maximum order reduction for $\widetilde{X}_1(\lambda)$ provided the descriptor system realization $(A - \lambda E, B_2, C, D_2)$ is *maximally observable*, i.e., the pair $(A + B_2 F - \lambda E, C + D_2 F)$ is observable for any F [93]. If this condition is not fulfilled, then the least order can be achieved after a preliminary order reduction, where a maximum number of unobservable eigenvalues are eliminated using a suitable choice of F. If $E = I$ and $D_2 = 0$, a numerically stable algorithm proposed in [122] to compute the maximal (A, B_2)-invariant subspace contained in the kernel of C can be employed for this purpose. If E is a general invertible matrix, then the same algorithm can be applied to the triple $(E^{-1}A, E^{-1}B_2, C)$, provided E is not too ill-conditioned. The case $D_2 \neq 0$ can be addressed using the extended system technique suggested in [6, p. 240]. $\qquad\square$

An important application of the above order reduction technique is to determine least-order combinations of a left nullspace basis vectors, which satisfy additional fault detectability conditions (see Sect. 7.5). In this case, we deal with a homogeneous equation $Q(\lambda)G(\lambda) = 0$ and find a suitable fault detection filter $Q(\lambda)$ in the form

$$Q(\lambda) = H N_l(\lambda) + Y(\lambda)N_l(\lambda), \tag{10.96}$$

where $N_l(\lambda)$ is a proper rational left nullspace basis of $G(\lambda)$ and H is a constant matrix (to be appropriately selected to fulfil the fault detectability condition). Assuming $N_l(\lambda)$ has the observable descriptor realization

$$N_l(\lambda) = \left[\begin{array}{c|c} A_l - \lambda E_l & B_l \\ \hline C_l & D_l \end{array}\right],$$

this leads to a *dual* problem to be solved in Sect. 7.5, which involves an observable pair $(A_l - \lambda E_l, \widetilde{C}_l)$ with invertible E_l and with a $\widetilde{C}_l$ matrix partitioned as

$$\widetilde{C}_l = \begin{bmatrix} HC_l \\ C_l \end{bmatrix}.$$

In this case, a matrix K is sought such that the pair $(A_l + KC_l - \lambda E_l, HC_l)$ is *maximally unobservable*. For this purpose, the algorithm described in this section can be applied to the controllable pair $(A_l^T - \lambda E_l^T, [\, HC_l^T \; C_l^T \,])$ to determine a suitable "state feedback" K^T, which cancels the maximum number of uncontrollable eigenvalues.

We denote $\overline{A} = E^{-1}A$, $\overline{B}_1 = E^{-1}B_1$, $\overline{B}_2 = E^{-1}B_2$, and also and $\overline{\mathcal{B}}_1 = \operatorname{span} \overline{B}_1$ and $\overline{\mathcal{B}}_2 = \operatorname{span} \overline{B}_2$. The problem to determine F which makes the pair $(A + B_2F - \lambda E, B_1)$ maximally uncontrollable is equivalent [175] to compute a subspace $\mathcal{V}$ of least possible dimension satisfying

$$(\overline{A} + \overline{B}_2 F)\mathcal{V} \subset \mathcal{V}, \quad \overline{\mathcal{B}}_1 \subset \mathcal{V}. \tag{10.97}$$

This subspace is the least-order $(\overline{A}, \overline{B}_2)$-invariant subspace which contains $\overline{\mathcal{B}}_1$ [175]. The above condition can be equivalently rewritten as a condition defining $\mathcal{V}$ as a *Type I* minimum dynamic cover [43, 74]

$$\overline{A}\mathcal{V} \subset \mathcal{V} + \overline{\mathcal{B}}_2, \quad \overline{\mathcal{B}}_1 \subset \mathcal{V}. \tag{10.98}$$

In this section we describe a computational method for determining minimal dynamic covers, which relies on the reduction of the descriptor system pair $(A - \lambda E, [B_1, B_2])$ to a particular condensed form, for which the solution of the problem (i.e., the choice of appropriate F) is simple. This reduction is performed in two stages. The first stage is the orthogonal reduction performed with the **Procedure GSCSF** presented in Sect. 10.4.1. In the second stage, additional zero blocks are generated in the reduced matrices using non-orthogonal transformations. With additional blocks zeroed via a specially chosen state feedback F, the least-order $(\overline{A}, \overline{B}_2)$-invariant subspace containing $\overline{\mathcal{B}}_1$ can be identified as the linear span of the leading columns of the resulting right transformation matrix. In what follows we present in detail the second reduction stage as well as the determination of F.

We assume that after performing the **Procedure GSCSF**, we obtained the orthogonal transformation matrices Q and Z, such that the transformed system triple

$$(\widehat{A} - \lambda\widehat{E}, [\, \widehat{B}_1 \; \widehat{B}_2 \,], \widehat{C}) := (Q^TAZ - \lambda Q^TEZ, [\, Q^TB_1 \; Q^TB_2 \,], CZ) \tag{10.99}$$

has the pair $(\widehat{A}, \widehat{B})$, with $\widehat{B} = [\, \widehat{B}_1 \; \widehat{B}_2, \,]$, in the staircase form (10.89) and the matrix $\widehat{E}$ in the block structured form (10.90). The dimensions of the first 2ℓ diagonal blocks of $\widehat{A}$ and $\widehat{E}$ are determined by the two sets of dimensions $\nu_{1,j}$ and $\nu_{2,j}$ for $j = 1, \ldots, \ell$, and define the dimensions $n_1 := \sum_{j=1}^{\ell} \nu_{1,j}$ and $n_2 := \sum_{j=1}^{\ell} \nu_{2,j}$. Additionally, partition the columns of the resulting $\widehat{C}$ in accordance with the column structure of $\widehat{A}$ in (10.89)

$$\widehat{C} = [\, C_1 \; C_2 \cdots C_{2\ell-1} \; C_{2\ell} \,]. \tag{10.100}$$

In the second reduction stage we use non-orthogonal upper triangular left and right transformation matrices W and U, respectively, to annihilate the minimum number of blocks in $\widehat{A}$ and $\widehat{E}$ which allows to solve the minimum cover problem. Assume W and U have block structures identical to $\widehat{E}$. By exploiting the full rank of submatrices $A_{2k,2k-2}$ we can introduce zero blocks in the block row $2k$ of $\widehat{A}$ by annihilating the blocks $A_{2k,2j-1}$, for $j = k, k+1, \ldots, \ell$. Similarly, by exploiting the invertibility of $E_{2j-1,2j-1}$, we can introduce zero blocks in the block row $2k - 2$ of E by annihilating the blocks $E_{2k-2,2j-1}$, for $j = k, k+1, \ldots, \ell$ of $\widehat{E}$. This computation is performed for $k = \ell, \ell - 1, \ldots, 2$. Let $\widetilde{A} := W\widehat{A}U$, $\widetilde{E} := W\widehat{E}U$, $[\widetilde{B}_1 \ \widetilde{B}_2] := W[\widehat{B}_1 \ \widehat{B}_2] = [\widehat{B}_1 \ \widehat{B}_2]$, and $\widetilde{C} = \widehat{C}U$ be the system matrices resulted after this (non-orthogonal) reduction. Define also the feedback matrix $\widetilde{F} \in \mathbb{R}^{m_2 \times n}$ partitioned column-wise compatibly with $\widehat{A}$

$$\widetilde{F} = [\, F_1 \ 0 \ F_3 \ \cdots \ 0 \ F_{2\ell-1} \ 0 \,],$$

where $F_{2j-1} \in \mathbb{R}^{m_2 \times \nu_{1,j}}$ are such that $A_{2,0}F_{2j-1} + A_{2,2j-1} = 0$ for $j = 1, \ldots, \ell$. With this feedback, we introduced ℓ zero blocks in the second block row of $\widetilde{A} + \widetilde{B}_2\widetilde{F}$. Finally, consider the permutation matrix defined by

$$P = \begin{bmatrix} I_{\nu_{1,1}} & 0 & 0 & 0 & \cdots & 0 & 0 \\ 0 & 0 & I_{\nu_{1,2}} & 0 & \cdots & 0 & 0 \\ \vdots & \vdots & \vdots & \vdots & \ddots & \vdots & \vdots \\ 0 & 0 & 0 & 0 & \cdots & I_{\nu_{1,\ell}} & 0 \\ 0 & I_{\nu_{2,1}} & 0 & 0 & \cdots & 0 & 0 \\ 0 & 0 & 0 & I_{\nu_{2,2}} & \cdots & 0 & 0 \\ \vdots & \vdots & \vdots & \vdots & \ddots & \vdots & \vdots \\ 0 & 0 & 0 & 0 & \cdots & 0 & I_{\nu_{2,\ell}} \end{bmatrix}. \tag{10.101}$$

If we define $L = PWQ^T$, $V = ZUP^T$ and $F = \widetilde{F}V^{-1}$, then overall we obtained the reduced system $(\breve{A} - \lambda\breve{E}, [\breve{B}_1 \ \breve{B}_2], \breve{C}, [D_1 \ D_2])$ defined with

$$\breve{A} - \lambda\breve{E} := L(A + B_2F - \lambda E)V = \left[\begin{array}{c|c} \breve{A}_{11} - \lambda\breve{E}_{11} & \breve{A}_{12} - \lambda\breve{E}_{12} \\ \hline 0 & \breve{A}_{22} - \lambda\breve{E}_{22} \end{array}\right],$$

$$[\breve{B}_1 \,|\, \breve{B}_2] := L[B_1 \,|\, B_2] = \left[\begin{array}{c|c} \breve{B}_{11} & \breve{B}_{12} \\ \hline 0 & \breve{B}_{22} \end{array}\right], \tag{10.102}$$

$$\breve{C} := (C + D_2F)V = [\breve{C}_1 \,|\, \breve{C}_2],$$

where, by construction, the pairs $(\breve{A}_{11} - \lambda \breve{E}_{11}, \breve{B}_{11})$ and $(\breve{A}_{22} - \lambda \breve{E}_{22}, \breve{B}_{22})$ are in controllable staircase form. Thus, by the above choice of F, we made n_2 of the n eigenvalues of the pencil $A + B_2 F - \lambda E$ uncontrollable via B_1. It is straightforward to show that the matrix V_1 formed from the the first n_1 columns of V satisfies

$$\overline{A} V_1 = V_1 \breve{E}_{11}^{-1} \breve{A}_{11} - \overline{B}_2 F V_1, \quad \overline{B}_1 = V_1 \breve{E}_{11}^{-1} \breve{B}_{11}.$$

Thus, according to (10.98), $\mathcal{V} := \operatorname{span} V_1$ is a dynamic cover of *Type I* of dimension n_1. It can be shown using the results of [74] that $\mathcal{V}$ has minimum dimension.

To illustrate the computational procedure, we consider the reduced system in Example 10.1. First, the following zero blocks are introduced: A_{65}, E_{45}, A_{43}, A_{45}, E_{23}, E_{25} (in this order). The resulting $\widetilde{A}$ and $\widetilde{E}$ are

$$\widetilde{A} = \begin{bmatrix} A_{11} & A_{12} & A_{13} & A_{14} & A_{15} & A_{16} \\ A_{21} & A_{22} & A_{23} & A_{24} & A_{25} & A_{26} \\ A_{31} & A_{32} & A_{33} & A_{34} & A_{35} & A_{36} \\ 0 & A_{42} & 0 & A_{44} & 0 & A_{46} \\ 0 & 0 & A_{53} & A_{54} & A_{55} & A_{56} \\ 0 & 0 & 0 & A_{64} & 0 & A_{66} \end{bmatrix}, \quad \widetilde{E} = \begin{bmatrix} E_{11} & E_{12} & E_{13} & E_{14} & E_{15} & E_{16} \\ 0 & E_{22} & 0 & E_{24} & 0 & E_{26} \\ 0 & 0 & E_{33} & E_{34} & E_{35} & E_{36} \\ 0 & 0 & 0 & E_{44} & 0 & E_{46} \\ 0 & 0 & 0 & 0 & E_{55} & E_{56} \\ 0 & 0 & 0 & 0 & 0 & E_{66} \end{bmatrix}.$$

Additional blocks are zeroed using the feedback $\widetilde{F}$ to obtain

$$\widetilde{A} + \widetilde{B}_2 \widetilde{F} = \begin{bmatrix} A_{11} & A_{12} & A_{13} & A_{14} & A_{15} & A_{16} \\ 0 & A_{22} & 0 & A_{24} & 0 & A_{26} \\ A_{31} & A_{32} & A_{33} & A_{34} & A_{35} & A_{36} \\ 0 & A_{42} & 0 & A_{44} & 0 & A_{46} \\ 0 & 0 & A_{53} & A_{54} & A_{55} & A_{56} \\ 0 & 0 & 0 & A_{64} & 0 & A_{66} \end{bmatrix}.$$

Finally, after block permutations, we obtained the controllable staircase forms

$$\left[\, \breve{B}_{11} \,\middle|\, \breve{A}_{11} - \lambda \breve{E}_{11} \,\right] = \left[\begin{array}{c|ccc} A_{1,-1} & A_{11} - \lambda E_{11} & A_{13} - \lambda E_{13} & A_{15} - \lambda E_{15} \\ 0 & A_{31} & A_{33} - \lambda E_{33} & A_{35} - \lambda E_{35} \\ 0 & 0 & A_{53} & A_{55} - \lambda E_{55} \end{array}\right],$$

$$\left[\, \breve{B}_{22} \,\middle|\, \breve{A}_{22} - \lambda \breve{E}_{22} \,\right] = \left[\begin{array}{c|ccc} A_{2,0} & A_{22} - \lambda E_{22} & A_{24} - \lambda E_{24} & A_{26} - \lambda E_{26} \\ 0 & A_{42} & A_{44} - \lambda E_{44} & A_{46} - \lambda E_{46} \\ 0 & 0 & A_{64} & A_{66} - \lambda E_{66} \end{array}\right].$$

The above approach to compute a minimum dynamic cover of Type I is the basis of **Procedure GRMCOVER1**, presented in what follows. This procedure determines, for a pair of generators $(X_1(\lambda), X_2(\lambda))$ with the descriptor realizations given in (10.92), explicit minimal realizations for $\widetilde{X}_1(\lambda)$ and $Y(\lambda)$ (see (10.93) and (10.95)) in the form

$$\widetilde{X}_1(\lambda) = \left[\begin{array}{c|c} \lambda \breve{E}_{11} - \breve{A}_{11} & \breve{B}_{11} \\ \hline \breve{C}_1 & D_1 \end{array}\right], \qquad Y(\lambda) = \left[\begin{array}{c|c} \lambda \breve{E}_{11} - \breve{A}_{11} & \breve{B}_{11} \\ \hline \breve{F}_1 & 0 \end{array}\right],$$

where $\widetilde{F} P^T =: \left[\,\breve{F}_1 \;\; \breve{F}_2\,\right]$, with $\breve{F}_1$ having n_1 columns.

Procedure GRMCOVER1: Order reduction using dynamic covers of Type I

Inputs : $X_1(\lambda) = (A - \lambda E, B_1, C, D_1)$ and $X_2(\lambda) = (A - \lambda E, B_2, C, D_2)$
Outputs: $\widetilde{X}_1(\lambda) = (\breve{A}_{11} - \lambda \breve{E}_{11}, \breve{B}_{11}, \breve{C}_1, D_1)$ and $Y(\lambda) = (\breve{A}_{11} - \lambda \breve{E}_{11}, \breve{B}_{11}, \breve{F}_1, 0)$ such that
$\qquad \widetilde{X}_1(\lambda) = X_1(\lambda) + X_2(\lambda)Y(\lambda)$ has least McMillan degree.

(1) Apply **Procedure GSCSF** to the system triple $(A - \lambda E, [\,B_1\ B_2\,], C)$ to
determine the orthogonally similar system triple $(\widehat{A} - \lambda \widehat{E}, [\,\widehat{B}_1\ \widehat{B}_2\,], \widehat{C})$ defined
in (10.99) and (10.100), and the dimensions $\nu_{1,j}$ and $\nu_{2,j}$ for $j = 1, \ldots, \ell$;
set $n_1 := \sum_{j=1}^{\ell} \nu_{1,j}$.

(2) With $\widehat{A}$ partitioned as in (10.89) and $\widehat{E}$ partitioned as in (10.90), perform the second stage of the
special reduction for Type I covers:

Set $W = I$, $U = I$, and partition W and U in blocks analogous to $\widehat{E}$ in (10.90).

for $k = \ell, \ell - 1, \ldots, 2$

 Comment. Annihilate blocks $A_{2k,2j-1}$, for $j = k, k+1, \ldots, \ell$.
 for $j = k, k+1, \ldots, \ell$

  Compute $U_{2k-2,2j-1}$ such that $A_{2k,2k-2}U_{2k-2,2j-1} + A_{2k,2j-1} = 0$.
  $A_{i,2j-1} \leftarrow A_{i,2j-1} + A_{i,2k-2}U_{2k-2,2j-1}, i = 1, 2, \ldots, 2k$.
  $E_{i,2j-1} \leftarrow E_{i,2j-1} + E_{i,2k-2}U_{2k-2,2j-1}, i = 1, 2, \ldots, 2k - 2$.
  $C_{2j-1} \leftarrow C_{2j-1} + C_{2k-2}U_{2k-2,2j-1}$.
  $U_{i,2j-1} \leftarrow U_{i,2j-1} + U_{i,2k-2}U_{2k-2,2j-1}, i = 1, 2, \ldots, 2\ell$.

 end

 Comment. Annihilate blocks $E_{2k-2,2j-1}$, for $j = k, k+1, \ldots, \ell$.
 for $j = k, k+1, \ldots, \ell$

  Compute $W_{2k-2,2j-1}$ such that $W_{2k-2,2j-1}E_{2j-1,2j-1} + E_{2k-2,2j-1} = 0$.
  $A_{2k-2,i} \leftarrow A_{2k-2,i} + W_{2k-2,2j-1}A_{2j-1,i}, i = 2j - 2, 2j - 1, \ldots, 2\ell$.
  $E_{2k-2,i} \leftarrow E_{2k-2,i} + W_{2k-2,2j-1}E_{2j-1,i}, i = 2j, 2j + 1, \ldots, 2\ell$.
  $W_{2k-2,i} \leftarrow W_{2k-2,i} + W_{2k-2,2j-1}W_{2j-1,i}, i = 1, 2, \ldots, 2\ell$.

 end

end

Denote $\widetilde{A} - \lambda \widetilde{E} := W\widehat{A}U - \lambda W\widehat{E}U$, $[\,\widetilde{B}_1\ \widetilde{B}_2\,] := W[\,\widehat{B}_1\ \widehat{B}_2\,]$, $\widetilde{C} := \widehat{C}U$.

(3) Compute $\widetilde{F} = [\,F_1\ 0\ F_3\ \cdots\ 0\ F_{2\ell-1}\ 0\,]$, where $F_{2j-1} \in \mathbb{R}^{m_2 \times \nu_1^{(j)}}$ are such that
$A_{2,0}F_{2j-1} + A_{2,2j-1} = 0$ for $j = 1, \ldots, \ell$.

(4) With P defined in (10.101), compute $\breve{A} - \lambda \breve{E} = P(\widetilde{A} + \widetilde{B}_2\widetilde{F} - \lambda \widetilde{E})P^T$,
$\breve{B}_1 = P\widetilde{B}_1$, $\breve{C} = (\widetilde{C} + D_2\widetilde{F})P^T$ and $\breve{F} = \widetilde{F}P^T$.

(5) Set $\widetilde{X}_1(\lambda) = \big(\breve{A}(1\!:\!n_1, 1\!:\!n_1) - \lambda \breve{E}(1\!:\!n_1, 1\!:\!n_1), \breve{B}_1(1\!:\!n_1, :), \breve{C}(:, 1\!:\!n_1), D_1\big)$
and $Y(\lambda) = \big(\breve{A}(1\!:\!n_1, 1\!:\!n_1) - \lambda \breve{E}(1\!:\!n_1, 1\!:\!n_1), \breve{B}_1(1\!:\!n_1, :), \breve{F}(:, 1\!:\!n_1), 0\big)$.

As stated in Sect. 10.4.1, the reduction of system matrices to the special control-lability form at Step (1) can be performed by using exclusively orthogonal similarity transformations. It can be shown that the computed condensed matrices $\widehat{A}$, $\widehat{E}$, $\widehat{B}_1$, $\widehat{B}_2$ and $\widehat{C}$ are exact for matrices which are nearby to the original matrices A, E, B_1, B_2 and C respectively. Thus this part of the reduction is *numerically backward stable*.

The computations performed at Step (2), representing the second stage of the special reduction, and the computation of the feedback matrix $\widetilde{F}$ at Step (3) involve the solution of many, generally overdetermined, linear equations. Therefore, these steps are generally not numerically stable. In spite of this, the numerical reliability of the overall computations can be guaranteed, as long as W and U, the block upper triangular transformation matrices employed at Step (2), have no excessively large condition numbers. The condition numbers can be approximated as $\kappa(L) \approx \|W\|_F^2$ and $\kappa(V) \approx \|U\|_F^2$. It follows that if these norms are relatively small (e.g., ≤ 10000) then practically there is no danger for a significant loss of accuracy due to performing non-orthogonal reductions. On contrary, large values of these norms provide a clear hint of potential accuracy losses. In practice, it suffices only to look at the largest magnitudes of the generated elements of W and U at Step (2) to obtain equivalent information. For the computation of $\widetilde{F}$, condition numbers for solving the underlying equations can be also easily estimated. However, a large norm of $\widetilde{F}$ is an indication of possible accuracy losses. For Step (2) of the reduction, a simple operation count is possible by assuming all blocks are 1×1, and this indicates a computational complexity of $\mathcal{O}(n^3)$. Thus, the overall computational complexity of **Procedure GRMCOVER1** is also $\mathcal{O}(n^3)$.

10.4.3 *Order Reduction Using Minimum Dynamic Covers of Type II*

The computational problem which we address in this section is the following: given the descriptor system pair $(A - \lambda E, B)$ with $A, E \in \mathbb{R}^{n \times n}$ and E invertible, $B \in \mathbb{R}^{n \times m}$, and B partitioned as $B = [\, B_1 \; B_2 \,]$ with $B_1 \in \mathbb{R}^{n \times m_1}$, $B_2 \in \mathbb{R}^{n \times m_2}$, determine the matrices F and G such that the pair $(A + B_2 F - \lambda E, B_1 + B_2 G)$ has maximal number of uncontrollable eigenvalues.

This computation is useful to determine least-order solutions of linear rational equations using state-feedback and feedforward techniques. For the compatible linear rational system of equations $G(\lambda)X(\lambda) = F(\lambda)$, considered also in Sect. 10.4.2, assume there exists a particular solution $X_1(\lambda)$ which is proper. Then, the general solution can be expressed as in (10.91), where $X_2(\lambda)$ is a proper rational basis of the right nullspace of $G(\lambda)$. The proper TFMs $X_1(\lambda)$ and $X_2(\lambda)$ thus generate all solutions of $G(\lambda)X(\lambda) = F(\lambda)$. Assume $X_1(\lambda)$ and $X_2(\lambda)$ have the controllable descriptor realizations in (10.92) with invertible E. Let F be a state-feedback gain and let G be a feedforward gain. Then, the TFMs defined as

$$\left[\, \widetilde{X}_1(\lambda)\ \widetilde{X}_2(\lambda) \,\right] := \left[\begin{array}{c|cc} A + B_2 F - \lambda E & B_1 + B_2 G & B_2 \\ \hline C + D_2 F & D_1 + D_2 G & D_2 \end{array}\right] \qquad (10.103)$$

generate also all solutions. It is straightforward to check that

$$\widetilde{X}_1(\lambda) = X_1(\lambda) + X_2(\lambda)Y(\lambda), \quad \widetilde{X}_2(\lambda) = X_2(\lambda)\widetilde{Y}(\lambda), \tag{10.104}$$

where $Y(\lambda)$ and $\widetilde{Y}(\lambda)$ have the descriptor system realizations

$$\left[\, Y(\lambda)\ \widetilde{Y}(\lambda)\,\right] = \left[\begin{array}{c|cc} A + B_2 F - \lambda E & B_1 + B_2 G & B_2 \\ \hline F & G & I \end{array}\right]. \tag{10.105}$$

It follows that $\widetilde{X}_1(\lambda)$ is another particular solution, while $\widetilde{X}_2(\lambda)$ is another right nullspace basis, because $\widetilde{Y}(\lambda)$ is invertible. If the gains F and G are determined such that the pair $(A + B_2 F - \lambda E, B_1 + B_2 G)$ is maximally uncontrollable, then the resulting realizations of $\widetilde{X}_1(\lambda)$ and $Y(\lambda)$ contain a maximum number of uncontrollable eigenvalues which can be eliminated using minimal realization techniques. Thus, $\widetilde{X}_1(\lambda)$ represents another particular solution with a reduced McMillan degree. An important application of the above order reduction technique addressed in Sect. 7.9 is to determine a least-order solution of the EMMP by solving a dual linear rational equation $G(\lambda) = X(\lambda)H(\lambda)$ using the techniques presented in Sect. 10.3.7.

The problem to determine the matrices F and G, which make the descriptor system pair $(A + B_2 F - \lambda E, B_1 + B_2 G)$ maximally uncontrollable, is essentially equivalent [93] to compute a subspace $\mathcal{V}$ having least possible dimension and satisfying

$$(\overline{A} + \overline{B}_2 F)\mathcal{V} \subset \mathcal{V}, \quad \mathrm{span}\,(\overline{B}_1 + \overline{B}_2 G) \subset \mathcal{V}, \tag{10.106}$$

where $\overline{A} = E^{-1}A$, $\overline{B}_1 = E^{-1}B_1$ and $\overline{B}_2 = E^{-1}B_2$. If we denote $\overline{\mathcal{B}}_1 = \mathrm{span}\,\overline{B}_1$ and $\overline{\mathcal{B}}_2 = \mathrm{span}\,\overline{B}_2$, then the above condition can be equivalently rewritten also as a condition defining a *Type II* minimum dynamic cover [43, 74] of the form

$$\overline{A}\mathcal{V} \subset \mathcal{V} + \overline{\mathcal{B}}_2, \quad \overline{\mathcal{B}}_1 \subset \mathcal{V} + \overline{\mathcal{B}}_2. \tag{10.107}$$

The computation of the minimal dynamic covers of Type II can be done in two stages using a similar technique as for the Type I covers presented in Sect. 10.4.2. The first stage is identical to the reduction performed for covers of Type I and is performed using **Procedure GSCSF**. Two orthogonal transformation matrices Q and Z are determined, such that the transformed system triple

$$(\widehat{A} - \lambda\widehat{E}, [\,\widehat{B}_2\ \widehat{B}_1\,], \widehat{C}) := (Q^T A Z - \lambda Q^T E Z, [\,Q^T B_2\ Q^T B_1\,], CZ) \tag{10.108}$$

has the pair $(\widehat{A}, \widehat{B})$, with $\widehat{B} = [\,\widehat{B}_2\ \widehat{B}_1,\,]$, in the staircase form (10.89) and the matrix $\widehat{E}$ in the block structured form (10.90). The dimensions of the first 2ℓ diagonal blocks of $\widehat{A}$ and $\widehat{E}$ are determined by the two sets of dimensions $\nu_{1,j}$ and $\nu_{2,j}$ for $j = 1, \ldots, \ell$, and define the dimensions $n_1 := \sum_{j=1}^{\ell} \nu_{1,j}$ and $n_2 := \sum_{j=1}^{\ell} \nu_{2,j}$. Additionally, partition the columns of the resulting $\widehat{C}$ in accordance with the column structure of $\widehat{A}$ in (10.89)

$$\widehat{C} = [\,C_1\ C_2\ \cdots\ C_{2\ell-1}\ C_{2\ell}\,]. \tag{10.109}$$

In the second reduction stage we use non-orthogonal upper triangular left and right transformation matrices W and U, respectively, to annihilate the minimum number of blocks in $\widehat{A}$ and $\widehat{E}$ which allows to solve the minimum cover problem. Assume W and U have block structures identical to $\widehat{E}$. By exploiting the invertibility of the diagonal blocks $E_{2j,2j}$, we can introduce zero blocks in the block row $2k-1$ of E by annihilating the blocks $E_{2k-1,2j}$, for $j = k, k+1, \ldots, \ell$ of $\widehat{E}$. Similarly, by exploiting the full rank of submatrices $A_{2k-1,2k-3}$, we can introduce zero blocks in the block row $2k-1$ of $\widehat{A}$ by annihilating the blocks $A_{2k-1,2j}$, for $j = k-1, k, \ldots, \ell$. Let $\widetilde{A} := W\widehat{A}U$, $\widetilde{E} := W\widehat{E}U$, $[\,\widetilde{B}_2\ \widetilde{B}_1\,] := W[\,\widehat{B}_2\ \widehat{B}_1\,] = [\,\widehat{B}_2\ \widehat{B}_1\,]$, and $\widetilde{C} = \widehat{C}U$ be the system matrices resulted after this (non-orthogonal) reduction.

Choose the feedforward matrix $G \in \mathbb{R}^{m_2 \times m_1}$ such that $A_{1,-1}G + A_{1,0} = 0$ and the feedback matrix $\widetilde{F} \in \mathbb{R}^{m_2 \times n}$ partitioned column-wise compatibly with $\widehat{E}$ as

$$\widetilde{F} = [\,0\ F_2\ \cdots\ F_{2\ell-2}\ 0\ F_{2\ell}\ 0\,],$$

where F_{2j} are such that $A_{1,-1}F_{2j} + A_{1,2j} = 0$ for $j = 1, \ldots, \ell$. With the permutation matrix

$$P = \left[\begin{array}{cc|cc|c|cc}
0 & I_{v_{2,1}} & 0 & 0 & \cdots & 0 & 0 \\
\hline
0 & 0 & 0 & I_{v_{2,2}} & \cdots & 0 & 0 \\
\hline
\vdots & \vdots & \vdots & \vdots & \ddots & \vdots & \vdots \\
\hline
0 & 0 & 0 & 0 & \cdots & 0 & I_{v_{2,\ell}} \\
\hline
I_{v_{1,1}} & 0 & 0 & 0 & \cdots & 0 & 0 \\
\hline
0 & 0 & I_{v_{1,2}} & 0 & \cdots & 0 & 0 \\
\hline
\vdots & \vdots & \vdots & \vdots & \ddots & \vdots & \vdots \\
\hline
0 & 0 & 0 & 0 & \cdots & I_{v_{1,\ell}} & 0
\end{array}\right], \tag{10.110}$$

we define $L = PWQ^T$, $V = ZUP^T$ and $F = \widetilde{F}V^{-1}$. Overall we obtain the reduced system $(\breve{A} - \lambda\breve{E}, [\,\breve{B}_2\ \breve{B}_1\,], \breve{C}, [\,\breve{D}_2\ \breve{D}_1\,])$ defined with

$$\breve{A} - \lambda\breve{E} := L(A + B_2F - \lambda E)V = \left[\begin{array}{c|c} \breve{A}_{11} - \lambda\breve{E}_{11} & \breve{A}_{12} - \lambda\breve{E}_{12} \\ \hline 0 & \breve{A}_{22} - \lambda\breve{E}_{22} \end{array}\right],$$

$$[\,\breve{B}_2\,|\,\breve{B}_1\,] := L\,[\,B_2\,|\,B_1 + B_2G\,] = \left[\begin{array}{c|c} 0 & \breve{B}_{12} \\ \hline \breve{B}_{21} & 0 \end{array}\right], \tag{10.111}$$

$$\breve{C} := (C + D_2F)V = [\,\breve{C}_1\,|\,\breve{C}_2\,],$$

$$[\,\breve{D}_2\,|\,\breve{D}_1\,] := [\,D_2\,|\,D_1 + D_2G\,],$$

where, by construction, the pairs $(\breve{A}_{11} - \lambda\breve{E}_{11}, \breve{B}_{12})$ and $(\breve{A}_{22} - \lambda\breve{E}_{22}, \breve{B}_{21})$ are in controllable staircase form. Thus, by the above choice of F and G, we made n_1

of eigenvalues of the pair $(A + B_2 F - \lambda E, B_1 + B_2 G)$ uncontrollable. The first n_2 columns V_1 of V satisfy

$$\overline{A} V_1 = V_1 \check{E}_{11}^{-1} \check{A}_{11} - \overline{B}_2 F V_1, \quad \overline{B}_2 G = V_1 \check{E}_{11}^{-1} \check{B}_{12} - \overline{B}_1$$

and thus, according to (10.107), span a *Type II* dynamic cover of dimension n_2 for the pair $(\overline{A}, [\overline{B}_1 \ \overline{B}_2])$. It can be shown using the results of [74] that the resulting *Type II* dynamic cover $\mathcal{V}$ has minimum dimension.

To illustrate the computational procedure, we consider the reduced system in Example 10.1. First, the following zero blocks are introduced: E_{56}, A_{54}, A_{56}, E_{34}, E_{36}, $A_{3,2}$, A_{34}, A_{36}, E_{12}, E_{14} and E_{16} (in this order). We obtain

$$[\tilde{B}_2 \ \tilde{B}_1 \mid \tilde{A}] = \begin{bmatrix} A_{1,-1} & A_{1,0} & A_{11} & A_{12} & A_{13} & A_{14} & A_{15} & A_{16} \\ 0 & A_{2,0} & A_{2,1} & A_{22} & A_{2,3} & A_{24} & A_{2,5} & A_{26} \\ 0 & 0 & A_{31} & 0 & A_{33} & 0 & A_{35} & 0 \\ 0 & 0 & 0 & A_{42} & A_{43} & A_{44} & A_{45} & A_{46} \\ 0 & 0 & 0 & 0 & A_{53} & 0 & A_{55} & 0 \\ 0 & 0 & 0 & 0 & 0 & A_{64} & A_{65} & A_{66} \end{bmatrix},$$

$$\tilde{E} = \begin{bmatrix} E_{11} & 0 & E_{13} & 0 & E_{15} & 0 \\ 0 & E_{22} & E_{23} & E_{24} & E_{25} & E_{26} \\ 0 & 0 & E_{33} & 0 & E_{35} & 0 \\ 0 & 0 & 0 & E_{44} & E_{45} & E_{46} \\ 0 & 0 & 0 & 0 & E_{55} & 0 \\ 0 & 0 & 0 & 0 & 0 & E_{66} \end{bmatrix}.$$

Additional blocks are zeroed using the feedback $\tilde{F}$ and feedforward gain G to obtain

$$[\tilde{B}_2 \ \tilde{B}_1 + \tilde{B}_2 G \mid \tilde{A} + \tilde{B}_2 \tilde{F}] = \begin{bmatrix} A_{1,-1} & 0 & A_{11} & 0 & A_{13} & 0 & A_{15} & 0 \\ 0 & A_{2,0} & A_{2,1} & A_{22} & A_{2,3} & A_{24} & A_{2,5} & A_{26} \\ 0 & 0 & A_{31} & 0 & A_{33} & 0 & A_{35} & 0 \\ 0 & 0 & 0 & A_{42} & A_{43} & A_{44} & A_{45} & A_{46} \\ 0 & 0 & 0 & 0 & A_{53} & 0 & A_{55} & 0 \\ 0 & 0 & 0 & 0 & 0 & A_{64} & A_{65} & A_{66} \end{bmatrix}.$$

Finally, after block permutations, we obtained the controllable staircase forms

$$[\check{B}_1 \mid \check{A}_1 - \lambda \check{E}_1] = \begin{bmatrix} A_{2,0} & A_{2,2} - \lambda E_{2,2} & A_{2,4} - \lambda E_{2,4} & A_{2,6} - \lambda E_{2,6} \\ 0 & A_{4,2} & A_{4,4} - \lambda E_{4,4} & A_{4,6} - \lambda E_{4,6} \\ 0 & 0 & A_{6,4} & A_{6,6} - \lambda E_{6,6} \end{bmatrix},$$

$$[\check{B}_2 \mid \check{A}_2 - \lambda \check{E}_2] = \begin{bmatrix} A_{1,-1} & A_{1,1} - \lambda E_{1,1} & A_{1,3} - \lambda E_{1,3} & A_{1,5} - \lambda E_{1,5} \\ 0 & A_{3,1} & A_{3,3} - \lambda E_{3,3} & A_{3,5} - \lambda E_{3,5} \\ 0 & 0 & A_{5,3} & A_{5,5} - \lambda E_{5,5} \end{bmatrix}.$$

The above approach to compute a minimum dynamic cover of Type II is the basis of **Procedure GRMCOVER2** presented in what follows. This procedure determines, for a pair of generators $(X_1(\lambda), X_2(\lambda))$ with the descriptor realizations given in (10.92), explicit minimal realizations for $\widetilde{X}_1(\lambda)$ and $Y(\lambda)$ (see (10.103) and (10.105))) in the form $\widetilde{X}_1(\lambda) = (\breve{A}_{11} - \lambda \breve{E}_{11}, \breve{B}_{12}, \breve{C}_1, \breve{D}_1)$ and $Y(\lambda) = (\breve{A}_{11} - \lambda \breve{E}_{11}, \breve{B}_{12}, \breve{F}_1, G)$, where $\widetilde{F} P^T =: \begin{bmatrix} \breve{F}_1 & \breve{F}_2 \end{bmatrix}$, with $\breve{F}_1$ having n_2 columns.

Procedure GRMCOVER2: Order reduction using dynamic covers of Type II

Inputs : $X_1(\lambda) = (A - \lambda E, B_1, C, D_1)$ and $X_2(\lambda) = (A - \lambda E, B_2, C, D_2)$
Outputs: $\widetilde{X}_1(\lambda) = (\breve{A}_{11} - \lambda \breve{E}_{11}, \breve{B}_{12}, \breve{C}_1, \breve{D}_1)$ and $Y(\lambda) = (\breve{A}_{11} - \lambda \breve{E}_{11}, \breve{B}_{12}, \breve{F}_1, G)$ such that
$\qquad \widetilde{X}_1(\lambda) = X_1(\lambda) + X_2(\lambda)Y(\lambda)$ has least McMillan degree.

(1) Apply **Procedure GSCSF** to the system triple $(A - \lambda E, [\, B_2 \ B_1 \,], C)$ to
determine the orthogonally similar system triple $(\widehat{A} - \lambda \widehat{E}, [\, \widehat{B}_2 \ \widehat{B}_1 \,], \widehat{C})$ defined
in (10.108) and (10.109), and the dimensions $\nu_{1,j}$ and $\nu_{2,j}$ for $j = 1, \ldots, \ell$;
set $n_2 := \sum_{j=1}^{\ell} \nu_{2,j}$.

(2) With $\widehat{A}$ partitioned as in (10.89) and $\widehat{E}$ partitioned as in (10.90), perform the second stage
of the special reduction for Type II covers:

    Set $W = I_n$, $U = I_n$, and partition W and U in blocks analogous to $\widehat{E}$ in (10.90).

    for $k = \ell, \ell - 1, \ldots, 1$

        Comment. Annihilate blocks $E_{2k-1,2j}$, for $j = k, k+1, \ldots, \ell$.
        for $j = k, k+1, \ldots, \ell$

            Compute $W_{2k-1,2j}$ such that $W_{2k-1,2j} E_{2j,2j} + E_{2k-1,2j} = 0$.
            $A_{2k-1,i} \leftarrow A_{2k-1,i} + W_{2k-1,2j} A_{2j,i}, i = 2j - 2, 2j - 1, \ldots, 2\ell$.
            $E_{2k-1,i} \leftarrow E_{2k-1,i} + W_{2k-1,2j} E_{2j,i}, i = 2j, 2j + 1, \ldots, 2\ell$.
            $W_{2k-1,i} \leftarrow W_{2k-1,i} + W_{2k-1,2j} W_{2j,i}, i = 1, 2, \ldots, 2\ell$.

        end

        if $k > 1$ **then**

            Comment. Annihilate blocks $A_{2k-1,2j}$, for $j = k - 1, k, \ldots, \ell$.
            for $j = k - 1, k, \ldots, \ell$

                Compute $U_{2k-3,2j}$ such that $A_{2k-1,2k-3} U_{2k-3,2j} + A_{2k-1,2j} = 0$.
                $A_{i,2j} \leftarrow A_{i,2j} + A_{i,2k-3} U_{2k-3,2j}, i = 1, 2, \ldots, 2k - 1$.
                $E_{i,2j} \leftarrow E_{i,2j} + E_{i,2k-3} U_{2k-3,2j}, i = 1, 2, \ldots, 2k - 3$.
                $C_{2j} \leftarrow C_{2j} + C_{2k-3} U_{2k-3,2j}$.
                $U_{i,2j} \leftarrow U_{i,2j} + U_{i,2k-3} U_{2k-3,2j}, i = 1, 2, \ldots, 2\ell$.

            end

        end if

    end

    Denote $\widetilde{A} - \lambda \widetilde{E} = W \widehat{A} U - \lambda W \widehat{E} U$, $[\, \widetilde{B}_2 \ \widetilde{B}_1 \,] = W [\, \widehat{B}_2 \ \widehat{B}_1 \,]$, $\widetilde{C} = \widehat{C} U$.

(3) Compute $\widetilde{F} = [0 \ F_2 \ \cdots \ F_{2\ell-2} \ 0 \ F_{2\ell} \ 0]$, where F_{2j} are such that $A_{1,-1} F_{2j} + A_{1,2j} = 0$ for
$j = 1, \ldots, \ell$; compute G such that $A_{1,-1} G + A_{1,0} = 0$.

(4) With P in (10.110), compute $\breve{A} - \lambda \breve{E} = P(\widetilde{A} + \widetilde{B}_2 \widetilde{F} - \lambda \widetilde{E}) P^T$, $\breve{B}_1 = P(\widetilde{B}_1 + \widetilde{B}_2 G)$,
$\breve{C} = (\widetilde{C} + D_2 \widetilde{F}) P^T$, $\breve{D}_1 = D_1 + D_2 G$ and $\breve{F} = \widetilde{F} P^T$.

(5) Set $\widetilde{X}_1(\lambda) = \Big(\breve{A}(1:n_2, 1:n_2) - \lambda \breve{E}(1:n_2, 1:n_2), \breve{B}_1(1:n_2, :), \breve{C}(:, 1:n_2), \breve{D}_1 \Big)$
and $Y(\lambda) = \Big(\breve{A}(1:n_2, 1:n_2) - \lambda \breve{E}(1:n_2, 1:n_2), \breve{B}_1(1:n_2, :), \breve{F}(:, 1:n_2), G \Big)$.

The numerical properties of **Procedure GRMCOVER2** are the same as those of **Procedure GRMCOVER1**, which are discussed in Sect. 10.4.2.

10.4.4   *Minimal Realization Using Balancing Techniques*

The aim of the algorithm presented in this section is to determine minimal order realizations of stable systems in a descriptor state-space form, by exploiting the concept of balanced realization. For a balanced realization, the controllability and observability properties are perfectly equilibrated. This is expressed by the fact that the controllability and observability Gramians are equal and diagonal. The eigenvalues of the Gramian of a balanced system are called the *Hankel singular values*. The largest singular value represents the *Hankel norm* of the corresponding TFM of the system, while the smallest one can be interpreted as a measure of the nearness of the system to a non-minimal one. Important applications of balanced realizations are to ensure minimum sensitivity to roundoff errors of real-time filter models or to perform model order reduction, by reducing large order models to lower order approximations. The order reduction can be performed by simply truncating the system state to a part corresponding to the "large" singular values, which significantly exceed the rest of "small" singular values. In what follows we present a procedure to compute minimal balanced realizations of stable descriptor systems. This procedure is instrumental in solving the Nehari approximation problem (see **Procedure GNEHARI** in Sect. 10.4.5).

For a stable state-space system $(A - \lambda E, B, C, D)$ with E invertible, the controllability Gramian P and observability Gramian Q satisfy appropriate generalized Lyapunov equations. In the continuous-time case P and Q satisfy

$$\begin{aligned} APE^T + EPA^T + BB^T &= 0, \\ A^T QE + E^T QA + C^T C &= 0, \end{aligned} \tag{10.112}$$

while in the discrete-time case

$$\begin{aligned} APA^T - EPE^T + BB^T &= 0, \\ A^T QA - E^T QE + C^T C &= 0. \end{aligned} \tag{10.113}$$

Since for a stable system both Gramians P and Q are positive semi-definite matrices, in many applications it is advantageous to determine these matrices directly in (Cholesky) factored forms as $P = SS^T$ and $Q = R^T R$, where both S and R can be chosen upper triangular matrices. Algorithms to compute directly these factors have been proposed in [62] for standard systems (i.e., with $E = I$) and extended to descriptor systems in [108]. The following minimal realization procedure proposed in [119] extends to descriptor systems the algorithms proposed in [120] for standard systems. This procedure determines for a stable system $(A - \lambda E, B, C, D)$ the minimal balanced realization $(\widetilde{A} - \lambda I, \widetilde{B}, \widetilde{C}, D)$ and the corresponding balanced diagonal Gramian matrix $\widetilde{\Sigma}$. The nonzero Hankel singular values are the decreasingly ordered diagonal elements of $\widetilde{\Sigma}$ and the largest Hankel singular value is $\|G(\lambda)\|_H$, the Hankel norm of the corresponding TFM $G(\lambda) = C(\lambda E - A)^{-1}B + D$.

Procedure GBALMR: Balanced minimal realization of stable systems

Input   : $(A - \lambda E, B, C, D)$ such that $\Lambda(A, E) \subset \mathbb{C}_s$
Outputs: Minimal realization $(\widetilde{A}, \widetilde{B}, \widetilde{C}, D)$, $\widetilde{\Sigma}$

(1) Compute the upper triangular factors S and R such that $P = SS^T$ and $Q = R^T R$ satisfy the appropriate Lyapunov equations (10.112) or (10.113), in accordance with the system type, continuous- or discrete-time, respectively.

(2) Compute the singular value decomposition

$$RES = \begin{bmatrix} U_1 & U_2 \end{bmatrix} \begin{bmatrix} \widetilde{\Sigma} & 0 \\ 0 & 0 \end{bmatrix} \begin{bmatrix} V_1^T \\ V_2^T \end{bmatrix},$$

where $\widetilde{\Sigma} > 0$.

(3) With the projection matrices $T_l = \widetilde{\Sigma}^{-1/2} U_1^T R$ and $T_r = SV_1 \widetilde{\Sigma}^{-1/2}$, compute the matrices of the minimal realization $(\widetilde{A}, \widetilde{B}, \widetilde{C}, D)$ with

$$\widetilde{A} = T_l A T_r, \quad \widetilde{B} = T_l B, \quad \widetilde{C} = C T_r.$$

Remark 10.10 The projection matrices satisfy $T_l E T_r = I$ and for a minimal standard system (A, B, C, D) we have $T_l = T_r^{-1}$. The reduction of a linear state-space model to a balanced minimal realization may involve the usage of ill-conditioned transformations (or projections) for systems which are nearly non-minimal or nearly unstable. This is why, for the computation of minimal realizations, the so-called *balancing-free* approaches, as proposed in [132] for standard systems and in [119] for descriptor systems, are generally more accurate. In this case, we can avoid any inversion by using at Step (3) the projection matrices $T_l = U_1^T R$ and $T_r = SV_1$ to obtain the descriptor minimal realization $(\widetilde{A} - \lambda \widetilde{E}, \widetilde{B}, \widetilde{C}, D)$ with the invertible $\widetilde{E} = T_l E T_r$.
□

10.4.5 Solution of Nehari Problems

In this section we consider the solution of the following optimal Nehari problem: Given $R(\lambda)$ such that $R^\sim(\lambda) \in \mathcal{H}_\infty$, find a $Y(\lambda) \in \mathcal{H}_\infty$ which is the closest to $R(\lambda)$ and satisfies

$$\|R(\lambda) - Y(\lambda)\|_\infty = \|R^\sim(\lambda)\|_H. \tag{10.114}$$

This computation is encountered in the solution of the AMMP formulated in Sect. 9.1.10. As shown in [54], to solve the Nehari problem (10.114), we can solve instead for $Y^\sim(\lambda)$ the optimal zeroth-order Hankel-norm approximation problem

$$\|R^\sim(\lambda) - Y^\sim(\lambda)\|_\infty = \|R^\sim(\lambda)\|_H\,. \tag{10.115}$$

In what follows, we only give a solution procedure for the solution of (10.114) in the continuous-time setting. The corresponding procedure for discrete-time systems is much more involved (see [61]) and therefore we prefer the approach based on employing a bilinear transformation as suggested in [54]. To solve the continuous-time Nehari problem (10.114), we solve the optimal zeroth-order Hankel-norm approximation problem to determine $Y(-s)$ such that

$$\|R(-s) - Y(-s)\|_\infty = \|R^\sim(s) - Y^\sim(s)\|_\infty = \|R^\sim(s)\|_H\,. \tag{10.116}$$

The following procedure is a straightforward adaptation of the general Hankel-norm approximation procedure proposed in [54, 114] for *square* $R(\lambda)$ with poles only in $\mathbb{C}_u$. Assuming $(A - \lambda E, B, C, D)$ is a state-space realization of $R(\lambda)$ (not necessarily minimal), this procedure computes the optimal stable Nehari approximation $Y(\lambda) = (\widetilde{A} - \lambda\widetilde{E}, \widetilde{B}, \widetilde{C}, \widetilde{D})$.

Procedure GNEHARI: Generalized optimal Nehari approximation

Input : $R(\lambda) = (A - \lambda E, B, C, D)$
Output: $Y(\lambda) = (\widetilde{A} - \lambda\widetilde{E}, \widetilde{B}, \widetilde{C}, \widetilde{D})$ such that $\|R(\lambda) - Y(\lambda)\|_\infty = \|R^\sim(\lambda)\|_H$.

(1) For a discrete-time system employ the bilinear transformation $z = \frac{1+s}{1-s}$:

$$(E, A, B, C, D) \leftarrow (E + A,\, A - E,\, \sqrt{2}B,\, \sqrt{2}C(E + A)^{-1}E,\, D - C(E + A)^{-1}B)\,.$$

(2) Compute using the **Procedure GBALMR** the balanced minimal realization $(\widehat{A}, \widehat{B}, \widehat{C}, D)$ of the system $(-A - sE, -B, C, D)$ and the corresponding diagonal Gramian $\widehat{\Sigma}$ of the balanced system satisfying $\widehat{A}\widehat{\Sigma} + \widehat{\Sigma}\widehat{A}^T + \widehat{B}\widehat{B}^T = 0$ and $\widehat{A}^T\widehat{\Sigma} + \widehat{\Sigma}\widehat{A} + \widehat{C}^T\widehat{C} = 0$.

(3) Partition $\widehat{\Sigma}$ in the form $\widehat{\Sigma} = \mathrm{diag}(\sigma_1 I, \widehat{\Sigma}_2)$, such that $\widehat{\Sigma}_2 - \sigma_1 I < 0$ and partition $\widehat{A}$, $\widehat{B}$, and $\widehat{C}$ conformably with $\widehat{\Sigma}$, as

$$\widehat{A} = \begin{bmatrix} \widehat{A}_{11} & \widehat{A}_{12} \\ \widehat{A}_{21} & \widehat{A}_{22} \end{bmatrix}, \quad \widehat{B} = \begin{bmatrix} \widehat{B}_1 \\ \widehat{B}_2 \end{bmatrix}, \quad \widehat{C} = [\,\widehat{C}_1 \;\; \widehat{C}_2\,]\,;$$

compute an orthogonal U such that $U\widehat{B}_1^T = -\widehat{C}_1$.

(4) Compute the descriptor system realization $(\widetilde{A} - \lambda\widetilde{E}, \widetilde{B}, \widetilde{C}, \widetilde{D})$ of $Y(s)$ as

$$\begin{aligned}
\widetilde{E} &= \widehat{\Sigma}_2^2 - \sigma_1^2 I\,, \\
\widetilde{A} &= -(\sigma_1^2\widehat{A}_{22} + \widehat{\Sigma}_2\widehat{A}_{22}\widehat{\Sigma}_2 - \sigma_1\widehat{C}_2^T U\widehat{B}_2^T)\,, \\
\widetilde{B} &= -(\widehat{\Sigma}_2\widehat{B}_2 + \sigma_1\widehat{C}_2^T U)\,, \\
\widetilde{C} &= \widehat{C}_2\widehat{\Sigma}_2 + \sigma_1 U\widehat{B}_2^T\,, \\
\widetilde{D} &= D - \sigma_1 U\,.
\end{aligned}$$

(5) For a discrete-time system employ the bilinear transformation $s = \frac{z-1}{z+1}$:

$$(\widetilde{E}, \widetilde{A}, \widetilde{B}, \widetilde{C}, \widetilde{D}) \leftarrow (\widetilde{E} - \widetilde{A},\, \widetilde{A} + \widetilde{E},\, \sqrt{2}\widetilde{B},\, \sqrt{2}\widetilde{C}(\widetilde{E} - \widetilde{A})^{-1}\widetilde{E},\, \widetilde{D} + \widetilde{C}(\widetilde{E} - \widetilde{A})^{-1}\widetilde{B})\,.$$

Remark 10.11 If $R(\lambda)$ is not square, then the **Procedure GNEHARI** can be applied to an augmented square $R_a(\lambda)$ formed by adding a sufficient number of zero rows or columns to $R(\lambda)$. From the resulting solution $Y_a(\lambda)$, we obtain the solution $Y(\lambda)$ of the original Nehari problem by removing the rows or columns corresponding to the added zero rows or columns in $R_a(\lambda)$. $\square$

10.5 Software Tools

The development of dedicated software tools for the synthesis of fault detection filters must fulfil some general requirements for robust software implementation, but also some requirements specific to the field of fault detection. In what follows, we shortly discuss some of these requirements (see also Sect. A.3 of the Appendix for additional aspects).

A basic requirement is the use of general, numerically reliable and computationally efficient numerical approaches as basis for the implementation of all computational functions, to guarantee the solvability of problems under the most general existence conditions of the solutions. Synthesis procedures suitable for robust software implementations are described in this book (see also [161]).

For beginners, it is important to be able to easily obtain preliminary synthesis results. Therefore, for ease of use, simple and uniform user interfaces are desirable for all synthesis functions. This can be achieved by relying on meaningful default settings for all problem parameters and synthesis options. Also, the synthesis functions to solve approximate synthesis problems should be applicable to solve the exact synthesis problems as well. On the other side, the solution of an exact problem for a system with noise inputs should provide a first approximation to the solution of the approximate synthesis problem.

For experienced users, an important requirement is to provide an exhaustive set of options to ensure the complete freedom in choosing problem-specific parameters and synthesis options. Among the frequently used synthesis options we mention: the number of residual signal outputs; stability degree for the poles of the resulting filters or the location of their poles; type of the employed nullspace basis (e.g., minimal proper, full-order observer based); performing least-order synthesis, etc.

In what follows we present an overview of the existing tools for the two software environments for scientific and numerical computing MATLAB and Julia.

10.5.1 *MATLAB Tools*

MATLAB is a widely used commercial software within both industry and academics, which provides several advanced tools for the analysis, synthesis and simulation of control systems as well as for visualization of simulation results. The *Control*

System Toolbox of MATLAB [88] provides an object-oriented framework to handle LTI systems in descriptor system representation. This feature was instrumental for the implementation of **DSTools**, a freely available collection of descriptor systems tools, whose functions are based on the algorithms described in this book (and also in its first edition [161]). This toolbox covers the basic computational needs to implement the synthesis procedures presented in Chaps. 5 and 6. The basic numerical linear algebra support for the implementation of the **DSTools** collection is provided by several core functions of MATLAB based on the linear algebra package LAPACK [3], jointly with a set of *mex*-functions based on subroutines of SLICOT [12], a subroutine library for control theory, based primarily on the basic vector-matrix operations library BLAS [36, 37, 78] and LAPACK. Currently, both **DSTools**[1] and SLICOT[2] are freely distributed via **GitHub**. See [162] for a comprehensive documentation of **DSTools**.

Following the publication of the first edition [161], **FDITools**, a freely available collection of MATLAB functions, has been implemented for the analysis and solution of fault detection problems along the previously formulated requirements. **FDITools** supports the basic synthesis approaches of linear residual generation filters for both continuous-time and discrete-time LTI systems. The underlying synthesis techniques rely on reliable numerical algorithms developed by the author and are described in the Chaps. 5–7 of this book. The functions of the **FDITools** collection rely on the system objects of the *Control System Toolbox* and the computational tools available in **DSTools** for the manipulation of rational TFMs via their descriptor system representations. The version V1.0 of the **FDITools** collection covers all synthesis procedures described in this book and, additionally, includes several useful analysis functions, as well as functions for an easy setup of synthesis models. Currently, **FDITools**[3] is freely distributed via **GitHub**; see [163] for a comprehensive documentation of this toolbox. A precursor of **FDITools** was the *Fault Detection Toolbox* for MATLAB, a proprietary software of the German Aerospace Center (DLR), developed between 2004 and 2014 (see [144] for the status of this toolbox around 2006).

10.5.2   *Julia Tools*

Julia established in the recent years as a powerful, innovative and flexible environment, suitable for the development of high-performance scientific and numerical computing tools, whose performance is comparable to similar tools implemented using traditional statically-typed languages such as Fortran or C (see https://julialang.org/.). As a dynamic programming language, Julia features optional typing, multiple dispatch and good performance, achieved using type inference and just-in-time com-

---

[1] https://github.com/andreasvarga/DescriptorSystemTools.

[2] https://github.com/SLICOT/SLICOT-Reference/.

[3] https://github.com/andreasvarga/FDITools.

pilation [14]. Julia is the basis of the implementation of several thousands packages, which form a vast software ecosystem covering a large range of application domains.

A notable recent development is **FaultDetectionTools**,[4] a free Julia package, which implements the complete functionality of **FDITools** in the Julia language. The required computational ingredients for the implementation of the **FaultDetectionTools** are provided by three supporting packages implemented in Julia, namely: **DescriptorSystems**[5] for handling descriptor system representations; **MatrixEquations**,[6] for solving various control-related matrix equations (Lyapunov, Sylvester, Riccati); and **MatrixPencils**,[7] for computations with linear matrix pencils. The supporting packages have been implemented to fulfil the following, nowadays widely accepted basic requirements for the development of high-performance numerical software:

- employing exclusively numerically stable or numerically reliable algorithms;
- ensuring high computational efficiency;
- enforcing robustness against numerical exceptions (overflows, underflows) and poorly scaled data;
- ensuring ease-of-use, high portability and high reusability.

There are several important differences between the collections of tools implemented in MATLAB and Julia. A first notable difference is that while the MATLAB implementation relies on a set of *mex*-function interfaces to Fortran subroutines available in SLICOT, the Julia implementations employ exclusively tools implemented in the Julia language. This implementation strategy was facilitated by the possibility offered by Julia to call directly particular LAPACK routines via suitable wrappers which underlie the linear algebra functions. Also of interest is to mention that several new wrappers to LAPACK functions have been added using the flexible interface provided by Julia to the languages C and Fortran. As consequence of these facts, the achieved implementations run nearly at the speed of the equivalent Fortran implementations and provide an increased functional flexibility compared to the rigid *mex*-function interfaces in MATLAB. An additional benefit is that the Julia implementations are independent of various computer architectures and operating systems (e.g., Windows, Linux, macOS) and not constrained to employ a particular set of compilers (of which some are commercial!) to build the MATLAB *mex*-functions. As replacements of *mex*-functions, the **MatrixPencils** package includes implementations of functions to compute various Kronecker-like forms, minimal realizations of descriptor systems, additive decompositions of descriptor systems, conversion between descriptor system and rational matrix representations, or the computation of system poles and zeros (see [165] for other applications). Similarly, the **MatrixEquations** provides a set of functions to solve matrix equations like various Lyapunov, Sylvester or Riccati equations. Since all these tools are implemented in Julia, a programming

---

⁴ https://github.com/andreasvarga/FaultDetectionTools.jl.

⁵ https://github.com/andreasvarga/DescriptorSystems.jl.

⁶ https://github.com/andreasvarga/MatrixEquations.jl.

⁷ https://github.com/andreasvarga/MatrixPencils.jl.

language that is both easy to use and fast, this fact illustrates the capability of Julia to successfully address the technical computing challenge known as the *two language problem* [14].

Another important difference is the provided user interfaces for the main computational functions. The MATLAB implementation strongly benefited of the already defined system objects in the *Control System Toolbox* (e.g., to handle descriptor systems or rational transfer function matrices), the availability of several basic computational functions (e.g., for the solution of generalized Riccati equations, computation of system norms) and the existence of advanced simulation and visualization tools. In contrast, the implementation of **DescriptorSystems** in Julia involved the definition of suitable system objects to represent state-space systems in descriptor or standard form jointly with functions which allow to operate on such representations (e.g., to perform various system couplings). Also, for handling input–output representations, a rational transfer function object has been defined (based on the polynomial and rational function definitions available in the **Polynomials**[8] package). The functions implemented in **FaultDetectionTools** rely on a set of custom system objects with inputs grouped in several categories, named as **controls**, **disturbances**, **faults** and **noise**. While such input grouping is a standard functionality provided in the *Control System Toolbox* of MATLAB, each of the system objects defined in Julia (e.g., synthesis models with the above input groups, fault detection filters in implementation and internal forms, etc.) requires a specific handling. An interesting side effect of the custom object definitions in Julia is that the system data depend on a generic *type parameter*, say **T**, for which the standard setting is **T = Float64**, to support double precision floating-point computations in all functions. However, occasionally, particular settings as **T = BigFloat** or **T = Measurement{Float64}** (defined in the **Measurements**[9] package) are also supported by some functions, to allow computations with arbitrary precision as in Example 7.1 or manipulation and simulation of systems with parametric uncertainties as in Example 5.16. In this context, the *multiple dispatch* mechanism implemented in Julia is a key feature to ensure the genericity and composability of software implementations.

We must also mention the differences in the provided plotting capabilities of the synthesis results. The MATLAB implementations fully benefit of a versatile set of dedicated plotting tools such as for the visualization of step and general time responses or of various frequency responses. The generated figures have publication quality and served for the illustrations presented in the first edition of this book [161]. A similar plotting functionality is only available, to a certain extent, in the **ControlSystems**[10] package of Julia. This package covers some of the functionality of the *Control System Toolbox* of MATLAB and provides several dedicated plotting functions using **Plots**,[11] the standard plotting package of Julia. For the

---

[8] https://github.com/JuliaMath/Polynomials.jl.

[9] https://github.com/JuliaPhysics/Measurements.jl.

[10] https://github.com/JuliaControl/ControlSystems.jl.

[11] https://github.com/JuliaPlots/Plots.jl.

production of publication quality plots for this book, we relied on an alternative plotting package, **Makie**,[12] via its backend **CairoMakie**.

Finally, we have to mention the difference in the user's experience regarding the speed of executions when working with MATLAB and Julia. While MATLAB produces often immediately the computational results, Julia needs time at the first execution to compile all needed functions. However, subsequent executions are usually at maximum speed, because Julia is inherently fast. Here are some runtime results obtained with MATLAB 2022b under Windows 11 to execute the **CS1_1.m** script [161, pp. 205–210] and with Julia 1.8.3 to execute the equivalent script **CS1_1.jl** (see Listings 8.1–8.4). The extended versions of scripts involve both numerical computations as well as plotting of some results. At the first execution, MATLAB clearly outperformed Julia, needing 12.3 s, while Julia 195.8 s. The second execution in MATLAB required 6.9 s and in Julia required 56.4 s, but starting with the third execution, both MATLAB and Julia needed only 6 s, and thus both exhibit similar runtime performance. The runtime results have been obtained on an Intel Core i7–10700 PC with 16 GB RAM running at 2.9–4.8 GHz.

In what follows, we succinctly describe available software tools in the Julia environment, which implement the numerically reliable algorithms discussed in this book. As already mentioned, the basic computational needs to implement the synthesis procedures presented in Chaps. 5 and 6 are covered by the functions of the **DescriptorSystems**. The basic numerical linear algebra support for the implementation of this collection is provided by several LAPACK-based functions of the **LinearAlgebra** package of Julia such as **svdvals**, **qr**, **schur** or **ordschur**, to mention only a few. Additionally, a layer of LAPACK-wrappers is accessible to users to implement efficient code which fully exploits the underlying problem structures. The need to solve generalized continuous-time and discrete-time algebraic Riccati equations, positive Lyapunov equations and generalized Sylvester equations is covered by the functions available in the **MatrixEquations** package. Similarly, the **MatrixPencil** package provides the basic functionality to compute system poles and zeros using Kronecker-like form-based reductions, to perform conversions between state-space and input–output system descriptions, to compute minimal order descriptor representations using orthogonal similarity transformation-based reductions, to determine additive decompositions based on various spectral splitting (e.g., finite-infinite, stable-unstable), to perform eigenvalue assignment using state-feedback and many others.

In Table 10.1 we present the list of functions of **DescriptorSystems**, which are explicitly called in the listings of examples presented in this book. A comprehensive list of functions and their detailed documentations is available in the online documentation of this package.[13] For an introduction to the provided basic functionality see also Sect. A.2 of the Appendix.

---

[12] https://github.com/MakieOrg/Makie.jl.

[13] https://andreasvarga.github.io/DescriptorSystems.jl/dev/.

Table 10.1 Functions of **DescriptorSystems** used in the examples of this book

Function	Description
append	Appending two descriptor systems by concatenating their inputs and outputs
dss2rm	Building the rational transfer function matrix of a descriptor system
dss	Constructing state-space descriptor system models
evalfr	Evaluating the transfer function matrix in a frequency point
eye	Building an identity matrix of given order
feedback	Forming the closed-loop system corresponding to a static output feedback
gir	Computation of generalized irreducible realizations
glcf	Generalized left coprime factorization
glinfnorm	Evaluation of the L_∞ norm of a descriptor system
glmcover1	Computation of left minimum dynamic cover of Type-1 based order reduction
glnull	Computation of minimal rational left nullspace bases
gpole	Computation of system poles (both finite and infinite)
iszero	Checking if the transfer function matrix is zero
order	Providing the order of a descriptor system
rss	Generation of random standard state-space systems
rtf	Constructing rational transfer functions

The functions of the **FaultDetectionTools** package cover the following main aspects:

- building fault detection and model detection related system objects;
- analysis of synthesis models;
- synthesis of fault detection and model detection filters;
- performance evaluation;
- computational utilities.

Accordingly, in Table 10.2 we present the list of explicitly called functions of the **FaultDetectionTools** package, in the listings of examples presented in this book. A complete list of functions and their detailed documentations is available in the online documentation of this package.[14] For an introduction to the provided basic functionality see also Sect. A.3 of the Appendix.

10.6   Notes and References

Section 10.1. The numerical linear algebra aspects related to the SVD, QR factorization, the real Schur and generalized real Schur decompositions are covered in several textbooks, of which we mention the works of Stewart [118] and of Golub and Van

---

[14] https://andreasvarga.github.io/FaultDetectionTools.jl/dev/.

Table 10.2 Functions of **FaultDetectionTools** used in the examples of this book

Constructors

fdimodset	Setup of synthesis models for solving fault detection and isolation problems
mdmodset	Setup of synthesis models for solving model detection problems
FDFilter	Building a fault detection filter object
FDFilterIF	Building a fault detection filter internal form object
FDIFilter	Building a fault detection and isolation filter object
FDIFilterIF	Building a fault detection and isolation filter internal form object
fdIFeval	Evaluation of the internal forms of fault detection and isolation filters

Analysis

fdigenspec	Generation of achievable fault detection specifications
mdgenspec	Generation of achievable model detection specifications

Synthesis

efdsyn	Exact synthesis of fault detection filters
efdisyn	Exact synthesis of fault detection and isolation filters
afdsyn	Approximate synthesis of fault detection filters
afdisyn	Approximate synthesis of fault detection and isolation filters
emmsyn	Exact model-matching-based synthesis of fault detection filters
ammsyn	Approximate model-matching-based synthesis of fault detection filters
emdsyn	Exact synthesis of model detection filters
amdsyn	Approximate synthesis of model detection filters

Performance

fditspec	Computation of the weak or strong structure matrix
fdisspec	Computation of the strong structure matrix
fdiscond	Computation of the fault detection sensitivity condition
fdif2ngap	Computation of the fault-to-noise gap
fdimmperf	Computation of the model-matching performance

Utilities

fditspec_	Computation of the weak/strong structure matrix of a descriptor system model

Loan [58]. The latter work, which also contains an up to date list of further references, served for the estimation of the computational efforts in terms of the required number of flops for the basic decompositions considered in Sect. 10.1. The book [63] is a modern reference for roundoff error analysis of floating-point computations. The computation of the controllability and observability staircase forms for standard and descriptor systems using orthogonal similarity transformations is addressed in [122]. The detailed algorithm underlying **Procedure GCSF** has been proposed by the author in [131] and several implementations in Julia covering the controllability and observability forms of both descriptor and standard systems are available in the **MatrixPencils** package. Algorithms for the computation of Kronecker-like forms of linear pencils, by using SVD-based rank determinations, and SVD-based row and column compressions, have been proposed in [28, 121]. Albeit numerically reliable, these algorithms have a computational complexity $\mathcal{O}(n^4)$, where n is the minimum of row or column dimensions of the pencil. More efficient algorithms of complexity $\mathcal{O}(n^3)$ have been proposed in [9, 101, 134], which rely on using QR factorizations with column pivoting for rank determinations, and row and column compressions. The **Procedure PREDUCE** is based on the method proposed in [101] and implementations of this algorithm in Julia are available in the **MatrixPencils** package, covering the reduction of both general linear matrix pencils as well as several linear system analysis related structured pencils.

Section 10.2. For a complete coverage of the topic of this section see [116]. The algorithms for the solution of linear matrix equations can be seen as extensions of the Bartels–Stewart method proposed for the solution of the Sylvester equation $AX + BX = C$ in [5]. This algorithm employs the reduction of A and B to RSFs and is considered a numerically reliable method. Further enhancements of this method and extensions to generalized Sylvester equations have been proposed in [57], where one of matrices (that with larger size) is reduced to a Hessenberg form, while the other is reduced to the RSF. Detailed algorithms for the solution of generalized Sylvester matrix equation are described in [50]. Similar algorithms with obvious simplifications can be employed to solve standard and generalized Lyapunov equations. An important algorithm for the solution of Lyapunov equations having positive semi-definite solutions has been proposed in [62], where the solution $X \geq 0$ is directly determined in a Cholesky-factored form $X = SS^T$. The extension of this algorithm to solve generalized Lyapunov equations has been proposed in [108]. The first numerically reliable algorithm to solve standard Riccati equations is the Schur method proposed in [77]. Enhancements of this method to cover discrete-time problems with singular state matrix followed in [106] and to address nearly singular problems in [90, 123]. In all these methods, however, the underlying Hamiltonian or symplectic structure of intervening matrix pencils is not exploited. Therefore, a new direction in developing algorithms for solving GCAREs and GDAREs are the structure exploiting and structure preserving methods to compute eigendecompositions of the Hamiltonian and symplectic pencils (see the book [90] and the recent survey [11]). A comprehensive collection of solvers for various Lyapunov, Sylvester and Riccati equations has been implemented in the Julia package **MatrixEquations**.

Section 10.3. The reliable numerical computation of irreducible realizations of descriptor systems has been considered in [122]. The orthogonal reduction-based algorithm to compute generalized controllability staircase forms, which underlies **Procedure GIR**, has been proposed in [131]. The algorithm to compute a rational nullspace basis of a rational matrix has been proposed in [138] and is related to the approach proposed in [8] to compute polynomial basis using pencil reduction techniques. For the computation of system zeros, an algorithm based on the Kronecker-like form has been proposed in [91]. The approach for the computation of the additive spectral decomposition employed in **Procedure GSDEC** has been proposed in [70]. The iterative pole dislocation techniques underlying the **Procedure GRCF** and **Procedure GRCFID** have been developed in the spirit of the approach described in [124] (see also [135, 166]). Alternative, non-iterative approaches to compute coprime factorizations with inner denominators have been proposed in [100, 102]. The methods presented in Sect. 10.3.6 to compute inner-outer factorizations of full column rank rational matrices are particular versions of the general methods for continuous-time systems proposed in [103] and for discrete-time systems proposed in [100]. The formulas for the complementary inner factors have been derived in [177]. The numerically reliable computational approach for solving linear rational equations, presented in Sect. 10.3.7, has been proposed in [140]. All algorithms discussed in this section have been implemented in the Julia package **DescriptorSystems**. These implementations strongly rely on the pencil manipulation techniques provided by the **MatrixPencils** package.

Section 10.4. The algorithm underlying **Procedure GSCSF** to compute the special controllability staircase form, employed in the methods to determine minimum dynamic covers, is a particular instance of the descriptor controllability staircase algorithm of [131]. An implementation of this procedure in Julia is available in the **MatrixPencils** package. This algorithm together with the computational methods of minimal dynamic covers have been developed in [142]. The minimal realization procedure, based on balancing techniques, has been proposed in [132] for standard systems. The extension of this technique to descriptor systems has been proposed in [119] and is the basis of **Procedure GBALMR**. The state-space method for the solution of the Nehari problem for continuous-time systems has been developed in [54]. The order reduction techniques based on dynamic covers, balancing techniques and Nehari approximation have been implemented in the Julia package **DescriptorSystems**.

Section 10.5. BLAS is a set of specifications for standard vector and matrix operations, which form the core of implementing numerical algebra algorithms. Three levels of abstraction served to define the functionality of BLAS. Level-1 BLAS basically covers operations with and on vectors [78] and served for the implementation of the widely used linear algebra package LINPACK [38]. Level 2 BLAS for matrix-vector operations [37] and Level-3 BLAS for matrix-matrix operations [36] formed the basic layer for implementing the high-performance linear algebra package LAPACK [3]. This package, originally written in Fortran 77, has been designed to run efficiently on a wide range of high-performance machines by using the BLAS, which can be optimized for each computing environment. Moreover, the use of BLAS, makes the subroutines portable and efficient across a wide range of computers. The technology for developing, testing and documenting LAPACK has been adopted by the developers of SLICOT [12, 126]. The initial version of the *Descriptor Systems Toolbox* for MATLAB is described in [136] (see also [126]). The first version of the *Fault Detection Toolbox* for MATLAB is described in [144], while the last version of this toolbox is described in [154].

Appendix A
Introduction to Julia Language and Packages

In this appendix we present basic information on using Julia together with the **DescriptorSystems** and **FaultDetectionTools** packages.

A.1 Julia Basics

Installing Julia. You have to install Julia locally on your computer. For this you have to download a 64-bit version of Julia for your platform (e.g., Windows, Linux or macOS) from the official Julia download site.[1] It is recommended to download the latest stable version of Julia, which at the time of writing is **v1.8.5**. Then follow the platform specific instructions[2] to perform installation and to launch Julia. You can also *optionally* install an editor, which supports Julia (e.g., Visual Studio Code— VSCode[3]), and activate the plug-ins for Julia integration. The full language manual is available online,[4] together with many tutorials and learning resources.[5]

Runing Julia. The easiest way to run Julia code is by starting the REPL command line interface. The *Read Evaluate Print Loop* (REPL) command line interface is a simple and straightforward way of using Julia. Once installed locally, Julia can be launched and the Julia REPL will appear within which Julia commands can be entered and executed following the prompt `julia>`. Alternatively, an *Integrated Development Environment* (IDE) such as VSCode (via the *Julia for VSCode* plug-in) can be employed, where a REPL command line interface is provided in a terminal window.

---

[1] https://julialang.org/downloads/.

[2] https://julialang.org/downloads/platform/.

[3] https://code.visualstudio.com/docs/languages/julia.

[4] https://docs.julialang.org.

[5] https://julialang.org/learning/.

When using the REPL alone or within an IDE (preferably VSCode), typically one will also work with Julia files, which have the **.jl** extension. The code listings in this book are stored in such files, which are available on the **GitHub** repository of the **FaultDetectionTools** package[6] (see subdirectory **Examples**). After starting Julia, help may be obtained for any command through the use of **?**. For example, try **? cd** to obtain the following output (truncated to the first seven lines).

```
help?> cd
search: cd Cdouble gcd gcdx secd asecd cld Cmd codeunit codeunits

  cd(dir::AbstractString=homedir())

  Set the current working directory.

  See also: pwd, mkdir, mkpath, mktempdir.
```

Installing packages. The Julia package manager **Pkg** is used to add (install) a Julia package. To enter the package manager system, type **]** to the Julia prompt. You'll then see the package manager prompt, **(@v1.8) pkg>**. You exit the package manager by typing Ctrl-C (Control-C) or by hitting the backspace key, which returns you to the **julia>** prompt. The following commands add the **DescriptorSystems** and **FaultDetectionTools** packages:

```
(@v1.8) pkg> add DescriptorSystems
(@v1.8) pkg> add FaultDetectionTools
```

To execute all code examples and case study examples presented in this book, other packages are necessary to be also installed in a similar way. To execute all code examples, the packages **Measurements**, **GenericLinearAlgebra**, **Polynomials**, **Makie**, **CairoMakie** and **LaTeXStrings** are also required. To execute all case study examples, the packages **JLD2** and **Optim** must be additionally installed. The status of the current Julia environment (i.e., the installed packages with their versions) can be queried with

```
(@v1.8) pkg> status
  [13f3f980] CairoMakie v0.10.1
  [a81e2ce2] DescriptorSystems v1.3.6
  [fbfc5fa0] FaultDetectionTools v1.2.1
  [14197337] GenericLinearAlgebra v0.3.5
  [033835bb] JLD2 v0.4.29
  [b964fa9f] LaTeXStrings v1.3.0
  [ee78f7c6] Makie v0.19.1
  [eff96d63] Measurements v2.8.0
  [429524aa] Optim v1.7.4
```

The packages can be updated to their latest versions using:

```
(@v1.8) pkg> update
```

For both the **status** and **update** commands, a package name can be additionally specified, to query the information for or to update a particular package, respectively.

---

[6] https://github.com/andreasvarga/FaultDetectionTools.jl.

Using packages. After a package is added to the Julia environment (which needs to be done only once), it can be loaded in Julia to access the functions contained in the packages. To import a package, the command **using**, followed by the package name, or a comma separated list of package names is used, as shown below

```
julia> using  FaultDetectionTools, DescriptorSystems
```

In the code examples in this book we also use other packages (e.g., plotting related packages such as **Makie**, **CairoMakie** and **LaTeXStrings**) for which we included the **using** statement in the code. By loading a package with **using** into the Julia kernel, all functions, objects, and types used in the package become accessible. Note that, while the installation of a package is a one-time operation, which must be performed before the package can be used, the use of **using** is needed in every session during which the package functionality is required.

Using the code examples. All code examples in this book as well as their variants can be easily executed from the Julia REPL. For example, to perform the computations of Example 7.1, the code snippet in Listing 7.1 can be copied and pasted into the REPL. Alternatively, the script **Ex7_1.jl** must be evaluated using the **include** command. In the following listing, first the **FaultDetectionTools** package is loaded with the help of command **using**. Then, the current directory is changed with **cd** to the subdirectory **Examples** of the package base directory determined with **pkgdir**. The script **Ex7_1.jl** defines a Julia module **Ex7_1**, which allows to access the computed results (e.g., the exact poles in **Ex7_1.pexact** and the computed poles[7] in **Ex7_1.poles**). The plot in the related Fig. 7.1 can be separately displayed by entering **Ex7_1.Fig7_1** or **Ex7_1.fig**.

```
julia> using FaultDetectionTools
julia> cd(joinpath(pkgdir(FaultDetectionTools), "Examples"))
julia> include("Ex7_1.jl")  # evaluate script
Example 7.1 with Fig7.1
 Exact poles   Computed poles
25×2 Matrix{ComplexF64}:
 -25.0+0.0im   -25.3924+0.0im
 -24.0+0.0im    -24.488-1.34474im
 -23.0+0.0im    -24.488+1.34474im
 -22.0+0.0im   -22.5233-2.65101im
 -21.0+0.0im   -22.5233+2.65101im
 -20.0+0.0im   -20.0104-3.41785im
 -19.0+0.0im   -20.0104+3.41785im
 -18.0+0.0im   -17.3526-3.54236im
 -17.0+0.0im   -17.3526+3.54236im
 -16.0+0.0im   -14.8476-3.13093im
 -15.0+0.0im   -14.8476+3.13093im
 -14.0+0.0im   -12.6443-2.3726im
 -13.0+0.0im   -12.6443+2.3726im
 -12.0+0.0im   -10.7726-1.44266im
 -11.0+0.0im   -10.7726+1.44266im
 -10.0+0.0im   -9.18481-0.464465im
  -9.0+0.0im   -9.18481+0.464465im
  -8.0+0.0im   -7.95697+0.0im
```

---

[7] The results have been computed with Julia 1.8.3 running under Windows 11.

```
 -7.0+0.0im   -7.00366+0.0im
 -6.0+0.0im   -5.99981+0.0im
 -5.0+0.0im   -5.00001+0.0im
 -4.0+0.0im        -4.0+0.0im
 -3.0+0.0im        -3.0+0.0im
 -2.0+0.0im        -2.0+0.0im
 -1.0+0.0im        -1.0+0.0im
```

To execute all code examples in the Chapters 2, 5, 6 and 7 of the book and all case study examples in Chapter 8 you can use the following commands:

```julia
julia> using FaultDetectionTools
julia> cd(joinpath(pkgdir(FaultDetectionTools), "Examples"))
julia> include("runbookexamples.jl")
julia> include("runcasestudies.jl")
```

This generates for each script a Julia module with the same name, which encapsulates all computed results and all generated figures in the respective script. For example, any internal variable, say **x**, in a script **Exi_j.jl** or **CSi_j.jl** can be accessed as **Exi_j.x** or **CSi_j.x**, respectively. Furthermore, each figure Fig**p.q** generated in **Exi_j.jl** or **CSi_j.jl** can be visualized by simply entering **Exi_j.Figp_q** or **CSi_j.Figp_q**, respectively.

Advanced usage. To have full access to the available software tools or to be able to modify and adapt to your needs the provided code examples, it is recommended to make a local copy of the **FaultDetectionTools** package on your machine. This can be easily done by downloading from **GitHub** a *zip*-ed version of the files of the repository, which you can then unpack to your preferred location on your machine. The work with Julia files is facilitated using a suitable IDE tool, like VSCode, which is highly recommended to be installed together with the plug-in for Julia support.

Issues to be aware of them. Working with Julia can be a fascinating experience due to the achievable high runtime speed and performance. While for users Julia may appear as an interpreted language, every new function and code snippet is in fact compiled using Julia's *Just-In-Time* (JIT) compiler before running them. Therefore, on a first execution of a code, runtime is slower than at the second and subsequent runs. Such delays are typically experienced when loading a package with the **using** command or at the first execution of some functions. Subsequent executions are usually at maximum speed, because Julia is inherently fast.

A.2 **DescriptorSystems** Basics

The **DescriptorSystems** collection provides functions, which allow the manipulation of linear time-invariant systems in their most general representations via state-space descriptor system models and rational transfer function matrices. This collection provides all computational functions necessary to efficiently and accurately solve synthesis problems of fault detection and isolation filters. The provided functionality of the package is also instrumental for the easy checking of various synthesis results in this context.

In the following subsections, we present succinctly the main categories of functions available in the **DescriptorSystems** package and provide several examples to illustrate the usage of functions, which are relevant to solve computational problems encountered in the synthesis of fault detection filters. To execute these examples in the Julia REPL, the **DescriptorSystems** package must be first loaded with the command:

```julia
julia> using DescriptorSystems
```

Other packages can be loaded in a similar way as the need arises. A detailed documentation of the implemented functions is available online.[8]

To check the package installation, it is advisable to execute first the provided interactive demonstration script, by executing the following commands:

```julia
julia> cd(joinpath(pkgdir(DescriptorSystems), "test"))
julia> include("DSToolsDemo.jl")
```

Warning: When executing the demonstration script for the first time, all involved functions are also compiled by the JIT-compiler and this usually involves some compilation times. However, you will not experience any delay when executing this script for a second time, when Julia's runtime corresponds to the expected maximum speed!

A.2.1   Building Models

Two system object types are defined within the **DescriptorSystems** package to represent *descriptor system state-space models* of the form (9.37) and *rational transfer functions* of the form (9.3) of continuous- or discrete-time single-input–single-output systems (9.1) or (9.2), respectiely. To build such models two basic constructors are provided, as listed in the following table, jointly with a function to extract matrix data from state-space representations:

Basic constructors	
dss	Construction of descriptor state-space models
rtf	Building rational transfer functions
dssdata	Extraction of matrix-data from descriptor state-space models

A state-space descriptor system model object can be constructed using the convenience constructor command **dss**. A descriptor state-space model of the form (9.37) is specified by the quintuple of matrices (A, E, B, C, D) with compatible dimensions and a sample time T_s, which for a continuous-time system is zero, while for a discrete-time system is either a positive number or is equal to -1, if the sample time

---

[8] https://andreasvarga.github.io/DescriptorSystems.jl/dev/.

is not explicitly specified. For a standard state-space model, the descriptor matrix is the identity matrix $E = I$. For example, the continuous-time descriptor system with the state-space realization

$$\left[\begin{array}{c|c} A - sE & B \\ \hline C & D \end{array}\right] = \left[\begin{array}{ccccc|cc} 1 & -s & 0 & 0 & 0 & 0 & 0 \\ 0 & 1 & -s & 0 & 0 & 0 & 0 \\ 0 & 0 & 1 & 0 & 0 & -1 & 0 \\ 0 & 0 & 0 & -1-s & 0 & 0 & 1 \\ 0 & 0 & 0 & 0 & -s & 0 & 1 \\ \hline 1 & 0 & 0 & -1 & 0 & 0 & 1 \\ 0 & 0 & 0 & 0 & 1 & 0 & 0 \end{array}\right]$$

can be entered using the following commands:

```
julia> A = [ 1 0 0 0 0;
             0 1 0 0 0;
             0 0 1 0 0;
             0 0 0 -1 0;
             0 0 0 0 0];
julia> E = [ 0 1 0 0 0;
             0 0 1 0 0;
             0 0 0 0 0;
             0 0 0 1 0;
             0 0 0 0 1];
julia> B = [ 0 0 -1 0 0; 0 0 0 1 1]';
julia> C = [ 1 0 0 -1 0; 0 0 0 0 1];
julia> D = [ 0 1; 0 0 ];
julia> sys = dss(A,E,B,C,D)
DescriptorStateSpace{Int64, Matrix{Int64}}

State matrix A:
5×5 Matrix{Int64}:
 1   0   0    0   0
 0   1   0    0   0
 0   0   1    0   0
 0   0   0   -1   0
 0   0   0    0   0

Descriptor matrix E:
5×5 Matrix{Int64}:
 0   1   0   0   0
 0   0   1   0   0
 0   0   0   0   0
 0   0   0   1   0
 0   0   0   0   1

Input matrix B:
5×2 Matrix{Int64}:
  0   0
  0   0
 -1   0
  0   1
  0   1

Output matrix C:
2×5 Matrix{Int64}:
```

```
 1   0   0   -1   0
 0   0   0    0   1

Feedthrough matrix D:
2×2 Matrix{Int64}:
 0   1
 0   0

Continuous-time state-space model.
```

The resulting system object **sys** is a Julia structure of type

DescriptorStateSpace{Int64, Matrix{Int64}}

with the fields **A**, **E**, **B**, **C**, **D** and **Ts**. All system matrices have in this example integer elements (i.e., are of type **Int64**) and can be accessed as **sys.A**, **sys.E**, **sys.B**, **sys.C** and **sys.D**. The sample time can be accessed as **sys.Ts**. The dimensions of the underlying vectors x, y and u in (9.37) can be queried as **sys.nx**, **sys.ny** and **sys.nu**, respectively.

A standard state-space model with $E = I$ can be constructed using the same constructor command in the form

```julia
julia> sysp = dss(A,B,C,D);
julia> typeof(sysp.E)
LinearAlgebra.UniformScaling{Bool}
```

where the trailing semicolon **;** suppresses the displaying of **sys**. In this case, the descriptor matrix E has the type **UniformScaling{Bool}**, which simply says that E is the identity matrix. The following shows that the system **sys** is indeed in a standard state-space form

```julia
julia> using LinearAlgebra
julia> sysp.E == I
true
```

A discrete-time system with a sample time $T_s = 10$ can be specified in the form

```julia
julia> sysd = dss(A,E,B,C,D,Ts = 10);
```

The above three different calling forms of **dss** illustrate the Julia multiple dispatch capabilities implemented for this command. To see the full range of possibilities to generate descriptor state-space models with **dss** see the documentation or enter **? dss**.

The **DescriptorSystems** collection exclusively deals with regular descriptor models, for which the pole pencil $A - \lambda E$ is regular. All functions guarantee that the computed results are regular descriptor systems, provided the input systems are regular. We have to stress, that for efficiency reasons, in most of functions of **DescriptorSystems**, the regularity condition for the input system data is only *assumed*, but it is not explicitly checked. Therefore, it is likely that some functions, even if they perform without issuing error messages, may still deliver nonsense results if the regularity assumption is not fulfilled by the input system descriptions.

A rational transfer function system model object can be constructed using the convenience constructor command **rtf**. A single-input single-output rational transfer function model of the form (9.3) is specified by the numerator and denominator polynomials $\alpha(\lambda)$ and $\beta(\lambda)$, respectively, of the rational transfer function $g(\lambda) = \alpha(\lambda)/\beta(\lambda)$ and a sample time T_s. The function **rtf** can be used to build a rational transfer function object from the numerator and denominator polynomials (constructed using the **Polynomials** package[9]), or from the roots of the numerator and denominator polynomials and a gain. To see the full range of possibilities to generate rational transfer function models with **rtf** see the documentation or enter **? rtf**.

An alternative convenient way to build rational transfer function models exploits the algebraic manipulation possibilities with rational functions, which are similar to the operations with real or complex numbers. Therefore, all valid operations with rational functions can be also performed with rational transfer functions sharing the same sample time. A simple technique to build the transfer function of a continuous-time system is to define first the indeterminate, say s, as a rational (improper) transfer function with denominator equal to one and then use the usual algebraic operations to build the transfer function. This construction is illustrated below to construct the transfer function $g(s) = s/(s + 1)$:

```
julia> s = rtf('s');
julia> g = s/(s+1)
RationalTransferFunction{Int64, :s, Polynomial{Int64, :s}, 0.0}
s // (s + 1)


Continuous-time rational transfer function model.
```

The transfer function $h(z) = z/(z - 2)$ of a discrete-time system with a sample time $T_s = 10$ sec can be built in a similar way:

```
julia> z = rtf('z',Ts = 10);
julia> h = z/(z-2)
RationalTransferFunction{Int64, :z, Polynomial{Int64, :z}, 10.0}
z // (z - 2)


Sample time: 10.0 second(s).
Discrete-time rational transfer function model.
```

To build transfer functions matrices for multiple-input multiple-output models of the form (9.4), we can simply rely on the capabilities of Julia to build and manipulate abstract arrays with elements of arbitrary types. For example, to enter the continuous-time improper transfer function matrix

$$G_c(s) = \begin{bmatrix} s^2 & \dfrac{s}{s+1} \\ 0 & \dfrac{1}{s} \end{bmatrix}, \tag{A.1}$$

the following commands can be used:

---

[9] https://github.com/JuliaMath/Polynomials.jl.

```
julia> s = rtf('s');                    # define the complex variable s
julia> Gc = [s^2 s/(s+1); 0 1/s] # define the 2×2 improper Gc(s)
2×2 Matrix{RationalTransferFunction{Int64, :s,
                           Polynomial{Int64, :s}, 0.0}}
From input 1 to output 1
s^2

From input 1 to output 2
0

From input 2 to output 1
s // (s + 1)

From input 2 to output 2
1 // s

Continuous-time rational transfer function matrix model.
```

Similarly, the discrete-time improper transfer function matrix

$$G_d(z) = \begin{bmatrix} z^2 & \dfrac{z}{z-2} \\ 0 & \dfrac{1}{z} \end{bmatrix},$$

with sample time equal to 0.5 can be entered using the commands:

```
julia> z = rtf('z',Ts=0.5);            # define the complex variable z
julia> Gd = [z^2 z/(z-2);0 1/z];  # define the 2×2 improper Gd(z)
```

Most of functions of **DescriptorSystems** accept only state-space system objects as inputs and therefore the input–output models must be converted to a standard or descriptor state-space form. For the above defined TFMs $G_c(s)$ and $G_d(z)$, these conversions can be done simply using the function **dss**:

```
julia> sysc = dss(Gc,minimal=true);
julia> sysd = dss(Gd,minimal=true);
julia> order(sysc) == order(sysd) == 5
true
```

Without using the keyword argument **minimal=true**, the resulting state-space realizations are usually non-minimal. Incidentally, both of the resulting minimal state-space realizations **sysc** and **sysd** have order 5. The function **dss** can be also called by providing the numerators and denominators as two polynomial matrices (see the demonstration script **DSToolsDemo.jl** for example).

Model building is additionally supported by several coupling/interconnecting functions listed in the following table:

Interconnecting descriptor system models	
append	Building aggregate models by appending the inputs and outputs
parallel	Connecting models in parallel (also overloaded with $+$)
series	Connecting models in series (also overloaded with $*$)
horzcat	Horizontal concatenation of descriptor system models (also overloaded with [* *])
vertcat	Vertical concatenation of descriptor system models (also overloaded with [*; *])

A.2.2 *Operations with Rational Transfer Functions*

The **DescriptorSystems** package implements the following operations on rational transfer functions:

Operations on rational transfer functions and matrices	
zpk	Computation of zeros, poles and gain of a rational transfer function
simplify	Pole–zero cancellation
normalize	Normalization of a rational transfer function to monic denominator
confmap	Applying a conformal mapping transformation to a rational transfer function or rational transfer function matrix
rtfbilin	Generation of common bilinear transformations and their inverses

Let $g(\lambda) = \alpha(\lambda)/\beta(\lambda)$ be a rational transfer function of the form (9.3). The function **zpk** can be used to compute the *zero-pole-gain* representation (9.7) of $g(\lambda)$. For the cancellation of common factors of the numerator and denominator polynomials $\alpha(\lambda)$ and $\beta(\lambda)$ of $g(\lambda)$ the function **simplify** can be used. The function **normalize** can be used to enforce a monic denominator polynomial $\beta(\lambda)$ via suitable scaling of the coefficients of the numerator and denominator polynomials. The function **rtfbilin** can be used to build the rational transfer functions of several commonly used bilinear transformations and their inverses. To apply a conformal mapping transformation to a rational transfer function or a transfer function matrix the function **confmap** can be used.

An interesting aspect worth to be mentioned is a powerful feature of the Julia language to allow the application of any operation on a scalar entity to the elements of arrays of such entities, using the so-called *dot syntax* implementing the Julia's *broadcasting* mechanism. For example, if **G** is a transfer function matrix object whose elements are rational transfer functions, then enforcing the cancellation of common factors in all numerator and denominator polynomials of **G** can be done by calling **simplify** in its *dot syntax* form **simplify.(G)**.

A.2.3 *System Analysis*

The analysis functions available in **DescriptorSystems** are listed in the following table:

System analysis	
gpole	Poles of a descriptor system
gpoleinfo	Poles and pole structure information of a descriptor system
gzero	Zeros of a descriptor system
gzeroinfo	Zeros and zero structure information of a descriptor system
gnrank	Normal rank of the transfer function matrix of a descriptor system
ghanorm	Hankel norm of a proper and stable descriptor system
gl2norm	L_2 norm of a descriptor system
gh2norm	H_2 norm of a descriptor system
glinfnorm	L_∞ norm of a descriptor system
ghinfnorm	H_∞ norm of a descriptor system
gnugap	ν-gap distance between two descriptor systems
freqresp	Frequency response of a descriptor system
timeresp	Time response of a descriptor system
stepresp	Step response of a descriptor system
isregular	Test whether a descriptor system has a regular pole pencil
isproper	Test whether a descriptor system is proper
isstable	Test whether a descriptor system is stable
iszero	Check if the transfer function matrix of a descriptor system is zero

Several functions can be used to analyse the properties of state-space models such as **isproper**, to check whether a system is proper or not, **isstable** to check stability, **iszero** to check if the transfer function matrix is zero based on [164, Method 5.], **gzero** to compute the system zeros, **gpole** to compute the system poles, or **gnrank** to determine the normal rank of the system transfer function matrix. The application of these functions to the already defined systems **sys**, **sysc** and **sysd** is illustrated below:

```
julia> isproper(sys)
false
julia> isstable(sysc)
false
julia> iszero(sys-sysc)
true
julia> gzero(sys)
4-element Vector{Float64}:
  0.0
 -0.9999999999999999
 -0.0
 Inf
julia> gpole(sys)
4-element Vector{Float64}:
 -1.0
  0.0
 Inf
 Inf
julia> gnrank(sys)
```

Since **sys** and **sysc** have the same transfer function matrix, it follows that **sys** is also a minimal realization of the transfer function matrix $G_c(s)$ in (A.1).

For the computation of the system norms discussed in Sect. 9.2.7, several functions are available such as **ghanorm**, to compute the Hankel norm of a stable system, **gl2norm/gh2norm**, to compute the L_2/H_2 norms, respectively, and **glinfnorm/ghinfnorm** to compute the L_∞/H_∞ norms, respectively. The function **gnugap** can be used to evaluate the ν-gap distance between two systems [169].

A.2.4 Conversions between Model Representations

In the following table the functions available in the **DescriptorSystems** package are listed, which can be used to simplify a given state-space model-based representation either exactly or approximately:

Simplification of descriptor system models	
gir	Irreducible realizations of descriptor systems
gminreal	Minimal realizations of descriptor systems
gbalmr	Reduced-order approximations of descriptor systems using balancing related methods
gss2ss	Conversions to SVD-like coordinate forms without non-dynamic modes
dss2ss	Conversion of descriptor systems to standard form

To compute minimal realizations, the function **gminreal** is available. Alternatively, the function **gir** can be used to compute irreducible realizations, which, for standard systems are also minimal. The functions **gir**, **gminreal**, **gss2ss** and **dss2ss** usually reduce the order of the initial state-space models by eliminating parts of the model, which are either uncontrollable, or unobservable, or contain no dynamics. The resulting reduced order models have exactly the same transfer function matrix as the initial system. Approximate order reduction can be performed using model reduction techniques implemented in the function **gbalmr**.

The conversion of a proper descriptor state-space form to the standard state-space form can be performed using the function **gss2ss**. An example is given below, where a proper system, used as input, is constructed using **dss(E,A,B,C,D)**, with the matrices **A** and **E** interchanged:

```
julia>   sys1 = gss2ss(dss(E,A,B,C,D))[1]
DescriptorStateSpace{Float64, UniformScaling{Bool}}

State matrix A:
4×4 Matrix{Float64}:
  0.0    1.0     0.0    0.0
  0.0    0.0     0.0    0.0
  0.0    0.0    -1.0    0.0
  1.0    0.0     0.0    0.0

Input matrix B:
4×2 Matrix{Float64}:
```

```
   0.0    0.0
  -1.0    0.0
   0.0    1.0
   0.0    0.0

Output matrix C:
2×4 Matrix{Float64}:
  0.0    0.0    1.0    1.0
  0.0    0.0    0.0    0.0

Feedthrough matrix D:
2×2 Matrix{Float64}:
  0.0    1.0
  0.0   -1.0

Continuous-time state-space model.
```

The calling form **sys1 = gss2ss(dss(E,A,B,C,D))[1]** simply selects the
first output computed by **gss2ss** to define the reduced system **sys1**.

A second category of functions for model conversions is listed in the following
table:

Basic conversions on descriptor system models	
c2d	Discretization of continuous-time descriptor systems
dss2rm	Rational transfer function matrix of a descriptor system
dss2pm	Polynomial transfer function matrix of a descriptor system
gbilin	Generalized bilinear transformation of a descriptor system

The function **c2d** implements several discretization methods of continuous-time
systems. Besides computing the discretized system, this function also computes the
mapped initial condition of the state of the discrete-time system, which is determined
from the initial conditions of the state and input of the continuous-time system. With
appropriate options selected via keyword arguments, this function can determine the
state-mapping matrices, which relate the continuous-time state to the discrete-time
state and inputs. The function **c2d** can be also applied to rational transfer functions.

For visualization purposes, often the more compact transfer function matrix models
are better suited than state-space models. The function **dss2rm** can be used to
convert a state-space model to a rational transfer function matrix representation. The
following example illustrates how **dss2rm** and **zpk** can be combined via the use
of the *dot syntax* to apply **zpk** to each element of the rational matrix computed with
dss2rm:

```
julia> zpk.(dss2rm(sys))
2×2 Matrix{Tuple{Vector{Float64}, Vector{Float64}, Float64}}:
 ([], [0.0, 0.0, 0.0], -1.0)    ([-2.0], [-1.0], 1.0)
 ([], [], 0.0)                  ([], [], -1.0)
```

The result is an array of *tuples*, where each tuple contains three elements: the vector
of zeros (can be empty), the vector of poles (can be empty) and the gain.

A.2.5   Operations with Rational Matrices

The **DescriptorSystems** package implements the following basic operations on rational matrices via their descriptor system representations:

Basic operations on rational matrices	
gdual	Building the dual of a descriptor system (also overloaded with **transpose**)
ctranspose	Building the conjugate of a system (also overloaded with **adjoint** and **'**)
inv	Inversion of a system
ldiv	Left division for two systems (also overloaded with ****)
rdiv	Right division for two systems (also overloaded with **/**)

These operations are described in Sect. 9.2.3 and play an important role in implementing many of the functions available in **DescriptorSystems**. Therefore, we discuss more in detail some aspects of using these functions.

Let $G(\lambda)$ be a rational transfer function matrix having the descriptor system realization $(A - \lambda E, B, C, D)$. We assume that the state-space system object **sys** has been defined to represent $G(\lambda)$ (e.g., using the **dss** constructor).

The *transpose* $G^T(\lambda)$ and the equivalent *dual* descriptor system realization $G^T(\lambda) = (A^T - \lambda E^T, C^T, B^T, D^T)$ can be simply computed as **transpose (sys)** or **gdual(sys)**, respectively. In some applications, it is useful to combine the construction of dual with the reverse permutation of state variables, which comes down to determine the realization of the dual as $G^T(\lambda) = (PA^T P - \lambda P E^T P, PC^T, B^T P, D^T)$, where P is the (orthogonal) permutation matrix with ones down on the secondary diagonal

$$ P = \begin{bmatrix} 0 & & 1 \\ & \ddots & \\ 1 & & 0 \end{bmatrix}. $$

This can be achieved by using **gdual(sys,rev = true)** to compute the dual system.

The *conjugate* (or *adjoint*) $G^\sim(\lambda)$ is defined in the continuous-time case as in (9.54) and in the discrete-time case as in (9.55) and can be computed in three equivalent ways as **ctranspose(sys)**, or **adjoint(sys)**, or simply **sys'**.

If $G(\lambda)$ is invertible, then an inversion free realization of the *inverse* $G^{-1}(\lambda)$ in the form (9.53) can be computed as **inv(sys)**, using the overloaded linear algebra function **inv**. Operations involving inverses can often be performed using the overloaded matrix operators **** (left divide) or **/** (right divide), as for example, to compute **sys1\\sys2** or **sys1/sys2** for two systems **sys1** and **sys2**. However, these operations do not avoid the explicit building of the inverses, as it can be seen by performing **sys1\\sys1** or **sys2/sys2**, which, instead of producing a non-dynamic systems with the direct feedthrough gain equal to the identity matrix, determine realizations of orders (at least) double of the orders of **sys1** or **sys2**. Alternative computation of **sys1\\sys2** can be done using the function **glsol**, while for the evaluation of **sys1/sys2**, the function **grsol** can be used. Both functions determine minimal order realizations of the solutions. These functions belong to a set of functions, which implement advanced operations on rational matrices such as the computation of additive spectral decompositions, nullspace bases, generalized inverses, or the solution of linear rational matrix equations. The list of these functions is presented in the following table:

Advanced operations on transfer function matrices	
gsdec	Additive spectral decompositions
grnull	Right nullspace basis of a transfer function matrix
glnull	Left nullspace basis of a transfer function matrix
grange	Range space basis of a transfer function matrix
gcrange	Coimage space basis of a transfer function matrix
grsol	Solution of the linear rational matrix equation $G(\lambda)X(\lambda) = F(\lambda)$
glsol	Solution of the linear rational matrix equation $X(\lambda)G(\lambda) = F(\lambda)$
grmcover1	Right minimum dynamic cover of Type 1-based order reduction
glmcover1	Left minimum dynamic cover of Type 1-based order reduction
grmcover2	Right minimum dynamic cover of Type 2-based order reduction
glmcover2	Left minimum dynamic cover of Type 2-based order reduction
ginv	Generalized inverses

The function **glnull** for left nullspace basis computation is central to solve synthesis problems of fault detection filters. In what follows, we present an example taken from [72, p. 459], to compute a rational left nullspace basis with specified poles. The example is the 3×4 rational matrix

$$G(s) = \begin{bmatrix} \dfrac{1}{s} & 0 & \dfrac{1}{s} & s \\ 0 & (s+1)^2 & (s+1)^2 & 0 \\ -1 & (s+1)^2 & s^2+2s & -s^2 \end{bmatrix},$$

with normal rank equal to 2, for which we determine the following stable rational left nullspace basis with a pole equal to -1

$$N_l(s) = \begin{bmatrix} \dfrac{s}{s+1} & -\dfrac{1}{s+1} & \dfrac{1}{s+1} \end{bmatrix}.$$

The computation of this basis can be performed with the following commands:

```julia
julia> s = rtf('s');
julia> G = [1/s 0 1/s s;
            0 (s+1)^2 (s+1)^2 0;
            -1 (s+1)^2 s^2+2*s -s^2];
julia> sys = dss(G,minimal=true);

julia> Nl = glnull(sys,poles = [-1])[1];
julia> println(dss2rm(Nl))
1×3 Matrix{RationalTransferFunction{Float64,:s,
          Polynomial{Float64, :s}, 0.0}}
From input 1 to output 1
```

```
(1.0000000000000002*s - 3.597533769998865e-16) // (1.0*s + 1.0)

From input 2 to output 1
-1.0000000000000007 // (1.0*s + 1.0)

From input 3 to output 1
1.0000000000000013 // (1.0*s + 1.0)

Continuous-time rational transfer function matrix model.

julia> iszero(N1*sys)
true
```

The use of the function **glnull** to solve a fault detection filter synthesis problem
is also shown in Listing 7.3.

The functions **grsol** and **glsol** can be used to address controller and filter
synthesis problems by solving *exact* model-matching problems. In conjunction with
the minimum dynamic cover techniques implemented in the functions **grmcover1**,
glmcover1, **grmcover2** and **glmcover2**, these functions can be used to compute solutions of least McMillan degree. In what follows, we show how to use the
function **glsol** to solve the synthesis problem of Example 5.3, as an exact model-
matching problem:

```
julia> s = rtf('s');
julia> Gu = [(s+1)/(s+2); (s+2)/(s+3)];   # enter Gu(s)
julia> Gd = [(s-1)/(s+2); 0];             # enter Gd(s)
julia> Gf = [Gu [0;1]];                   # define Gf(s)
julia> mu = 1; md = 1; p = 2; mf = mu+1;  # set dimensions
julia> Mr = [(s+2)/(s+3) 1];              # enter Mr(s)

julia> Ge = dss([Gu Gd Gf; eye(mu,mu+md+mf)]);
julia> Fe = dss([zeros(1,mu+md) Mr]);
julia> Q = glsol(Ge,Fe,mindeg = true)[1];
julia> order(Q)
1
julia> iszero(Q*Ge-Fe)
true
```

The use of the function **glnull** in conjunction with **glmcover1** to solve a least
order fault detection filter synthesis problem is shown in Listing 7.4.

A.2.6 *Factorizations of Rational Matrices*

The **DescriptorSystems** package implements several functions for the factorization of rational matrices described in Sects. 9.2.6 and 9.2.8:

Factorizations of transfer function matrices	
grcf	Right coprime factorization with proper and stable factors
glcf	Left coprime factorization with proper and stable factors
grcfid	Right coprime factorization with inner denominator
glcfid	Left coprime factorization with inner denominator
gnrcf	Normalized right coprime factorization
gnlcf	Normalized left coprime factorization
giofac	Inner-outer and QR-like factorizations of a transfer function matrix
goifac	Co-outer–coinner and RQ-like factorizations of a transfer function matrix
grsfg	Right spectral factorization of $\gamma^2 I - G^{\sim}(\lambda)G(\lambda)$
glsfg	Left spectral factorization of $\gamma^2 I - G(\lambda)G^{\sim}(\lambda)$

The functions **grcf/glcf** to compute stable coprime factorizations are important to solve stabilization problems for the synthesis of controllers and fault detection filters. The functions **grcfid/glcfid** to compute coprime factorizations with inner denominators are important to implement the functions for norm computations and spectral factorizations. The functions **gnrcf/gnlcf** to compute normalized coprime factorizations play an important role in the evaluation of the ν-gap metric [169]. And finally, the functions **giofac/goifac** to compute inner-outer factorizations and **grsfg/ glsfg** to compute spectral factorizations are instrumental to solve approximate model-matching problems (see next section).

Due to its relevance for solving stabilization problems of fault detection filters, we illustrate the use of the function **glcf** for the computation of a stable left coprime factorization of an improper transfer function matrix with unstable poles. We consider the 2×2 rational matrix $G_c(s)$ in (A.1) with the poles $\{-1, 0, \infty, \infty\}$, whose (minimal) state-space realization is the previously computed **sysc**. A stable left coprime factorization $G_c(s) = M^{-1}(s)N(s)$, with the stable and proper factors

$$N(s) = \begin{bmatrix} -\dfrac{s^2}{(s+1)(s+2)} & -\dfrac{s}{(s+1)^2(s+2)} \\ 0 & \dfrac{1}{s+3} \end{bmatrix}, \quad M(s) = \begin{bmatrix} -\dfrac{1}{(s+1)(s+2)} & 0 \\ 0 & \dfrac{s}{s+3} \end{bmatrix}$$

can be computed using the following commands, which also check the validity of the computed results:

```julia
julia> N, M = glcf(sysc,mininf=true,mindeg=true,evals=[-1,-2,-3]);

julia> iszero(M*sysc-N)        # check the factorization
true
julia> gpole(N)                # check poles of N
4-element Vector{Float64}:
 -3.0
 -2.0
 -1.0
 -1.0
julia> gpole(M)                # check poles of M
3-element Vector{Float64}:
 -3.0
 -2.0
```

```
 -1.0
julia> gzero(gir([N M]))   # check coprimeness (no zeros)
Float64[]
```

A.2.7 Solving Approximate Model-Matching Problems

For the solution of approximate model-matching problems the functions listed in the following table can be used:

Solution of approximate model-matching problems	
grasol	Approximate solution of the linear rational matrix equation $G(\lambda)X(\lambda) = F(\lambda)$
glasol	Approximate solution of the linear rational matrix equation $X(\lambda)G(\lambda) = F(\lambda)$
glinfldp	Solution of the least distance problem $\min_{X(\lambda)} \|[\,G_1(\lambda) - X(\lambda)\,G_2(\lambda)\,]\|_\infty$
gnehari	Generalized Nehari approximation

The functions **grasol** and **glasol** compute the approximate solutions, which minimize the $\mathcal{L}_2$- or $\mathcal{L}_\infty$-norms of the corresponding residuals $E(\lambda) := G(\lambda)X(\lambda) - F(\lambda)$ and $\widetilde{E}(\lambda) := X(\lambda)G(\lambda) - F(\lambda)$, respectively (see Sect. 9.1.10). These functions reduce the model-matching problems to least distance problems, which are then solved using the functions **glinfldp** and **gnehari**.

The following commands illustrate the γ-iteration-based solution of the $\mathcal{H}_\infty$ model-matching problem of the example used in [46, p. 111]:

```
julia> s = rtf('s');
julia> W = (s+1)/(10s+1);
julia> T1 = [W; 0];
julia> T2 = [-(s-1)/(s^2+s+1)*W; (s^2-2*s)/(s^2+s+1)*W];
julia> G = dss(T2, minimal=true); F = dss(T1);
julia> atol = 1.e-7;

julia> Xopt, info = grasol(G, F; mindeg = true, atol);
julia> println("Optimal error: $(info.mindist)")
Optimal error: 0.2521288923421184

julia> ghinfnorm(G*Xopt-F)[1] ≈ info.mindist
true

julia> Xsub, infosub = grasol(G, F, 0.2729; mindeg = true, atol);
julia> println("Suboptimal error: $(infosub.mindist)")
Suboptimal error: 0.25536759751557475

julia> ghinfnorm(G*Xsub-F)[1] ≈ infosub.mindist
true
```

A.3 `FaultDetectionTools` Basics

The **FaultDetectionTools** package provides a collection of functions for the synthesis of linear fault detection and model detection filters. The implementation of the functions included in this collection follows several principles, which have been consequently enforced when implementing these functions. These principles have been partly formulated in the Sect. 10.5 and, together with new aspects, are discussed below:

- *Using general, numerically reliable and computationally efficient numerical approaches as basis for the implementation of all computational functions, to guarantee the solvability of problems under the most general existence conditions of the solutions*. The main consequence for the users is that the implemented synthesis methods provide a solution whenever one exists. These methods are extensively described in this book, which thus forms the methodological and computational basis of all implemented analysis and synthesis functions.
- *Support for the most general model representation of linear time-invariant systems in form of generalized state-space descriptions, also known as descriptor systems*. All analysis and synthesis functions are applicable to both continuous- and discrete-time systems. The basis for the implementation of all functions is the **DescriptorSystems** package, which serves to define suitable synthesis models, to solve efficiently and accurately filter synthesis problems, to perform various pre- and post-synthesis analysis, as well as to check various synthesis results.
- *Providing simple user interfaces*. All synthesis functions rely on default settings of problem parameters and design options, which allow to easily obtain preliminary synthesis results. Also, all functions to solve a class of problems (e.g., fault detection) use the same input model.
- *Providing an exhaustive set of options*. This ensures the complete freedom in choosing problem specific parameters and synthesis options.
- *Guaranteeing the reproducibility of results*. This feature is enforced by employing the so-called *design matrices*. These are internally used to build linear combinations of left nullspace basis vectors and are occasionally randomly generated (if not explicitly provided). The values of the employed design matrices are returned as additional information by all synthesis functions. The use of design matrices also represents a convenient mean to perform an optimization-based tuning of these matrices to achieve specific performance characteristics for the resulting filters.

In this section, we present succinctly the main functionality provided by the **FaultDetectionTools** package to solve synthesis problems of fault detection and model detection filters, starting with the building of synthesis models, then performing preliminary analysis, filter synthesis and evaluation of achieved performance. We provide simple examples to illustrate the usage of several functions. To run the examples, the **FaultDetectionTools** and **DescriptorSystems** packages must be first loaded with the command:

```julia
julia> using FaultDetectionTools, DescriptorSystems
```

Other packages can be loaded in a similar way if the need arises. A detailed documentation of the implemented functions is available online.[10]

Recall that, all examples and case studies presented in this book can be reproduced using a collection of scripts, which is part of the **FaultDetectionTools** package. Most of the provided scripts have two versions, a *compact* one, which calls directly the functions of the **FaultDetectionTools**, and an equivalent *extended* version, which calls mainly functions of the **DescriptorSystems** package and does not perform plotting. For example, to execute the script corresponding to Example 5.4, any of the following last two commands can be used:

```
julia> cd(joinpath(pkgdir(FaultDetectionTools), "Examples"))
julia> include("Ex5_4c.jl")    # compact version
julia> include("Ex5_4.jl")     # extended version
```

A.3.1 Solving Fault Detection Problems

Building Synthesis Models

The **FaultDetectionTools** package relies on custom synthesis model objects, which support the grouping of inputs according to a predefined naming convention. For the solution of fault detection problems, the synthesis models are state-space models of the form

$$E\lambda x(t) = Ax(t) + B_u u(t) + B_d d(t) + B_f f(t) + B_w w(t) + B_a v(t),$$
$$y(t) = Cx(t) + D_u u(t) + D_d d(t) + D_f f(t) + D_w w(t) + D_a v(t). \tag{A.2}$$

This model exhibits five input groups, which are named as follows: **controls** for the control inputs $u(t)$, **disturbances** for the disturbance inputs $d(t)$, **faults** for the fault inputs $f(t)$, **noise** for the noise inputs $w(t)$, and **aux** for the auxiliary inputs $v(t)$.

A typical situation is when a plant model is available for fault monitoring purposes, having the form

$$E\lambda x(t) = Ax(t) + B\widetilde{u}(t),$$
$$y(t) = Cx(t) + D\widetilde{u}(t), \tag{A.3}$$

where $\widetilde{u}(t)$ is a global input, whose components are the control inputs, disturbance inputs and noise inputs. Some fault inputs can be already present and additional fault inputs can be defined as actuator and sensor faults. The building of the synthesis model is just the operation to define the above groupings of various categories of input components, using the synthesis model setup function **fdimodset**.

As an example, consider a plant model **sys** of the form (A.3) with the input vector $\widetilde{u}(t)$ having 3 components, the output vector $y(t)$ having 3 components and a 3-dimensional state vector $x(t)$. Assume that the first two inputs are control inputs,

---

[10] https://andreasvarga.github.io/FaultDetectionTools.jl/dev/.

which are both susceptible to actuator faults, and the third input is a disturbance input, which is not measurable and therefore is considered as an unknown input. All outputs are measurable, and assume that the first output is susceptible to sensor fault. The following commands generate a synthesis model with additive faults as described above:

```
julia> A = [0.75 -2.26 0.71; 0.05 -1.8 -0.47; 1.37 0.49 2.37];
julia> B = [0.26 2.13 1.01; -1.08 1.86 -0.68; -0.1 0.63 0.03];
julia> C = [-0.56 2.22 0.03; -0.38 1.14 -0.32; -0.85 1.08 -3.16];
julia> D = [0.8 0.56 0.47; 0.97 0.02 0.25; 0.37 0.45 0.8];
julia> sys = dss(A,B,C,D);
julia> sysf = fdimodset(sys,c = 1:2,d = 3,fa = 1:2,fs = 1)
FDIModel{Float64}
DescriptorStateSpace{Float64, LinearAlgebra.UniformScaling{Bool}}

State matrix A:
3×3 Matrix{Float64}:
 0.75   -2.26    0.71
 0.05   -1.8    -0.47
 1.37    0.49    2.37

Input matrix B:
3×6 Matrix{Float64}:
  0.26    2.13    1.01    0.26    2.13    0.0
 -1.08    1.86   -0.68   -1.08    1.86    0.0
 -0.1     0.63    0.03   -0.1     0.63    0.0

Output matrix C:
3×3 Matrix{Float64}:
 -0.56    2.22    0.03
 -0.38    1.14   -0.32
 -0.85    1.08   -3.16

Feedthrough matrix D:
3×6 Matrix{Float64}:
 0.8     0.56    0.47    0.8     0.56    1.0
 0.97    0.02    0.25    0.97    0.02    0.0
 0.37    0.45    0.8     0.37    0.45    0.0

Continuous-time state-space model.

Input groups:
Name           Channels
controls       1:2
disturbances   3:3
faults         4:6
```

The function **rss** has been used to generate the above state-space model **sys** with randomly generated coefficient matrices with elements rounded to two decimal digits. However, to allow the reproducibility of the results, we entered explicitly the state-space system matrices A, B, C and D and built the corresponding state-space model using the **dss** constructor. The function **fdimodset** uses the system model **sys** to build a synthesis model using keyword arguments, which allow to specify the grouping of system inputs. For this case, the keyword arguments **c** and **d** define the indices of control and disturbance inputs, respectively, and the keyword arguments **fa** and **fs** specify the indices of inputs and outputs susceptible to actuator and sensor faults, respectively. The structures of the resulting input and feedthrough matrices reflect this definition of faults inputs in agreement with (2.4).

The resulting synthesis model **sysf** has the custom type **FDIModel** and is a structure whose field names can be inspected using the function **fieldnames**:

```
julia> fieldnames(FDIModel)
(:sys, :mu, :md, :mf, :mw, :ma)
```

Thus, the underlying descriptor system model (A.2) can be accessed as **sysf.sys** and the dimensions of the control, disturbance, fault, noise and auxiliary vectors are contained in **sysf.mu**, **sysf.md**, **sysf.mf**, **sysf.mw** and **sysf.ma**, respectively. The ranges of indices of control, disturbance, fault, noise and auxiliary inputs can be queried as **sysf.controls**, **sysf.disturbances**, **sysf.faults**, **sysf.noise** and **sysf.aux**, respectively.

Preliminary Analysis

To perform fault detection and isolation related analysis, the functions listed in the following table are available:

Analysis of FDI synthesis models	
fdigenspec	Generation of achievable FDI specifications
fdichkspec	Feasibility analysis of a set of FDI specifications

The function **fdigenspec** can be used to generate the achievable weak and strong FDI specifications to be used for solving synthesis problems of FDI filters, while the function **fdichkspec** performs the analysis of the feasibility of a set of FDI specifications. For the definitions of *isolability* related concepts see Sect. 3.4.

To determine the achievable strong FDI specifications for the generated synthesis model **sysf** with respect to *constant* faults (i.e., for frequency equal to 0), the function **fdigenspec** can be used as follows:

```
julia> S = fdigenspec(sysf,FDfreq = 0)
4×3 BitMatrix:
 1  1  1
 0  1  1
 1  0  1
 1  1  0
```

For the above example, the possible fault signatures are contained in the resulting structure matrix

$$S = \begin{bmatrix} 1 & 1 & 1 \\ 0 & 1 & 1 \\ 1 & 0 & 1 \\ 1 & 1 & 0 \end{bmatrix},$$

which indicates that the EFDP and AFDP are solvable (first row of S) and the EFDIP and AFDIP are also solvable using the specifications contained in the rows of

$$S_{FDI} = \begin{bmatrix} 0 & 1 & 1 \\ 1 & 0 & 1 \\ 1 & 1 & 0 \end{bmatrix}. \tag{A.4}$$

Solving FDI Synthesis Problems

The available functions to solve FDI synthesis problems are listed in Table 10.2. Typically, each synthesis function has as inputs a synthesis model object, possible optional inputs (e.g., desired structure matrix, reference filter model) and synthesis options, which are specified via keyword arguments (e.g., desired stability degree of filter poles). All optional inputs and keyword arguments have meaningful default values, which allow to use each synthesis function via simple calls. On the other hand, the provided keyword arguments ensure the highest flexibility in solving fault detection problems. The results computed by the synthesis functions are a filter object, which is provided in both implementation and internal forms, and some synthesis related information. In what follows, we present typical synthesis commands to solve exact synthesis problems. Similar commands can be used to solve approximate synthesis problems.

Solving Fault Detection Problems

For the solution of an EFDP, the function **efdsyn** can be used. To ensure a stability degree -2 for the filter poles, the simplest form of using this function is

```julia
julia>   Q, R, info = efdsyn(sysf, sdeg = -2);
```

where **Q** is the resulting fault detection filter object of type **FDFilter** in the implementation form, **R** is the resulting filter object of type **FDFilterIF** in the internal form, and **info** is a *named-tuple*, which contains additional synthesis related information.

The filter object **Q** contains the resulting filter in a standard state-space form, which corresponds to the implementation form (3.3). The underlying filter can be obtained via **Q.sys** and the dimensions of the partitioned filter input vectors as **outputs** and **controls**, can be accessed as the integers contained in **Q.ny** and **Q.mu**, respectively. The ranges of the indices of the output and control inputs can be accessed as **Q.outputs** and **Q.controls**, respectively. The resulting filter poles can be computed using the overloaded function **gpole**:

```julia
julia>   gpole(Q)
1-element Vector{Float64}:
 -1.9999999999999993
```

The filter object **R** contains the resulting filter in a standard state-space form, which corresponds to the internal form (3.4), with additional auxiliary inputs. The internal form can be used for post-synthesis performance analysis as well as for possible further tuning of the fault detection filter (e.g., using optimization techniques) to ensure a desired filter behaviour. The field names associated with this structure can be inspected with

```julia
julia> fieldnames(FDFilterIF)
(:sys, :mu, :md, :mf, :mw, :ma)
```

Thus, the underlying filter model can be accessed as **R.sys** and the dimensions of the control, disturbance, fault, noise and auxiliary input vectors are contained in

R.mu, **R.md**, **R.mf**, **R.mw** and **R.ma**, respectively. The ranges of indices of control, disturbance, fault, noise and auxiliary inputs can be queried as **R.controls**, **R.disturbances**, **R.faults**, **R.noise** and **R.aux**, respectively. A particular feature of the generated object **R.sys** is related to using the nullspace method as first step in all synthesis functions. Since the transfer function matrices $R_u(\lambda)$ and $R_d(\lambda)$ in (3.4) are always zero, by convention, the corresponding input dimensions are set to zero, such that only the state-space realization of $R_f(\lambda)$ is explicitly provided for this example. This can be checked by looking to the resulting dimensions:

```julia
julia> ( R.mu, R.md, R.mf, R.mw, R.ma )
(0, 0, 3, 0, 0)
```

Another aspect of most synthesis techniques is that the implementation form and internal form state-space representation share the state and output matrices as in (7.14).

The full internal form corresponding to the computed **Q** and synthesis model **sysf** can be still computed with the function **fdIFeval**, as shown below

```julia
julia> Rf = fdIFeval(Q, sysf; minimal = true, atol = 1.e-7);
julia> iszero(Rf.sys[:,Rf.controls],atol=1.e-10) &&
       iszero(Rf.sys[:,Rf.disturbances],atol=1.e-10)
true
```

where a minimal realization of the resulting filter is determined using the keyword argument **minimal = true** jointly with **atol = 1.e-7** for setting the absolute tolerance, and the nullspace synthesis condition (i.e., $R_u(\lambda) = 0$ and $R_d(\lambda) = 0$) is explicitly checked.

The information additionally provided in the named tuple **info** is

```julia
julia> info
(tcond=3.0, degs=[1, 2], S=Bool[1 1 1; 1 1 1], HDesign=[0.0 1.0])
```

where:

info.tcond   is the maximum of the condition numbers of the employed nonorthogonal transformation matrices for which a large value is an indication of possible loss of accuracy in the computed results;

info.degs   is an integer vector containing the increasingly ordered degrees of a left minimal polynomial nullspace basis of $G(\lambda) := \left[\begin{smallmatrix} G_u(\lambda) & G_d(\lambda) \\ I & 0 \end{smallmatrix}\right]$, which can be seen as tentative orders for least order syntheses;

info.S   is the binary structure matrix of $\overline{G}_f(\lambda) := N_l(\lambda)\left[\begin{smallmatrix} G_f(\lambda) \\ 0 \end{smallmatrix}\right]$ corresponding to the computed left rational nullspace basis $N_l(\lambda)$ of $G(\lambda)$;

info.HDesign   is the design matrix H in (7.40) employed for the synthesis of the fault detection filter. Using the resulting design matrix as synthesis option entered via the keyword argument **HDesign**, the computed results can be exactly reproduced. Moreover, optimally tuned values of the design matrix can be determined, for example, by maximizing the value of the achieved fault sensitivity condition (5.20) or (5.21) (see the optimal tuning example at the end of this subsection).

Solving Fault Detection and Isolation Problems

For the solution of an EFDIP using the structure matrix S_{FDI} in (A.4), the function **efdisyn** can be used. To ensure a stability degree -2 for the filter poles, the simplest form of using this function is

```julia
julia>   Q1, R1, info1 = efdisyn(sysf, S[2:end,:], sdeg = -2);
```

The resulting filter object **Q1** contains a bank of 3 scalar output filters, each in a standard state-space form, which corresponds to the partitioned form of the filter in (3.14). The underlying vector of filters can be obtained via **Q1.sys** and the dimensions of the partitioned filter input vectors as **outputs** and **controls**, can be accessed as the integers contained in **Q1.ny** and **Q1.mu**, respectively. The ranges of the indices of the output and control inputs can be queried as **Q1.outputs** and **Q1.controls**, respectively.

The filter object **R1** contains the resulting vector of filters in standard state-space form, which corresponds to the internal form in (5.40) with additional auxiliary inputs. Thus, the underlying vector of filters can be accessed as **R1.sys** and the dimensions of the control, disturbance, fault, noise and auxiliary input vectors are contained in **R1.mu**, **R1.md**, **R1.mf**, **R1.mw** and **R1.ma**, respectively. The ranges of indices of control, disturbance, fault, noise and auxiliary inputs can be queried as **R1.controls**, **R1.disturbances**, **R1.faults**, **R1.noise** and **R1.aux**, respectively. In the resulting **FDIFilterIF** object **R1**, by convention, the control and disturbance vector dimensions are set to zero, the corresponding transfer function matrices being null.

The full internal form corresponding to the computed **Q1** and synthesis model **sysf** can be computed with the function **fdIFeval**, as shown below

```julia
julia>  Rf1 = fdIFeval(Q1, sysf; minimal = true, atol = 1.e-7);
julia>  indcd = [Rf1.controls; Rf1.disturbances];
julia>  iszero(vcat(Rf1.sys...)[:,indcd])
true
```

where minimal realizations of all component filters are determined by using the keyword arguments **minimal = true** and **atol = 1.e-7**, and the nullspace synthesis condition (i.e., $R_u(\lambda) = 0$ and $R_d(\lambda) = 0$) is explicitly checked by building the systems corresponding to their stacked transfer function matrices. Here, the use of Julia's *splat operator* "**...**" jointly with the vertical concatenation operator **vcat** acting on a vector of systems (the internal forms) allows to build vertically concatenated systems and to extract the part corresponding to the control and disturbance inputs in a very compact way.

Solving Model-Matching Problems

We illustrate how to solve an EMMP using the previously computed results to solve the EFDP and EFDIP. To solve the EMMP we use as reference filter, the internal form **R** computed with **efdsyn** to solve the EFDP. As it is checked, the following code computes essentially the same results as **efdsyn**:

```julia
julia>   Q2, R2, info2 = emmsyn(sysf, R);
julia>   iszero(Q2.sys-Q.sys) && iszero(R2.sys-R.sys)
true
```

Here, **Q2**, of type **FDFilter**, is the resulting fault detection filter object in the implementation form, **R2**, of type **FDFilterIF**, is the resulting filter object in the internal form, and **info2** is a *named-tuple*, which contains additional synthesis related information.

To illustrate the solution of an EMMP with a block structured reference filter we can use as reference filter, the internal form **R1** computed with **efdisyn** to solve the EFDIP. As checked in the following code, the computed results are essentially the same as those determined with **efdisyn**:

```julia
julia> Q3, R3, info3 = emmsyn(sysf, R1, atol = 1.e-7);
julia> iszero(vcat(Q3.sys...)-vcat(Q1.sys...)) &&
       iszero(vcat(R3.sys...)-vcat(R1.sys...))
true
```

Here, **Q3**, of type **FDIFilter**, is the resulting fault detection and isolation filter object in the implementation form, **R3**, of type **FDIFilterIF**, is the resulting filter object in the internal form, and **info3** is a *named-tuple*, which contains additional synthesis related information.

Analysis of Synthesis Results

The functions listed in the following table can be used for assessment purposes:

Performance evaluation of FDI filters	
fditspec	Computation of the weak or strong structure matrix
fdisspec	Computation of the strong structure matrix
fdiscond	Computation of the fault detection sensitivity condition
fdif2ngap	Computation of the fault-to-noise gap
fdimmperf	Computation of the model-matching performance

For example, to assess the achieved structure matrix from the results computed with **efdisyn**, we can use the function **fdisspec**:

```julia
julia>  fdisspec(R1)
3×3 BitMatrix:
 0  1  1
 1  0  1
 1  1  0
```

For the assessment of the achieved sensitivity condition from the results computed with **efdsyn**, we can use the function **fdiscond**:

```julia
julia>  fdiscond(R)[1]
0.5183693425036974
```

To improve this value, several synthesis trials can be performed with **efdsyn** using different randomly generated design matrices specified via the keyword argument **HDesign** as shown below

```julia
julia> temp = efdsyn(sysf, sdeg = -2; HDesign=randn(1,2));
julia> fdiscond(temp[2])[1]
```

This example illustrates the way Julia returns the outputs in the tuple **temp**, with three components, where the second component **temp[2]** is the internal form of the fault detection filter. For an acceptable value near enough to 1 of the sensitivity condition, the corresponding value of the design matrix can be obtained from the third output parameter as **temp[3].HDesign**. An optimum value can be determined using optimization-based tuning. A typical setup for this is the following:

```
julia> using Optim
julia> # define the function to be minimized
julia> f(x) = -fdiscond(efdsyn(sysf, sdeg = -2, HDesign=x)[2])[1]
f (generic function with 1 method)
julia> scond_init = -f(info.HDesign)      # initial value
0.5183693425036974
julia> result = optimize(f, info.HDesign, NelderMead());
julia> scond_fin = -f(result.minimizer) # final value
0.923950737703156
```

The function **fdimmperf** can be used to check the model-matching performance, by computing the norm of the error between the desired reference filter and the achieved internal form:

```
julia> fdimmperf(R,R2)
1.9170232274390572e-16
```

This check indicates an exact match, as to be expected for an exact synthesis.

A.3.2 Solving Model Detection Problems

Building Synthesis Models

For the solution of model detection problems, the synthesis models are collections of N state-space models of the form

$$E^{(i)}\lambda x^{(i)}(t) = A^{(i)}x^{(i)}(t) + B_u^{(i)}u(t) + B_d^{(i)}d^{(i)}(t) + B_w^{(i)}w^{(i)}(t) + B_a^{(i)}v^{(i)}(t),$$
$$y^{(i)}(t) = C^{(i)}x^{(i)}(t) + D_u^{(i)}u(t) + D_d^{(i)}d^{(i)}(t) + D_w^{(i)}w^{(i)}(t) + D_a^{(i)}v^{(i)}(t),$$

$$(A.5)$$

where $u(t) \in \mathbb{R}^{m_u}$ is the control input (common to all systems) and for $i = 1, \ldots, N$, $x^{(i)}(t) \in \mathbb{R}^{n^{(i)}}$, $y^{(i)}(t) \in \mathbb{R}^p$, $d^{(i)}(t) \in \mathbb{R}^{m_d^{(i)}}$, $w^{(i)}(t) \in \mathbb{R}^{m_w^{(i)}}$, and $v^{(i)}(t) \in \mathbb{R}^{m_a^{(i)}}$ are, respectively, the state vector, output vector, disturbance input vector, noise input vector and auxiliary input vector of the i-th system. These models exhibit four input groups, which are named as follows: **controls** for the common control inputs $u(t)$, **disturbances** for the disturbance inputs $d^{(i)}(t)$, **noise** for the noise inputs $w^{(i)}(t)$, and **aux** for the auxiliary inputs $v^{(i)}(t)$.

A typical situation is when N plant models are available for model detection purposes, with the i-th model having the form

$$E^{(i)}\lambda x^{(i)}(t) = A^{(i)}x^{(i)}(t) + B^{(i)}\widetilde{u}^{(i)}(t),$$
$$y^{(i)}(t) = C^{(i)}x^{(i)}(t) + D^{(i)}\widetilde{u}^{(i)}(t),$$

$$(A.6)$$

where $\widetilde{u}^{(i)}(t)$ is a global input for the i-th system, whose components are the common control inputs, disturbance inputs and noise inputs. The building of the synthesis model is just the operation to define the above groupings of various categories of input components, using the synthesis model setup function **mdmodset**.

As an example, consider $N = 3$ stable plant models in a vector of standard system models **sys** of the form (A.6) with the input vector $\tilde{u}(t)$ having 3 components, the output vector $y(t)$ having two components and a one-dimensional state vector $x(t)$. Assume that the first two inputs are control inputs and the third input is a disturbance input. The following command generates a synthesis model for model detection purposes:

```julia
julia> A1 = [-0.22;;]; B1 = [0.77 1.1 -0.23];
julia> C1 = [0.37; 0.55;;]; D1 = [0.56 0.16 0.3; 0.34 0.84 0.87];
julia> sys1 = dss(A1,B1,C1,D1);
julia> sys = [sys1/10^(i-1) for i in 1:3];
julia> sysm = mdmodset(sys,c = 1:2, d = 3);
```

The basic model **sys1** has been generated using the **rss** function, with the entries of the state-space matrices rounded to two decimal digits. Observe the syntax used to enter **A1** as an 1×1 matrix and **C1** as a 2×1 matrix. (Entering **A1 = [-0.22]** and **C1 = [0.37; 0.55]** would generate vectors instead matrices!) The Julia language construct

```julia
sys = [sys1/10^(i-1) for i in 1:3]
```

builds a vector **sys** of three systems, whose input–output maps are differently scaled (e.g., to model different levels of loss of efficiency faults of actuators or sensors). This construct is called an *array comprehension* and is a concise replacement for the longer **for**-loop version.

The resulting synthesis model **sysm** has the custom type **MDModel** and is a structure whose field names can be inspected using the function **fieldnames**:

```julia
julia> fieldnames(MDModel)
(:sys, :mu, :md, :mw, :ma)
```

Thus, the underlying descriptor system models (A.5) can be accessed as **sysm.sys**, the dimension of the control vector is contained in **sysm.mu**, the dimensions of disturbance, noise and auxiliary vectors are contained in the vectors **sysm.md**, **sysm.mw** and **sysm.ma**, respectively. The i-th component model can be accessed as **sysm.sys[i]** and the corresponding dimensions of disturbance, noise and auxiliary vectors are **sysm.md[i]**, **sysm.mw[i]** and **sysm.ma[i]**, respectively.

An alternative way to build a synthesis model is to build first a vector of component synthesis models, where each component model is built with the constructor function **MDModel**. For the above example, the same result can be achieved using the following command sequence:

```julia
julia> syscm = [MDModel(sys1/10^(i-1),mu=2,md=1)
                     for i in 1:3];
julia> sysm = mdmodset(syscm);
```

Preliminary Analysis

To perform model detection-related analysis, the functions listed in the following table are available:

Analysis of model detection synthesis models	
mdgenspec	Generation of achievable model detection specifications
mddist	Computation of distances between component models
mddist2c	Computation of pairwise distances between two sets of component models

The function **mdgenspec** generates the achievable structure matrix to solve a model detection problem for a given set of synthesis models. The function **mddist** can be used to compute the pairwise distances between a collection of component models, using several distance functions such as the ν-gap metric, $\mathcal{H}_\infty$-norm or $\mathcal{H}_2$-norm. The function **mddist2c** can be used to compute the pairwise distances between the component models of two collections of synthesis models.

For a control input-based discrimination of component models, a necessary condition for the solvability of a model detection problem is that the pairwise distances between the control channels of the component models are nonzero. This condition can be checked for the generated synthesis model **sysm** with the function **mddist** as shown below

```
julia> mdists = mddist(sysm)[1]
3×3 Matrix{Float64}:
 0.0       0.818181   0.969787
 0.818181  0.0        0.404058
 0.969787  0.404058   0.0
```

As expected, the distance between the third and first models is larger than the distance between the second and first models. The solution of the model detection problem targets a similar distance mapping to be achieved by the resulting model detection filters.

The model detection problem is solvable as can be read out of the achievable structure matrix for step control inputs characterized by the frequency 0:

```
julia> S = mdgenspec(sysm, MDfreq = 0)
3×3 BitMatrix:
 0  1  1
 1  0  1
 1  1  0
```

Solving Model Detection Problems

For the solution of an EMDP, the function **emdsyn** can be used. To ensure a stability degree -2 for the filter poles, this function can be called with the keyword arguments **sdeg = -2** (to place the poles to have real parts at most -2) and **smarg = -2** (to shift the stability margin to -2 for the real parts of acceptable poles):

```
julia> Q, R, info = emdsyn(sysm, sdeg = -2, smarg = -2);
```

where **Q** is the resulting bank of model detection filters object of type **MDFilter** in the implementation form, **R** is the resulting array of model detection filters object of type **MDFilterIF** in the internal form, and **info** is a *named-tuple*, which contains additional synthesis related information.

The filter object **Q** contains the resulting bank of model detection filters in standard state-space forms, which represent the implementation forms of the filters as in (4.2) with the corresponding block structured global filter in (4.3). The underlying vector of filters can be obtained via **Q.sys** and the dimensions of the partitioned filter input vectors as **outputs** and **controls**, can be accessed as the integers contained in **Q.ny** and **Q.mu**, respectively. The ranges of the indices of the output and control inputs can be queried as **Q.outputs** and **Q.controls**, respectively.

The filter object **R** contains the resulting internal forms of model detection filters, which represent the implementation forms of the filters as in (4.4). The underlying array of filters can be accessed as **R.sys** and the dimension of the control vector is contained in **R.mu**, the dimensions of disturbance, noise and auxiliary vectors are contained in the vectors **R.md**, **R.mw** and **R.ma**, respectively. The (i, j)-th component internal form can be accessed as **R.sys[i,j]** and the corresponding dimensions of disturbance, noise and auxiliary vectors are **sysm.md[j]**, **sysm.mw[j]** and **sysm.ma[j]**, respectively.

The resulting poles of the model detection filters can be computed using:

```julia
julia> gpole(Q)
3-element Vector{Vector{Float64}}:
 [-2.0]
 [-2.0000000000000004]
 [-2.0]
```

where the i-th component of the resulting vector contains the poles of the i-th filter.

The internal form corresponding to the computed **Q** and synthesis model **sysm** can be computed with the function **mdIFeval**, as shown below

```julia
julia>    Rf = mdIFeval(Q,sysm;minimal = true, atol = 1.e-7);
julia>    all(iszero.(R.sys .- Rf.sys,atol=1.e-7))
true
```

where minimal realizations are determined and the equality of computed results with the explicitly determined internal form is explicitly checked. Note the use of Julia's broadcasting approach by which all function calls have a "dot" call version (here the overloaded function **iszero**), while the "dot" operation "**.-**" involves all pairs of elements of the two-dimensional arrays of system models **R.sys** and **Rf.sys**.

The information provided in the named tuple **info** for each individual filter is similar to that obtained for the function **efdsyn**. A useful additional information is the achieved distance-mapping performance contained in **info.MDperf**, representing the peak gains of the associated internal representations of the model detection filters. The resulting values can be displayed with:

```julia
julia> info.MDperf
3×3 Matrix{Float64}:
  0.0  1.0  1.1
 10.0  0.0  1.0
 11.0  1.0  0.0
```

It is easy to observe that the resulting gains qualitatively map the distances between the models, as previously determined with the function **mddist**.

Analysis of Synthesis Results
The functions listed in the following table can be used for assessment purposes:

Performance evaluation of model detection filters	
mdspec	Computation of the weak structure matrix
mdsspec	Computation of the strong structure matrix
mdperf	Computation of the distance-matching performance
mdmatch	Computation of the distance-matching performance to a component model
mdgap	Computation of the noise gaps

For example, to assess the achieved strong structure matrix from the results computed with **emdsyn**, we can use the function **mdsspec** as follows:

```
julia>  mdsspec(R)
3×3 BitMatrix:
 0  1  1
 1  0  1
 1  1  0
```

which is the targeted structure matrix to solve model detection problems.

We can check the achieved distance-matching performance using the function **mdperf**:

```
julia> mdperf(R)[1] ≈ info.MDperf
true
```

A.4   Notes and References

Section A.1. A good general introduction to the Julia language is provided in [79]. Introductory material on linear algebra focusing on vector and matrix computations in Julia is available in [20], the Julia companion to the textbook [19]. Chapter 1 of [96] served as inspiration for our succinct presentation of Julia's basic issues.

Section A.2. A detailed documentation of the **DescriptorSystems** package has been prepared with the **Documenter**[11] package of Julia and is available online.[12]

Section A.3. A detailed documentation of the **FaultDetectionTools** package, prepared with the **Documenter** package of Julia, is available online.[13]

---

[11] https://github.com/JuliaDocs/Documenter.jl.

[12] https://andreasvarga.github.io/DescriptorSystems.jl/dev/.

[13] https://andreasvarga.github.io/FaultDetectionTools.jl/dev/.

References

1. H. Alwi, C. Edwards, and C. P. Tan. *Fault Detection and Fault-Tolerant Control Using Sliding Modes*. Springer-Verlag, London, 2011.
2. B. D. O. Anderson, T. S. Brinsmead, F. De Bruyne, J. Hespanha, D. Liberzon, and A. S. Morse. Multiple model adaptive control. I. Finite controller coverings. *Int. J. Robust Nonlin. Control*, 10:909–929, 2000.
3. E. Anderson, Z. Bai, J. Bishop, J. Demmel, J. Du Croz, A. Greenbaum, S. Hammarling, A. McKenney, S. Ostrouchov, and D. Sorensen. *LAPACK User's Guide, Third Edition*. SIAM, Philadelphia, 1999.
4. Y. Baram. An information theoretic approach to dynamical systems modeling and identification. *IEEE Trans. Automat. Control*, 23:61–66, 1978.
5. R. H. Bartels and G. W. Stewart. Algorithm 432: Solution of the matrix equation AX+XB=C. *Comm. ACM*, 15:820–826, 1972.
6. G. Basile and G. Marro. *Controlled and Conditioned Invariants in Linear System Theory*. Prentice-Hall, Inc., 1991.
7. M. Basseville and I. V. Nikiforov. *Detection of Abrupt Changes – Theory and Application*. Prentice-Hall, Inc., 1993.
8. T. Beelen. *New algorithms for computing the Kronecker structure of a pencil with applications to systems and control theory*. Ph. D. Thesis, Eindhoven University of Technology, 1987.
9. T. Beelen and P. Van Dooren. An improved algorithm for the computation of Kronecker's canonical form of a singular pencil. *Linear Algebra Appl.*, 105:9–65, 1988.
10. V. Belevitch. *Classical Network Theory*. Holden Day, San Francisco, 1968.
11. P. Benner. Theory and numerical solution of differential and algebraic Riccati equations. In P. Benner, M. Bollhöfer, D. Kressner, C. Mehl, and T. Stykel, editors, *Numerical Algebra, Matrix Theory, Differential-Algebraic Equations and Control Theory*, pages 67–105. Springer, 2015.
12. P. Benner, V. Mehrmann, V. Sima, S. Van Huffel, and A. Varga. SLICOT – a subroutine library in systems and control theory. In B. N. Datta, editor, *Applied and Computational Control, Signals and Circuits*, volume 1, pages 499–539. Birkhäuser, 1999.
13. P. Benner, V. Sima, and M. Voigt. L_∞-norm computation for continuous-time descriptor systems using structured matrix pencils. *IEEE Trans. Automat. Control*, 57:233–238, 2012.
14. J. Bezanson, A. Edelman, S. Karpinski, and V.B. Shah. Julia: A fresh approach to numerical computing. *SIAM Review*, 59:65–98, 2017.
15. M. Blanke, M. Kinnaert, J. Lunze, and M. Staroswiecki. *Diagnosis and Fault-Tolerant Control*. Springer-Verlag, Berlin, 2003.
16. P. M. M. Bongers, O. H. Bosgra, and M. Steinbuch. L_∞-norm calculation for generalized state space systems in continuous and discrete time. In *Proceedings of the American Control Conference, Boston, MA, USA*, pages 1637–1639, 1991.
17. J. D. Bošković and R. K. Mehra. Multiple-model adaptive flight control scheme for accommodation of actuator failures. *J. Guid. Control Dyn.*, 25:712–724, 2002.
18. S. Boyd and V. Balakrishnan. A regularity result for the singular values of a transfer matrix and a quadratically convergent algorithm for computing its L_∞-norm. *Syst. Control Lett.*, 15:1–7, 1990.
19. S. Boyd and L. Vandenberghe. *Introduction to Applied Linear Algebra – Vectors, Matrices, and Least Squares*. Cambridge University Press, 2018.
20. S. Boyd and L. Vandenberghe. *Introduction to Applied Linear Algebra – Vectors, Matrices, and Least Squares. Julia Language Companion*. Online draft version: https://web.stanford.edu/~boyd/vmls/vmls-julia-companion.pdf, 2021.
21. N.A. Bruinsma and M. Steinbuch. A fast algorithm to compute the H_∞-norm of a transfer function. *Syst. Control Lett.*, 14:287–293, 1990.
22. S. L. Campbell. *Singular Systems of Differential Equations*. Pitman, London, 1980.
23. J. Chen and R. J. Patton. *Robust Model-Based Fault Diagnosis for Dynamic Systems*. Kluwer Academic Publishers, London, 1999.

24. J. Chen, R. J. Patton, and G.-P. Liu. Optimal residual design for fault diagnosis using multi-objective optimization and genetic algorithms. *Int. J. Syst. Sci.*, 27:567–576, 1996.
25. E. Y. Chow and A. S. Willsky. Analytical redundancy and the design of robust failure detection systems. *IEEE Trans. Automat. Control*, 29:603–614, 1984.
26. L. Dai. *Singular Control Systems*, volume 118 of *Lecture Notes in Control and Information Sciences*. Springer Verlag, New York, 1989.
27. J. Demmel and B. Kågström. Stably computing the Kronecker structure and reducing subspaces of singular pencils $A - \lambda B$ for uncertain data. In J. Cullum and R. A. Willoughby, editors, *Large Scale Eigenvalue Problems*, volume 127 of *Mathematics Studies Series*, pages 283–323. North-Holland, Amsterdam, 1986.
28. J. Demmel and B. Kågström. The generalized Schur decomposition of an arbitrary pencil $A - \lambda B$: robust software with error bounds and applications. Part I: Theory and algorithms. Part II: Software and applications. *ACM Trans. Math. Software*, 19:160–174, 175–201, 1993.
29. S. X. Ding. *Model-based Fault Diagnosis Techniques, 2nd Edition*. Springer Verlag, Berlin, 2013.
30. S. X. Ding. *Data-driven Design of Fault Diagnosis and Fault-tolerant Control Systems*. Springer Verlag, Berlin, 2014.
31. S. X. Ding, P. M. Frank, E. L. Ding, and T. Jeinsch. A unified approach to the optimization of fault detection systems. *Int. J. Adapt. Control Signal Process.*, 14:725–745, 2000.
32. S. X. Ding and L. Guo. Observer-based fault detector. In *Proceedings of the IFAC World Congress, San Francisco, CA, USA*, volume N, pages 187–192, 1996.
33. S. X. Ding, L. Guo, and T. Jeinsch. A characterization of parity space and its application to robust fault detection. *IEEE Trans. Automat. Control*, 44:337–343, 1999.
34. X. Ding and P. M. Frank. Fault detection via factorization. *Syst. Control Lett.*, 14:431–436, 1990.
35. X. Ding and P. M. Frank. Frequency domain approach and threshold selector for robust model-based fault detection and isolation. In *Proceedings of the IFAC Symposium SAFEPROCESS, Baden-Baden, Germany*, 1991.
36. J. J. Dongarra, J. Du Croz, S. Hammarling, and I. Duff. A set of level 3 basic linear algebra subprograms. *ACM Trans. Math. Software*, 16:1–17, 1990.
37. J. J. Dongarra, J. Du Croz, S. Hammarling, and R. Hanson. An extended set of Fortran basic linear algebra subprograms. *ACM Trans. Math. Software*, 14:1–17, 1988.
38. J. J. Dongarra, C. B. Moler, J. R. Bunch, and G. W. Stewart. *LINPACK User's Guide*. SIAM, Philadelphia, 1979.
39. G.-R. Duan. *Analysis and Design of Descriptor Linear Systems*, volume 23 of *Advances in Mechanics and Mathematics*. Springer, New York, 2010.
40. A. Edelmayer and J. Bokor. Optimal H_∞ scaling for sensitivity optimization of detection filters. *Int. J. Robust Nonlin. Control*, 12:749–760, 2002.
41. A. Edelmayer, J. Bokor, and L. Kevitzky. An H_∞-filtering approach to robust detection of failures in dynamic systems. In *Proceedings of the IEEE Conference on Decision and Control, Lake Buena Vista, FL, USA*, pages 3037–3039, 1994.
42. A. Emami-Naeini, M. M. Akhter, and S. M. Rock. Effect of model uncertainty on failure detection: The threshold selector. *IEEE Trans. Automat. Control*, 33:1106–1115, 1988.
43. E. Emre, L. M. Silverman, and K. Glover. Generalized dynamic covers for linear systems with applications to deterministic identification and realization problems. *IEEE Trans. Automat. Control*, 22:26–35, 1977.
44. S. Fekri, M. Athans, and A. Pascoal. Issues, progress and new results in robust adaptive control. *Int. J. Adapt. Control Signal Process.*, 20:519–579, 2006.
45. G. D. Forney. Minimal bases of rational vector spaces with applications to multivariable linear systems. *SIAM J. Control*, 13:493–520, 1975.
46. B. A. Francis. *A Course in H^∞ Theory*, volume 88 of *Lecture Notes in Control and Information Sciences*. Springer-Verlag, New York, 1987.
47. P. M. Frank and X. Ding. Frequency domain approach to optimally robust residual generation and evaluation for model-based fault diagnosis. *Automatica*, 30:789–804, 1994.

48. E. Frisk and M. Nyberg. A minimal polynomial basis solution to residual generation for fault diagnosis in linear systems. *Automatica*, 37:1417–1424, 2001.

49. F. R. Gantmacher. *Theory of Matrices*, volume 2. Chelsea, New York, 1959.

50. J. D. Gardiner, A. J. Laub, J. J. Amato, and C. B. Moler. Solution of the Sylvester matrix equation $AXB^T + CXD^T = E$. *ACM Trans. Math. Software*, 18:223–231, 1992.

51. J. Gertler. *Fault Detection and Diagnosis in Engineering Systems*. Marcel Dekker, New York, 1998.

52. J. J. Gertler and R. Monajemy. Generating directional residuals with dynamic parity relations. *Automatica*, 31:627–635, 1995.

53. J. J. Gertler and D. Singer. A new structural framework for parity equation-based failure detection and isolation. *Automatica*, 26:381–388, 1990.

54. K. Glover. All optimal Hankel-norm approximations of linear multivariable systems and their L^∞-error bounds. *Int. J. Control*, 39:1115–1193, 1984.

55. K. Glover and A. Varga. On solving non-standard $\mathcal{H}_-/\mathcal{H}_{2/\infty}$ fault detection problems. In *Proceedings of the IEEE Conference on Decision and Control, Orlando, FL, USA*, pages 891–896, 2011.

56. D. E. Goldberg, P. Lancaster, and L. Rodman. *Invariant Subspaces of Matrices with Applications*. John Wiley & Sons, New York, 1986.

57. G. H. Golub, S. Nash, and C. Van Loan. A Hessenberg–Schur method for the problem $AX + XB = C$. *IEEE Trans. Automat. Control*, 24:909–913, 1979.

58. G. H. Golub and C. F. Van Loan. *Matrix Computations, 4th Edition*. John Hopkins University Press, Baltimore, 2013.

59. P. Goupil. Oscillatory failure case detection in the A380 electrical flight control system by analytical redundancy. *Control Eng. Pract.*, 18:1110–1119, 2010.

60. P. Goupil. AIRBUS state of the art and practices on FDI and FTC in flight control system. *Control Eng. Pract.*, 19:524–539, 2011.

61. D. W. Gu, M. C. Tsai, S. D. O'Young, and I. Postlethwaite. State-space formulae for discrete-time H^∞ optimization. *Int. J. Control*, 49:1683–1723, 1989.

62. S. J. Hammarling. Numerical solution of the stable, non-negative definite Lyapunov equation. *IMA J. Numer. Anal.*, 2:303–323, 1982.

63. N. J. Higham. *Accuracy and Stability of Numerical Algorithms*. SIAM, Philadelphia, 1996.

64. R. A. Horn and C. R. Johnson. *Matrix Analysis, 2nd Edition*. Cambridge University Press, New York, 2013.

65. M. Hou. Fault detection and isolation for descriptor systems. In R. J. Patton, P. M. Frank, and R. N. Clark, editors, *Issues of Fault Diagnosis for Dynamic Systems*, pages 115–144. Springer Verlag, London, 2000.

66. M. Hou and P. C. Müller. Fault detection and isolation observers. *Int. J. Control*, 60:827–846, 1994.

67. M. Hou and R. J. Patton. An LMI approach to $\mathcal{H}_-/\mathcal{H}_\infty$ fault detection observers. In *Proceedings of the UKACC International Conference on Control, Exeter, UK*, pages 305–310, 1996.

68. R. Isermann. *Fault-Diagnosis Systems, An Introduction from Fault Detection to Fault Tolerance*. Springer, Berlin, 2006.

69. I. M. Jaimoukha, Z. Li, and V. Papakos. A matrix factorization solution to the $\mathcal{H}_-/\mathcal{H}_\infty$ fault detection problem. *Automatica*, 42:1907–1912, 2006.

70. B. Kågström and P. Van Dooren. Additive decomposition of a transfer function with respect to a specified region. In M. A. Kaashoek, J. H. van Schuppen, and A. C. M. Ran, editors, *Signal Processing, Scattering and Operator theory, and Numerical Methods, Proceedings of the International Symposium on Mathematical Theory of Networks and Systems 1989, Amsterdam, The Netherlands*, volume 3, pages 469–477. Birhhäuser, Boston, 1990.

71. B. Kågström and L. Westin. Generalized Schur methods with condition estimators for solving the generalized Sylvester equation. *IEEE Trans. Automat. Control*, 34:745–751, 1989.

72. T. Kailath. *Linear Systems*. Prentice Hall, Englewood Cliffs, 1980.

73. S. Kim, J. Choi, and Y. Kim. Fault detection and diagnosis of aircraft actuators using fuzzy-tuning IMM filter. *IEEE Trans. Aerosp. Electron. Syst.*, 44:940–952, 2008.

74. G. Kimura. Geometric structure of observers for linear feedback control laws. *IEEE Trans. Automat. Control*, 22:846–855, 1977.

75. M. Kinnaert and Y. Peng. Residual generator for sensor and actuator fault detection and isolation: a frequency domain approach. *Int. J. Control*, 61:1423–1435, 1995.

76. P. Kunkel and V. Mehrmann. *Differential-Algebraic Equations. Analysis and Numerical Solution*. EMS Publishing House, Zurich, 2006.

77. A. Laub. A Schur method for solving the algebraic Riccati equation. *IEEE Trans. Automat. Control*, 24:913–921, 1979.

78. C. L. Lawson, R. J. Hanson, D. R. Kincaid, and F. T. Krogh. Basic linear algebra subprograms for Fortran usage. *ACM Trans. Math. Software*, 5:308–323, 1979.

79. B. Lauwens and A.B. Downey. *Think Julia: How to Think Like a Computer Scientist*. O'Reilly Media, 2019.

80. D. J. N. Limebeer and G. D. Halikias. A controller degree bound for $\mathcal{H}^\infty$-optimal control problems of the second kind. *SIAM J. Control Optim.*, 26:646–677, 1988.

81. N. Liu and K. Zhou. Optimal solutions to multi-objective robust fault detection problems. In *Proceedings of the IEEE Conference on Decision and Control, New Orleans, LA, USA*, pages 981–988, 2007.

82. N. Liu and K. Zhou. Optimal robust fault detection for linear discrete time systems. *J. Control Science Eng.*, vol. 2008, Article ID 829459, 16 pages, https://doi.org/10.1155/2008/829459, 2008.

83. M. Lovera, C. Novara, P. L. dos Santos, and D. Rivera. Guest editorial special issue on applied LPV modeling and identification. *IEEE Trans. Control Syst. Technol.*, 19:1–4, 2011.

84. D. G. Luenberger. An introduction to observers. *IEEE Trans. Automat. Control*, 16:596–602, 1971.

85. M.-A. Massoumnia. *A geometric approach to failure detection and identification in linear systems*. Ph. D. Thesis, Department of Aeronautics and Astronautics, Massachusetts Institute of Technology, Cambridge, MA, 1986.

86. M.-A. Massoumnia and W. E. Vander Velde. Generating parity relations for detecting and identifying control system component failures. *J. Guid. Control Dyn.*, 11:60–65, 1986.

87. M.-A. Massoumnia, G. C. Verghese, and A. S. Willsky. Failure detection and identification. *IEEE Trans. Automat. Control*, 34:316–321, 1988.

88. MathWorks. Control System Toolbox (R2015b-R2022b), User's Guide. The MathWorks Inc., Natick, MA, 2015–2022.

89. P. S. Maybeck. Multiple model adaptive algorithms for detecting and compensating sensor and actuator/surface failures in aircraft flight control systems. *Int. J. Robust Nonlin. Control*, 9(14):1051–1070, 1999.

90. V. L. Mehrmann. *The Autonomous Linear Quadratic Control Problem*, volume 163 of *Lectures Notes in Control and Information Sciences*. Springer Verlag, Berlin, 1991.

91. P. Misra, P. Van Dooren, and A. Varga. Computation of structural invariants of generalized state-space systems. *Automatica*, 30:1921–1936, 1994.

92. C. B. Moler and C. F. Van Loan. Nineteen dubious ways to compute the exponential of a matrix. *SIAM Rev.*, 20:801–836, 1978.

93. A. S. Morse. Minimal solutions to transfer matrix equations. *IEEE Trans. Automat. Control*, 21:131–133, 1976.

94. R. Murray-Smith and T. A. Johansen, editors. *Multiple Model Approaches to Modelling and Control*. Taylor & Francis, London, 1997.

95. K. S. Narendra and J. Balakrishnan. Adaptive control using multiple models. *IEEE Trans. Automat. Control*, 42:171–187, 1997.

96. Y. Nazarathy and H. Klok. *Statistics with Julia. Fundamentals for Data Science, Machine Learning and Artificial Intelligence*. Springer Nature, Switzerland, 2021.

97. H. Niemann and J. Stoustrup. Design of fault detectors using H_∞ optimization. In *Proceedings of the IEEE Conference on Decision and Control, Sydney, Australia*, pages 4237–4238, 2000.

98. M. Nyberg. Criterions for detectability and strong detectability of faults in linear systems. *Int. J. Control*, 75:490–501, 2002.

99. M. Nyberg and E. Frisk. Residual generation for fault diagnosis of systems described by linear differential-algebraic equations. *IEEE Trans. Automat. Control*, 51:1995–2000, 2006.

100. C. Oară. Constructive solutions to spectral and inner-outer factorizations with respect to the disk. *Automatica*, 41:1855–1866, 2005.

101. C. Oară and P. Van Dooren. An improved algorithm for the computation of structural invariants of a system pencil and related geometric aspects. *Syst. Control Lett.*, 30:39–48, 1997.

102. C. Oară and A. Varga. Minimal degree coprime factorization of rational matrices. *SIAM J. Matrix Anal. Appl.*, 21:245–278, 1999.

103. C. Oară and A. Varga. Computation of general inner-outer and spectral factorizations. *IEEE Trans. Automat. Control*, 45:2307–2325, 2000.

104. D. Ossmann and A. Varga. Optimization-based tuning of LPV fault detection filters for civil transport aircraft. In C. Vallet, D. Choukroun, C. Philippe, G. Balas, A. Nebylov, and O. Yanova, editors, *Progress in Flight Dynamics, Guidance, Navigation, Control, Fault Detection and Avionics*, volume 6 of *EUCASS Advances in Aerospace Sciences*, pages 263–280. TORUS Press, 2013.

105. D. Ossmann and A. Varga. Detection and identification of loss of efficiency faults of flight actuators. *Int. J. Appl. Math. Comput. Sci.*, 25:53–63, 2015.

106. T. Pappas, A. J. Laub, and N. R. Sandell Jr. On the numerical solution of the discrete-time algebraic Riccati equation. *IEEE Trans. Automat. Control*, 25:631–641, 1980.

107. R. J. Patton and M. Hou. Design of fault detection and isolation observers: a matrix pencil approach. *Automatica*, 34(9):1135–1140, 1998.

108. T. Penzl. Numerical solution of generalized Lyapunov equations. *Adv. Comput. Math.*, 8:33–48, 1998.

109. M. S. Phatak and N. Viswanadham. Actuator fault detection and isolation in linear systems. *Int. J. Syst. Sci.*, 19:2593–2603, 1988.

110. K. Poolla, P. Khargonekar, A. Tikku, J. Krause, and K. Nagpal. A time-domain approach to model validation. *IEEE Trans. Automat. Control*, 39:951–959, 1994.

111. Z. Qiu and J. Gertler. Robust FDI systems and H_∞ optimization. In *Proceedings of the IEEE Conference on Decision and Control, San Antonio, TX, USA*, pages 1710–1715, 1993.

112. M. L. Rank and H. Niemann. Norm based design of fault detectors. *Int. J. Control*, 72:773–783, 1999.

113. A. Saberi, A. A. Stoorvogel, P. Sannuti, and H. Niemann. Fundamental problems in fault detection and identification. *Int. J. Robust Nonlinear Control*, 10:1209–1236, 2000.

114. M. G. Safonov, R. Y. Chiang, and D. J. N. Limebeer. Optimal Hankel model reduction for nonminimal systems. *IEEE Trans. Automat. Control*, 35:496–502, 1990.

115. M. G. Safonov, E. A. Jonckheere, M. Verma, and D. J. N. Limebeer. Synthesis of positive real multivariable feedback systems. *Int. J. Control*, 45:817–842, 1987.

116. V. Sima. *Algorithms for Linear-quadratic Optimization*. Marcel Dekker, New York, 1996.

117. R. S. Smith and J. Doyle. Model invalidation: a connection between robust control and identification. In *Proceedings of the American Control Conference, Pittsburgh, PA, USA*, volume 2, pages 1435–1440, 1989.

118. G. W. Stewart. *Introduction to Matrix Computations*. Academic Press, New York, 1974.

119. T. Stykel. Gramian based model reduction for descriptor systems. *Math. Control Signals Syst.*, 16:297–319, 2004.

120. M. S. Tombs and I. Postlethwaite. Truncated balanced realization of a stable non-minimal state-space system. *Int. J. Control*, 46:1319–1330, 1987.

121. P. Van Dooren. The computation of Kronecker's canonical form of a singular pencil. *Linear Algebra Appl.*, 27:103–141, 1979.

122. P. Van Dooren. The generalized eigenstructure problem in linear systems theory. *IEEE Trans. Automat. Control*, 26:111–129, 1981.

123. P. Van Dooren. A generalized eigenvalue approach for solving Riccati equations. *SIAM J. Sci. Stat. Comput.*, 2:121–135, 1981.

124. P. Van Dooren. Rational and polynomial matrix factorizations via recursive pole-zero cancellation. *Linear Algebra Appl.*, 137/138:663–697, 1990.

125. P. Van Dooren. Numerical Linear Algebra for Signals, Systems and Control. University of Louvain, Louvain la Neuve, Belgium, Draft notes prepared for the Graduate School in Systems and Control, 2003.

126. S. Van Huffel, V. Sima, A. Varga, S. Hammarling, and F. Delebecque. High-performance numerical software for control. *IEEE Control Syst. Mag.*, 24:60–76, 2004.

127. J. Vandewalle and P. Dewilde. On the irreducible cascade synthesis of a system with real rational transfer matrix. *IEEE Trans. Circuits Syst.*, 24:481–494, 1977.

128. A. I. G. Vardulakis and N. Karcanias. Proper and stable, minimal MacMillan degrees bases of rational vector spaces. *IEEE Trans. Automat. Control*, 29:1118–1120, 1984.

129. A. Varga. A Schur method for pole assignment. *IEEE Trans. Automat. Control*, 26:517–519, 1981.

130. A. Varga. The numerical stability of an algorithm for pole assignment. In G. Leininger, editor, *Computer Aided Design of Multivariable Technological Systems, Proceedings of the Second IFAC Symposium, West Lafayette, IN, USA*, pages 117–122. Pergamon Press, Oxford, 1983.

131. A. Varga. Computation of irreducible generalized state-space realizations. *Kybernetika*, 26:89–106, 1990.

132. A. Varga. Minimal realization procedures based on balancing and related techniques. In F. Pichler and R. Moreno Díaz, editors, *Computer Aided Systems Theory – EUROCAST'91, A Selection of Papers from the Proceedings of the Second International Workshop on Computer Aided System Theory, Krems, Austria*, volume 585 of *Lecture Notes in Computer Science*, pages 733–761. Springer Verlag, Berlin, 1992.

133. A. Varga. On stabilization of descriptor systems. *Syst. Control Lett.*, 24:133–138, 1995.

134. A. Varga. Computation of Kronecker-like forms of a system pencil: Applications, algorithms and software. In *Proceedings of the IEEE International Symposium on Computer-Aided Control System Design, Dearborn, MI, USA*, pages 77–82, 1996.

135. A. Varga. Computation of coprime factorizations of rational matrices. *Linear Algebra Appl.*, 271:83–115, 1998.

136. A. Varga. A DESCRIPTOR SYSTEMS toolbox for MATLAB. In *Proceedings of the IEEE International Symposium on Computer-Aided Control System Design, Anchorage, AK, USA*, pages 150–155, 2000.

137. A. Varga. Computational issues in fault-detection filter design. In *Proceedings of the IEEE Conference on Decision and Control, Las Vegas, NV, USA*, volume 4, pages 4760–4765, 2002.

138. A. Varga. On computing least order fault detectors using rational nullspace bases. In *Proceedings of the IFAC Symposium SAFEPROCESS, Washington D.C., USA*, 2003.

139. A. Varga. Reliable algorithms for computing minimal dynamic covers. In *Proceedings of the IEEE Conference on Decision and Control, Maui, HI, USA*, pages 1873–1878, 2003.

140. A. Varga. Computation of least order solutions of linear rational equations. In *Proceedings of the International Symposium on Mathematical Theory of Networks and Systems, Leuven, Belgium*, 2004.

141. A. Varga. New computational approach for the design of fault detection and isolation filters. In M. Voicu, editor, *Advances in Automatic Control*, volume 754 of *The Kluwer International Series in Engineering and Computer Science*, pages 367–381. Kluwer Academic Publishers, Dordrecht, 2004.

142. A. Varga. Reliable algorithms for computing minimal dynamic covers for descriptor systems. In *Proceedings of the International Symposium on Mathematical Theory of Networks and Systems, Leuven, Belgium*, 2004.

143. A. Varga. Numerically reliable methods for optimal design of fault detection filters. In *Proceedings of the IEEE Conference on Decision and Control, Seville, Spain*, pages 2391–2396, 2005.

144. A. Varga. A FAULT DETECTION toolbox for MATLAB. In *Proceedings of the IEEE Conference on Computer Aided Control System Design, Munich, Germany*, pages 3013–3018, 2006.

145. A. Varga. Fault detection and isolation of actuator failures for a large transport aircraft. In *Proceedings of the First CEAS European Air and Space Conference, Berlin, Germany*, 2007.

146. A. Varga. On designing least order residual generators for fault detection and isolation. In *Proceedings of the 16th International Conference on Control Systems and Computer Science, Bucharest, Romania*, pages 323–330, 2007.

147. A. Varga. General computational approach for optimal fault detection. In *Proceedings of the IFAC Symposium SAFEPROCESS, Barcelona, Spain*, pages 107–112, 2009.

148. A. Varga. Least order fault and model detection using multi-models. In *Proceedings of the Conference on Decision and Control, Shanghai, China*, pages 1014–1019, 2009.

149. A. Varga. Monitoring actuator failures for a large transport aircraft – the nominal case. In *Proceedings of the IFAC Symposium SAFEPROCESS, Barcelona, Spain*, 2009.

150. A. Varga. The nullspace method – a unifying paradigm to fault detection. In *Proceedings of the Conference on Decision and Control, Shanghai, China*, pages 6964–6969, 2009.

151. A. Varga. On computing achievable fault signatures. In *Proceedings of the IFAC Symposium SAFEPROCESS, Barcelona, Spain*, pages 935–940, 2009.

152. A. Varga. Integrated algorithm for solving $\mathcal{H}_2$-optimal fault detection and isolation problems. In *Proceedings of the Conference on Control and Fault-Tolerant Systems (SysTol), Nice, France*, pages 353–358, 2010.

153. A. Varga. Integrated computational algorithm for solving $\mathcal{H}_\infty$-optimal FDI problems. In *Proceedings of the IFAC World Congress, Milano, Italy*, pages 10187–10192, 2011.

154. A. Varga. Linear FDI-Techniques and Software Tools. FAULT DETECTION Toolbox V1.0 – Technical Documentation, German Aerospace Center (DLR), Institute of Robotics and Mechatronics, 2011.

155. A. Varga. On computing minimal proper nullspace bases with applications in fault detection. In P. Van Dooren, S. P. Bhattacharyya, R. H. Chan, V. Olshevsky, and A. Routray, editors, *Numerical Linear Algebra in Signals, Systems and Control*, volume 80 of *Lecture Notes in Electrical Engineering*, pages 433–465. Springer Verlag, Berlin, 2011.

156. A. Varga. Descriptor system techniques in solving $\mathcal{H}_{2/\infty}$-optimal fault detection and isolation problems. In L. T. Biegler, S. L. Campbell, and V. Mehrmann, editors, *Control and Optimization with Differential-Algebraic Constraints*, volume 23 of *Advances in Design and Control*, pages 105–125. SIAM, 2012.

157. A. Varga. New computational paradigms in solving fault detection and isolation problems. *Annu. Rev. Control*, 37:25–42, 2013.

158. A. Varga, S. Hecker, and D. Ossmann. Diagnosis of actuator faults using LPV-gain scheduling techniques. In *Proceedings of the AIAA Guidance, Navigation, and Control Conference, Portland, OR, USA*, 2011.

159. A. Varga and D. Ossmann. LPV model-based robust diagnosis of flight actuator faults. *Control Eng. Pract.*, 31:135–147, 2014.

160. A. Varga, D. Ossmann, and H.-D. Joos. A fault diagnosis based reconfigurable longitudinal control system for managing loss of air data sensors for a civil aircraft. In *Proceedings of the IFAC World Congress, Cape Town, South Africa*, pages 3489–3496, 2014.

161. A. Varga. *Solving Fault Diagnosis Problems, Linear Synthesis Techniques, 1st Edition*. vol. 84 of Studies in Systems, Decision and Control, Springer International Publishing, 2017.

162. A. Varga. *Descriptor System Tools (DSTOOLS) User's Guide*. https://arxiv.org/pdf/1707.07140, 2018.

163. A. Varga. *Fault Detection and Isolation Tools (FDITOOLS) User's Guide*. https://arxiv.org/pdf/1703.08480, 2018.

164. A. Varga. On checking null rank conditions of rational matrices. "https://arxiv.org/abs/1812.11396", 2018.

165. A. Varga. On computing the Kronecker structure of polynomial and rational matrices using Julia. "https://arxiv.org/abs/2006.06825", 2020.

166. A. Varga. On recursive computation of coprime factorizations of rational matrices. *Linear Algebra and its Applications*, 623:478–502, 2021.

167. G. Verghese, P. Van Dooren, and T. Kailath. Properties of the system matrix of a generalized state-space system. *Int. J. Control*, 30:235–243, 1979.

168. M. Vidyasagar. *Control System Synthesis: A Factorization Approach*. The MIT Press, Cambridge, MA, 1985.

169. G. Vinnicombe. *Uncertainty and Feedback: H_∞ Loop-Shaping and the v-gap Metric*. Imperial College Press, London, 2001.

170. N. Viswanadham and K. D. Minto. Robust observer design with application to fault detection. In *Proceedings of the American Control Conference, Atlanta, GA, USA*, 1988.

171. D. Wang and K.-Y. Lum. Adaptive unknown input observer approach for aircraft actuator fault detection and isolation. *Int. J. Adapt. Control Signal Process.*, 21:31–48, 2007.

172. M. E. Warren and A. E. Eckberg. On the dimension of controllability subspaces: A characterization via polynomial matrices and Kronecker invariants. *SIAM J. Contr.*, 13:434–445, 1975.

173. J. H. Wilkinson. The perfidious polynomial. In G. H. Golub, editor, *Studies in Numerical Analysis*, volume 24 of *Studies in Mathematics*, pages 1–28. Mathematical Association of America, 1984.

174. A. S. Willsky. Detection of abrupt changes in dynamic systems. In M. Basseville and A. Benveniste, editors, *Detection of Abrupt Changes in Signals and Dynamical Systems*, volume 77 of *Lecture Notes in Control and Information Science*, pages 27–49. Springer-Verlag, Berlin, 1986.

175. W. M. Wonham and A. S. Morse. Feedback invariants of linear multivariable systems. *Automatica*, 8:93–100, 1972.

176. P. Zhang and S. X. Ding. An integrated trade-off design of observer based fault detection systems. *Automatica*, 44:1886–1894, 2008.

177. K. Zhou, J. C. Doyle, and K. Glover. *Robust and Optimal Control*. Prentice Hall, Upper Saddle River, 1996.

178. A. Zolghadri, D. Henry, J. Cieslak, D. Efimov, and P. Goupil. *Fault Diagnosis and Fault-Tolerant Control and Guidance for Aerospace Vehicles, From Theory to Application*. Springer-Verlag, London, 2014.